DOUBLE LAYERS AND OTHER NONLINEAR POTENTIAL STRUCTURES IN PLASMAS

Fourth Symposium on

DOUBLE LAYERS AND OTHER NONLINEAR POTENTIAL STRUCTURES IN PLASMAS

Innsbruck, Austria

July 6 – 8, 1992

Editor

Roman W. Schrittwieser

Institute for Ion Physics
University of Innsbruck
Austria

World Scientific
Singapore • New Jersey • London • Hong Kong

Published by

World Scientific Publishing Co Pte Ltd
5 Toh Tuck Link, Singapore 596224
USA office: 27 Warren Street, Suite 401-402, Hackensack, NJ 07601
UK office: 57 Shelton Street, Covent Garden, London WC2H 9HE

British Library Cataloguing-in-Publication Data
A catalogue record for this book is available from the British Library.

**DOUBLE LAYERS AND OTHER
NONLINEAR POTENTIAL STRUCTURES IN PLASMAS**

ISBN-13 9789810214326
ISBN-10 9810214324

This meeting
and all the associated publications
are dedicated
to the memory of the late

Christoph K. Görtz
Dwight R. Nicholson
Robert Alan Smith
Linhua Shan

Fourth Symposium on Double Layers and Other Nonlinear Potential Structures in Plasmas

July 6 to 8, 1992, Innsbruck Austria

Local Organizer:

Roman Schrittwieser
Institute for Ion Physics, University of Innsbruck,
Technikerstr. 25/III, A-6020 Innsbruck, Austria,

International Scientific Committee:

Acknowledgements:

This symposium was supported by the following institutions and companies:

Österreichische Forschungsgemeinschaft, Vienna, under grant No. 06/1723,
Tiroler Landesregierung, Innsbruck,
Der Bürgermeister der Stadt Innsbruck,
Firma Balzers, Wien,
Firma Leybold, Wien,
Landeshypothekenbank Tirol.

The editor wishes to thank the following persons for their kind help in the organisation:

Paul Eichberger,
Carin Schrittwieser,
Armin Siebenförcher,
Carsten Winkler,
Martin Zimmerling.

Christoph K. Görtz

Dwight R. Nicholson

Robert Alan Smith

Linhua Shan

Dedication:

We all remember the tragic events on November 1, 1991, when four of our colleagues from the Physics and Astronomy Department of the University of Iowa were shot to death on campus. All of them were plasma physicists and each made important contributions either directly to double layer research or to closely related fields, so the organizer of this first double layer symposium since their deaths wishes to dedicate the meeting and all the associated publications to their memory. Due to their research interest, they were known to most of the participants of this meeting either personally or by their scientific achievements. Many of us have lost not only close scientific colleagues but also good friends.

Christoph Görtz, professor of physics, did research on Jupiter's moon Io, electron beams in auroral field lines, the spokes of Saturn's B ring and the magnetospheres of Earth, Saturn and Jupiter. A native of Germany, he obtained his doctorate from Rhodes University in South Africa in 1972 and went to Iowa the following year. He was senior editor of the *Journal of Geophysical Research - Space Physics*.

Dwight R. Nicholson, chair of the Department of Physics and Astronomy and professor of physics. He was a plasma theorist who studied Langmuir turbulence, stability properties, solitons, nonlinear phenomena and neural networks. He wrote a widely used textbook, *Introduction to Plasma Theory* (Wiley, 1983). He obtained his PhD in plasma physics from the University of California, Berkeley, in 1975 and joined the Iowa faculty in 1978.

Robert Alan Smith, associate professor of physics, was an expert on double layers, the Jovian magnetosphere and Jupiter's moon Io. He worked at the Naval Research Lab for 11 years and received his PhD in physics from the University of Maryland in 1973.

Linhua Shan, a research investigator in the physics department, received his PhD from Iowa in December 1990. His thesis, a Monte Carlo simulation evaluating the origin of Saturn's B, ring suggested that the B ring is not much older than 100 million years.

These obituaries have partly been taken from the December issue of *Physics Today*. The kind permission of the editor of *Physics Today* to reproduce this information is grateful acknowledged.

Roman W. Schrittwieser

Preface:

This book contains the full manuscripts of the invited and contributed papers, accepted for presentation at the Fourth Symposium on Double Layers and Other Nonlinear Potential Structures in Plasmas, which took place in Innsbruck, Austria, from July 6 to 8, 1992. This was the week which followed the 1992 International Conference on Plasma Physics. The meeting was the fourth in a series of symposia which started in 1982 with the "Symposium on Plasma Double Layers", organized by the plasma physics group at the Risø National Laboratory near Roskilde, Denmark. The "Second Symposium on Plasma Double Layers and Related Topics" took place in 1984 in Innsbruck. In 1987 the plasma physics group of the "Alexandru-Ioan-Cuza" University in Iasi, Romania, organized the Third Symposium on Plasma Double Layers in Bucharest. All three meetings were organized following one of the International Conferences on Plasma Physics (1982, Gothenburg, Sweden; 1984, Lausanne, Switzerland; 1987, Kiev, Ukraine). No double layer symposium followed the 1989 International Conference on Plasma Physics, New Delhi, India. A workshop on "Double Layers in Astrophysics" took place in Huntsville, Alabama, USA, in 1986. Now, after a period of more than five years since the last double layer symposium, another meeting of this kind was long overdue.

The 63 manuscripts were directly reproduced without editing from the authors' originals. Thus the resposibility for typing or other errors rests entirely with the authors. The editor would like to thank all authors for their kind collaboration.

The editor gratefully acknowledges the help of the International Scientific Committee in the choice of invited speakers and the evaluation of the contributions. Besides the 14 invited papers, 23 experimental papers, 20 theoretical papers, 3 computer simulation papers, 2 space related papers and 1 "other" paper were submitted in the form of one-page abstracts and accepted by the International Scientific Committee. The one-page abstracts have each been reviewed by two members of the International Scientific Committee. In the cases of conflicting opinions, an adjudicator made the final decisions. In total 2 abstracts had to be rejected and in three cases the referees suggested that 2 abstracts should be combined into one. One post-deadline abstract was review by just one member of the International Scientific Committee. Several contributions have, however, not been presented during the Symposium and no final manuscripts have been received by the editor. This is the reason why there are several gaps in the consecutive numbering of the contributions.

The following abbreviations are used to distiguish the contributions:

In .... Invited Papers
Ex .... Contributed Papers Experiments
Th .... Contributed Papers Theory
Si .... Contributed Papers Simulation
Sp .... Contributed Papers Space
Ot .... Contributed Papers Others

Contents:

Invited Lectures:

Contributed Papers:

Experiments:

Contributed Papers:

Theory:

Contributed Papers:

Simulation:

Contributed Papers:

Space:

Contributed Papers:

Others:

List of participants:

ALLEN, John Edward
Dept. of Engineering Science, Univ. of Oxford, Parks Road,
Oxford, OX1 3PJ, *UK*

ARMSTRONG, Richard John
Institute of Mathematical & Physical Sciences, University of Tromsø,
P.O.Box 953
N-9001 Tromsø, *Norway*

AXNÄS, Ingvar
Plasma Physics Department, Alfvén Laboratory, Royal Institute of Technology ,
S-10044 Stockholm, *Sweden*

BABOOLAL, Satyanand
Dept.of Computer Science, University of Durban-Westville, Privat Bag X54001,
4000 Durban, *South Africa*

BINGHAM, Robert
Rutherford Appleton Laboratory, Chilton, Didcot,
Oxon X11 0QX, *UK*

BIRDSALL, Charles K.
University of California, EECS Department, Cory Hall,
Berkeley, CA 94720, *USA*

BONHOMME, Gerard
Lab.P.M.I., Universitè de Nancy I, BP 239,
F-54506 Vandoevre-Les-Nancy Cedex, *France*

BOSTRÖM, Rolf
Swedish Institute of Space Physics, Uppsala Division,
S-75591 Uppsala, *Sweden*

BRYANT, Duncan Alan
Rutherford Appleton Laboratory, Chilton, Didcot,
OXON OX11 0QX, *UK*

BULANOV, Sergei Vladimir
General Physics Institute, Plasma Physics Department,
Russian Academy of Sciences, Vavilova 38
117942 Moscow, *Russia*

CALLEBAUT, Dirk
Natural Sciences Department, University of Antwerp, Universiteitsplein 1,
B-2610 Antwerp, *Belgium*

CARLQVIST, Per
Plasma Physics Department, Alfvén Laboratory, Royal Institute of Technology,
S-10044 Stockholm, *Sweden*

CARNEIRO, Lindolff Thadeu
Laboratorio de Fisica de Plasma, Universidade Federal Fluminense,
Niteroi, RJ 24.020, *Brazil*

CARPENTER, Tom
Department of Physics and Astronomy, University of Iowa,
Iowa City, IA 52242, *USA*

CERCEK, Milan
Plasma Physics Laboratory, Reactor Physics Department,
Institute Jozef Stefan, Jamova 39, P.O. Box 100,
SLO-61111 Ljubljana, *Slovenia*

CHIUEH, Tzihong
National Central University of Taiwan, Physics Department,
Chung-li, Taiwan 32054, *Republic of China*

DEUTSCH, Rudolf
Institut für Plasmaforschung der Universität Stuttgart, Pfaffenwaldring 31, DW-7000 Stuttgart 80, *Germany*

EICHBERGER, Paul
Institute for Ion Physics, University of Innsbruck, Technikerstr. 25, A-6020 Innsbruck, *Austria*

FÄLTHAMMAR, Carl-Gunne
Plasma Physics Department, Alfvén Laboratory, Royal Institute of Technology, S-10044 Stockholm, *Sweden*

HATAKEYAMA, Rikizo
Department of Electronic Engineering, Tohoku University, Aramaki Aza-Aoba, Sendai 980, *Japan*

HELLBERG, Manfred A.
Plasma Physics Research Institute, University of Natal, King George V Avenue Durban 4001, *South Africa*

HERSHKOWITZ, Noah
Nuclear Engineering Department, University of Wisconsin, 1500 Johnson Drive, Madison, WI 53706-1687, *USA*

HONZAWA, Tadao
Dept. of Electric & Electronic Engineering, Utsunomiya University, Ishii-machi 2753, Utsunomiya 321, *Japan*

IANOPOL, Bogdan
Faculty of Physics, University "Al. I.Cuza", Blvd. Copou Nr. 11, RO-6600 IASI, *Romania*

JASPERSE, John R.
Geophysical Directorate/GIM, Phillips Laboratory, Hanscom AFB, Bedford, MA 01731-5000, *USA*

JELIC, Nicola
Plasma Physics Laboratory, Reactor Physics Department, Institute Jozef Stefan, Jamova 39, P.O. Box 100 SLO-61111 Ljubljana, *Slovenia*

JENSEN,Vagn O.
Optics and Fluid Dynamics Department, Risø National Laboratory DK-4000 Roskilde, *Denmark*

JOHN, P. I.
Institute for Plasma Research, Near Indira Bridge, Bhat, Gandhinagar 382 424, *India*

KAW, Predhiman K.
Institute for Plasma Research, Near Indira Bridge, Bhat, Gandhinagar 382 424, *India*

KLINGER, Thomas
Institute for Experimental Physics, University of Kiel, Olshausenstr. 40, DW-23 Kiel 1, *Germany*

KNORR, Georg
Department of Physics and Astronomy, University of Iowa, Iowa City, IA 52242, *USA*

KOEPKE, Mark E.
Physics Department, Hodges Hall, University of West Virginia, Morgantown, WV 26506, *USA*

KRISHAN, Vinod
Indian Institute of Astrophysics,
Bangalore 560034
India

KRUMM, Peter
Plasma Physics Research Institute, University of Natal, King George V Avenue,
Durban 4001, *South Africa*

KUHN, Siegbert
Institute for Theoretical Physics, University of Innsbruck, Technikerstr. 25
A-6020 Innsbruck, *Austria*

LUCA, Dumitru
Faculty of Physics, University "Al. I.Cuza", Blvd. Copou Nr. 11,
RO-6600 IASI, *Romania*

LUNDBERG, Jonas
Department of Plasma Physics, Umea University,
S-90187 Umea, *Sweden*

MÄLKKI, Anssi
Geophysics Department, Finnish Meteorological Institute, P.O. Box 503,
SF-00101 Helsinki, *Finland*

MANFREDI,Giovanni
PMMS/CNRS, 3A Av. de la Recherche Scientifique,
F-45071 Orleans Cedex, *France*

MARTIN, Pablo
Departmento de Fisica, Universidad Simon Bolivar, Apartado 89000
Caracas 1080A, *Venezuela*

MASE, Hiroshi
Department Electric and Electronic Engineering,
Ibaraki University, Naka Narusawa-cho,
Hitachi 316, *Japan*

MASLOV, Vassily I.
Kharkov Institute of Physics and Technology, Academichskaya 1,
310108 Kharkov, *Ukraine*

MATSUMOTO, Kazunori
Department of Electronics and Informatics, Toyama Prefectural University,
Kosugi-machi, Imizu-gun,
Toyama 939-03, *Japan*

MERLINO, Robert L.
Department of Physics and Astronomy, University of Iowa,
Iowa City, IA 52242, *USA*

MIENO, Tetsu
Department of Physics, Shizuoka University, 836 Ohya,
Shizuoka 422, *Japan*

MOLA, Stefano
PMMS/CNRS, 3A Av. de la Recherche Scientifique
F-45071 Orleans Cedex, *France*

NAMBU, Mitsuhiro
College of General Education, Ropponmatsu, Kyushu University,
Fukuoka 810, *Japan*

OERTL, Michael
Institute for Zoophysiology, University of Innsbruck, Technikerstr. 25,
A-6020 Innsbruck, *Austria*

PÉCSELI, Hans L.
Department of Physics, Unviversity of Oslo, P.O.Box 1048, Blindern,
N-0315 Oslo, *Norway*

PIEL, Alexander
Institute for Experimental Physics, University of Kiel, Olshausenstr. 40-60,
DW-2300 Kiel, *Germany*

POPA, Gheorghe
Faculty of Physics, University "Al. I.Cuza", Blvd. Copou Nr. 11,
RO-6600 IASI, *Romania*

RAADU, Michael A.
Plasma Physics Department, Alfvén Laboratory, Royal Institute of Technology,
S-10044 Stockholm, *Sweden*

RAPOZO,Candido da Cunha
Laboratorio de Fisica de Plasma, Universidade Federal Fluminense,
Niteroi, RJ 24.020, *Brazil*

RASMUSSEN, Jens Juul
Optics and Fluid Dynamics Department, Risø National Laboratory,
DK-4000 Roskilde, *Denmark*

SAKHAROV,Alexander
General Physics Institute, Plasma Physics Department,
Russian Academy of Sciences, Vavilova 38,
117942 Moscow, *Russia*

SANDULOVICIU, Mircea
Faculty of Physics, University "Al. I.Cuza", Blvd. Copou Nr. 11,
RO-6600 IASI, *Romania*

SATO, Naoyuki
Department of Electrical and Electronic Engineering,
Ibaraki University, 4-12-1 Nakanarusawa,
Hitachi, Ibaraki 316, *Japan*

SATO, Noriyoshi
Department of Electronic Engineering, Tohoku University, Aramaki Aza-Aoba,
Sendai 980, *Japan*

SATO, Teruyuki
Plasma Science Center, Nagoya University, Furoh-cho
Nagoya 464-01, *Japan*

SAXENA, Yogesh C.
Institute for Plasma Research, Near Indira Bridge, Bhat,
Gandhinagar 382 424, *India*

SCHAMEL, Hans
Physical Institute, University of Bayreuth, Universitätsstrasse,
DW-8580 Bayreuth, *Germany*

SCHLÜTER, Hans-Wilhelm
Institute for Experimental Physics II, NB 5/132,
Ruhr-Universität, Universitätstrasse 150,
DW-4630 BOCHUM 1, *Germany*

SCHRITTWIESER, Roman W.
Institute for Ion Physics, University of Innsbruck, Technikerstr. 25,
A-6020 Innsbruck, *Austria*

SEN, Abhijit
Institute for Plasma Research, Near Indira Bridge, Bhat,
Gandhinagar 382 424, *India*

SHIVAROVA, Antoniya P.
Faculty of Physics, Sofia University, 5.A.Ivanov Blvd.,
BG-1126 Sofia, *Bulgaria*

SIEBENFÖRCHER, Armin
Institute for Ion Physics, University of Innsbruck, Technikerstr. 25,
A-6020 Innsbruck, *Austria*

SKORIK, Max
Department of Engineering Sciences, University of Oxford, Parks Road,
OXFORD, OXI 3PJ, *UK*

STENFLO, Lennart
Department of Plasma Physics, Umea University,
S-90187 UMEA, *Sweden*

STENZEL, Reiner
Department of Physics, University of California, 405 Hilgard Ave,
Los Angeles, CA 90024-1547, *USA*

TAKEDA, Susumu
1-12-19, Gakunancho,
Okayama 700, *Japan*

TAKEDA, Yasumasa
Department of Physics, College of Science and Technology, Nihon University,
1-8, Kanda Surugadai - Chiyoda-ku,
Tokyo 101, *Japan*

TALASMAN, Sorin J.
Faculty of Physics, University "Al. I.Cuza", Blvd. Copou Nr. 11,
RO-6600 IASI, *Romania*

TANIKAWA, Takao
Institute of Research and Development, Tokai University,
2-8-1-705, Tsurumaki-kita, Hadano,
Kanagawa 257, *Japan*

TANIZUKA, Noboru
Integrated Arts & Sciences, University of Osaka Prefecture, Gakuen-cho 1-1,
Sakai 593, *Japan*

TATARONIS, John A.
Department of Electrical and Computer Engineering,
University of Wisconsin, 1500 Johnson Drive
MADISON, WI 53706-1687, *USA*

TORVÉN,Staffan
Plasma Physics Department, Alfvén Laboratory, Royal Institute of Technology,
S-10044 Stockholm, *Sweden*

UEHARA, Mituo
Department de Fisica di ITA, Universidade do Vale do Paraiba, H-23A Nr.102,
12225 Sao José dos Campos, *Brazil*

VALENTINI, Hans-Burkhard
Institut für Physikalische Hochtechnologie, Helmholtzweg 4,
DO-6900 Jena, *Germany*

VERHEEST, Frank
Institute for Theoretical Mechanics, Rijksuniversiteit Gent, Krijgslaan 281
B-9000 Gent, *Belgium*

VOLWERK, Martin
Sternekundig Instituut Utrecht, P.O.Box 80.000
NL-3508 TA - Utrecht, *Netherlands*

WANG, En-Yao
Southwestern Institute of Physics, P.O.Box 432, Chengdu,
Sichuan 610041, *China*

WINKLER, Carsten
Institute for Ion Physics, University of Innsbruck, Technikerstr. 25,
A-6020 Innsbruck, *Austria*

ZHYKHARSKY, Aleksandr V.
Scientific-Technical Centre "Kinetika", P.O.Box 36,
630078 Novosibirsk, *Russia*

ZIMMERLING, Martin
Institute for Ion Physics, University of Innsbruck, Technikerstr. 25,
A-6020 Innsbruck, *Austria*

ZOLER, David
Department of Physics and Astronomy, Tel-Aviv University,
69978 Tel-Aviv, *Israel*

Author Index:

INVITED LECTURES

THE FORMATION OF DOUBLE LAYERS IN A MERCURY ARC DISCHARGE

J.E. Allen and H.S. Maciel,[1] Department of Engineering Science, University of Oxford, Oxford OX1 3PJ, U.K.

A study has been carried out of free double layers formed within the plasma volume of a low–pressure mercury–arc discharge at high current densities. The layer is observed to form in the later stages of current limitation, the onset of which occurs for a ratio of drift to thermal speed of electrons of about unity. The free 'double layer' is found to have a triple or quadruple layer structure. Electrical energy is converted by the layer into kinetic energy of the charged particles.

1. INTRODUCTION: A double layer (DL) is a region of non–neutral plasma consisting of two adjacent space charge layers, one positively charged and one negatively charged. Recently, it has become clear that DLs are only a special case of a more general multiple–space–charge–layer phenomenon. These structures are observed to form in computer simulated and laboratory experiments in current–carrying and voltage–driven plasmas as well as in plasmas driven by beam–particle injection or wave launching. Additional experimental evidence is provided by measurements in the auroral zone.

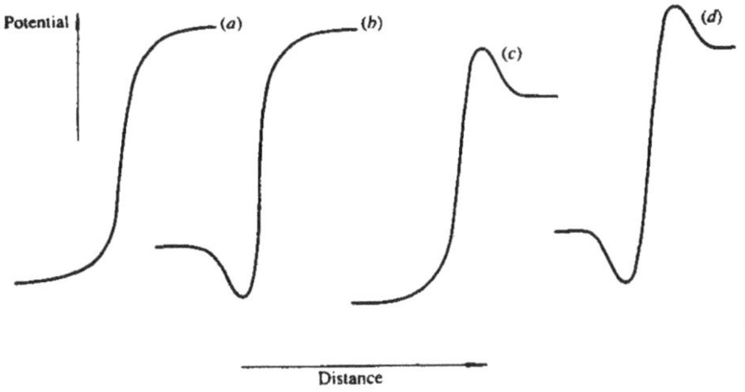

Figure 1 – Schematic plots of the various potential profiles of multiple layers: profile (a) corresponds to a double layer, profiles (b) and (c) correspond to triple layers and profile (d) corresponds to a quadruple layer.

Figures 1(*a*–*d*) show the topologies of potential profiles of double and multiple layers that have been observed in laboratory and in computer experiments. An ordinary DL necessarily has a monotonic potential profile (figure 1*a*), whereas a triple layer contains a potential dip (figure 1*b*) or a potential hump (figure 1*c*) and a quadruple layer contains a potential dip and a potential hump (figure 1*d*). Intense electric fields, localized low conductivity, acceleration of electrons and ions to appreciable energies, turbulence and current disruption in certain cases are all important features of DLs; see the reviews by Torvén,[1] Schamel,[2] Hershkowitz[3] and Allen.[4]

In gas discharges DLs can be formed as a matching agency between a cathode and the positive column, or they form to control the ionization and the flow of charged par-

[1] Present address: Departamento de Física ITA/CTA–12225. São José dos Campos. São Paulo, Brazil.

ticles at a constriction, or, more spectacularly, they form in discharges with uniform cross–section as a result of processes that lead to current limitation in the discharge. This latter DL, which was named the 'free double layer' by Levine & Crawford,[5] is the main subject of this lecture which is based on the work of Maciel and Allen.[6]

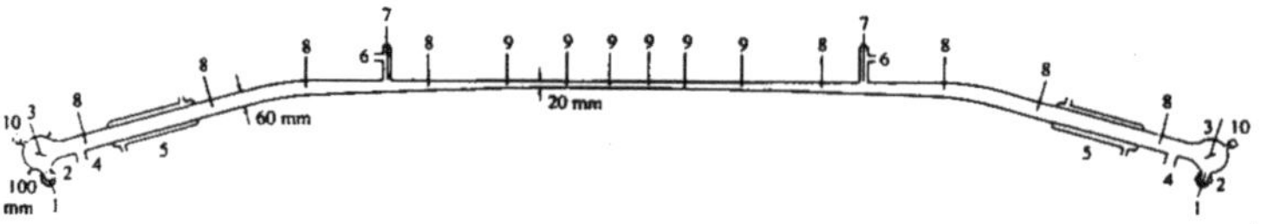

Figure 2 – Experimental apparatus: 1, tungsten electrode; 2, cooling jacket surrounding the mercury pool; 3, umbrella; 4, port to diffusion pump; 5, cooling jacket; 6, ionization (pressure) gauge; 7, auxiliary electrode; 8, fixed probe; 9, mobile probes P1, ..., P6; 10, filler.

2. EXPERIMENTAL APPARATUS: The aims of the design of the experimental apparatus were, first, to achieve the conditions for the formation of a free DL and, secondly, to obtain a sufficient degree of control over the physical variables that a proper experimental verification of the DL formation and its internal structure would be possible under a range of conditions of the discharge. The discharge tube used by Stangeby & Allen[7] was chosen as the basis for the design. It is long (4.7 m) in order to avoid the disturbing effects of the end electrodes over the central part, which constitutes the experimental region. In fact, for the positive column to be truly independent of the electrode regions, the discharge must be several metres long, since only then is the 'average electron' more likely to be lost to the column wall than to the anode (a vapour pressure of the order of millitorr is assumed).[8] Sharp bends and abrupt constrictions in the overall tube are avoided. Not only would they act as a source of noise and fluctations in the discharge, but also a constriction DL may well form to aid current continuity. This kind of DL should be avoided because our intention is to study the formation of a free DL in the absence of any disturbances. An overall view of the discharge tube is given in figure 2. It comprises three main sections: a central section 1.2 m long and of 28 mm external diameter, a left–hand section and a right–hand section. Both end sections are similar and are connected to the central one by gradually and smoothly reducing their diameter from 60 mm to 28 mm over a length of 30 cm, thus avoiding abrupt constriction which could induce undesirable constriction–DL formation. The whole tube system is symmetrical and the discharge can be run in either direction. Further details can be found in the paper by Maciel & Allen.[6]

3. EXPERIMENTAL RESULTS:

3.1. *The voltage–current characteristic of the discharge*: The voltage–current (V-I) characteristic of the discharge is measured under conditions such that layer formation can be achieved. These conditions refer mainly to the mercury–vapour pressure, which is in the range (1.2 – 3.0)×10⁻⁴ Torr. The impurity level must be below 10 % and the cathode spot well anchored. We observed that discharge survival during the layer formation is very critical, and excess impurities and spot motion not only hinder layer formation but may lead to a current cut–off as the discharge is driven towards layer formation.

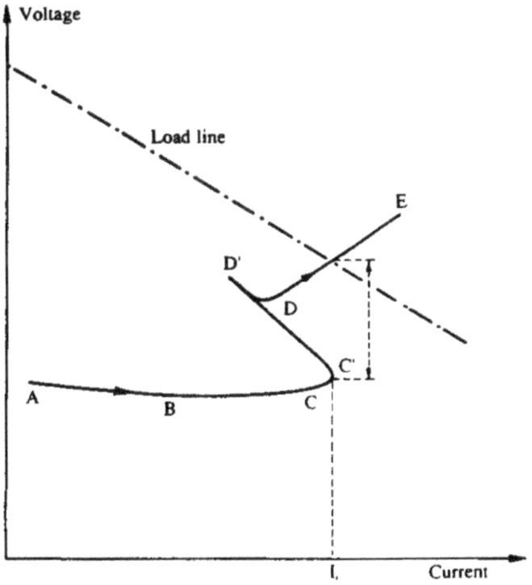

Figure 3 – Schematic diagram of the d.c. voltage–current characteristic, showing the different modes of operation of the discharge: A to B, non-oscillatory mode; B to C, stable oscillatory mode; C′ to D′, unstable oscillatory mode. I_c is the critical current at which current limitation takes place.

The schematic picture of figure 3 shows the general features of the d.c. V-I characteristic of the discharge at low pressure. The whole characteristic is divided into regions, which can be described as follows.

The low–current region from point A to point B corresponds to the standard arc characteristic in which the voltage decreases slightly with increasing current. At higher current (region B to C) a positive characteristic is seen, in which weak oscillations are observed, the current ripple being below 1 % and the frequency a few tens of kilohertz. C′ is a critical point at which the operational point moves spontaneously along the load line towards higher voltages and lower currents. The transition from C′ to D′ occurs in a turbulent way, and the current shows spikes corresponding to ripple as high as 80 %. This turbulent process ends with the appearance of a visible layer across the central column. After the layer has been established, the current may be increased further (region D to E). In that part of the characteristic, the discharge performs a well–defined relaxation oscillation of approximately 200 kHz and the current ripple is about 20 %.

An experimental d.c. V-I characteristic is shown in figure 4. The d.c. voltage and current were recorded on an X–Y plotter while the external resistance was reduced gradually in order to increase the current. The temperature of the cooling jackets was kept constant and the vapour pressure was sufficiently low (about 2×10⁻⁴ Torr) that the characteristic was slightly positive even for a current as low as 3 A. Weak oscillations are observed in the stable oscillatory mode with a ripple of about 0.4 % at about 30 kHz. At the critical point $V \simeq 90$ V, $I \simeq 6$ A a strong turbulent process is triggered and the unstable oscillatory mode begins. This mode progresses spontaneously, with the operational point moving up the load line towards higher voltage and lower current. At the point $V \simeq$ 140 V, $I \simeq 5$ A the visible layer shows up in the central column, ending the turbulent process and starting the layer mode. The operational point still moves back slightly along the same load line owing to the layer motion, but throughout the layer mode the discharge voltage is significantly higher than in the stable oscillatory mode. Most of this voltage difference is concentrated along the layer. The oscillations in the layer mode are of relaxation type with a current ripple of about 20 % at about 200 kHz, which is a low frequency compared with the ion–plasma frequency $f_{pi} \simeq 3$ MHz.

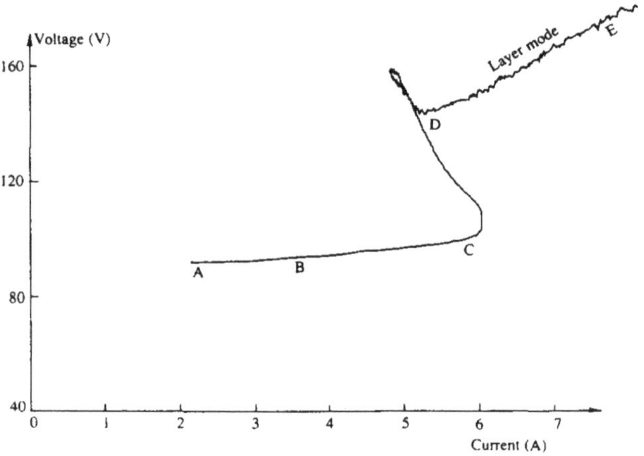

Figure 4 – Experimental plot of the voltage–current characteristic of the discharge. Vapour pressure 1.5×10^{-4} Torr.

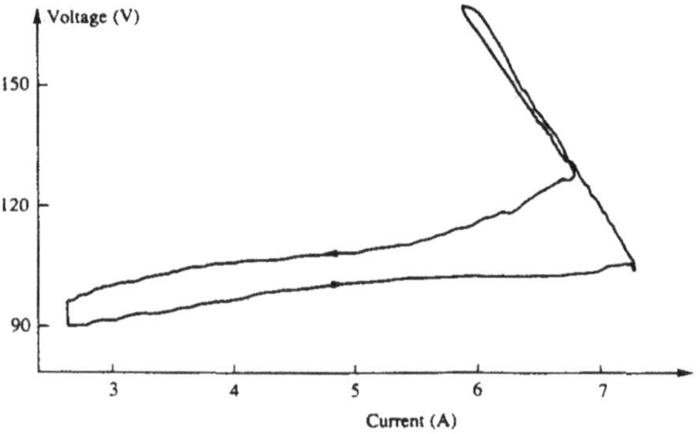

Figure 5 – Hysteresis loop formed by increasing the current until the free double layer is formed and then reducing the current until the layer disappears.

In the layer mode the current may be varied – not only increased but also decreased. In the latter case the V-I characteristic describes a hysteresis curve (see figure 5). The loop is closed when the current is reduced so much that the layer becomes unstable and can no longer survive.

3.2. *Layer formation and current limitation*: The best condition for the formation of the layer is that in which the pressure in both ends of the central column are balanced, but it can be formed for small imbalances, in contrast with the findings of Levine & Crawford.[5] We have seen that the layer always formed in the final stage of the unstable oscillatory mode, which is triggered after a certain critical current has been reached. At that critical point (point C′ in figure 3) the incremental resistance of the discharge, as indicated by the derivative dV/dI, tends to infinity. This apparent limit to the discharge current has been known as 'current limitation' in the gas discharge literature. It is clear that the current limitation is not the result of double–layer formation but rather the converse is true.

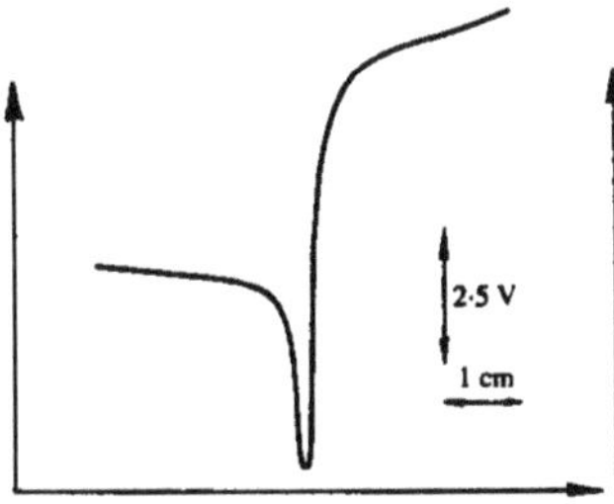

Figure 6

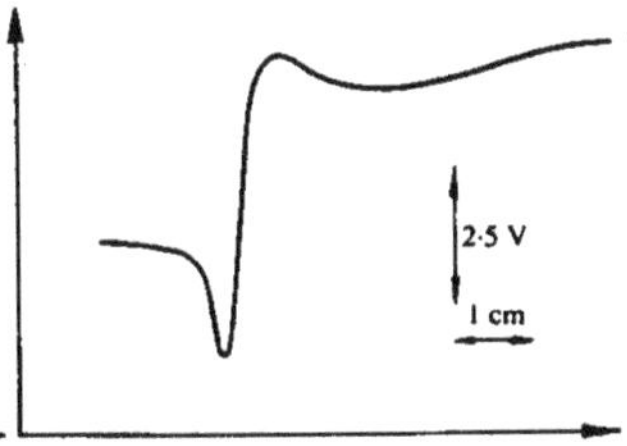

Figure 7

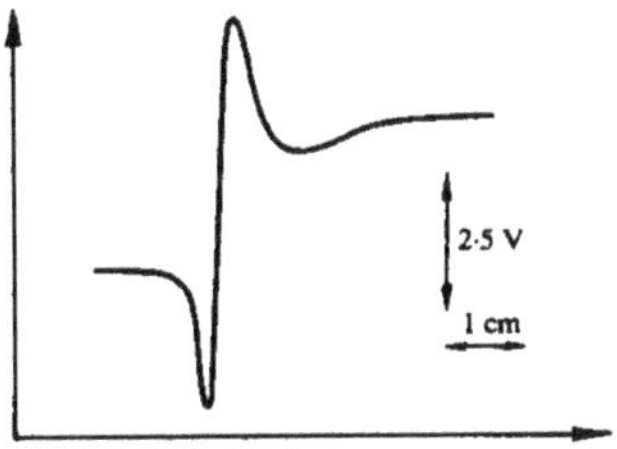

Figure 8

Figure 6 – Experimental plot of the most common potential profile of the free double layer.
Figure 7 – Experimental potential profile of the free double layer with a slight potential drop on the anode side.
Figure 8 – Experimental profile of the layer, showing the formation of a potential hump. In this case the free double layer has a quadruple–layer structure.

3.3. *Potential profiles of the layer*: In the present work the layer is produced in such a condition that it moves against the electron flux with approximately constant velocity. During its transit it crosses the probes P1, ..., P6 placed along the central column (see figure 2). The d.c. floating potential profile of the layer is recorded with an X–Y plotter in the time–base mode. The time co–ordinate is converted into axial position to within an error of not more than 20 %. The probes are of tungsten and flat–faced, with the unshielded surface parallel to the axis. In this way the effect of electron drift velocity on the floating potential is minimized. The probes are radially movable so that the axial distribution of potential can be found at different radial positions. Such measurements are necessary for determining the three–dimensional structure of the layer.

Basically there are two kinds of axial potential profiles. Figure 6 shows the most typical one, which consists of a potential jump preceded by a potential dip. The axial electric field in the anode–side plasma is higher than that in the cathode–side plasma but much smaller than the axial electric field within the layer. The presence of the potential dip preceding the potential jump is the main difference from the usual monotonic profile of a double layer. This feature is common to any potential profile of free double layers measured in a mercury–arc discharge. One variation of the typical profile is shown in figure 7. It retains the principal features of the typical profile but shows a slight drop of the potential on the anode side, after which the potential picks up again. Figure 8 shows a significant modification to the profile. In this case, instead of a slight potential drop, there is formation of a perfect potential hump bordering the anode plasma. We have observed that the height of the potential hump can be even bigger than the depth of the potential dip.

We refer to figures 6, 7 and 8 as profiles type I, II and III respectively. All of these structures may be formed during the same transit of the layer; for instance, the type III profile with a small potential hump at P2, type II at P3, type I at P4, type II at P5 and type III with a higher potential hump at P6. This shows that the layer structure changes as its moves in different regions of the discharge column. It is clear that the structure of the type II profile represents a transitional stage between the type I and type III structures.

A good experimental condition for this kind of measurement is obtained with a difference in the cooling temperature of both jackets of not more than 2°C. The effect of increasing the coolant temperature and thus the vapour pressure is to reduce the potential differences across the whole layer. The type III profile is hardly ever found for coolant temperature above 5°C.

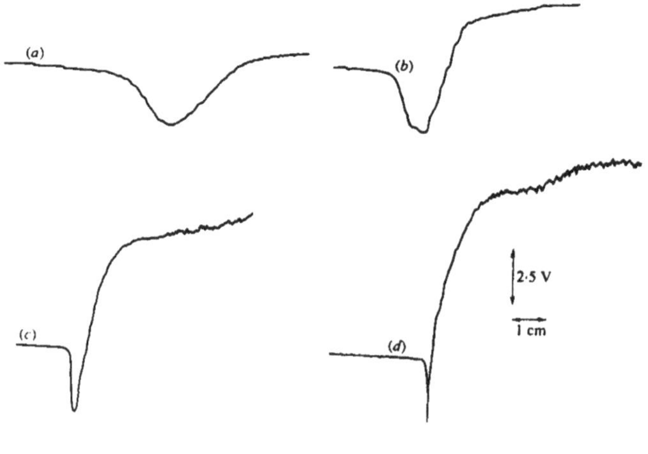

Figure 9 – Axial potential profiles across the free double layer, taken with the probe at different positions from the walls: (a) wall; (b) 2 mm; (c) 10 mm; (d) 12.5 mm; pressure 2×10^{-4} Torr.

While the motion of the type I and type II layers is not disturbed very much during crossing of the probes, the type III layer is very sensitive to the presence of the probe and may move back and forth if the probe is too deep inside the discharge column. Measurement of this profile with the probe on the axis has not been successful.

The three–dimensional structure of the layer can be found from the set of axial potential profiles taken for various radial positions. Figure 9 shows one set of these profiles for the typical layer. Owing to the probe disturbance described above, we have not obtained a complete set of profiles for the type III layer.

Figure 9(*a*–*d*) show that the potential well of the profile becomes deeper and narrower near the axis while the potential jump becomes higher and steeper.

3.4. *Measurement of the electron drift velocity*: An important parameter for the description of phenomena concerned with plasma instability and free–double–layer formation in both theoretical and experimental work is the ratio between the electron drift speed and the electron thermal speed. Torvén & Babić[9] observed that layer formation in a mercury discharge occurred for values of this ratio in the range 0.9 – 1.1. Levine & Crawford[5] reported a much lower value of about 0.3 in a discharge in argon. In all the early work the electron drift velocity was estimated from the discharge current and the electron density.

We have developed a new technique for obtaining a better measurement of the ratio of drift to thermal speed of the electrons. We use a directional probe,[10] which is a plane probe that can be rotated so that the perpendicular to its exposed surface can assume any angle with respect to the electron flow direction (axial). The probe is placed on the axis but can also be moved to other radial positions.

The following equation is a good approximation for small values of γ, where γ is the angle between the normal to the probe surface and the radial direction:

$$F(\gamma) = \left[\frac{I(\gamma)}{I_0} + \frac{\pi}{4} - 1\right]^{\frac{1}{2}} - \left(\frac{\pi}{4}\right)^{\frac{1}{2}} = \frac{R \sin \gamma}{2^{\frac{1}{2}}}, \tag{1}$$

$I(\gamma)$ is the electron saturation current for angle γ and $I_0 = I(0)$. The function $F(\gamma)$ can be constructed from the probe characteristic taken at different angles. The ratio R of the drift and thermal velocities of the electrons can be obtained from the graph $F(\gamma)$ versus $\sin \gamma$ which is a straight line. The error involved in the above approximation is not critical in the range $-20^0 < \gamma < 20^0$ because only terms of order higher than $\sin^2\gamma$ have been ignored.

The advantage of this method is that it does not involve intermediate calculations of other parameters like electron temperature and electron density and does not rely on the measurement of a single value of the electron saturation current. It still requires determination of the plasma potential, but it is not too sensitive to the experimental error involved in this determination. However, this method demands more care with respect to the experimental procedure, for example it is very important to have good probe orientation. This can be obtained by looking at the floating potential of the probe: the probe will be at $\gamma = \pi/2$ when the floating potential reaches its maximum absolute value. An orientation with uncertainty less than 2^0 is obtainable when the drift velocity is high.

Figure 10 shows a set of the directional probe characteristics.

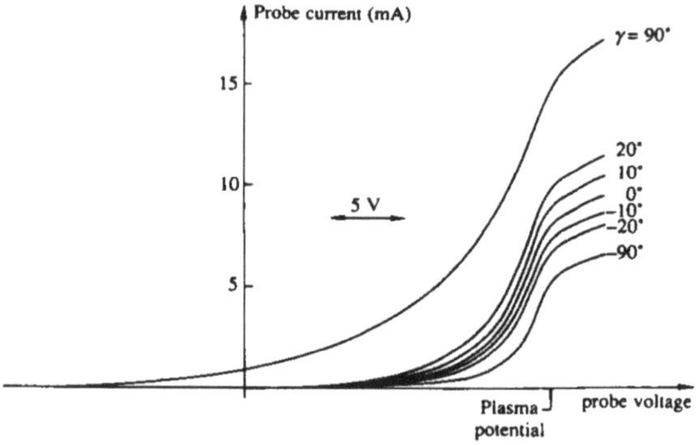

Figure 10 – Current–voltage characteristics of the directed probe for different angles γ.

3.5. *Results of probe measurements*: Measurements of T_e, N_{0e} and R were made on the axis of the column, while the discharge was driven towards current limitation by keeping the discharge current at the constant value $I = 5.6$ A, and gradually decreasing the coolant temperature T_0. We have determined this set of discharge parameters for a fixed value of $T_0 = 25^0$C and again for every 5^0C drop of T_0 until the discharge shows signs of being near current limitation. The results are given in table 1.

The experimental error involved in the determination of T_e, N_{0e} and R are estimated to be about 2 %, 5 % and 5 % respectively. T_0 has an uncertainty of 0.2^0C. The number density on the axis was deduced from measurements made with the a wall probe.

$T_0(^0C)$	T_e(eV)	$N_{0e}(10^{16}\ m^{-3})$	R
25	1.9	47	0.40
20	2.1	37	0.45
15	2.4	32	0.48
10	3.1	22	0.65
5	8.2	7.4	0.99

Table 1

Figure 11 shows the plot of R versus T_0. The observed behaviour of the discharge is also indicated in the figure. We can see that current limitation is reached for temperatures below 5^0C, for which the drift velocity is of the order of the electron thermal velocity. This agrees with the values estimated by Torvén and Babić.[9] Levine and Crawford[5] estimated a low value $R \simeq 0.3$, possibly because they ignored the effect of electron drift on the probe characteristic, from which the electron density was obtained.

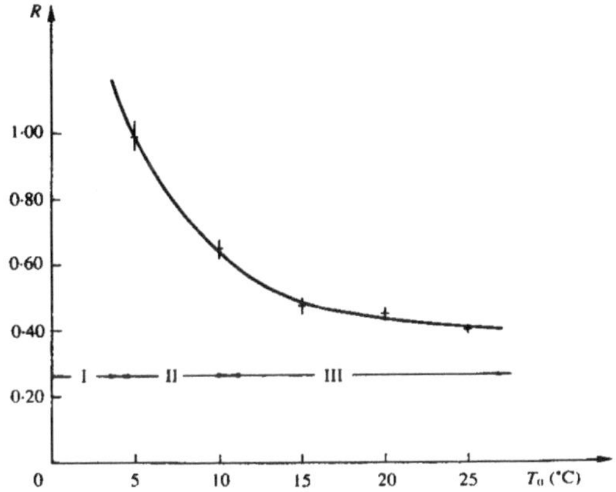

Figure 11 – Variation of the ratio of drift and thermal speeds of the electrons as the discharge is driven towards current limitation by lowering the temperature of the coolant. The discharge current is held at 5.6 A.

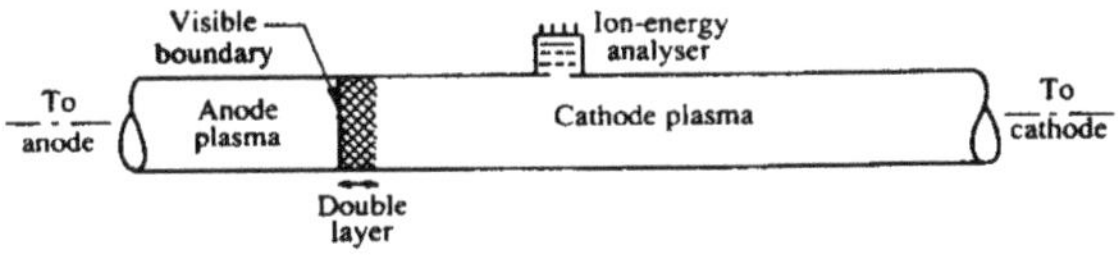

Figure 12 – Schematic diagram showing the experimental arrangement for measurements with the ion–energy analyser.

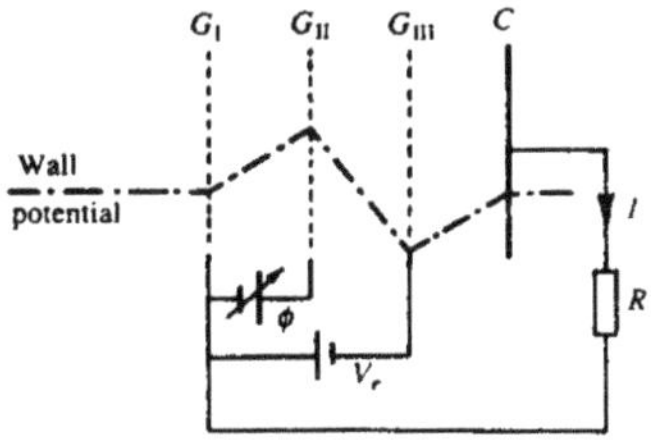

Figure 13 – Schematic diagram showing the polarization of the grids of the analyser.

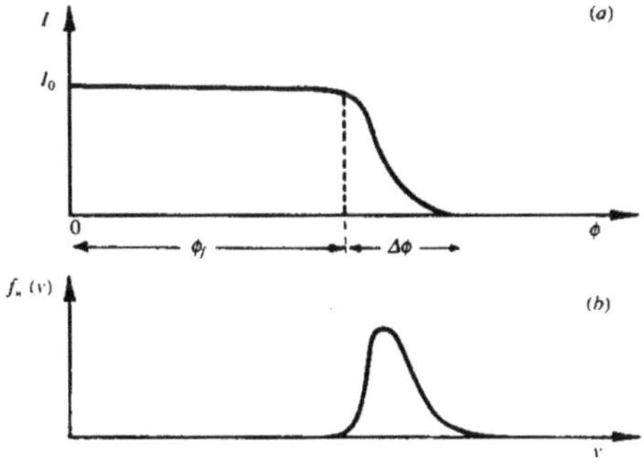

Figure 14 – Schematic diagrams of (a) ion current versus retarding voltage and (b) ion distribution function of radial velocity at the wall; $v = (2e\Phi/m)^{\frac{1}{2}}$.

3.6. *Measurement of the ion energies*: For measurement of ion energies we have built a small multigrid electrostatic energy analyser, which is placed flush with the wall. It can be used to measure the radially directed energy of the ions. Figure 12 depicts the experimental arrangement. The ion–energy analyser (figure 13), has three grids made of stainless–steel mesh with a transparency of 50 %. The holes in the mesh have dimension 0.09 mm, which is less than the Debye length, approximately 0.2 mm. The grid GI is at the potential of the wall. A retarding voltage Φ is applied to grid GII. At GIII a voltage is applied that is high enough to reflect all the electrons. This polarization reduces the current due to secondary electrons emitted from the collector C. The ion current is measured from the potential drop across the small resistance R. This potential drop is less than 1 mV; the collector therefore stays practically at the wall potential.

The ions enter the wall sheath edge with a distribution function of velocities obeying the Bohm condition, and travel from the sheath edge towards the wall in a free flight under the influence of the accelerating electric field. If $f_w(v)$ is the distribution of the ion radial velocities at the wall then the ion current at the collector is

$$I = I_0 - eA \int_0^{v^*} v f_w(v)\, dv. \tag{2}$$

where $v^* = (2e\Phi/m_i)^{\frac{1}{2}}$, I_0 is the ion current when there is no ion reflection and A is the area of the collector. Differentiation with respect to Φ gives the distribution function at the wall:

$$f_w\left[\left[\frac{2e\Phi}{m_i}\right]^{\frac{1}{2}}\right] = -\frac{m_i}{Ae^2}\frac{dI}{d\Phi}. \tag{3}$$

Figure 14(a) is a schematic plot of the ion current I versus retarding voltage Φ, and figure 14(b) shows the corresponding ion velocity distribution function at the wall. The ion current remains constant for retarding voltage up to $\Phi = \Phi_f$. At higher voltage the current decreases owing to the reflection of ions, and falls to zero at $\Phi = \Phi_f + \Delta\Phi$. The voltage Φ_f accounts for the drop in potential across the sheath and $\Delta\Phi$ for the distributed energy of the ions at the wall. According to the positive–column theory, the expected values of Φ_f and $\Delta\Phi$ are about $5.5\ kT_e/e$ and kT_e/e respectively.

Figure 15 shows the experimental plot of the I versus Φ characteristic in the three possible cases. The discharge current is about 5.6 A and the pressure is about 2.0×10^{-4} Torr in all cases. In the case of curve I, the discharge is capable of sustaining a double layer but the formation of the latter has not yet taken place. After the layer is formed, the curve II is taken at the cathode side and curve III is taken just at the anode side of the layer. The characteristic at the cathode side remains unaltered for different distances between the double layer and the analyser, but this is not true on the anode side; that is, as the double layer moves away, the characteristic changes from curve III to curve IV.

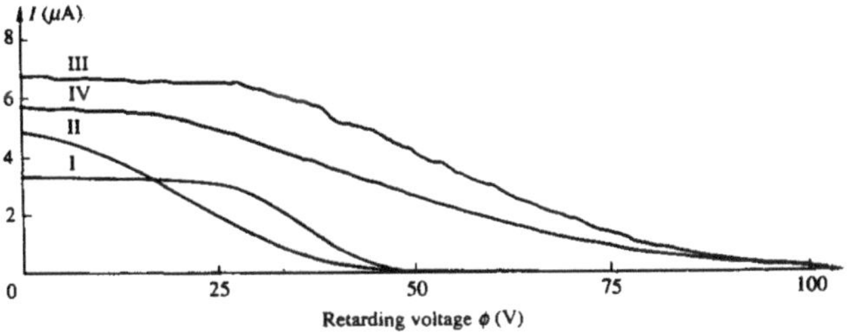

Figure 15 – Experimental current versus retarding voltage characteristic of the ion–energy analyser: (I) before layer formation; (II) on cathode side of the layer; (III) just on the anode side of the layer; (IV) on anode side but 30 cm from the layer. The discharge current is held at 5.6 A for all the curves.

Comparisons of curve I (before layer formation) with curve II (at the cathode side of the layer) and curve III (at the anode side of the DL) show a remarkable effect of the layer on the distributions of ion energies in the surrounding plasmas. There is a strong 'heating' of the ions at the anode side and a 'cooling' on the cathode side. The extraordinarily high energies that the ions can achieve (above 100 eV) in the anode plasma are to be explained in terms of transfer of energy from the electrons accelerated by the DL.

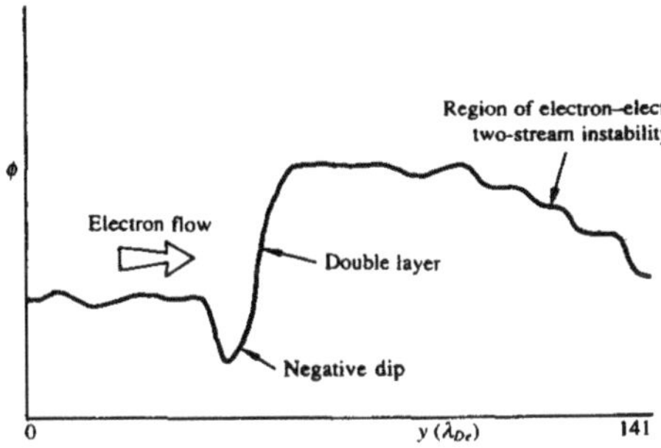

Figure 16 – Schematic diagram of a double layer with a potential dip as seen in the simulations of Ref. 11. The plot is taken from Ref. 13.

4. CONCLUDING REMARKS ON THE CHARACTER OF THE FREE DL: Our experiments on the free DL have led to two main results. First, the layer has a non–monotonic potential profile and instead corresponds to a triple or a quadruple layer rather than a double layer as has been supposed in previous work. Secondly, the layer is formed in the later stages of current limitation in the arc discharge. The first result comes from probe measurements and the second from observations of the V-I characteristic of the discharge, which show that, after current limitation occurs, the discharge evolves towards a new mode of operation – the layer mode – which is inaugurated by the layer formation. The potential dip on the cathode side of the layer was not quite noticable in the

results reported by Levine & Crawford,[5] who first attempted to make probe measurements inside a free DL. The first time the DL was observed to form with a dip on the low–potential side was in the numerical experiments performed by DeGroot *et al.*[11] Interest in this potential dip accompanying the DL potential transition arose following the numerical simulations of Sato & Okuda,[12] who observed that this kind of DL was shown to form from ion–acoustic fluctuations in a plasma carrying current.

Figure 16 shows the non–monotonic potential profile first obtained in a simulation (see the review by Borovsky[13]). This DL was observed to form with a scale length of (20 – 60)λ_D in a plasma of cold ions and warm electrons when the electron drift velocity was of the order of or greater than the electron thermal velocity. To make a comparison with our measurements, figure 17 shows a typical potential profile of the free DL. It has the same topology and comparable scales of length and potential as the DL seen in the simulation of DeGroot *et al.*[11]

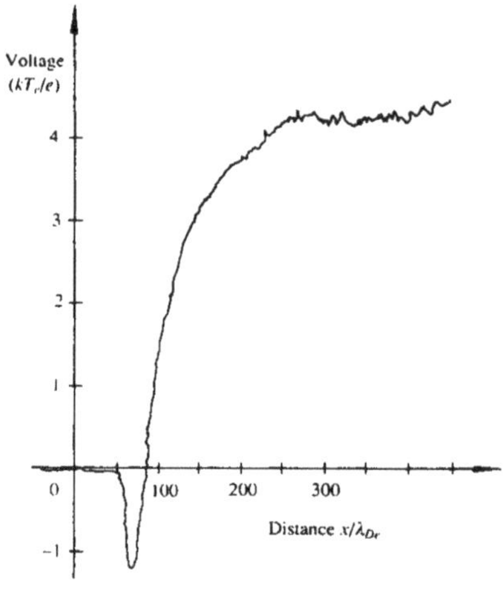

Figure 17 – Experimental potential profile of the free double layer, showing the approximate scales of distance and voltage.

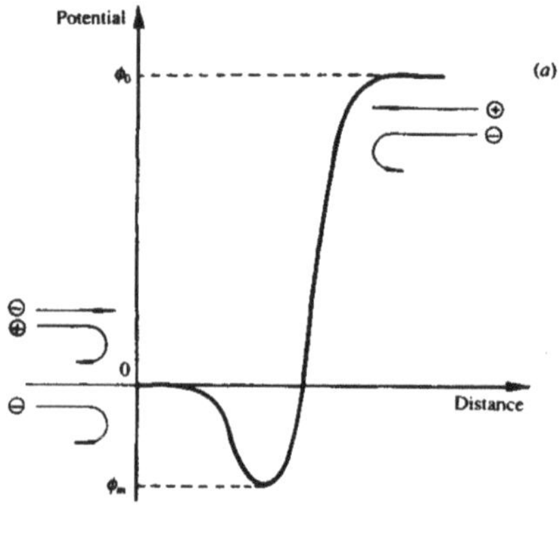

Figure 18 – Typical profile of a layer (triple layer) and sample electrons (–) and ions (+) that are reflected and transmitted through the layer.

The potential hump with which the free DL sometimes appears has been observed by Singh & Schunk [14] in numerical simulations. They associate the hump in the high–potential side of the DL with the development of an electron hole due to the excitation of an electron–electron two–stream instability in that region.

One of the important characteristics of the layer model concerns the charged–particle populations that build up the layer space charge. A typical potential profile of the layer that is measured experimentally is shown in figure 18. It is a potential profile of a triple layer separating the cathode plasma at zero potential from the anode plasma at a higher potential Φ_0. The potential minimum is Φ_m. Ions that originated in the cathode plasma are reflected back at some point of the layer potential jump. They constitute a population of ions trapped at the cathode side. On the other hand, the ions that originated in the anode plasma may fall through the layer, thereby constituting the free–ion population. We deduce the existence of an 'ion hole' in the ion phase space. In fact, layers that have similar topology have been named 'asymmetric ion holes'.[2]

The electrons that originate in the cathode plasma are partly reflected by the barrier of the potential dip and partly transmitted through the layer. Those reflected constitute the population of electrons trapped at the cathode side and those transmitted constitute the free–electron population. The electrons that originated in the anode plasma are mostly reflected by the layer. They constitute a population of electrons trapped at the anode side.

Although there are electron trajectories that cross the layer from the anode to the cathode side, they are in a minority because to do this they must have an energy at least equal to $e(\Phi_0 + |\Phi_m|)$. Their contribution to the free population can therefore be neglected. Similar arguments apply to ions that cross the layer from the cathode to the anode side.

In summary, the triple layer comprises five particle populations:

f_{etc}, electrons trapped at the cathode side;

f_{eta}, electrons trapped at the anode side;

f_{ef}, free electrons;

f_{itc}, ions trapped at the cathode side;

f_{if}, free ions.

Here the f_s are distribution functions and the subscripts refer to electrons (e) or ions (i), trapped (t) or free (f) in the cathode side (c) or anode side (a). Calculations based on this model have been reported elsewhere.[15]

REFERENCES:

1. S. Torvén, in *Wave Instabilities in Space Plasmas* (ed. P.J. Palmadesso & K. Papadopoulos), Reidel, (1979) p. 109.
2. H. Schamel, in *Proceedings of the 1st Symposium on Plasma Double Layers* (ed. P. Michelsen & J.J. Rasmussen), Risø National Laboratory, Roskilde, Denmark (1982), p. 13.

3. N. Hershkowitz, *Space Sci. Rev.* **41** (1985), 351.
4. J.E. Allen, *Plasma Phys. Contr. Fusion* **27** (1985), 1343.
5. J.S. Levine and F.W. Crawford, *J. Plasma Phys.* **24** (1980), 359.
6. H.S. Maciel and J.E. Allen, *J. Plasma Phys.* **42** (1989), 321.
7. P.C. Stangeby and J.E. Allen, *J. Phys.* D **6** (1973), 224.
8. F.W. Crawford and S.A. Self, *Int. J. Electron.* **18** (1965), 569.
9. S. Torvén and M. Babić, in *Proceedings of the 12th Int. Conf. on Phenomena in Ionized Gases* (ed. J.G.A. Hülscher & D.C. Schram), Eindhoven, Netherlands, Elsevier (1975), p. 124.
10. J.E. Allen and F. Magistrelli, *C.N.E.N. Report* RTI/FI 62 (1962), 7.
11. J.S. DeGroot, C. Barnes, A.E. Walstead and O. Buneman, *Phys. Rev. Lett.* **38** (1977), 1283.
12. T. Sato and H. Okuda, *Phys. Rev. Lett.* **44** (1980), 740.
13. J.E. Borovsky, in *Proceedings of the 2nd Symposium on Plasma Double Layers and Related Topics* (ed. R. Schrittwieser & G. Eder), University of Innsbruck (1984), p. 33.
14. N. Singh and R.W. Shunk, *Plasma Phys. Contr. Fusion* **26** (1984), 859.
15. H.S. Maciel and J.E. Allen, in *Proceedings of the 2nd Symposium on Plasma Double Layers and Related Topics* (ed. R. Schrittwieser & G. Eder), University of Innsbruck (1984), p. 218.

STRONGLY DRIVEN SHEATHS

C.K. Birdsall and V. Vahedi
Plasma Theory and Simulation Group
Department of Electrical Engineering
and Computer Sciences
University of California at Berkeley
Berkeley, CA 94720

Abstract

It is remarkable that a 1 eV bounded plasma may be driven by a 100 or 1000 volt signal (DC or RF) on an electrode, yet remain a bounded plasma, under rather ordinary conditions. An isolated charge or a single species, of course, would be accelerated away. Our talk presents some examples of strongly driven plasmas and makes some estimates of why the plasmas are "contained" and not destroyed. The examples are drawn from computer experiments,[1] starting from ideal models and adding more physical features, until a "complete" model is obtained.[2] (The "complete" models compare well with laboratory experiments.)

The main example is a voltage driven RF discharge. The initial model has warm electrons but immobile ions, the so-called Godyak homogeneous model. We will present this model in detail. Then we will add, in steps, mobile ions, background gas, ionization, excitation, scattering, and charge exchange, noting the effects obtained with each addition. The full model will be a current driven discharge.

Acknowledgments

Work supported by the U.S. Navy Office of Naval Research, Contract No. N00014–85–K0809 and the U.S. Department of Energy, Contract No. DE–FG03–90ER 54079.

References

1. C.K. Birdsall, "Particle-in-cell charged particle simulations, plus Monte-Carlo collisions with neutral atoms, PIC-MCC," *IEEE Trans. Plasma Science* **19**(2), pp. 65-85, April 1991.
2. V. Vahedi, M.A. Lieberman, M.V. Alves, J.P. Verboncoeur, and C.K. Birdsall, "A one-dimensional collisional model for plasma-immersion ion implantation," *J. Appl. Phys.* **69**(4), pp. 2008-2015, 15 February 1991.
3. V.A. Godyak and R.B.Piejak, "Abnormally low electron energy ad heating-mode transition in a low-pressure argon rf discharge at 13.56 MHz," *Phys. Rev. Lett.* *65*(8), 996, 20 August 1990.
4. B.P. Wood, Chapter 3 in "Sheath heating in low pressure capacitive radio frequency discharges," Ph.D. thesis, University of California, Berkeley, CA, November 1991.

A Simple Voltage Driven RF Discharge with Mobile Electrons and Fixed Uniform Ions (Godyak Homogeneous Model)

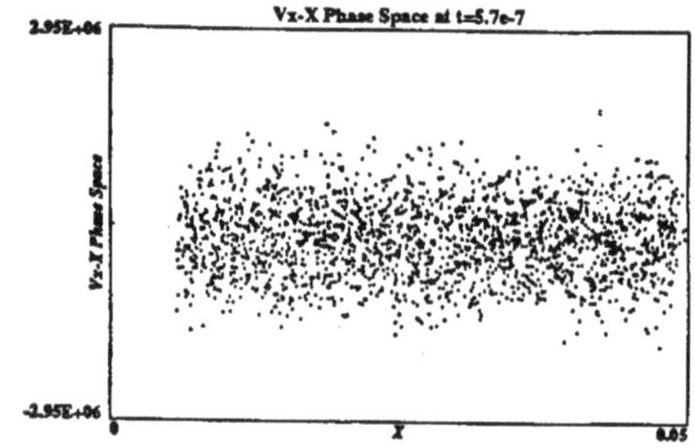

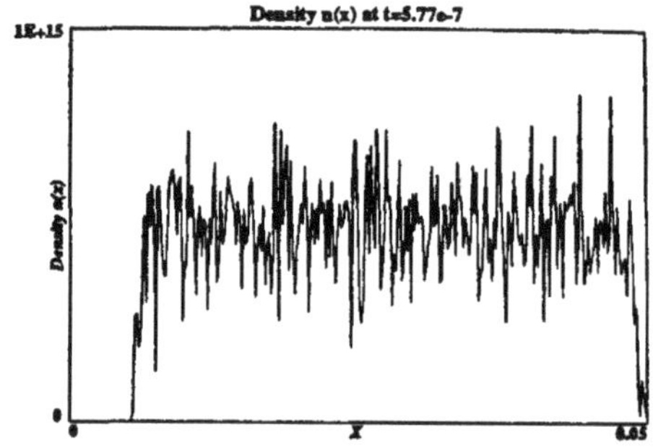

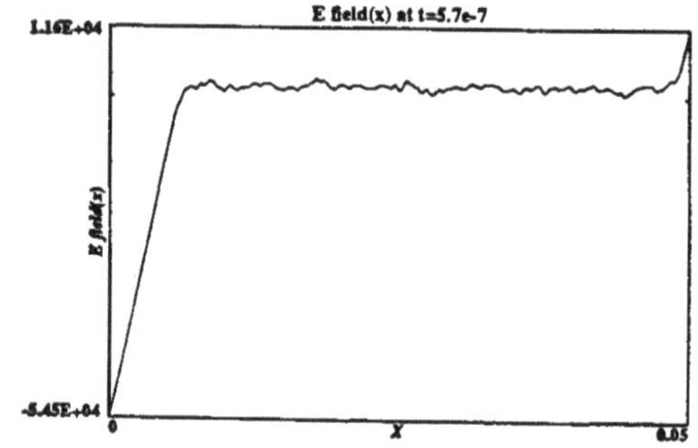

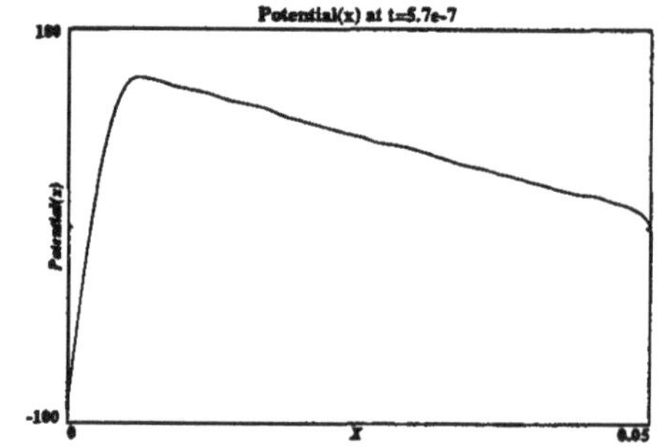

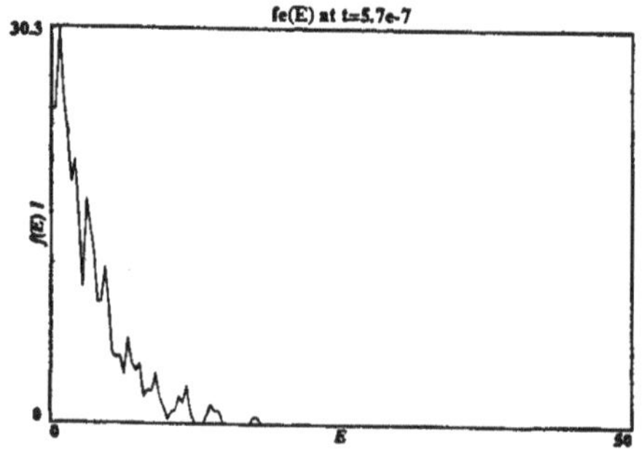

These are snapshots of $v_x - x$ phase space, density $n(x)$, $E_x(x)$, $\phi(x)$ and $f_e(E)$ at the left (target) electrode. Initially $T_e = 1\,eV$, $n_o = 5 \times 10^{14}/m^3$. RF drive is 100 Volts at $f_o = 13.56$ MHz. Electrode spacing is 0.05 m. With uniform immobile ions, $E_x(x)$ is larger than with mobile ions.

A Simple Voltage Driven RF Discharge with Mobile Electrons and Fixed Uniform Ions (Godyak Homogeneous Model)

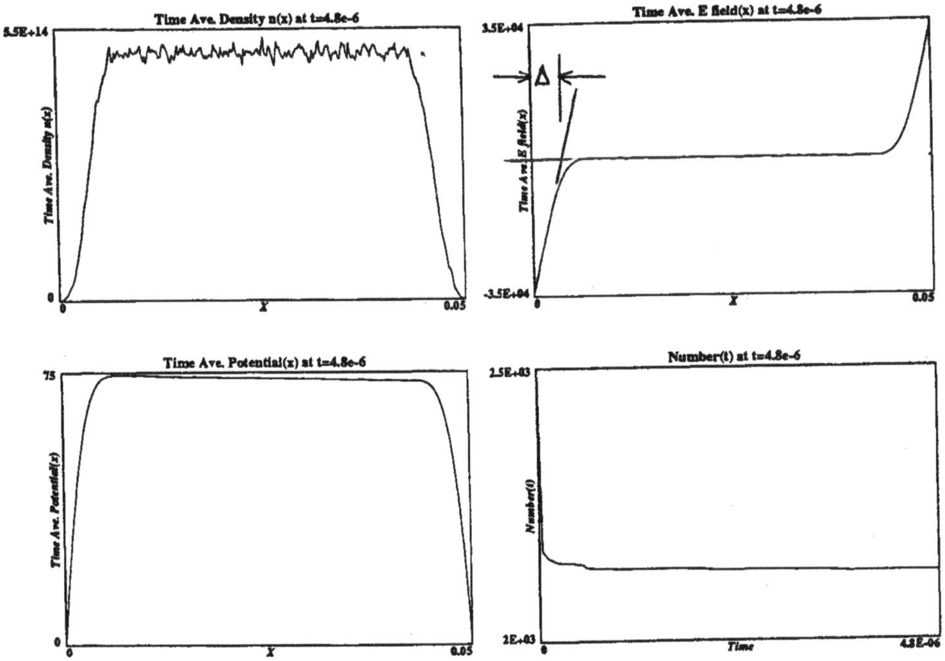

Three of these are time averaged plots, of $n(x)$, $E_x(x)$ and $\phi(x)$. The "vacuum gap," Δ, is estimated from the slope of $E_x(0)$, to be 0.0035 m (roughly equal to $10\ \lambda_D$) which predicts $f_{series} = f_{plasma}\sqrt{2\Delta/L_x} = 75.6$MHz. The total number of (computer) electrons decreases from 2500 at $t = 0$ to 2126 after about one RF cycle; the electrons are trapped thereafter.

On the next page, the spectra of $I(f)$ shows peaks at f_0 and f_{series}, noted as 1,4. $V_{mid}(f)$ shows peaks at f_0, $2f_0$, $f_{series} \pm f_0$, $2f_{series}$, and f_{plasma}, (peaks 1, 2, 3, 5, 6, 7). Power (f) has peaks at $2f_0$ and $f_{series} \pm f_0$ (peaks 2, 3, 5).

Raising n_0 to $5 \times 10^{15}/m^3$ changes these spectra to some extent, namely with a more distinct response in $V_{mid}(f)$ at f_{plasma}.

A Simple Voltage Driven RF Discharge with Mobile Electrons and Fixed Uniform Ions (Godyak Homogeneous Model)

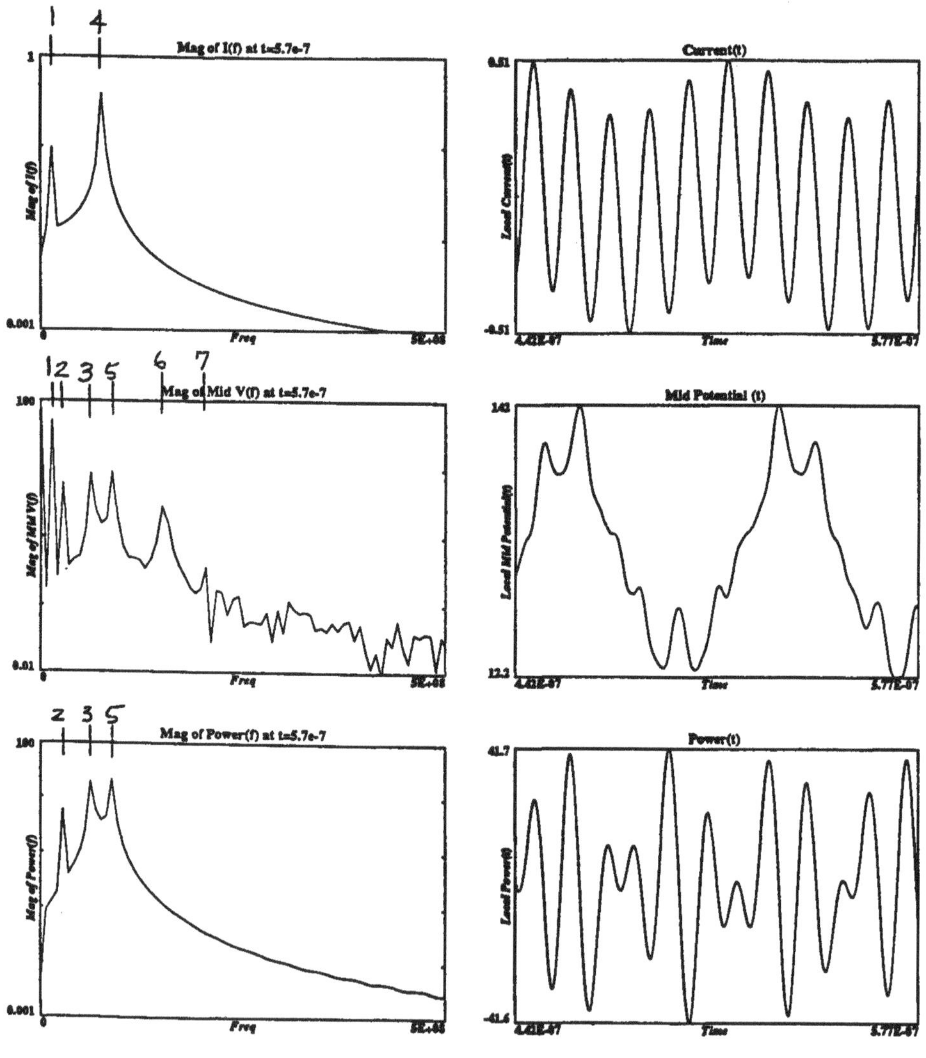

Voltage Driven RF Discharge with Mobile Electrons and Ions, without Electron and Ion-Neutral Collisions

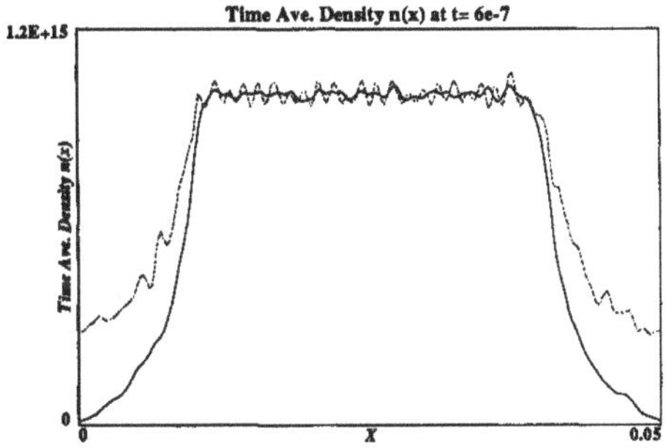

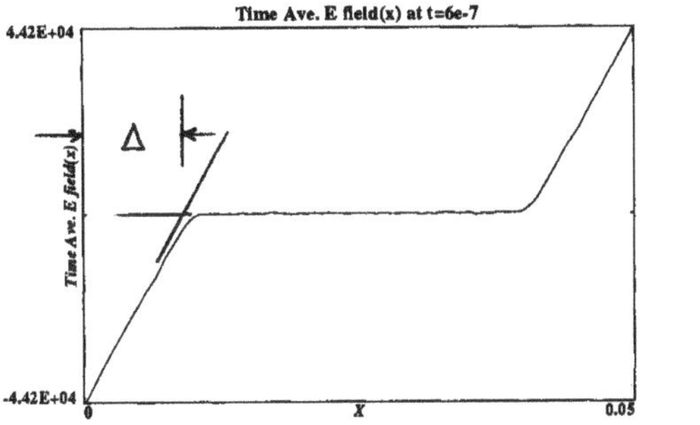

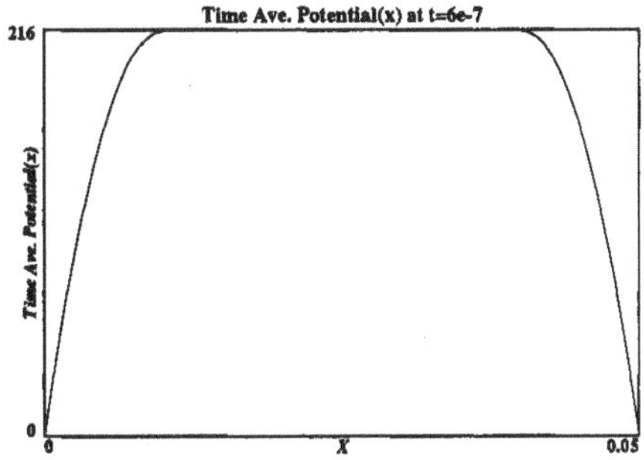

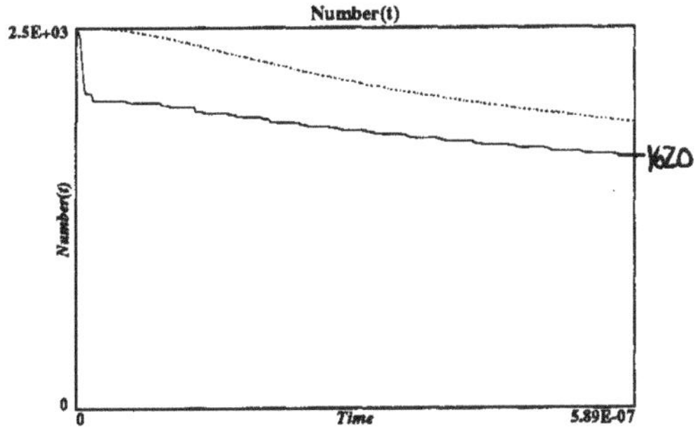

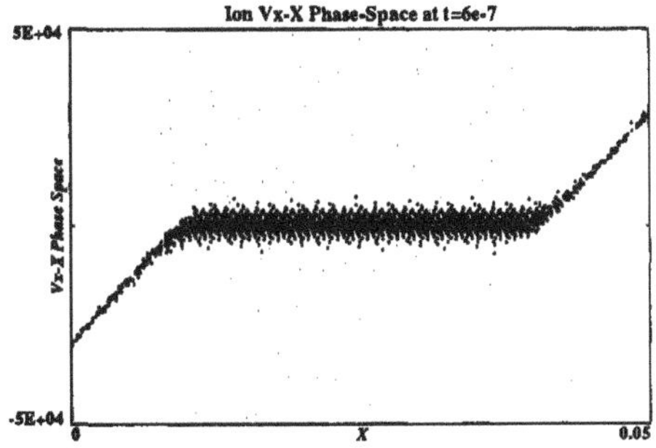

The argon ions now move, producing the time-average densities $n_e(x)$ (solid curve), $n_i(x)$ (dashed curve), field $E_x(x)$ (now much closer to zero in the bulk), and potential $\phi(x)$. Also now the ions escape (as shown) so the electrons also leave. The ions, with $f_{pi} = 1.05$ MHz $\ll f_0 = 13.56$ MHz move with the time average $E_x(x)$, accelerated to both electrodes. f_{series} and f_{plasma} appear in $I(f)$, $V_{mid}(f)$, but are much less pronounced.

Particle Density of Electrons and Ions in an Argon Current-Driven RF Discharge (Gap= 2cm, I= 0.3 A rms)

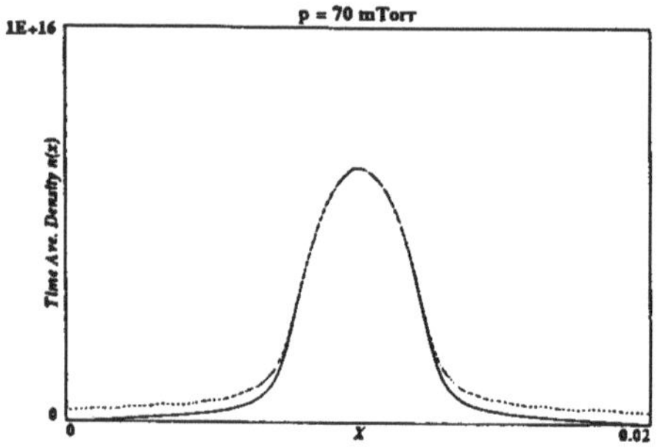

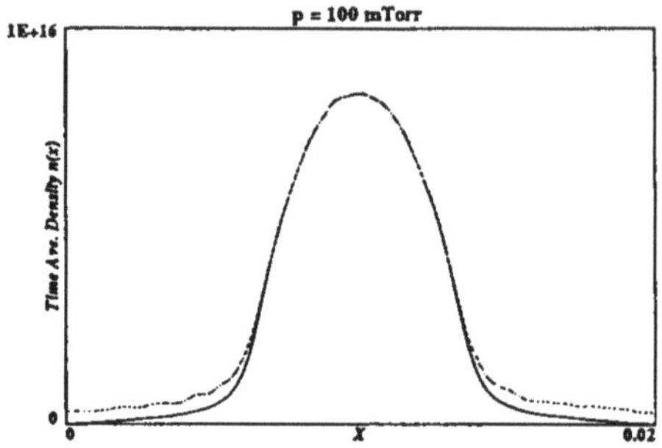

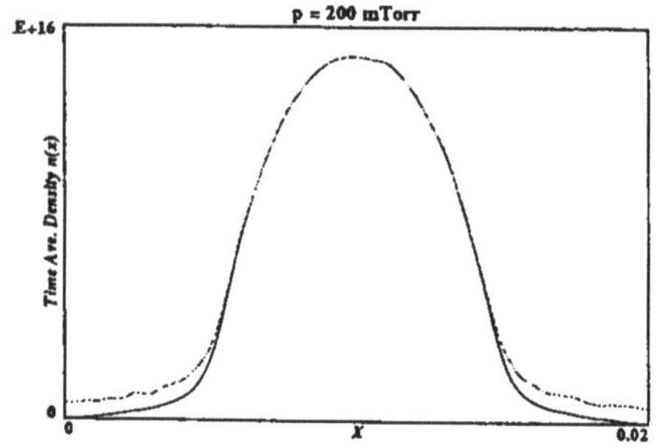

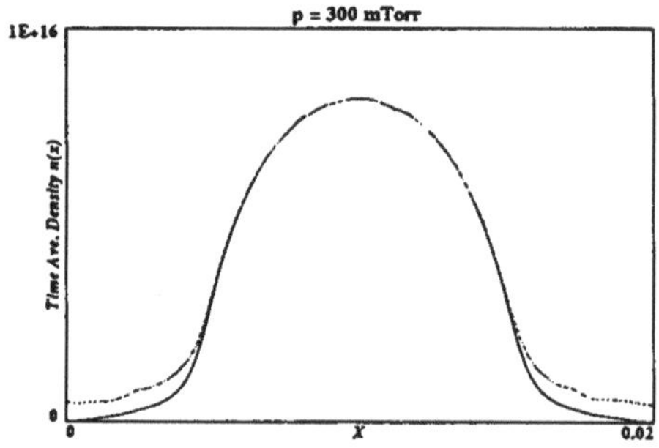

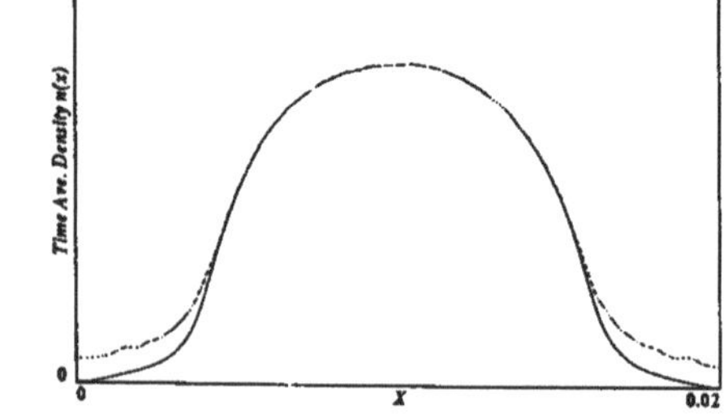

Next, electron and ion collisions with neutrals (scatter, excitation, ionization, charge exchange) are included. The examples shown are current driven at 13.56 MHz at 70, 100, 200, 300 and 500 milliTorr pressure. The discharge is self-sustaining.

This set of time-average density n(x) plots shows the ion density reaching both walls (dashed line), with much smaller electron density; dynamically $n_e(x,t)$ shifts from left to right at f_0.

Time Averaged Potential in an Argon Current-Driven RF Discharge (Gap= 2cm, I= 0.3 A rms)

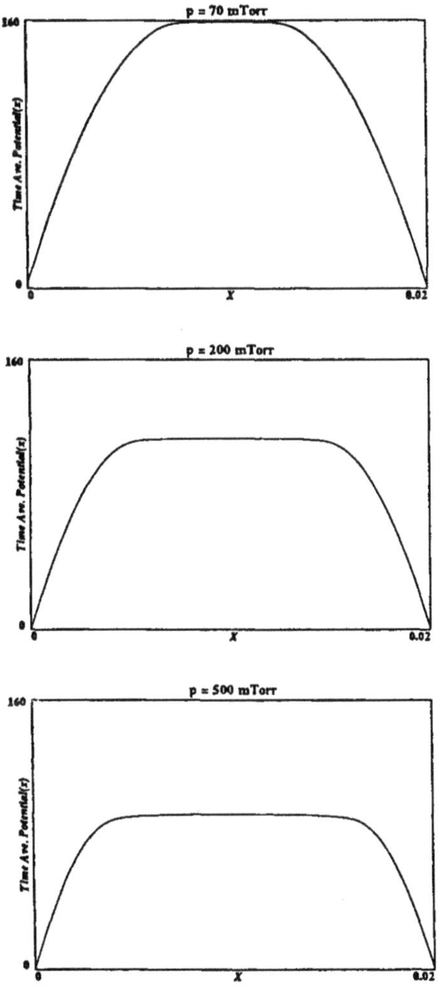

The time-average $\phi(x)$ shows similar behavior, with large E(x) in the sheaths, and E(x) very near zero in the bulk.

Spatial Profile of Average J.E in an Argon Current-Driven RF Discharge (Gap= 2cm, I= 0.3 A rms)

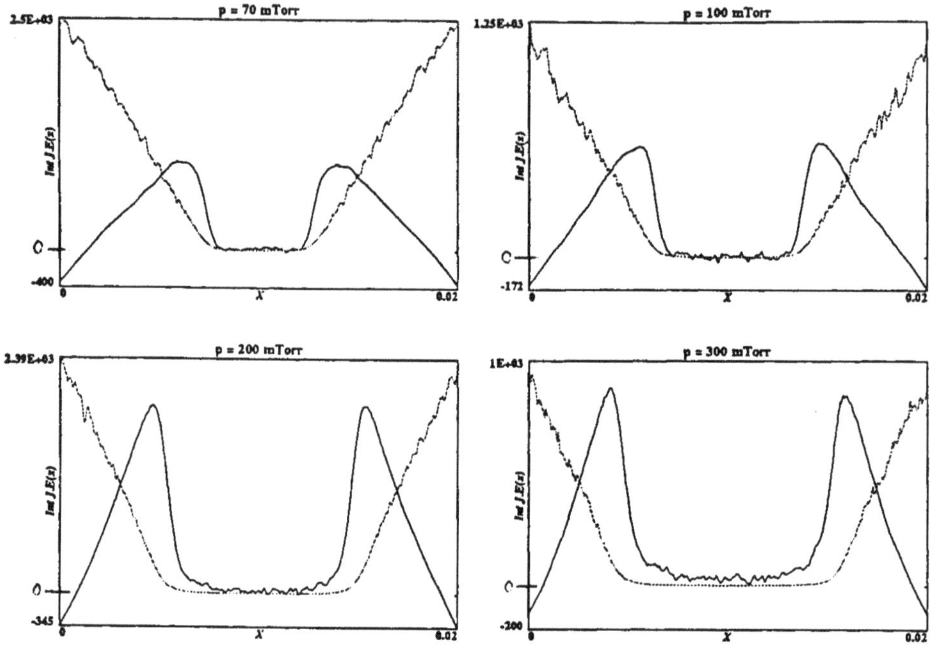

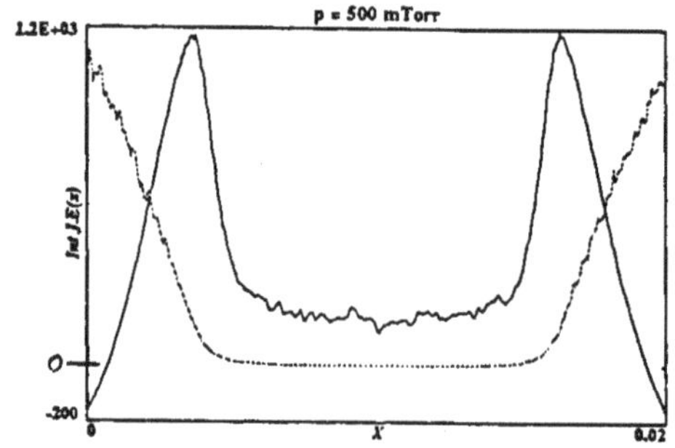

These plots are the time-average values of $\underline{J} \cdot \underline{E}$, the power delivered by the field to the particle (work done by $\underline{E}$ on $\underline{J}$), over the RF diode. The ions (dashed line) are clearly accelerated in the sheaths, with no work done on the ions in the bulk plasma. The electrons are heated in (and by) the sheaths, with little work done in the bulk plasma at low pressures, but with increased (ohmic) heating done in the bulk at high pressures.

Ionization Profile in an Argon Current-Driven RF Discharge (Gap= 2 cm , I = .3 A rms)

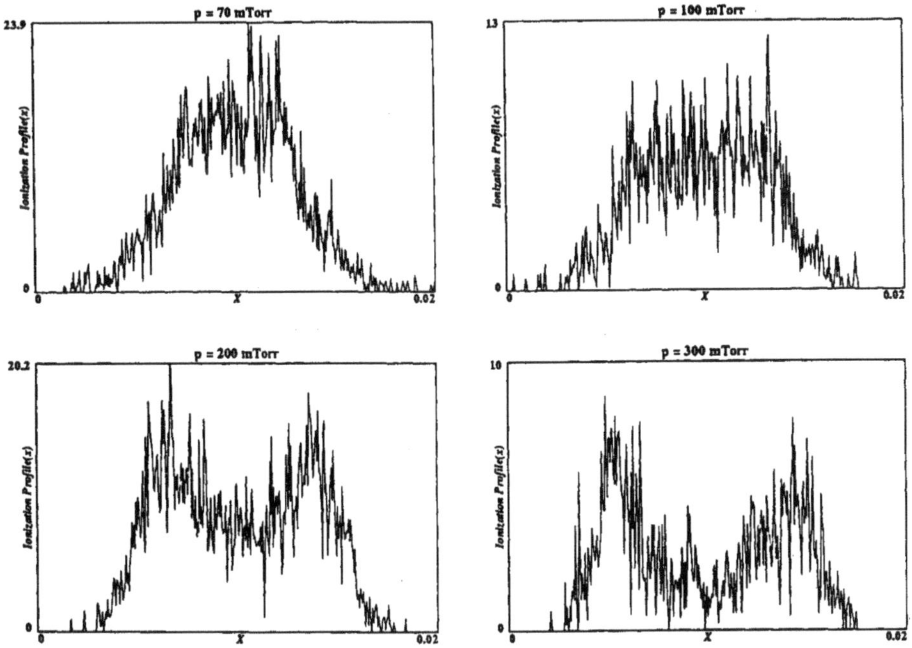

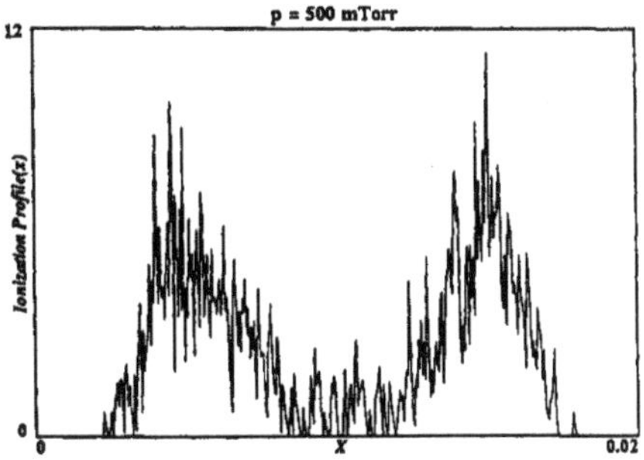

The ionization profiles follow the density profiles at low pressure, but peak at about the sheath edges at high pressures.

Electron Energy Distribution Function in an Argon Current-Driven RF Discharge (Gap= 2 cm, I= 0.3 A rms)

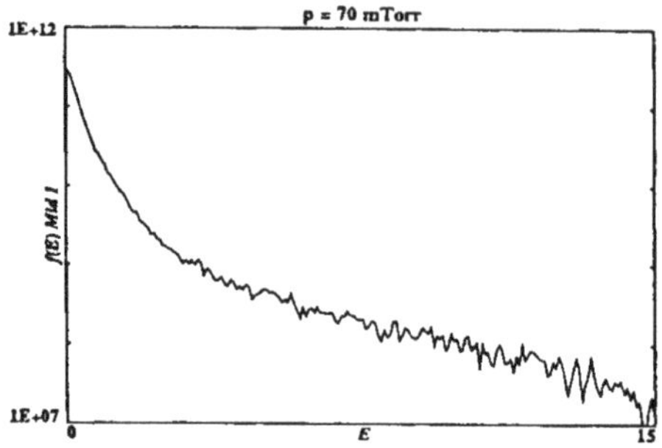

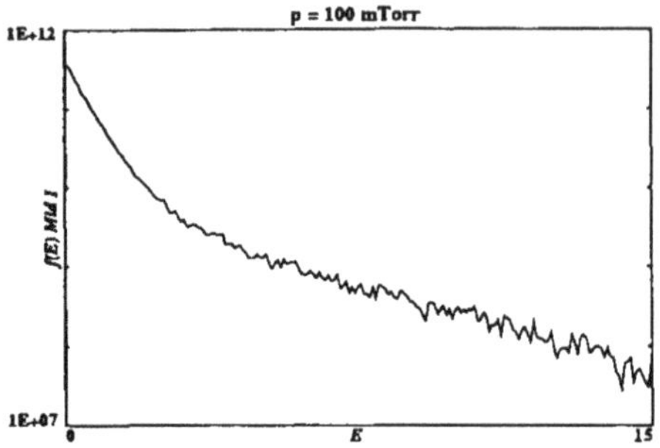

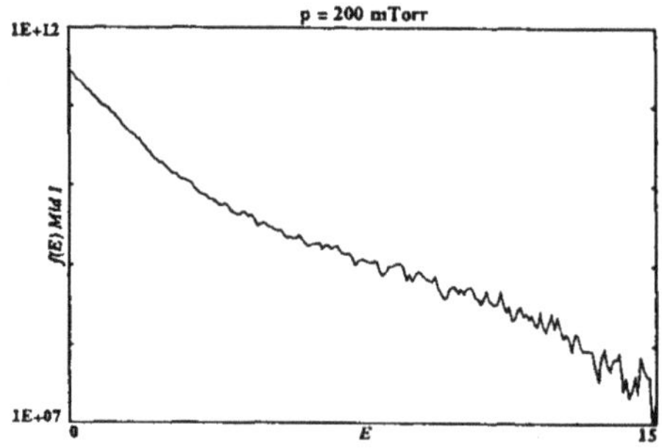

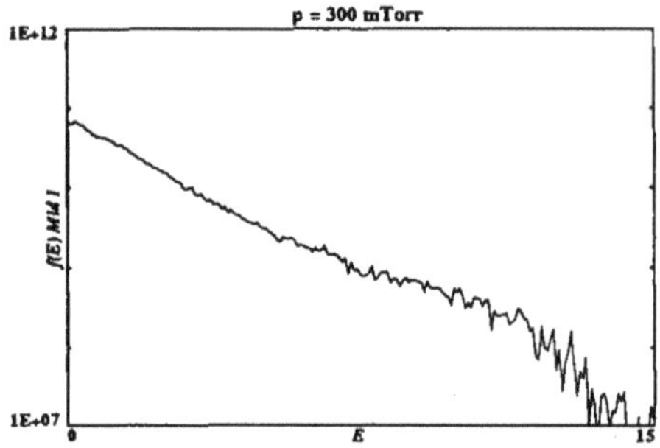

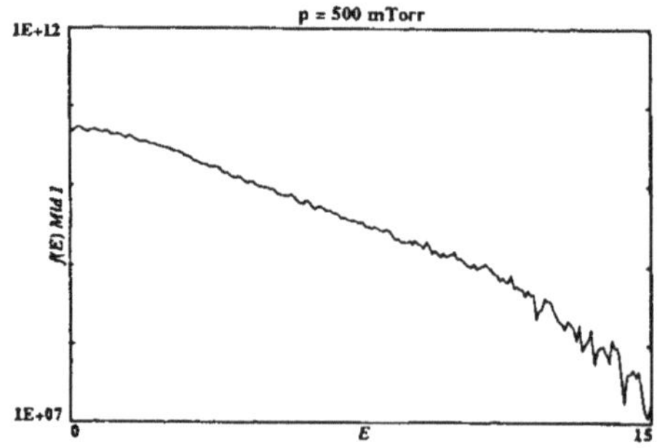

The electron energy distribution function, $f_e(E)$, shows a cool bulk temperature (below 1 eV) and a much warmer tail (few eV), at low pressures, but a single T_e at higher pressures, consistent with the $\underline{J} \cdot \underline{E}$ picture, increasing bulk heating at 500 mT. Such behavior is also observed in laboratory experiments.[3] Dynamically, the hot tail appears at the left-hand sheath ($v_x < 0$) and then at the right-hand sheath ($v_x > 0$), with velocity sign lost on these plots of $E = \frac{1}{2}mv^2$. In addition, at low pressures, one observes fast electrons (above ionization energies) having beam like behavior in the bulk, bouncing between the sheaths.[4]

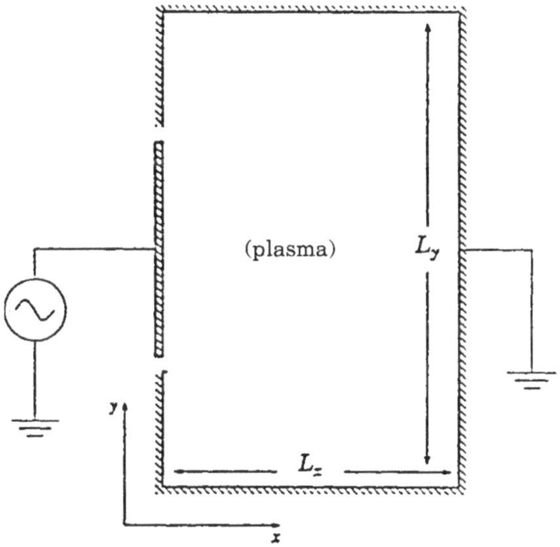

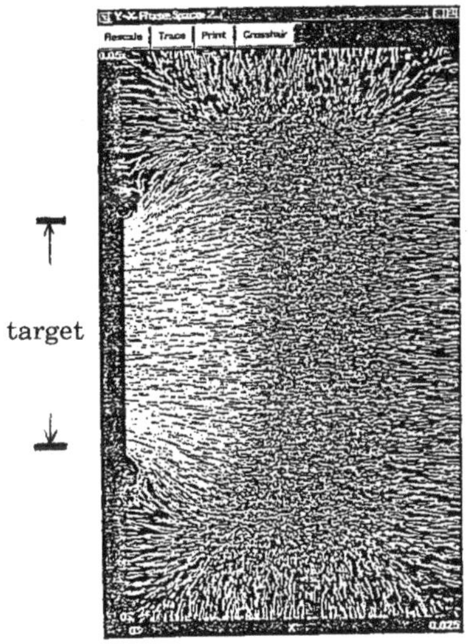

A two-dimensional RF discharge model is shown here. The driven electrode (target) has one-sixth the area of the grounded electrode, driven at $V_0 = 500\sin(2\pi f_0 t) - 400$ where $f_0 = 13.56$ Hz. (The latter replaces the self bias that would occur if the external circuit had a blocking capacitor.) The electrodes are planar, $L_x = 2.5$cm, $L_y = 5$ cm; there are no variations in z. The gas is argon, at 100 mT pressure.

The right hand plot shows ion trajectories, accelerated into the target, with a variety of angles, and non-uniform. Smaller fluxes go to the grounded electrode.

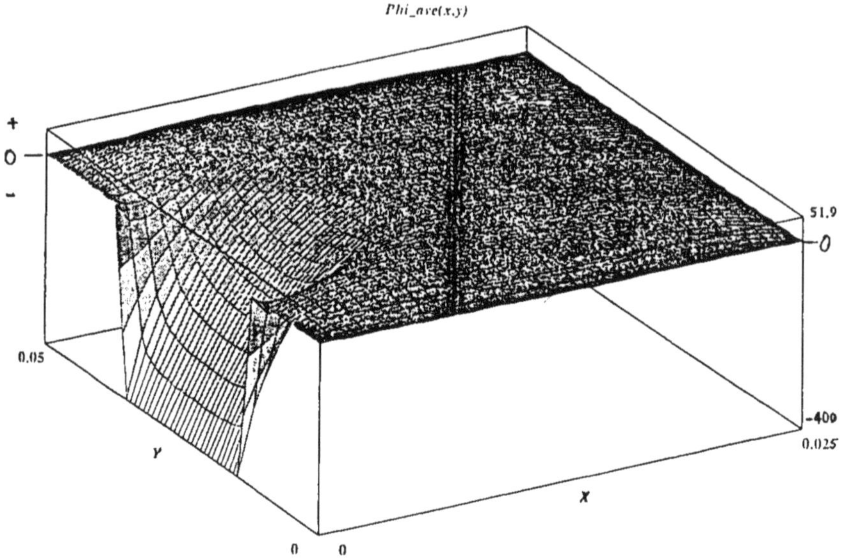

$\phi(x, y)$

This is the time average potential $\phi(x,y)$, for the 2d model in the previous figure. There is a very large potential drop at the target, with a far smaller sheath drop around the grounded wall.

THE OCCURRENCE AND ROLE OF AURORAL DOUBLE LAYERS

ROLF BOSTRÖM, PER-OLA DOVNER, ANDERS I. ERIKSSON, GUNNAR HOLMGREN,
Swedish Institute of Space Physics, Uppsala Division, S-755 91 Uppsala, Sweden;
ANSSI MÄLKKI, Finnish Meteorological Institute, Department of Geophysics,
P. O. Box 503, SF-00101 Helsinki, Finland

Abstract. In the auroral acceleration region, solitary structures frequently occur which constitute plasma density rarefactions (up to 50 %) in a thermal plasma background, are negatively charged (local potential minima of up to a few volts), and have a spatial scale of the order of 100 m. Some of the structures have a net potential drop, up to a few volts, predominantly directed upwards, and thus constitute Weak Double Layers (WDLs). The average upward potential drop for each structure is small (about 0.1 V), but possibly a large number of WDLs may uphold a significant, large scale potential drop and contribute to the observed upward acceleration of ions, and the downward acceleration of the electrons which produce the aurora as they enter the atmosphere. The structures also seem to occur in regions of upward flowing electric current, so the WDLs with upward directed net potential drops constitute loads on the circuit where electric potential energy from a magnetospheric generator mechanism is converted to kinetic energy of the accelerated particles.

1. INTRODUCTORY DISCUSSION: The plasma in the magnetosphere on the magnetic field lines connected to the auroral regions exhibits two remarkable features. It develops small scale (Debye and ion gyro radius scale) solitary nonlinear structures, including double layers (DLs). It also redistributes energy among the particles, so that a subset develops with energies much larger than that of the bulk. Possibly, the two phenomena are related in that DLs powered by a magnetospheric current generator may account for the acceleration of particles.

In the auroral acceleration region (see Figure 1), at altitudes of 5000 – 12000 km, electrons are accelerated downwards, along the geomagnetic field lines, to keV energies, sufficient to produce auroral displays as they precipitate into the earth's atmosphere. Positive ions (Hydrogen and Oxygen), of ionospheric origin, are also accelerated to similar energies, but in the opposite direction. These observations are most easily explained as an acceleration in a quasi static electric field. Mechanisms that can uphold static electric fields along the geomagnetic field lines are thus sought[1]. DLs constitute such a mechanism, and their importance have particularly been advocated by Alfvén and his Stockholm school[2–5].

The development of DLs may be an intrinsic property of any current carrying plasma with

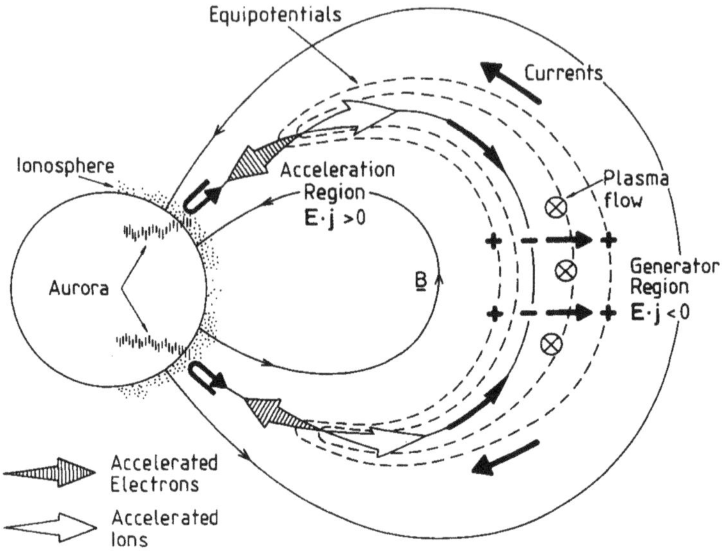

Figure 1: Schematic of the auroral acceleration region with its equipotential structure and current flow.

a certain level of wave excitation or turbulence. A fraction of the thermal electrons, those of low energy, will be reflected from local potential minima associated with the waves or turbulence. If the electrons start to carry an ordered, directed current, this reflection will be asymmetric, with more electrons reflected on the upstream (for the electrons) side of the potential well. An excess of negative charge thus develops , and an electric field builds up in the direction of the current ($\mathbf{E} \cdot \mathbf{j} > 0$); a Weak Double Layer (WDL; with voltage drop of the order of kT_e/e) is formed. This field necessarily accelerates those electrons that have enough energy to overcome the potential minimum, and current continuity is ensured. Also ions, flowing in the opposite direction, are accelerated. Thus, the WDL structures are dissipative elements in which electric potential energy of the electrons and ions is converted to kinetic energy. The WDLs can properly be termed particle accelerators. The mechanism of electron reflection is most interesting if the momentum exchange leads to a growth of the amplitude of the structures[6].

The Viking satellite, which in 1986 studied the auroral acceleration region, was well instrumented to observe small scale structures, like WDLs. The experimental results presented here are all from Viking, even if other spacecraft have also contributed substantially to the view of WDLs[7] and the acceleration region. WDLs were seen frequently in this region[8–10], particularly in association with beams of energized, upward flowing ions. It must be emphasized, however, that also other mechanisms might account for the acceleration. So far, we lack a quantitative evaluation of the relative importance of the various possible mechanisms. In this paper, we describe the *characteristics* of the WDLs, and show that they *must* contribute to *some* particle acceleration (which, on erroneous grounds, has been questioned[11]), and that, *in principle*, numerous WDLs can account for substantial acceleration. Even if, upon further analysis, it should turn out that the observed WDLs do not account for the major part of the acceleration, they constitute a basic plasma physics phenomenon of interest as such.

2. OBSERVATIONS AND INTERPRETATIONS: The low frequency wave experiment of Viking had four spherical probes, of diameter 10 cm, mounted on 40 m long wire booms in the spin plane of the spacecraft (spin rate 3 rpm). The spin plane was in the orbital plane (cartwheel mode). This nearly coincided with the magnetic meridian plane as the satellite was in a near polar orbit. Two of the probes could be operated either as density probes or as differential voltage probes, and the other two only as voltage probes. In the density mode the probes were operating at a fixed positive bias voltage relative to the spacecraft, and the current was measured by a logarithmic amplifier, and the AC output provided a measurement of relative density fluctuations ($\Delta n/n$). In voltage mode, the probes were given a fixed current bias so that they operated on the steep part of the Langmuir characteristic and the differential voltage fluctuations were measured, corresponding to differential potential fluctuations in the plasma. Two data channels, sampled at 853 samples/s, were available for data transfer. Thus, we could either measure $\Delta n/n$ in two points in space 80 m apart ($\Delta n_1/n$ and $\Delta n_2/n$), or two differential voltages between probes extended in perpendicular directions ($\Phi_1 - \Phi_2$ and $\Phi_3 - \Phi_4$), or $\Delta n/n$ at one point and the differential potential between two other points 80 m apart ($\Phi_3 - \Phi_4$).

Figure 2a shows 400 ms of data with two probes in $\Delta n/n$-mode, at this moment separated nearly along the geomagnetic field line as shown by the insert. We note localized density depletions of up to 40 %, seen first by the lower probe ($\Delta n_2/n$), and then by the upper one ($\Delta n_1/n$). The fact that the structures are *consistently* seen first by the lower and then by the upper probe is a strong indication that we are observing a true geophysical phenomenon, and not an effect of any disturbance caused by the spacecraft. From the time lag, we infer the propagation speed, and from the speed and the length of time that we observe a structure, we can derive the spatial extent of the structures. A statistical study of 416 cases shows that the structures consistently propagate upwards, along the magnetic field lines, with an average velocity of 20 km/s, and average extent along the field lines of 175 m. The extent of the solitary structures in the direction transverse to the magnetic field can not be directly measured. However, as we generally see the same structure by both probes even when these are separated transversly to the magnetic field,

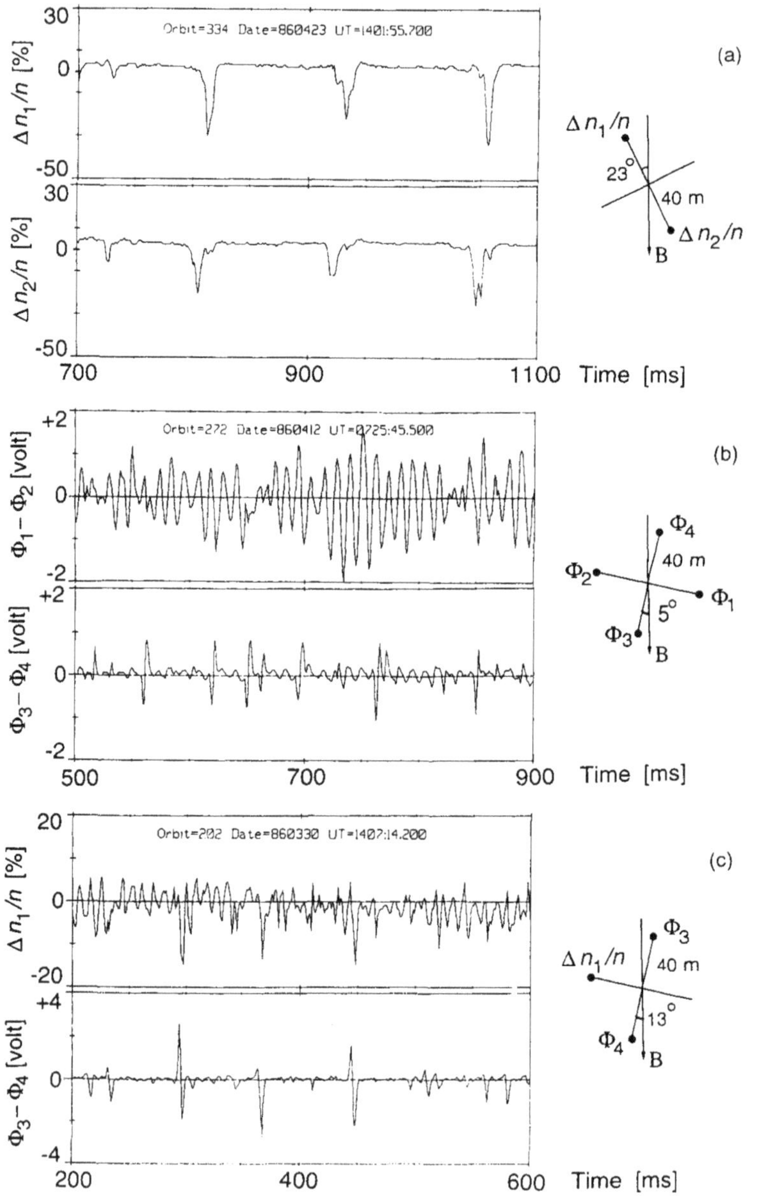

Figure 2: Observations of WDLs with three different modes of operation of the Viking probes.

we conclude that the transverse dimension is typically larger than the transverse probe separation of up to 80 m. From data with the probe configuration as shown in Figure 2a, we can also draw some conclusions as to the life time of the solitary structures. By comparing the amplitudes as seen by the two probes, we find a characteristic time for the growth or decay of amplitude on the order of 10 ms.

The velocities are in the ion acoustic range, suggesting that we observe a solitary ion wave. The structures, however, do not seem to be typical ion acoustic solitons, as they are rarefactive rather than compressive. Furthermore, we have not been able to statistically find any correlation between amplitude, spatial scale, and velocity[8] as suggested by the Korteweg – de Vries relations for one dimensional ion acoustic solitons.

Figure 2b, with two pairs of probes in differential potential mode, shows large amplitude EIC waves in the direction nearly perpendicular to the magnetic field, while the measurement of potential difference nearly along the magnetic field again displays solitary wave pulses, where the potential difference between the lower and upper probe first goes negative, then positive. Similar signatures have earlier been seen by the S3-3 satellite[7]. Assuming that these are the potential signatures of structures like those seen in Figure 2a, propagating upwards and passing first probe 3 and then probe 4, we conclude that the structures are associated with a local potential *minimum* of up to about 1 V. A statistical study of a large number of cases shows that the structures are consistently negatively charged. The large amplitude EIC waves are frequently seen in the same regions as the solitary structures, but there is no one-to-one correlation.

From potential (Langmuir) sweeps of the probes, performed in areas close to those where the solitary structures are observed, we conclude that there exists an electron population with characteristics like a thermal population with a temperature of some electron volts. What we measure in density mode (Figure 2a) is the depletion of this *thermal electron density*. For the structures to be negatively charged the *ion density* must be even more depleted. Thus we conclude that we have "holes" both in the ion and electron spatial distribution. The fact that the structures propagate with velocities in the ion acoustic range suggests that we have a *thermal ion population* present (in addition to the ion beam).

Figure 2c, with one probe in $\Delta n/n$-mode and one pair of probes in differential potential mode, shows that the solitary potential structures, like those of Figure 2b (although of reversed polarity as probes 3 and 4 have now interchanged position), occur at the same time as strong plasma density depletions, like those of Figure 2a. (The record of $\Delta n_1/n$ is here somewhat noisy as large potential variations of EIC waves couple into the density measurements particularly when the probe and spacecraft are separated perpendicular to the magnetic field, i. e., in the general direction of the electric field of the EIC waves. This effect is both qualitatively and quantitatively well understood as it agrees with predictions of classical probe theory.) From data of the type of Figure 2c, we can compare the magnitude of potential variations with that of density variations. Although the measurements are not made at the same point, and in fact separated by a distance of the same order as the expected extent of the structures, we find a positive correlation, consistent with what we would expect if the two were linked by a Boltzmann relation for electrons of a temperature in the range 2 – 20 eV. It seems reasonable that the electrons are in an equilibrium where the electrostatic force from the potential gradient of the minimum is balanced by the pressure gradient. For the ions, however, no such equilibrium is possible, as both the potential and pressure gradients act to refill the ion hole. The ions are thus in a dynamic state where the inertial forces come into play, and then the ion hole is expected to have a life time determined by the ion dynamics (the time it takes a thermal ion to propagate across the structure). The experimentally inferred characteristic time for temporal development of the order of 10 ms is consistent with this interpretation.

Particularly interesting, in the potential difference measurements along the magnetic field (lower traces of Figure 2b and 2c), is that some of the positive and negative excursions are not symmetric. This implies a net potential drop across the structure, up to a few volts in the examples of Figure 2c. The structures then constitute weak double layers (WDLs), which we thus regard as a particular form of the more general feature that we refer to as (solitary) structures. Most likely, the WDLs and the other solitary structures are not qualitatively different phenomena, but different phases of evolution of the structures, or manifestations of different boundary conditions. Most observed structures do not show any discernible net potential, and the average over a large number of structures is only about 0.1 V. However, what is interesting is that the potential drops predominantly point in the upward direction[16], and therefore a large number of WDLs can collectively build up a substantial potential difference along the magnetic field lines that might be important for the acceleration of auroral electrons and ions. It is also important to note that the predominant direction of the net potential drop is in the direction required to be consistent with the observed acceleration.

Assuming an average linear packing density of the structures (including those with no net voltage drop) of 10 %, consistent with Figure 2, an average net upward directed potential drop of 0.1 V per structure, and an extent in altitude of 5000 km, we find a total field-aligned potential drop of 0.5 kV, which is on the order of the typical energy for the accelerated ions. We do not always see solitary structures when the satellite is in the auroral acceleration region, and thus the assumed average packing density of 10 % of the structures may seem too high. On the other hand, we sometimes experience regions where the potential variations are very rapid, and which in principle could involve WDLs with a packing density approaching 100 % which can not be experimentally resolved. In fact, the development of multiple WDLs may be an intrinsic property of a current carrying turbulent plasma. It is also possible that the cases with relatively well behaved and well separated WDLs, like those shown in Figure 2 and which we have been able to study, correspond to cases of marginal instability of some process leading to the formation of WDLs and turbulence. In fact, it is very difficult to derive a good experimental value for the packing density, but it is only through such a number that the actual importance of WDLs for particle acceleration can be assessed.

From the data discussed above, we have constructed the schematic picture shown in Figure 3, illustrating how we envisage that a few structures start to collectively build up a large scale equipotential pattern like that of Figure 1. Some of the structures shown have negligible net voltage drop, and are here referred to as Solitary Waves (SWs), while those with a net potential are referred to as WDLs. The large scale equipotential pattern that results from a large number of WDLs will be one with a potential minimum in the center and upper part of the region where the WDLs occur.

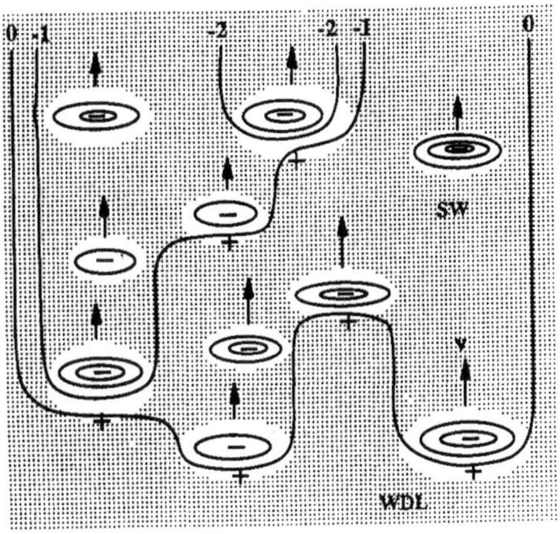

Figure 3: Schematic of how multiple WDLs build up a large scale equipotential pattern.

With the other instruments on Viking, it has been possible to establish the general background in which the solitary structures and WDLs appear. Figure 4 shows a composite of such data for a time period including and surrounding that for which the data shown in Figure 2a was recorded. Panel 4a shows the electron flux and Panel 4b the ion flux, both gray scale coded as functions of time and energy. The detectors on the spinning

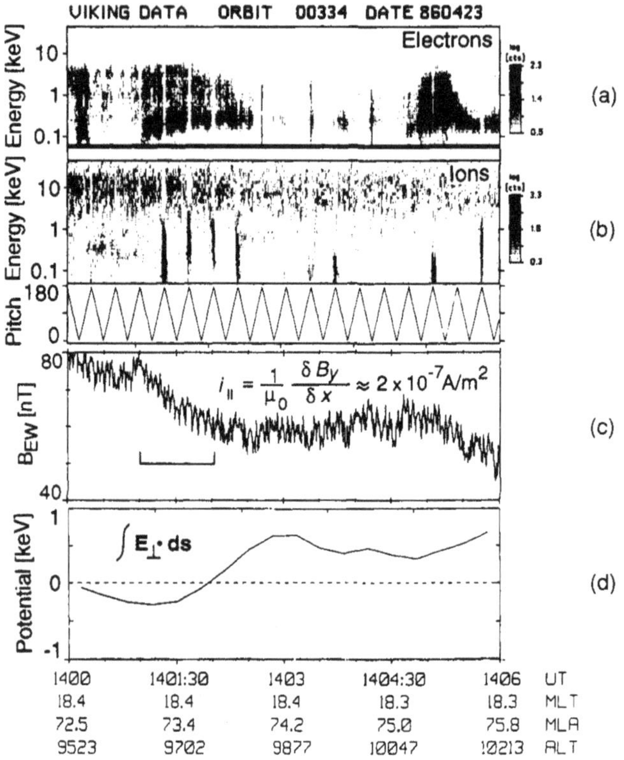

Figure 4: Particle and field measurements from Viking in the auroral acceleration region around the time of the observations of Figure 2a.

spacecraft measure fluxes in different directions at different times. The direction relative to the magnetic field is shown by the sawtooth curve marked Pitch, and 180° refers to particles moving up along the field lines. The most characteristic feature of the data is the beam of upward moving ions with energies of up to about 1 keV, which occurs around the time of the data of Figure 2a (1401:56 UT). Data from several other orbits of Viking show that the presence of an ion beam is a characteristic of regions where WDLs are observed. The most intense flux of electrons have relatively low energy as the measurements are presumably made in the upper part of the acceleration region where the electrons have not yet passed a large potential increase. The electrons, however, show a feature indicative of a potential difference between the satellite and the ionosphere, namely that the loss cone around 180° is wider for the lowest energies.

Panel 4c shows the east-west component of the variations of the background magnetic field along the satellite trajectory, indicating an upward field-aligned current intensity of about 2×10^{-7} A/m^2 in the region of interest. This is relatively moderate, and below estimates of the intensity required for a linear instability generating ion acoustic or EIC waves. Most important to note is that the current is *upward* directed so that $\mathbf{E} \cdot \mathbf{j} > 0$ in the WDLs, as required for them to operate as particle accelerators. Panel 4d shows the line integral of the electric field measured on Viking along the trajectory across the region of interest. We find a potential minimum of about 0.5 – 1 kV in the area where the WDLs are observed, consistent with the potential pattern of Figure 1 and with the observed energies of the accelerated ions.

3. DOUBLE LAYERS AND THE GENERATOR: WDLs by themselves are no sources of free energy, they just provide a mechanism to uphold potential drops along the magnetic field lines and to convert electric potential energy of the particles to kinetic energy. This potential energy must be provided by the external system, or more specifically the electric generator of the current circuit in which the DLs appear. A discussion of this generator, sometimes referred to as a dynamo, must thus be an integral part of a model for how the WDLs are formed, and for understanding their role in the system, particularly as particle accelerators. The generator is a region where $\mathbf{E} \cdot \mathbf{j} < 0$, i. e., a region where there exists a force which moves charged particles against the electric force $q\mathbf{E}$, so that they gain electric potential energy. This force may be referred to as an "electromotive force, emf." However, it must be noted that the existence of such an emf in no way implies that $\oint \mathbf{E} \cdot \mathbf{ds}$ has to be non-zero. On the contrary, it is quite feasible to construct a purely electrostatic model for the generator and the particle acceleration.

The generator may be loosely described as a region where mechanical energy is converted to electric energy, or where a mechanical force acts. The analogy with a concrete electrotechnical machine must not be carried too far, but it is evident that in the plasma there exists mechanical energy in the form of kinetic energy of the particles, both in their convective and gyro motions. Standard MHD-models for the generation of magnetospheric current systems[12–14] are based on these. It is this kinetic energy, or, in single particle terminology, the inertial force or inertial drift, that drives the particles in the opposite direction to the electric force. Evidently, the ions are here much more important than the electrons.

The generator is a region where a very large number of ions each loose a small amount of kinetic energy by moving a short distance across the equipotentials towards higher potential. The result of this, shown schematically in Figure 5, is the required current flow through the generator, and that a smaller number of ions and electrons become available

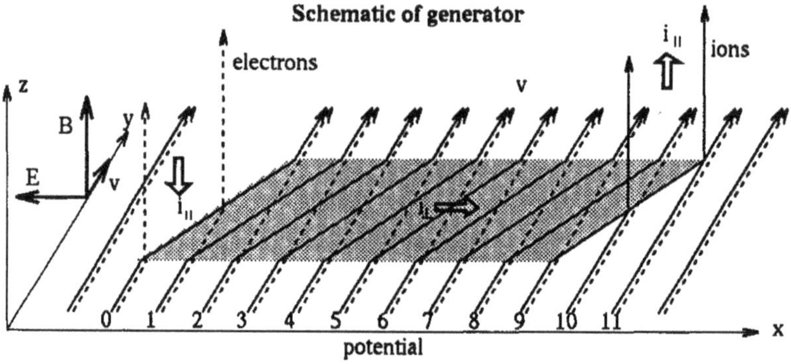

Figure 5: Schematic of MHD generator where a large number (here only 10) of ions each lose a small amount of energy (here 2 units) to produce some particles (here 2 ions and 2 electrons) at a high potential energy (here 10 units).

with a high potential energy at the terminals of the generator. The MHD generator is thus a device which converts the kinetic energy of a very large number of particles to substantial potential energy of a smaller number of particles. It is this potential energy that is used to drive the auroral current system, with the dissipative elements of the acceleration region, in the form of multiple WDLs, as well as the dissipative ionospheric segment of the current system. It is interesting to note that the whole system can be described in purely electrostatic terms.

4. CONCLUDING DISCUSSION: With the Viking wave experiment we have established that solitary structures *exist* in large numbers in the auroral acceleration region of the magnetosphere, having *all relevant characteristics* of WDLs, including their dissipative nature ($\mathbf{E} \cdot \mathbf{j} > 0$), and a predominant polarity such that they *must* contribute to the auroral particle acceleration. It has *not* been experimentally established whether they are the most important acceleration process, and other mechanisms must also be assessed.

Bryant et al.[11] have, incorrectly, criticized the DL model for particle acceleration, and claim that an electrostatic model for particle acceleration is *in principle* impossible to construct. The criticism seems to be partly based on a discussion of an isolated DL, not powered by an external system, which, as evident from our discussion above, would not work as an accelerator (and would not even exist). The criticism also seems to be based on the fact that a single particle, moved around a closed circuit loop in an electrostatic field, cannot gain more kinetic energy at the expense of potential energy in one part of the system than it has to lose in another part. However, with more than one particle some can gain energy at the expense of others, and no wave-particle interaction is *required* for this. Also, while the currents along the magnetic field lines are carried mainly by electrons, the transverse currents of the generator and of the ionospheric segment of the circuit (see Figure 1) are carried mainly by ions. There is no need to bring particles of one species around a closed loop.

Another incorrect view of the WDL model has been that the total voltage drop along the field lines is determined by the generator, and thus the voltage by which the particles will be accelerated bears no relation to the WDLs, and the WDLs could then hardly be said to be responsible for the acceleration. This is not so, as there are other resistive elements of the system, regarded as a circuit, such as the ionosphere and the possible internal resistance of the generator (the generator will be nearly a perfect voltage generator if it is the plasma convection that drives it, but more nearly a perfect current generator if it is the pressure gradients, i. e., the gyro motions). What is important then, using a lumped circuit analogy, is how the open circuit voltage, or the emf, provided by the generator, in the closed current circuit is split on the various loads; a possible voltage drop along the field lines, which accelerates particles, and resistive voltage drops in the current flow transverse to the field lines, which do not contribute to the acceleration. That is why it is important to find mechanisms, such as WDLs, electrostatic shocks or mirror forces, that can uphold parallel potential drops in the plasma[1].

WDLs remain a potential candidate for particle acceleration, but other models must also be considered. Electrostatic models, with potential drops along the magnetic field lines, caused by WDLs or other mechanisms, can explain many features of the observed particle spectra. The models do not imply that all particles are accelerated to the same energy, as the generator region may extend into the acceleration region and particles may enter the system from the side at different positions along the field lines, i. e., at different potentials. Also the WDLs are short lived and the total potential drop will fluctuate, producing a spread of energies.

The WDLs are clearly dynamic phenomena and it may seem inconsistent to suggest that they support a "static" electric field along the magnetic field lines. However, by "static" we do not mean a truly time independent field, but rather that we can use electrostatics to describe the field. The WDLs may support a non-zero *average* parallel electric field, and it is this field that may be important for the auroral acceleration. This is to discriminate from what we call wave theories for the aurora. Such theories also rely on parallel electric fields (at least for electron acceleration), but at non-zero frequencies. Wave theories generally work with Landau resonant electron acceleration, or "brute force" stochastic mechanisms relying on broad band waves with high amplitudes.

It may well be misleading to discriminate too strictly between wave models and static models of the auroral acceleration. One alternative mechanism for sustaining the parallel electric field needed in a static theory is the so-called "anomalous resistivity" caused by wave processes. The solitary waves and weak double layers discussed in this report are indeed wave phenomena themselves. Also, the dynamic coupling of a generator and a load in a "static" theory as the one discussed in this paper is mediated by Alfvén waves[15]. Thus, plasma waves are of central interest even in a static model, and it may well be that those waves also contribute to the auroral acceleration, in accordance with some wave theory, together with the static electric field.

To resolve the issue of the actual importance of the different mechanisms (WDLs, shocks, various wave mechanisms) that have been presented, we need both an assessment of the *efficiency* of the various mechanisms, and a *quantitative* evaluation of the *occurrence frequency* for each phenomenon. The latter point has often been overlooked. Data from which such evaluations of occurrence frequency can be performed are available, but difficult and laborious to analyse. A first attempt at such an evaluation for the WDLs has been done by Mälkki et al.[16], but we must note the problem mentioned above in deriving a good experimental measure for the packing density of the WDLs.

Acknowledgement. The analysis of the data from the Viking wave experiment is the result of team work at the Uppsala Division of the Swedish Institute of Space Physics, involving also co-investigators from Cornell University, and the Finnish Meteorological Institute. The data of Figure 4 have been provided by Dr. J. Woch of Kiruna Division, Swedish Institute of Space Physics, Dr. L. Zanetti of Applied Physics Laboratory, Johns Hopkins University, and Prof. L. Block of Department of Plasma Physics, Royal Institute of Technology.

REFERENCES:

1. C.-G. Fälthammar, *ESA Journal* **7** (1983), 385.
2. C. Jacobsen and P. Carlqvist, *Icarus* **3** (1964), 270.
3. H. Alfvén and P. Carlqvist, *Solar Physics* **1** (1967), 220.
4. P. Carlqvist and R. Boström, *J. Geophys. Res.* **75** (1970), 7140.
5. L. P. Block, *Cosmic Electrodynamics* **3** (1972), 349.
6. D. Tetreault, *J. Geophys. Res.* **96** (1991), 3549.
7. M. Temerin, K. Cerny, W. Lotko, and F. S. Mozer, *Phys. Rev. Lett.* **48** (1982), 1175.
8. R. Boström, G. Gustafsson, B. Holback, G. Holmgren, H. Koskinen, and P. Kintner, *Phys. Rev. Lett.* **61** (1988), 82.
9. R. Boström, B. Holback, G. Holmgren, and H. Koskinen, *Physica Scripta* **39** (1989), 782.

10. R. Boström, Observations of weak double layers on auroral field lines, *IEEE Trans. Plasma Sci.* (accepted for publication, 1992).
11. D. A. Bryant, R. Bingham, and U. de Angelis, *Phys. Rev. Lett.* **68** (1992), 37.
12. V. M. Vasyliunas, in *Particles and Fields in the Magnetosphere*, edited by B. M. McCormac (D. Reidel Publ. Co., Dordrecht, 1970), p. 60.
13. R. Boström, in *Physics of the Hot Plasma in the Magnetosphere*, Proceedings of the 30th Nobel Symposium, edited by B. Hultqvist and L. Stenflo (Plenum Press, N. Y., 1975), p. 341.
14. A. Hasegawa and T. Sato, *Space Plasma Physics* (Springer-Verlag, 1989), p. 165 – 168.
15. R. L. Lysak, *Space Sci. Rev.* **52** (1990), 33.
16. A. Mälkki, R. Boström, and G. Holmgren, *these proceedings* (1992).

Problems with Double Layers as Particle Accelerators

D A Bryant, R Bingham, and U de Angelis[(a)]

Rutherford Appleton Laboratory, Chilton, Didcot, Oxfordshire, United Kingdom OX11 0QX

Abstract

It is pointed out that the continuing advocacy of electrostatic double layers as particle accelerators in the aurora and other space and astrophysical plasmas is fundamentally unsound. It is suggested furthermore that there is little reason to invoke static or quasi-static electric fields as the cause of auroral electron acceleration. Stochastic acceleration by wave turbulence appears to present a natural explanation for electron acceleration in collisionless plasmas in general. This report is an expanded version of a recent paper [1] on this subject, taking into account subsequent comments and discussions held at the present meeting and elsewhere.

INTRODUCTION

Double layers — "two equal but oppositely charged, essentially parallel but not necessarily plane, space charge layers"[2] — have long been advocated[3] as accelerators for charged particles in the aurora[4], solar flares[5] and, by implication, other space and astrophysical plasmas. We wish to point out, however, that double layers (at least, as so defined) cannot, in fact, serve as particle accelerators, in the sense of providing energy, since the electric fields associated with any static configuration of space charge are conservative. Furthermore, since the line integral of the electric field outside such a double layer exactly balances the line integral inside, (equipotential contours being closed surfaces) it cannot even promote a net (ie more than a temporary and local) exchange of potential and kinetic energy, and is thus not an accelerator in this limited sense either.

The Tower of Pisa, invoked by adherents of acceleration by potential differences, can be considered an accelerator for stones falling from its top only if we are not interested in how the stones ascended the Tower in the first place. But if we start with all the stones on the ground our problem is to explain how some of them were raised to the top, before we can consider the Tower of Pisa combined with a lifting mechanism (perhaps even Galileo himself) as an accelerator. After all, Isaac Newton's apple wasn't accelerated by the tree! It should be stressed here that the term "particle acceleration" is used as a synonym of "particle energization" in the literature on plasmas[6] and in the context of high-energy particle acceleration in general. Nothing is learned, to employ another example, if the kinetic energy at the mid point of the swing of a pendulum is attributed simply to the potential energy at the extrema, when we wish to know what set the pendulum swinging.

Any model advocating potential differences (arising from double layers or any other source) to explain the measured kinetic energies of auroral particles must, therefore, solve the real problem of particle energization; or, if the model restricts itself just to the exchange of potential for kinetic energy, it should not be claimed to be an explanation of the aurora.

ORIGIN OF POTENTIAL-DIFFERENCE THEORY

Let us attempt to trace the steps by which the belief summarized in the opening sentences, so at variance with basic physics, appears to have arisen and why it continues to be promulgated, as the solution to the problem, fifty years after the "serious energy difficulty"[7] with this type of theory was pointed out. The pioneering observation was that the electrons responsible for producing a bright active auroral arc were concentrated within such a narrow band of energies that they were termed "monoenergetic"[8]. Since, at that time, acceleration processes of a statistical nature were thought to be incapable of generating a peak in a distribution function, an acceleration process involving (static) electric fields seemed to be the only possibility. However, measurements from higher flying rockets with more advanced detectors were able to demonstrate that velocity-space densities of the electron streams responsible for auroral arcs and other forms of discrete aurorae rise again towards low energies[9], making the original description of the distribution no longer apt. The term continues to be used in some quarters to describe the peak alone, though this is inappropriate, even misleading, when the peak-to-valley velocity-space-density ratio is typically of order 2 only[10], and there is commonly as much as a factor of 2 in velocity between the peak and the valley on the low-velocity flank. On occasions there is only a plateau or even just an inflexion (note that the earliest examples of "monoenergetic" distributions[11], when translated from count rate to velocity-space density were actually of this last type). The prime reason for seeking a process that could create a monoenergetic distribution has, therefore, long disappeared.

Most of the current literature advocating double layers in the aurora seems to concentrate on the trivial role of these structures to transform energy from potential to kinetic, whereas a prerequisite for any interpretation of particle acceleration by double layers, or any other quasi-static potential structure, is a quantitative model for a "dynamo" or "generator" that is also consistent with observations. This aspect of the problem is clearly recognized[12], and a model for an magnetohydrodynamic dynamo has recently been presented[13], but it is only barely sketched, and exhibits a number of inconsistencies[14].

PROBLEMS WITH THE POTENTIAL-DIFFERENCE THEORY

Conflict with Observations

Despite these developments many observations have been, and continue to be, put forward as revealing, directly or indirectly, the presence of static or quasi-static (relative to particle transit time) electric fields, both parallel and perpendicular to the local magnetic field. In the absence of information to the contrary, we understand that these fields should be treated as naturally occurring and not spacecraft induced. It is clear, incidentally, from the tens-of-minutes durations of moderate and active aurora[15] — a class including auroral arcs, from the lack of periodic oscillations or reversals in the acceleration process, and from the relative stability of the magnetic field, that induction can be neglected, at least as a primary agent. What evidence there is for a significant potential difference is at best highly equivocal, with a wide range of ad hoc assumptions being necessary to account for individual discrepancies. Taken together these create a substantial catalogue of evidence against the hypothesis[16]. Among the more telling are (i) a significant difference between the energies gained by electrons and positive ions travelling in opposite directions through the acceleration region[17], (ii) the frequent occurrence of counterstreaming electrons[18], and simultaneously upward flowing electrons and positive ions[19], (iii) a factor of up to 5 between energies gained by upward flowing ions of different mass (H^+ and O^+) though carrying the same charge[20], (iv) a lack of association between the behaviour of Ba^+ ions introduced, in active experiments, into the acceleration region and the aurora directly below[16].

The claim[21] that the characteristic kinetic energy is found to be equal to the potential difference obtained from electric field measurements has no foundation whatsoever, since the measured electric fields relate, in the model used, to potential differences *below* the spacecraft, and the observed electron distributions, in the same model, to potential differences *above* the spacecraft[22,23]. Moreover, even the estimation of potential drops below the spacecraft from measurements along the path is strongly qualified in the references cited by the declarations that "Experimental problems such as electric field saturation effects, threshold limitations, and sensitivity difficulties make this conclusion tentative at best"[22], and "...it is to be expected that the energy of maximum ion flux would be only a rough indicator of the potential below the satellite"[23]. Reservations have been expressed, too, about the quality of fit between observed and predicted velocity-space-density contours of electrons flowing upward in and near the so-called widened loss cone[16,24].

External Field

Many analytical studies of double layers in space plasmas have been restricted effectively to a single dimension in which the charge sheets are treated as being infinite in lateral extent[3]. This simplification is able to overlook the vital fact for a real double layer of an external as well as as internal field and, consequently, the vanishing line integral. Reference to double layers as "regions of a single electric field polarity"[25], a definition as "strong electric fields"[21], and an assessment that they behave as loads[12] , all seem to indicate neglect of the external field. The view that numerous 'weak double layers' in series would add to produce a significant combined potential difference[12] also requires the external field to vanish (in the same way that the potential difference across a row of batteries or capacitors becomes the sum only when the external fields are removed with conducting linkages). Note that a definition requiring a vanishing electric field on each side of the layer[26] does not satisfy Poisson's equation.

The same restrictions apply, of course, in the laboratory. Within a discharge tube a charged particle may, of course, exchange potential for kinetic energy to a degree determined by the boundary conditions. A double layer will help to determine where the exchange takes place, but cannot, of course, affect its magnitude[27,28]. The discharge tube as a whole could be considered to be an accelerator, but only in the same sense as the charged dome of a Van de Graaff generator. Energy is provided in these instances by the battery, or equivalent source of energy, across the discharge tube, and by the mechanically driven charging belt of the generator. The net potential difference resulting from the one-dimensional treatment does not, therefore, apply in practice.

'Generators' and 'Dynamos'

Related and equally disqualifying shortcomings are to be found in the many models that appeal to open-ended representations of electrostatic equi-potentials[29]. If they are closed, eg in the opposite hemisphere, the potential well has to be replenished, and the fundamental generator problem, mentioned above, immediately arises. Theories[30], and numerical simulations[31] that simply endow electrons with the potential energy later to appear as kinetic energy must be considered just as unsatisfactory. The solution to the problem of an energy source for particle acceleration in space cannot, we think, be solved either by drawing a few lines and calling them a "cosmic dynamo" with no indication of, or reference to the crucial non-conservative element, or to the mechanical constraints and intricacies of real dynamos (this is the state of affairs in current publications on the subject, decades after Alfvén made the then original and intriguing suggestion[4] before the advent of in-situ space observations). It should also be added, that almost fifty years of research have established beyond any doubt that a plasma can choose its behaviour from a near infinity of possibilities, the least likely being emulation of a simple circuit[32].

Length of Acceleration Region

A further major problem has been made even more acute recently by the announcement at this meeting[33] that typical inferred potential drops across double layers are $\approx 0.1V$ (rather than the previous $\approx 1V$). The double layers are estimated[12] to have a vertical extent of $\approx 100m$, which, together with the published data, implies a vertical separation of $\approx 2.5km$. The number of such double layers needed to give the characteristic $\approx 10keV$ peak energy (assuming that they could add) is then $\approx 10^5$. Since only 1 in 5 of these potential structures is deduced to have a non-zero potential difference[12], the total number required is $\approx 5 \times 10^5$. At the above spacing the length of the series of layers is, then, $\approx 1.25 \times 10^6 km$, or approximately 200 Earth radii. This is to be compared with the $\approx 10^4 km$, of order 2 Earth radii, length of the acceleration region confirmed from numerous satellite measurements. Even if the double layers are packed tightly with no spacing, the region would need to be 8 Earth radii in extent. This hardly seems a viable option. We would also have to accept that there is a high probability for hundreds of thousands of double layers to be generated synchronously along the electron paths. We think it more likely that this probability is negligibly small.

WAVE THEORY

Waves are well known to be very effective agents for transferring energy between different components within collisionless plasmas. It is also well established that the auroral region is turbulent and that it exhibits waves and wave-particle interactions of many kinds[34]. For such reasons it has been suggested for many years[35] that auroral-electron acceleration is due to wave-particle interactions. If a configuration of space charge is in motion, eg. as an electrostatic wave packet, or even a double layer moving at a suitable (resonant) velocity, reflexion of a charged particle by a potential barrier, although conservative in the wave frame, is accompanied by an energy exchange $\pm mv\Delta v$, where m is the mass of the reflected particle, v is the wave-packet velocity, and Δv is the velocity difference. This will be recognised as a (non-relativistic) electrostatic form of Fermi acceleration[36]. The same principle is used in celestial mechanics to speed a spacecraft as it passes a moving planet. When $v \rightarrow 0$ we recover the degenerate, conservative case. Consistent and characteristic features of the auroral acceleration region are highly irregular or "spiky" electrostatic fields[37] and broadband electrostatic waves[38]. When these are treated as an ensemble of electrostatic wave packets of the lower-hybrid type, and the constraints set by phase velocity, group velocity and amplitude governing stochastic interaction with electrons are taken into account in a Monte-Carlo model, the characteristic peak is found to evolve naturally[39]. Lower-hybrid turbulence is readily generated in a plasma whenever ion beams, ion conics, or ring distributions are present[32], or where there are inhomogeneities in the plasma[40]. Such waves are renowned for their effectiveness in accelerating or heating the electron component, by acting as the energy-transfer mechanism[41]. Other waves with similar electric-field properties would be equally effective[42], as a recent analysis of the effect of Alfvén waves confirms[43]. Acceleration of electrons is thus an expected consequence of turbulence on auroral field lines.

It should be mentioned at this point that double layers in the laboratory (to date the only well documented double-layers) are always associated with instabilities, wave activity and even turbulence[44]. This was pointed out by Alfvén when he wrote that noise in double layers is so important that "... theories which do not take it into consideration run some risk of being irrelevant"[45]. Had this timely caution been heeded, a double layer would have been seen to be a most unlikely candidate for the source of potential gradient supposed to be responsible for what was once thought to be an ordered particle acceleration. Given the great variety of situations where wave-turbulence energizes charged particles, a wave-particle interaction seems to be the natural first choice in a wide range of applications[46]. The electron accelerator in space plasmas may thus be seen, not as the equivalent of a Van de Graaff generator with unspecified drive mechanism, but as a stochastic linear accelerator governed by fundamental properties of the plasma[39].

CONCLUSION

We suggest, therefore, that there is no further reason to credit double layers, or any other source of static or quasi-static electric field, with the acceleration of electrons in the aurora or other space or astrophysical plasmas. All present evidence is overwhelmingly in favour of acceleration by electrostatic wave turbulence of the lower-hybrid or equivalent type.

REFERENCES

(a) Permanent address, Department of Physical Sciences, University of Naples.

1 D A Bryant, R Bingham and U deAngelis , Phys. Rev Letters, 68 37 1992

2 L P Block, Astrophys. and Space Science, 55, 59 (1978); see also 3 and 4

3 eg C K Goertz, Rev. Geophys. and Space Phys., 17, 418 (1979); Double layers in Astrophysics eds. Williams and Moorehead, NASA Conference Publication 2469, (1987).

4 H Alfvén, Tellus, 10, 104 (1958).

5 H Alfvén, and P Carlqvist, Sol. Phys. 1, 220 (1967).

6 D Le Queau, A Roux,and D Gresillon, (eds.) Collective Acceleration in Collisionless Plasmas, Cargese Workshop 1991, June 1992

7 T G Cowling, Terr. Mag. and Atmos. Elec., 47, 209 (1942).

8 C E McIlwain, J. Geophys. Res. 65, 2727 (1960)

9 D L Reasoner and C R Chappell, J. Geophys, Res. 78, 2176 (1973).

10 D A Bryant, in "Physics of Auroral Arc Formation" ed. Akasofu and Kan, AGU Geophysical monograph, 25, 103 (1981).

11 D S Evans, J. Geophys. Res. 73, 2315 (1968). Note that to derive a quantity proportional to velocity-space density from the count rate of an electrostatic analyser, one simply divides (in the non relativistic approximation) by the square of kinetic energy.

12 R Boström, Collective Acceleration in Collisionless Plasmas, Cargese Workshop 1991, 222 1992; see also R Boström, G Gustafsson, B Holback, G Holmgren, H Koskinen, and P Kintner, Phys. Rev. Lett. 61, 82 (1988).

13 R Lundin and L Eliasson, Ann. Geophysicae 9 202 1991

14 D A Bryant, Ann. Geophysicae 9 224

15 A P Krukonis and J A Whalen, J. Geophys. Res. 85 119 1980

16 D A Bryant, Physica Scripta, T30, 215 (1990); and "Solar and Planetary Plasma Physics" ed. Buti, World Scientific Press, p58, (1990); see also 39.

17 see measurements by P H Reiff, H L Collin, J D Craven, J L Burch, J D Winningham, E G Shelley, L A Frank and M A Freedman, J. Geophys. Res. 93, 7441 (1988).

18 R D Sharp, E G Shelley, R G Johnson and A G Ghielmetti, J. Geophys. Res. 85, 92 (1981).

19 B Hultqvist, R Lundin, K Stasiewicz, L Block, P-A Lindqvist, G Gustafsson, H Koskinen, A Bahnsen, T Potemra and L Zanetti, J. Geophys. Res. 93, 9765 (1988).

20 H L Collin, R D Sharp, E G Shelley, and R G Johnson, J. Geophys. Res. 86, 6820 (1981); R Lundin and B Hultqvist, J. Geophys. Res. 94, 6665 (1989).

21 J E Borovsky, Phys. Rev. Letters, in press

22 P F Mizera, J F Fennel D R Croley Jr and D J Gorney J. Geophys. Res. 86 7566 1981

23 M Temerin, M H Boehm and F S Mozer, Geophys. Res Lett. 8 799 1981

24 D R Croley Jr, P F Mizera and J F Fennell, J. Geophys. Res. 83 2701 1978

25 M Temerin, K Cerny, W C Lotko and F S Mozer, Phys. Rev. Letters, 48 1175 1982

26 e.g., H Alfvén, Cosmic Plasma, D Reidel, p29, (1981).

27 see for example Fig. 10 of P Leung, A Y Wong, and B H Quon, in "Relation between laboratory and Space Plasmas" ed. Kikuchi, D Reidel, 1981.

28 eg C Chan, in "Double layers in Astrophysics" eds. Williams and Moorehead, NASA Conference Publication 2469, p35 (1987); see also 27.

29 eg F S Mozer, C W Carlson, M K Hudson, R B Torbert, B Parady and J Yatteau, Phys. Rev. Lett. 38, 292 (1977); D A Gurnett, in "Critical Problems of Magnetospheric Physics" ed. Dyer, IUCSTP Washington DC, (1972); see also 3.

30 D W Swift, J Geophys. Res., 80, 2096 (1975); and M Hesse, J Birn and K Schindler, J Geophys. Res., 95, 18,929 (1990). For recognition of the problem, though, see D W Swift, J Geophys. Res., 81, 3935 (1976); D A Bryant, in "High-Latitude Space Plasma Physics" ed. Hultqvist and Hagfors, Plenum Press, p295 (1983); and 35.

31 eg Y T Chui and M Schulz, J. Geophys. Res. 83, 629 (1978); and see comment on invariability of this practice by T J Birmingham, Physica Scripta T18, 203 (1987).

32 A B Mikhailovskii, Theory of Plasma Instabilities, Consultants Bureau, N.Y. London (1974).

33 R Boström, These proceedings.

34 see Auroral Physics, C-I Meng, M J Rycroft and L A Frank (eds) Cambridge University Press 1991

35 D A Bryant, D S Hall and D R Lepine, Planet Space Sci., 26, 81 (1976); and 29.

36 E Fermi, Phys. Rev. 75, 1169 (1949); and Astrophys. J, 119, 1 (1954).

37 G-G Fálthammar, L P Block, P-A Lindqvist, G Marklund, A Pedersen and F S Mozer, Annales Geophysicae, 5A, 171 (1987).

38 R Pottelette, M Malingre, A Bahnsen, L Eliasson, K Stasiewicz, R E Erlandson and G Marklund Ann. Geophys. 6, 573 (1988).

39 D A Bryant, A C Cook, Z -S Wang, U de Angelis and C H Perry, J.Geophys. Res. 96, 13,829 (1991); see also 16.

40 eg R C Davidson, N T Gladd, C S Wu and D Huba Phys. Fluids 20, 301 (1977); and J H Hsia, S M Chiu, M F Hsia, R L Chou and C S Wu, Phys. Fluids 22, 1737 (1979).

41 J B McBride, E Ott, J P Boris, and J H Orens, Phys. Fluids, 15, 2367 (1972); P J Barrett, B D Fried, C F Kennel, J H Sellen, and R J Taylor, Phys. Rev. Lett., 28, 337 (1972); C F F Karney and N J Fisch, Phys. Fluids, 22, 1817 (1979).

42 D A Bryant, D S Hall and R Bingham, p119 in Auroral Physics, C-I Meng, M J Rycroft (eds.) Cambridge University Press 1991

43 C -H Hui and C E Seyler, J. Geophys. Res., 97 3953 1992

44 P Leung, in "Double Layers in Astrophysics", eds. Williams and Moorehead, NASA Conference Publication 2469, p89 (1987); see also 27.

45 H Alfvén, keynote address in "Double Layers in Astrophysics", eds. Williams and Moorehead NASA Conference Publication 2469, 1 (1987).

46 (i) solar-wind ion releases, bow shock, aurora — D A Bryant, Plasma Phys. and Controlled Fusion, 27, 1369 (1985); (ii) solar-wind ion releases — D S Hall, D A Bryant, C P Chaloner, D R Lepine and R Bingham, J. Geophys. Res., 91, 1320 (1986); (iii) Martian bow shock — D A Bryant, D S Hall, R Bingham and N M Shutte, Nature, 347, 236 (1990).

STAIRSTEP DOUBLE LAYERS, MAGNETIZED SHEATHS AND MAGNETIZED DOUBLE LAYERS

N. Hershkowitz, T. Intrator, D. Diebold, G-H. Kim and J. Meyer

University of Wisconsin, Department of Nuclear Engineering and Engineering Physics, 1500 Johnson Drive, Madison, WI 53706-1687 USA

Abstract

Laboratory measurements of double layers and sheaths in collisionless and collisional, magnetized and unmagnetized systems are discussed. The transition from single to two to three or more step double layers is reviewed together with the possible connection to BGK double layer solutions. The characteristic lengths of sheaths and presheaths in a uniform B field is shown to depend on the ion mean free path, the ion gyroradius, the ion sound gyroradius, and the angle between the magnetic field and the normal to a boundary (or the electric field). The first measurements of magnetic and charge exchange presheaths in ECR plasmas and magnetic presheaths in inductive magnetized plasmas are presented. Very recent experimental observations of multiple step 2 dimensional, time average, double layers, oriented with respect to the magnetic field, in plasmas produced by inductive sources in uniform magnetic fields are also discussed.

Introduction

Many laboratory double layer experiments have been motivated by reports of the presence of magnetospheric double layers.[1] It is not yet certain whether auroral electron acceleration takes place at one large potential step or at a series of much smaller steps joined together.[2,3,4,5,6] Single step double layers have been an active area of investigation for many years.[7] On the other hand, double layers consisting of a series of small steps have been demonstrated in only a few laboratory plasmas.[8,9,10,11] While such "stairstep" structures can be the result of ionization, e.g., striations in high pressure systems, experiments also find such structures in "collisionless" systems. Computer simulations of double layers have also found evidence for multiple structures but periodic boundary conditions often employed in such simulations do not permit structures with different potentials at boundaries.[12] Magnetospheric double layers are accompanied by dc magnetic fields so it is important to investigate the new physics associated with the addition of magnetic fields. Unfortunately, even the simpler problem of the effects of magnetic fields on boundary sheaths have not been investigated thoroughly. In this paper we describe laboratory experiments which investigate stairstep double layers, magnetized sheaths and magnetized double layers with selected examples from collisionless and collisional plasmas.

Most laboratory experiments[13,14] have been carried out with "collisionless" unmagnetized plasmas, parallel to the magnetic field in magnetized plasmas or parallel to the axis in axisymmetric magnetized plasmas. In those experiments, all collision lengths were comparable or greater than device dimensions. Such double layers can be described by BGK[15] solutions to the Maxwell's equations and Newton's laws.[16] The potential structures are entirely determined by specifying the electron and ion distribution functions at the device boundaries. This approach does a good job of describing single step potential transitions. Such double layer thicknesses are measured in Debye lengths.

Collisionless two or more step double layers have been reported in unmagnetized systems. The transition from single to two step double layers and from single to two to three or more step double layers will be reviewed here. In addition, the role of possible symbiotic effects[17] between plasma turbulence and electron distribution functions and the connection to BGK double layer solutions will be explored. New results are also presented on the formation of nonaxisymmetric sheaths and multiple double layer structures in magnetized plasmas.

Recently, interest in double layers and sheaths has developed in the plasma processing community. These double layers and sheaths differ from those previously investigated in two significant ways. In rf plasma processing, situations often exist in which the ions see a time averaged potential while the electrons respond to a time varying rf field. Double layer potential structures then exist on a transient basis during part of the discharge.[18] The other important difference is that plasma processing applications normally employ plasmas with collision lengths that are smaller than device dimensions and cannot be considered collisionless. These short collision lengths set the scale for presheath lengths.

Experimental Technique

All double layer experiments were carried out in one of two triple plasma devices -- DOLI-I and DOLI-II. We have used these devices in our laboratory for more than 10 years. Over the years a variety of different types of plasma sources and magnetic field configurations have been employed. At the present time DOLI-I is operating with $B = 0$ and with hot filament sources while DOLI-II operates with inductive plasma sources and with a uniform axial magnetic field of up to 350 gauss.

The DOLI devices each consists of two source chambers which bound a target chamber. The target chamber is separated from each source chamber by two biased grids which allow some measure of separate control of the electron and ion distribution functions at each of the two target end boundaries. Filaments are normally operated within the surface magnetic field produced by permanent magnets. This can trap the ionizing electrons and prevent them from reaching the target plasma. Inductive fields are produced by spiral inductive antennas mounted at the ends of the source chambers. The inductive antennas are normally operated at 13.56 MHz.

Measurements of magnetic presheaths were carried out in two devices. The first, specifically constructed for these studies, employed an inductive source, similar to that used in DOLI-II, to produce plasma which was separated by a grid from the magnetized plasma in contact with the tilted plate. The plasma in contact with the tilted plate first diffused across the magnetic field and grid. The second device was an electron cyclotron resonant plasma source which was being employed for plasma etching. Plasma was resonantly produced in a mirror cell and allowed to flow out along an expanding magnetic field where it encountered a tilted, grounded plate.

Plasma potential was measured using emissive probes either in strong emission,when the probe position was scanned, or by the inflection point technique,[19,20] when point by point measurements are made. The latter technique has a spatial resolution of 1 mm, can have a plasma potential uncertainty as small as 0.2 V and has been operated at plasma densities ranging from $< 10^4 cm^{-3}$ to as high as $10^{13} cm^{-3}$.[21]

Experimental Results

Double Layers and Sheaths

A sheath is a non-neutral potential step which occurs near a plasma boundary. It usually involves "trapped electrons", i.e., electrons which enter the sheath from the plasma and reflect back to the plasma and "free" ions which are not reflected by the sheath. A double layer can be viewed as a sheath which occurs separated from a plasma boundary. It differs from the usual sheath in that it usually contains at least three separate species, e.g. free and trapped electrons and free ions. Once the potential step is removed from the wall, it is not at all obvious that there should be one or two or more potential steps.

Multiple - Stairstep Double Layers

An experiment by Chan and Hershkowitz[10] in a triple plasma device reported that increasing the ratio of the system length L to the Debye length λ_D resulted in a transition from one to two double layer transitions. Subsequent experiments did not support this result and found that the number of double layer steps had a strong sensitivity to boundary conditions. In a more recent experiment,[11] the evolution of multiple double layers from a single double layer was investigated. It was suggested that symbiotic effects between plasma turbulence and electron distribution functions[17] and BGK solutions resulted in double layers.

A good example of a two step "collisionless" double layer is given in Fig. 1. This double layer was graphed by recording the floating potential of an emissive probe in strong emission as the probe was scanned along the axis of the target chamber of a triple plasma device. It is important to note that the potential at which the second double layer began to grow is not associated with an imposed boundary potential. Note that the structure resembles two BGK type double layers joined together.

Data showing the dependence of the multiple double layer structure on the east target grid (located at the right) potential ϕ_t are given in Fig. 2. For low value of ϕ_t, only one potential step is observed. As ϕ_t was increased two and then more steps appear. These steps, indicated by arrows, are only weakly apparent and the structure resembles a potential step caused by anomalous resistivity. This is not accidental because it is likely that that the potential structure is a consequence of interactions of ions and electrons with the fluctuating electric fields resulting from streaming instabilities. Data showing electric potential fluctuations versus position (see Fig. 3) show strong correlations with the dc potential structures. Earlier experiments[8] had also reported strong correlations between density fluctuations and potential structures.

Borovsky[17] argued that a symbiotic effect could exist between double layer potential steps and plasma turbulence. The argument suggests that the plasma turbulence associated with the first double layer plays the role of modifying the electron or ion distribution function to make it a suitable distribution function for the boundary of a second double layer transition.

Magnetized Structures

The characteristic lengths of sheaths and presheaths in a uniform B field depend on the Debye length, the ion mean free path λ_{in}, the ion gyroradius r_{ci}, the ion sound gyroradius r_{cs}, and $\sin\psi$, where ψ is the angle between the magnetic field and the normal to a boundary or the electric field. The first measurements of magnetic and charge exchange presheaths in ECR plasmas and magnetic presheaths in inductive magnetized plasmas are presented here.

Theory of Boundary Sheaths in Tilted Magnetic Field

First consider a B field oriented at an angle ψ with respect to an equipotential boundary. Unlike the $\psi = 0$ problem, for which the solution is well understood, very little experimental or theoretical work has been carried out with regard to this problem.

For $\psi = 0$, there is an electrostatic sheath whose thickness can be expressed

$$s \propto (e\phi/T_e)^{\alpha} \lambda_D$$

where $\alpha = 3/4$, $2/3$, 0.6 and 0.5 for collisionless, mobility limited with constant mobility or constant cross section and for frozen ions respectively. For the parameters of experiments reported here, electrostatic sheath thicknesses were several mm or less. The sheath is preceded by a Bohm presheath.

It is well known (but not well verified experimentally) that ions entering a one dimensional electrostatic sheath must satisfy a Bohm criteria[22,23,24] i.e. they must have a directed "Bohm velocity" $v \geq c_s$, where c_s is the ion sound velocity $c_s \equiv \sqrt{\frac{Te}{mi}}$. The presheath thickness depends on the system dimensions or on the ion collision mean free path. Representative plasma potential axial profile data, obtained with the ECR source, are given in Fig. 4. It is apparent that as the neutral pressure is increased, the presheath length is decreased. It can be shown that the presheath length is approximately proportional to the charge exchange length which is the shortest ion collision mean free path length.

For $\psi \neq 0$, ions gyrating about the magnetic field encounter the wall before their guiding center hits the wall. The resulting ion loss gives rise to a corresponding decrease in plasma potential which we identify as a "magnetic presheath". The magnetic presheath is comparable to the ion gyroradius in thickness. Generally, the electrostatic sheath thickness $s \ll r_{ci}$, the ion gyroradius.

Chodura[25] has derived an expression for the magnetic presheath thickness

$$s_m = \beta \frac{c_s}{\omega_{ci}} \sin \psi \equiv \beta r_{cs} \sin \psi. \tag{1}$$

Here, ω_{ci} is the ion gyrofrequency and β is a parameter found to be $\sqrt{6}$. Note that if $\beta = 1$, the magnetic sheath thickness is the guiding center position at which the first encounter occurs with the boundary. Equation (1) was derived under the assumption that ions had a Bohm velocity before encountering the magnetic presheath boundary. The experimental data suggest that the magnetic presheath is a quasineutral region with constant E field, it must be so if

$$\lambda_D \ll r_{cs}.$$

Note that there are now three characteristic lengths, λ_D, λ_c and r_{cs}.

We have investigated the consequences of Eq. (1) for two situations, a collisionless inductive plasma and an outflowing ECR plasma. In both cases a conducting plate was oriented at an angle ψ with respect to a magnetic field.

Magnetized Sheath Data

Representative data for plasma potential versus position, measured in the inductive source, are given in Fig. 5. The magnetic presheath is apparent as a region with an approximately constant electric field, near the boundary. The magnetic presheath length boundary is identified by an arrow in Fig. 5. Using similar data obtained for several values of magnetic field (up to 170 gauss), electron temperature (4-5 eV) and angle (30-80°), the magnetic presheath width is shown to be proportional to $\beta r_{cs} \sin \psi$ in Fig. 6. For these data we find $\beta = 1$.

Experimental results showing a combination of a charge exchange and magnetic sheath are given in Fig. 7. These data were obtained using a plasma flowing out of an electron cyclotron resonant (ECR) source into an expanding magnetic flux tube with a boundary consisting of a grounded conducting plate that is tilted with respect to the magnetic field at an angle ψ. Equipotential contours were obtained by making a series of point plasma potential measurements with an emissive probe using the inflection point technique.[19] Representative parameters are B = 150 gauss, T_e = 3-4 eV, N_2^+ ions, n = 10^{10}cm^{-3} and pressure = 0.75-2 mTorr. For these parameters $\lambda_D = 5 \times 10^{-3}$ cm, λ_c = 1.5-4 cm and r_{cs} = 5 cm.

It can be shown that for $\psi = 0$, the presheath length $s_p = \alpha\lambda_c$.

The equipotential contours in Fig. 7 show qualitative agreement with Eq. (1). For positions $x \gg \lambda_c, r_{cs}$, (with x measured from the plate along the magnetic field), the contours are approximately perpendicular to the B field, i.e. the E field is parallel top the field. This is consistent with an expanding magnetic flux tube, flux conservation, a decreased density and ion flux conservation.

Magnetized Double Layers

In recent experiments we have investigated multiple step two dimensional double layers, oriented with respect to the magnetic field. Double layers were studied in a triple plasma device with a uniform axial magnetic field of up to 350 gauss. Plasma was produced with inductive sources located at the far end of each source chamber (see Fig. 8). The resulting potential structures exhibit many new features. Representative equipotential structures are given in Fig. 8. Scans of the plasma potential parallel to the device axis have the interesting property that they show either relatively thick single transitions or two or more step transitions. It is apparent from these data that both of these structures are two aspects of the same structure. This only becomes apparent when the two dimensional structures are explored. The characteristic dimension of these structures appears to be proportional to λ_D in the parallel direction and forms a distribution in the perpendicular direction $L_\perp$, bounded by the ion gyroradius r_{ci} and r_{cs}. This result appears to be consistent with those reported by Lennartsson.[26]

The potential seen by the electrons and ions result because an inductive source is used to produce the plasma. Because the rf frequency ω is greater than the ion plasma frequency but much less than the electron plasma frequency, the ions see a time averaged potential while the electrons responds to the time varying rf field. It is then likely that the double layer potential structures only exist on a transient basis during part of the discharge. Such structures could accelerate ions and electrons in the same direction during part of the rf cycle. Our experiments have not yet clearly determined the size of the time varying rf potential in the target chamber so the importance of this effect is not yet clear. It may be that such a phenomena has been observed in space.[27]

Conclusions

Collisioneless stairstep double layers have been demonstrated in the laboratory both in one and two dimensions and with and without magnetic fields. It appears that collisionless stairstep double layers may be related to BGK double layers modified by plasma turbulence. Experimental measurements are also presented of time average double layers, oriented with respect to the magnetic field, in plasmas with inductive sources. The characteristic dimensions perpendicular to the B field are bounded by r_{ci} and r_{cs}.

The fist experimental measurements have been presented of presheaths in collisional plasmas and of the combination of magnetic presheaths and collisional presheaths. The characteristic dimensions are found to be $r_{cs} \sin\Psi$ and λ_c.

Acknowledgement

This work is supported by NSF Grant ATM-9017451.

References

1. M. Temerin, K. Cerney, W. Lotko, and F.S. Mozer, "Observation of double layers and solitary waves in the auroral plasma," Phys. Rev. Lett., vol. 48, 1175 (1982).

2. R. Boström, G. Gustafsson, B. Holback, G. Holmgren, H. Koskinen, and P. Kintner, "Characteristics of solitary waves and weak double layers in the magnetospheric plasma," Phys Rev. Lett., vol. 61, 82 (1988).

3. R. Boström, "Observations of weak double layers on auroral field lines," Abstracts of the Workshop on Plasma Experiments in the Laboratory and in Space, Alpbach, Austria, July 1-5, 1991, p.30 (1991).

4. R. Boström, "Are weak double layers important for auroral plasma dynamics?" Abstracts of the AGU Chapman Conference on Auroral Plasma Dynamics, convened by R.L. Lysak, Minneapolis, Oct. 21-25, 1991, p.25 (1991) and the discussion immediately following the above talk (1991).

5. F.S. Mozer, "Questions about the electric fields in the auroral acceleration region," Abstracts of the Workshop on Plasma Experiments in the Laboratory and in Space, Alpbach, Austria, July 1-5, 1991, p. 25 (1991).

6. F.S. Mozer, "Parallel electric fields," Abstracts of the AGU Chapman Conference on Auroral Plasma Dynamics, convened by R.L. Lysak, Minneapolis, Oct. 21-25, 1991, p.25 (1991).and the discussion immediately following the above talk (1991).

7. N. Hershkowitz, "Review of recent laboratory double layer experiments," Space Sci. Rev., vol. 41, pp. 351-391 (1985).

8. Guyot, M. and Hollenstein, Ch., Phys. Fluids 26, 1596 (1983).

9. Sato, N., Hatakeyama, R. Iizuka, S., Mieno, T., Saeki, K., Juul Rasumssen, J., and Michelson, P., Phys. Rev. Letters 45, 1330 (1981).

10. C. Chan, and N. Hershkowitz, "Transition from single to multiple double layers," Phys. Fluids, vol. 25, 2135 (1982).

11. A. Bailey III, and N. Hershkowitz, "Three step double layers in the laboratory," Geophys. Res. Lett., vol. 15, pp. 99-102 (1988).

12. C. Barnes, M.K. Hudson, and W. Lotko, "Weak double layers in ion-acoustic turbulence," Phys. Fluids 28, 1055 (1985).

13. Proceedings of Symposium on Plasma Double Layers, Riso National Laboratory, June 16-18, 1982, vol. Riso-R-472.

14. Proceedings on Second Symposium on Plasma Double Layers and Related Topics, July 5-6, 1984, Innsbruck, Austria.

15. I.B. Bernstein, J.M. Greene, and M.D. Kruskal, Phys. Rev., vol. 108, 546 (1957).

16. G. Knorr and C.K. Goertz, Astrophys. Space Sci. 31, 209 (1974).

17. J.E. Borovsky, "Parallel electric fields in extragalactic jets: double layers and anomalous resistivity in symbiotic relationships," Astrophys. Journal, 306, 451 (1986).

18. R.A. Gottscho, G.R. Scheller, D. Stoneback and T. Intrator, "The effect of electrode area ratio on low-frequency glow discharges," J. Appl. Phys 66, 2 (1989).

19. J.R. Smith, N Hershkowitz, and P. Coakley, "The inflection point method of interpreting emissive probe characteristics," Rev. Sci. Instrum., vol. 50, 210 (1979).

20. N. Hershkowitz, Plasma Diagnostics, Vol. 1, edited by O. Auciello and D.L. Flamm, Academic Press, San Diego, 1989, chapter 3, pp.113-184.

21. M.H. Cho, C. Chan, N. Hershkowitz, and T. Intrator, "Measurement of vacuum space potential by an emissive probe," Rev. Sci. Instrum., vol. 55, 631 (1984).

22. K.-U. Riemann, J. Phys. D, 24, 493 (1991).

23. F.F. Chen, Plasma Diagnostic Techniques, edited by R.H. Huddlestone and S.L. Leonard. New York: Academic Press, 1965, p. 134.

24. S. Meassick, M-H. Cho, and N. Hershkowitz, "Measurement of plasma presheath," IEEE Trans. Plasma Sci., vol. PS-13, 115 (1985).

25. R. Chodura, Phys. Fluids, vol. 25, 1628 (1982).

26. Lennartsson, W., "On the Spatial Scale of Oblique Double Layers," Proc. of Chapman Conf. on Ion Acceleration in the Magnetosphere, Wellesley, MA, p.323 (1985).

27. B. Hultqvist, L. Block, P.A. Lindqvist, G. Gustafsson, H. Koskinen, A. Bahnsen, T.A. Potemra, and L.J. Zanetti, "Simultaneous observation of upward moving field aligned energetic electrons and ions on auroral field lines," EOS Trans. Am. Geophys. Union, 69, 446 (1988).

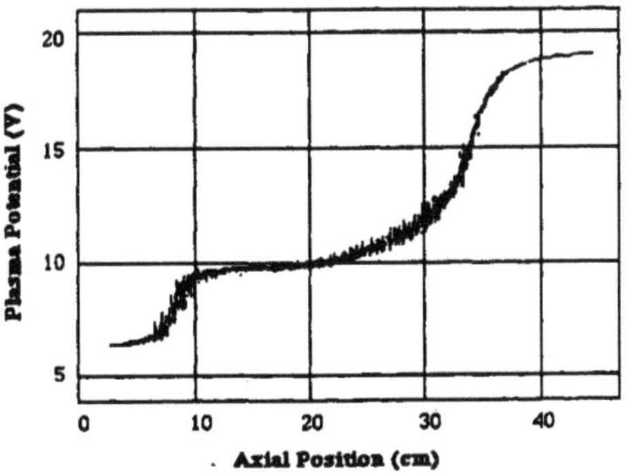

Figure 1. Representative Two Step Double Layer.

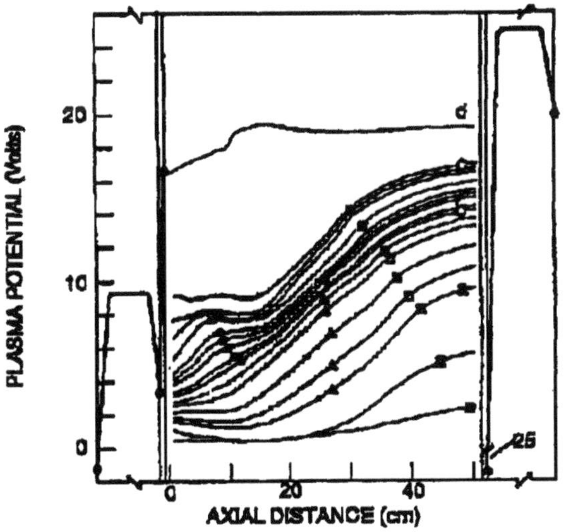

Figure 2. Evolution of multiple double layer with changes in east target grid bias voltage. (n_{west} ~ 8 x 10^8, n_{east} ~ 2.8 x 10^8 cm^{-3}. From bottom to top the curves correspond to grid bias voltages equal to 0, 5.0, 10.0 (labelled a), 11.25, 12.5, 13.75, 14.3, 14.75 (b), 15.0, 15.5, 15.75, 16.0, 16.5, 17.0, 17.25, 17.50 (c) and 17.75 V (d).

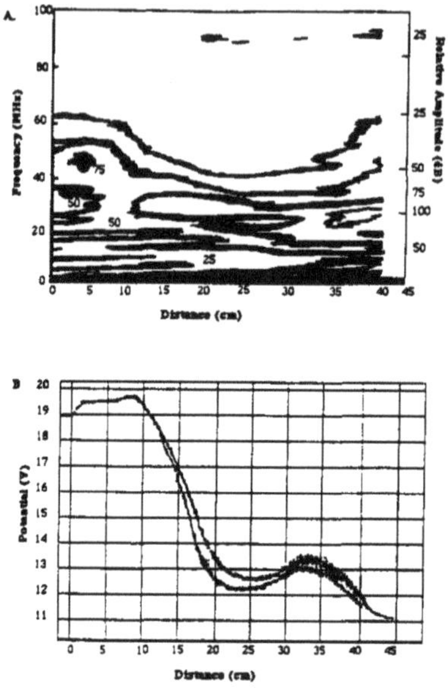

Figure 3. Capacitive probe measurements of the high frequency floating potential fluctuations associated with a DL. (a) Equi-amplitude contours (marked in db) in the frequency (MHz), distance (cm) domain. (b) Plasma potential (volts) vs. distance (cm) of the corresponding DL.

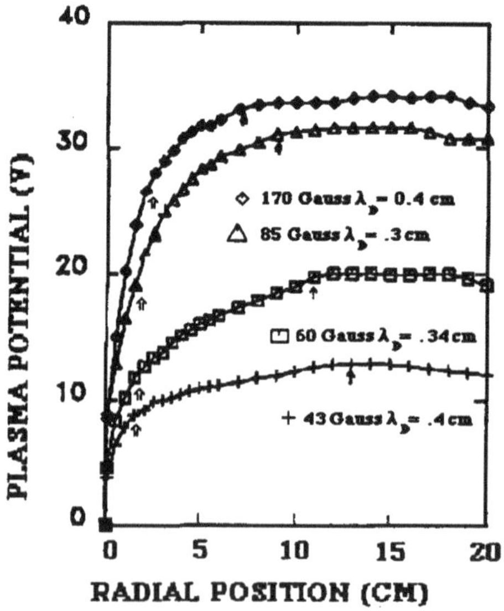

Figure 4. Plasma potential (volts) vs. radial position (cm) of the sheaths near a planar electrode in a He plasma for various applied magnetic fields and plasma conditions. The angle ψ between the applied magnetic fields and the normal to the planar electrode was 60 degrees. The solid arrows indicate the beginning of the magnetic presheath and the open arrows indicate the beginning of the electrostatic sheath.

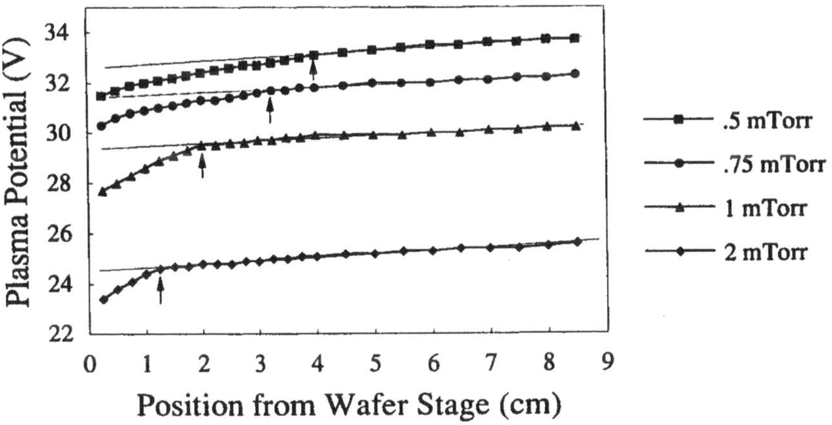

Figure 5. Axial scans near the boundary of an ECR device as a function of N_2 plasma pressure.

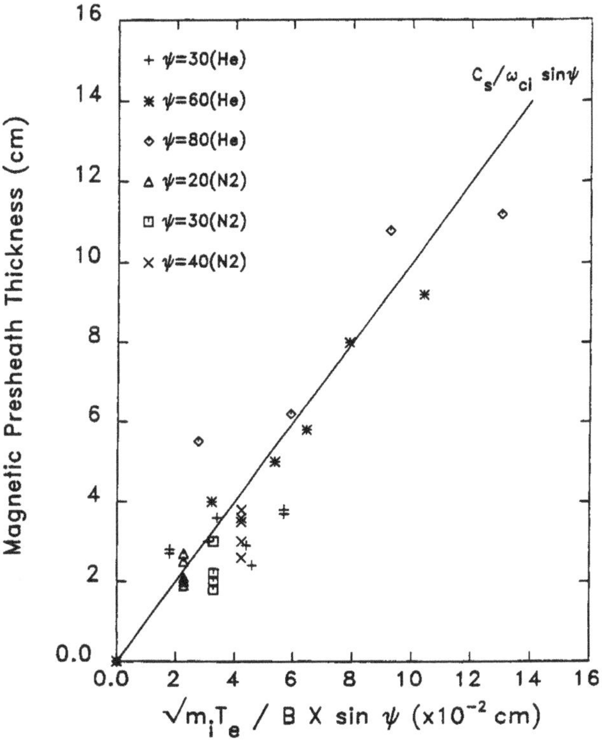

Figure 6. Measured magnetic presheath width vs. r_{cs} x sin ψ. Data are included for helium (inductive source) and nitrogen (ECR) plasma.

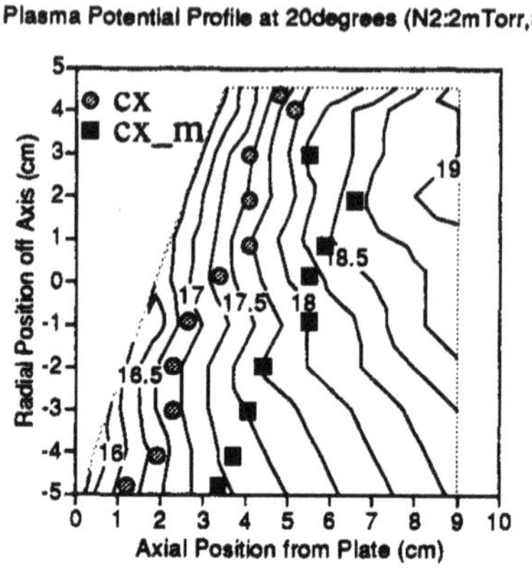

Figure 7. (a) Equi-potential controus near a boundary tilted at 20 degrees.

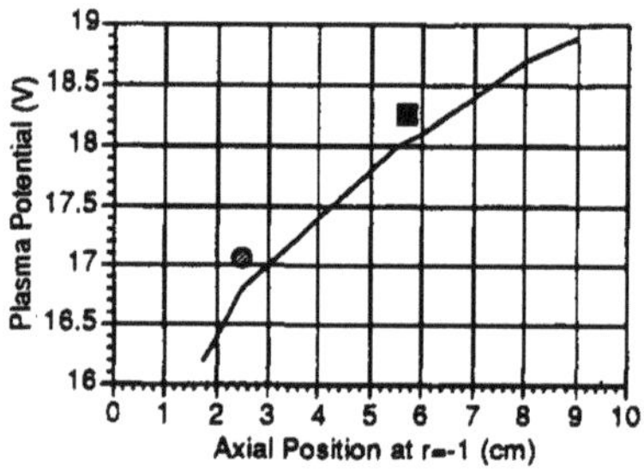

Figure 7. (b) Axial plasma potential scan at r = -1 cm.

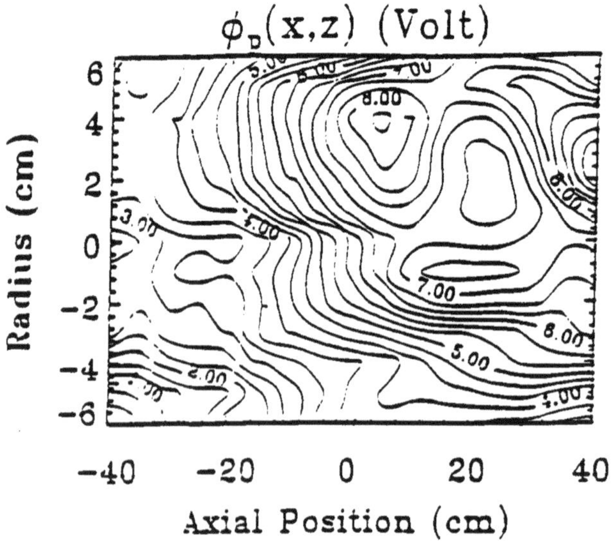

Figure 8. Equi-potential contours (in volts) in the radius (cm), axial position (cm) domain. The applied magnetic field was 360 gauss and in the axial direction.

KINETIC MODELING OF BOUNDED PLASMAS BY ANALYTIC–NUMERICAL METHODS: ADVANCING TO TWO–DIMENSIONAL AND COLLISIONAL SYSTEMS

S. Kuhn

Plasma and Energy Physics Group
Institute for Theoretical Physics, University of Innsbruck,
A–6020 Innsbruck, Austria

ABSTRACT. Realistic kinetic modeling of bounded plasmas is of interest in both fundamental plasma research and applied disciplines such as plasma technology and fusion research. Particle simulation has by now become a routine tool for such studies, but this paper deals with an alternative approach termed "Analytic–Numerical Plasma Simulation (ANPS)". This approach, which represents a combination of analytic and numerical methods closely related to the method of characteristics, naturally allows for realistic ion–electron mass ratios and high resolution. It either yields analytic solutions or, where these are impossible, can be advantageously applied in the form of a semi–analytic method called "trajectory simulation".
In the paper, the basic theory elements of the ANPS approach are outlined, previous results on 1d collisionless bounded plasma systems are reviewed, recent dc solutions for a 1d collisional and a 2d collisionless model of the single–ended Q–machine are presented, and some perspectives for the future development of ANPS are discussed.

1. INTRODUCTION AND SUMMARY

Realistic kinetic modeling of bounded plasmas is not only of interest in fundamental plasma research but is becoming increasingly important also in applied disciplines, such as plasma technology and fusion research. *Particle simulation* has by now become a routine tool for such studies, as is most impressively demonstrated by the widespread and ever increasing usage of the bounded–plasma particle codes PDW1 (/89Law/, /92Bul/) and PDx1 (/91Vahe/) developed at U.C. Berkeley, and similar codes (/79Dec/, /92Bau/). Nevertheless alternative, more theory–oriented methods are still desirable and needed because the simulations in question represent a special kind of *experiments* ("computer experiments") and it is widely accepted that theory is an essential prerequisite for understanding experiments.

This paper deals with such an alternative approach, to be termed *"Analytic–numerical plasma simulation (ANPS)"*, which has been under development for some time and in several places but still does not seem to be as well known as it would deserve. The ANPS approach represents a *combination of analytic and numerical methods closely related to the method of characteristics*, the characteristics of any kinetic plasma equation being identical with the trajectories of collisionless single–particle motion. It naturally allows for realistic electron–to–ion mass ratios and high resolution in both configuration and velocity spaces, and either yields *analytic solutions* or, where such are impossible, can be advantageously applied in the form of a semi–analytic method, to be called "*trajectory simulation*".

The bulk of the paper is organized as follows: In Sec. 2, the basic theory elements of the ANPS approach are briefly outlined. Section 3 contains a brief review of ANPS results on *(1d,1v) collisionless* bounded plasma systems (cartesian, cylindrical, spherical; partly time–dependent). In Sec. 4 we describe a new iteration scheme which heavily relies on trajectory integration and has already been used successfully for the self–consistent calculation of time–independent (dc) states of a cartesian (1d,2v) single–ended Q–machine model *including charge-exchange collisions* (Sec. 5), and of a cartesian (2d,3v) *collisionless* single–ended Q–machine model (Sec. 6), both applications representing significant advances over previous ANPS results. Finally, in Sec. 7, we comment on the present status of ANPS and identify some perspectives for its future development.

Clearly, the notation "(m d, n v)" indicates a system described by m space and n velocity coordinates. In fact, in the context of this paper it is no longer sufficient to simply speak of, e.g., "one–dimensional" systems because a system that is one–dimensional in configuration space (1d) may be two–dimensional in velocity space (2v) and hence must be denoted as (1d,2v), cf. Sec. 5.

2. BASIC THEORY ELEMENTS OF THE ANALYTIC–NUMERICAL SIMULATION (ANPS) APPROACH

2.A. Basic equations and boundary conditions

In the present section we briefly summarize the basic elements of the analytic–numerical approach to kinetic bounded–plasma modeling. For illustration (and in accordance with the present state of development of ANPS), we restrict ourselves to *electrostatic systems with given magnetic fields*. For these, the basic equations in the plasma region are (i) the kinetic equations for the velocity distribution functions and (ii) Poisson's equation for the electrostatic potential.

In integro–differential form, the kinetic equation for particle species s is given by

$$\left[\frac{\partial}{\partial t} + \mathbf{v}\cdot\frac{\partial}{\partial \mathbf{x}} + \frac{e_s}{m_s}(\mathbf{E} + \mathbf{v}\times\mathbf{B})\cdot\frac{\partial}{\partial \mathbf{v}}\right] f_s(\mathbf{x},\mathbf{v},t) \equiv \mathcal{C}_s(ff\,|\,\mathbf{x},\mathbf{v},t) \equiv \sum_p \mathcal{C}_{sp}(f_s f_p\,|\,\mathbf{x},\mathbf{v},t)\,, \quad (2.A.1)$$

with $\mathbf{E}(\mathbf{x},t)$ and $\mathbf{B}(\mathbf{x},t)$ the macroscopic electric and magnetic fields, e_s the charge and m_s the mass of a species–s particle, f_s the distribution function and $\mathcal{C}_s$ the total collision term of species s, and $\mathcal{C}_{sp}$ the partial collision term between species s and p. By writing the total and partial collision terms in the particular form given in (2.A.1) we wish to indicate that they are, as is usually assumed, *bilinear* functionals of the distribution functions but ultimately represent functions of $\mathbf{x}$, $\mathbf{v}$ and t.

With the usual relation $\mathbf{E} = -\nabla\Phi$, Poisson's equation reads

$$\nabla^2\Phi(\mathbf{x},t) = -\frac{1}{\epsilon_0}\rho(\mathbf{x},t) = -\frac{1}{\epsilon_0}\sum_s e_s n_s(\mathbf{x},t)\,, \quad (2.A.2)$$

where ρ is the total space–charge density and

$$n_s(\mathbf{x},t) = \int d^3v\ f_s(\mathbf{x},\mathbf{v},t) . \tag{2.A.3}$$

is the number density of particle species s.

Equations (2.A.1) and (2.A.2) must be complemented by appropriate boundary conditions. To be specific, let us assume here that we are dealing with a bounded spatial region where both the "emission" distribution functions and the electric potential are given at the boundaries:

$$f_s(\mathbf{x}_b,\mathbf{v},t) = f_{sb}(\mathbf{x}_b,\mathbf{v},t) \qquad [f_{sb} \text{ given for } \mathbf{v}\cdot\mathbf{n}(\mathbf{x}_b) > 0] \tag{2.A.4}$$

$$\Phi(\mathbf{x}_b,t) = \Phi_b(\mathbf{x}_b,t) \qquad [\Phi_b \text{ given}] , \tag{2.A.5}$$

where $\mathbf{x}_b$ is a "boundary point", i.e., a point lying infinitesimally close to the bounding surface, and $\mathbf{n}(\mathbf{x}_b)$ is the local normal to that surface pointing into the plasma region. This situation is schematically illustrated in Fig. 2.A.1.

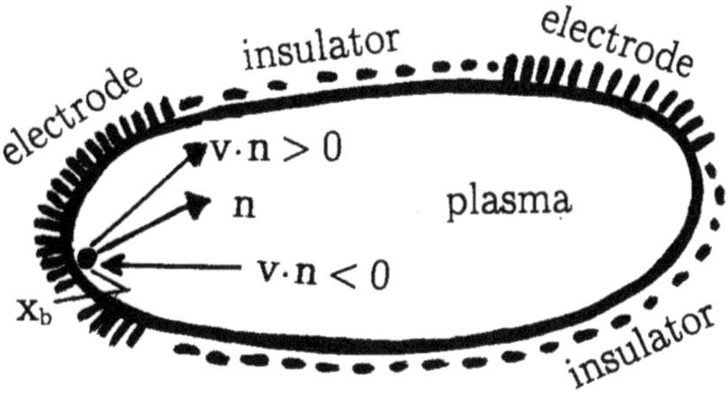

FIG. 2.A.1. Schematic of the bounded–plasma problem considered.

2.B. Trajectory integration of the kinetic equations

1. Collisionless trajectories and Lagrangian derivative. The species–s kinetic equation (2.A.1) can be rewritten in the form

$$\frac{d_s}{dt} f_s(\mathbf{x},\mathbf{v},t) = \mathscr{C}_s(ff\,|\,\mathbf{x},\mathbf{v},t) , \tag{2.B.1}$$

where

$$\frac{d_s}{dt} \equiv \frac{\partial}{\partial t} + \mathbf{v}\cdot\frac{\partial}{\partial \mathbf{x}} + \frac{e_s}{m_s}(\mathbf{E} + \mathbf{v}\times\mathbf{B})\cdot\frac{\partial}{\partial \mathbf{v}} \tag{2.B.2}$$

denotes the "Lagrangian" (or "total" or "substantial") time derivative along a collisionless species–s particle trajectory (or "characteristic") $[\mathbf{x}(t),\mathbf{v}(t)]$, i.e., a trajectory satisfying the equations of motion

$$\frac{d_s}{dt}\mathbf{x} = \mathbf{v} , \qquad m_s \frac{d_s}{dt}\mathbf{v} = e_s[\mathbf{E}(\mathbf{x},t) + \mathbf{v}\times\mathbf{B}(\mathbf{x},t)] . \tag{2.B.3a,b}$$

Note that in (2.B.1) to (2.B.3), $\mathbf{x}$ and $\mathbf{v}$ are generally shorthand for $\mathbf{x}(t)$ and $\mathbf{v}(t)$, respectively. Only when applying the operator (2.B.2), $\mathbf{x}$, $\mathbf{v}$ and t must be considered as independent variables first and assigned their current values $\mathbf{x}(t)$, $\mathbf{v}(t)$ and t only afterwards.

In all cases considered by us, we will be interested in trajectories passing through the "current point" $P \equiv [\mathbf{x},\mathbf{v},t]$, at which the distribution function is to be calculated. The generic point along such a trajectory will henceforth be written as $\hat{P} \equiv [\hat{\mathbf{x}},\hat{\mathbf{v}},\hat{t}] \equiv [\mathbf{x}(\hat{t}),\mathbf{v}(\hat{t}),\hat{t}]$, Generally, (2.B.3a,b) are integrated between a suitable starting point $\hat{P}_0$, i.e., with the "initial conditions"

$$[\mathbf{x}(\hat{t}=\hat{t}_0),\mathbf{v}(\hat{t}=\hat{t}_0),\hat{t}=\hat{t}_0] = [\hat{\mathbf{x}}_0,\hat{\mathbf{v}}_0,\hat{t}_0] \equiv \hat{P}_0 , \tag{2.B.4}$$

and the current point P, corresponding to the "final conditions"

$$[\mathbf{x}(\hat{t}=t),\mathbf{v}(\hat{t}=t),\hat{t}=t] = [\mathbf{x},\mathbf{v},t] = \hat{P} , \tag{2.B.5}$$

with $\hat{t}_0 \leq t$, cf. Fig. 2.B.1. More precisely, the standard problem of bounded–plasma trajectory simulation requires tracing the trajectory back in time from the current point to a suitable starting point – e.g., one where the distribution function is known.

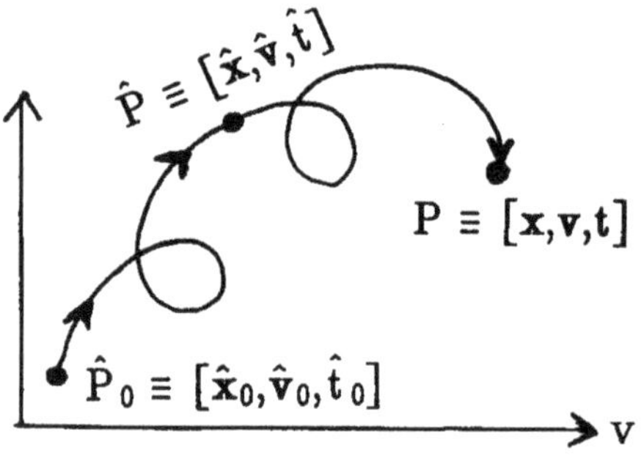

FIG. 2.B.1. Current, generic and initial points on a sample trajectory.

2. Formal trajectory integration. By "trajectory integration of the species–s kinetic equation" we mean integrating (2.B.1) along a specific species–s trajectory, which formally leads to the "trajectory–integrated kinetic equation"

$$f_s(\mathbf{x},\mathbf{v},t) = f_s(\hat{\mathbf{x}}_0,\hat{\mathbf{v}}_0,\hat{t}_0) + \int_{\hat{t}_0}^{t} d\hat{t}\ \mathscr{C}_s(ff\,|\,\hat{\mathbf{x}},\hat{\mathbf{v}},\hat{t}) . \tag{2.B.6}$$

Note that for both collisional *and* collisionless systems, we integrate along the *collisionless* particle trajectories, so that in any numerical evaluation it is crucial to have good methods for routinely solving (2.B.3) and (2.B.6) for given fields with the necessary precision.

Considering, in particular, an initial–value problem for $t \geq 0$, $f_s(\hat{\mathbf{x}}_0,\hat{\mathbf{v}}_0,\hat{t}_0)$ is (i) a boundary value if $\hat{t}_0 \geq 0$ and $\hat{P}_0$ is a boundary point, and (ii) an initial value if $\hat{t}_0 = 0$ and $\hat{P}_0$ is a point in volume (/84Kuhn1,2/), cf. Fig. 2.B.2. Hence, (2.B.6) is equivalent to the equations of motion plus boundary and initial conditions, and thus represents a very compact and powerful formal solution of the kinetic equations.

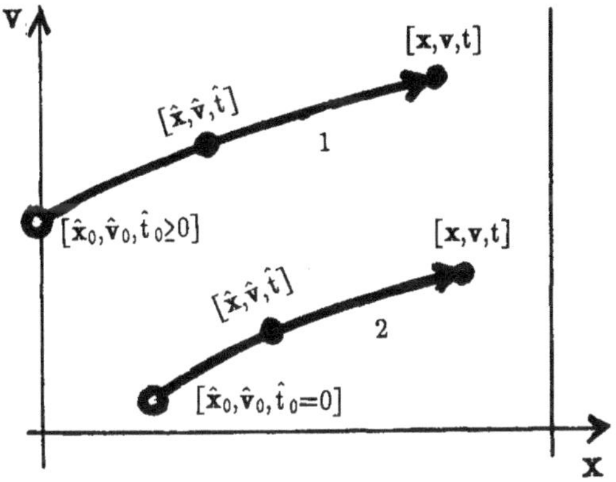

FIG. 2.B.2. The basic types of trajectories for the bounded–plasma initial–value problem: Trajectory 1 originates at a boundary point, while trajectory 2 originates at an initial point.

*3. **Practical procedure.*** We are now in a position to formulate the following procedure of evaluating the trajectory–integrated kinetic equation (2.B.6) for practical purposes. To calculate f_s at the current point $P = [\mathbf{x},\mathbf{v},t]$, we first trace the collisionless species–s trajectory passing through it back in time, trying to hit a point $\hat{P}_0 = [\hat{\mathbf{x}}_0,\hat{\mathbf{v}}_0,\hat{t}_0]$ where f_s is known. If we find such a point, we can proceed as follows: In the case of a *collisionless* particle species ($\mathscr{C}_s \equiv 0$), for which the kinetic equation (2.B.1) assumes the particularly simple form of the "Vlasov equation",

$$\frac{d_s}{dt} f_s(\mathbf{x},\mathbf{v},t) \equiv 0 , \tag{2.B.7}$$

Eq. (2.B.6) immediately yields the distribution function in the *explicit* form

$$f_s(\mathbf{x},\mathbf{v},t) = f_s(\hat{\mathbf{x}}_0,\hat{\mathbf{v}}_0,\hat{t}_0) \qquad (\text{for } \mathscr{C}_s \equiv 0) , \tag{2.B.8}$$

while in the case of a *collisional* species, for which $\mathscr{C}_s$ does not vanish identically, (2.B.6) is an integral equation whose further treatment requires evaluation of the trajectory integral

$$I(\hat{P}_0,P) \equiv \int_{t_0}^{t} d\hat{t}\; \mathscr{C}_s(ff|\hat{\mathbf{x}},\hat{\mathbf{v}},\hat{t}) . \tag{2.B.9}$$

Except for particularly simple cases allowing for fully analytic treatment (/80Kuhn/), this latter situation will usually require an iterative scheme in which some or all of the f's needed in the (usually numerical) evaluation of (2.B.9) will be given but approximate ones (e.g, those obtained in the previous iteration), cf. Secs. 4 and 5.

However, it may happen that no point $\hat{P}_0$ exists where f_s is known, in which case $f_s(P)$ must be determined from other considerations. If, e.g., in a dc situation the trajectory remains within a restricted domain of configuration space and closes on itself, we are dealing with trapped particles (/92Zhy4/), the distribution of which may be the result of subtle long–time evolution (/91Crys/).

2.C. Classification of methods

Solving kinetic bounded–plasma problems *always* requires *some* amount of numerical evaluation. Still, we find in appropriate and convenient to introduce the following global distinctions.

1. By ***numerical*** methods we mean ones where the underlying equations are solved numerically right from the beginning, such as in *particle simulation* or in schemes based on the *direct numerical integration* of the basic equations [e.g., *Vlasov simulation* (/85Law/)].

2. A method will be called ***analytic–numerical*** if much analytic groundwork is involved but a comparable portion of the work still has to rely on the computer. As an example let us mention the trajectory–simulation method outlined in Secs. 2.B, 3.D and 4 to 6.

3. ***Analytic*** methods are ones where most of the necessary operations are performed analytically and numerical evaluation is required only at a relatively late stage. This is the case, e.g., in the standard method for treating *(1d,1v), collisionless, dc* systems, cf. Secs. 2.D, 3.A.2 and 3.C.

2.D. Analytic solutions

1. To the present author's knowledge, *exact* analytic solutions to kinetic bounded–plasma problems have been obtained only for collisionless systems. The reason is that for these the velocity distribution functions can be written as functions of the constants of collisionless single–particle motion, $c_{s1}(\mathbf{x},\mathbf{v},t)$, $c_{s2}(\mathbf{x},\mathbf{v},t)$, ... :

$$f_s = f_s[c_{s1}(\mathbf{x},\mathbf{v},t),\, c_{s2}(\mathbf{x},\mathbf{v},t),\, \ldots] \tag{2.D.1}$$

with

$$\frac{d_s}{dt}\, c_{s1}(\mathbf{x},\mathbf{v},t) = 0\,, \quad \frac{d_s}{dt}\, c_{s2}(\mathbf{x},\mathbf{v},t) = 0\,, \ \ldots, \tag{2.D.2}$$

and if one succeeds in finding analytic expressions for c_{s1}, c_{s2}, ..., any function of the form (2.D.1) clearly represents an exact analytic solution to the collisionless kinetic equation (2.B.7). Again to the best of the author's knowledge, as far as bounded–plasma theory is concerned this has been achieved to date only for *(1d,1v) dc* problems, cf. Secs. 3.A.2 and 3.C.

2. However, there is a school of theorists from the former Soviet Union who have succeeded in treating *1d collisional* bounded–plasma systems in an approximate manner but with a remarkable degree of analyticity, including elastic collisions between charged and neutral particles (/64Baksht/, /64Kuch/, /74Bab/), charge–exchange collisions (/64Baksht/, cf. also Sec. 5), and Coulomb collisions (/62Kuch/).

3. As regards time–*dependent* bounded–plasma problems, a predominantly analytic treatment of small–amplitude periodic oscillations in the Pierce diode (Sec. 3.A.3) has been given in /90Hoer/. This method is based on a Fourier–series approach and could, in principle, be advantageously combined with the concept of trajectory integration.

4. All in all, however, it appears that our capability of finding analytic solutions to any but the simplest bounded–plasma problems is still very limited, and it would be a major achievement and breakthrough to find analytic solutions also for more complex (e.g., 2d) situations.

3. REVIEW OF (1d,1v) COLLISIONLESS ANPS RESULTS

3.A. The cartesian (1d,1v) model

1. General model description. In the cartesian (1d,1v) model we are dealing with a situation in which all macroscopic quantities depend only on x and t (such as the electrostatic potential $\Phi(x,t)$ or the particle number densities $n_s(x,t)$), both the electric field and the electric current density have only an x–component ($E(x,t)$ and $j(x,t)$, respectively), and the distribution functions are of the form $f_s(x,v_x,t)$. For this model, which actually represents a 1d approximation to configurations which in reality are essentially 2d, (2.A.1) and (2.A.2) become

$$\left[\frac{\partial}{\partial t} + v_x\frac{\partial}{\partial x} + \frac{e_s}{m_s}E\frac{\partial}{\partial v_x}\right]f_s = 0 \qquad \text{and} \qquad \frac{\partial E}{\partial x} = \sum_s \frac{e_s}{\epsilon_0}\int_{-\infty}^{+\infty} dv_x\, f_s\,, \tag{3.A.1,2}$$

respectively. At certain values of x, "electrodes" are placed in the form of infinite planes perpendicular to the x–direction, at which appropriate boundary conditions for the particles (emission, absorption, reflection, ...) and for the electric field or potential (e.g., $\Phi(x_{electrode},t) = \text{const}$) are prescribed.

The most widely used system of this type is the "plane–diode" model, which is characterized by the presence of just two electrodes bounding the intervening plasma, in which case (3.A.1) and (3.A.2) must be complemented by the field and particle boundary conditions at both electrodes, and by the external–circuit equation (/84Kuhnl/, /87Kuhnl/). While this model has been reviewed extensively in /87Kuhnl/, our present focus is to show very briefly why it lends itself particularly well to analytic treatment.

2. Analytic calculation of time–independent (dc) states. In a dc situation, the single–particle energy W_s is a constant of the motion (cf. Sec. 2.D), so that along the species–s particle trajectory passing through $[x,v_x]$ we have

$$W_s(\Phi(x),v_x) \equiv \frac{m_s v_x^2}{2} + e_s\Phi(x) = \frac{m_s\hat{v}_x^2}{2} + e_s\Phi(\hat{x}) = \frac{m_s\hat{v}_{x0}^2}{2} + e_s\Phi(\hat{x}_0)\,. \tag{3.A.3}$$

From these equalities we readily obtain

$$v_x = \sigma\sqrt{2[W_s - e_s\Phi(x)]/m_s}\,, \quad \hat{v}_x = \hat{\sigma}\sqrt{2[W_s - e_s\Phi(\hat{x})]/m_s}\,, \tag{3.A.4a,b}$$

$$\hat{v}_{x0} = \hat{\sigma}_0 \sqrt{2[W_s - e_s\Phi(\hat{x}_0)]/m_s}\,, \tag{3.A.4c}$$

where σ, $\hat{\sigma}$ and $\hat{\sigma}_0$ are the signs (± 1) of the velocities at the current, generic and initial points and must be chosen consistently with the situation considered. To this end, a global qualitative assumption must be made about the shape of the potential distribution (e.g., monotonically decreasing from left to right or with with a single maximum in front of the left–hand electrode, etc.)

Combining (2.B.8) with (3.A.4c) and assuming that the velocity distribution function at $\hat{x}_0$ is a *known* function f_{s0} of the velocity, we find

$$f_s(x,v_x) = f_s(\hat{x}_0,\hat{v}_{x0}) = f_{s0}(\hat{v}_{x0}) = f_{s0}\left[\hat{\sigma}_0 \sqrt{2[W_s(\Phi,v_x) - e_s\Phi(\hat{x}_0)]/m_s}\right]$$
$$\equiv g_s[W_s(\Phi,v_x)]\,, \tag{3.A.5}$$

which means that $f_s(x,v_x)$ has been obtained as a *known* function $g_s[W_s(\Phi,v_x)]$. Inserting this and $E(x) = -d\Phi(x)/dx$ into (3.A.2) yields Poisson's equation in the form

$$\frac{d^2\Phi}{dx^2} = -\frac{\rho(\Phi)}{\epsilon_0}\,, \tag{3.A.6}$$

with ρ a *known* function of Φ. Multiplying both sides by $d\Phi/dx$ and integrating between some suitable position x_1 and the current position x we obtain

$$\frac{d\Phi}{dx} = \tilde{\sigma}\sqrt{E_1^2 + P(\Phi,\Phi_1)}\,, \qquad \text{with} \qquad P(\Phi,\Phi_1) \equiv -\frac{2}{\epsilon_0}\int_{\Phi_1}^{\Phi} d\Phi'\rho(\Phi')\,, \tag{3.A.7a,b}$$

where the sign function $\tilde{\sigma}$ must be chosen consistently with the situation considered.

In many situations, and especially when the emission distribution functions $f_{s0}(v_x)$ are truncated Maxwellians (3.B.1), the functions $g_s(\Phi,v_x)$, $\rho(\Phi)$ and $P(\Phi,\Phi_1)$ can be obtained analytically. The last step, however, namely integrating (3.A.7a) for the potential profile $\Phi(x)$, usually requires *numerical* evaluation.

3. Probably the simplest and most fundamental bounded plasma system is the ***Pierce diode***, in which a *cold* collisionless electron beam propagates between two electrodes, the interelectrode space being filled with a neutralizing immobile ion background (/44Pier/, /84Kuhn2/, /84Kol/, /85Crys/, /90Hoer/, /92End/, and many others). This model, which even allows for *analytic time–dependent nonlinear solutions* (Sec. 2.D.3), has been reviewed extensively in /87Kuhn1/. One of its best–known features is the "Pierce instability", an electron instability playing an important role in bounded–plasma dynamics, cf. Sec. 3.D.

3.B. The single–emitter plasma diode (KDSI) model.

1. Model description. A particularly important special case of the plane–diode models addressed in Sec. 3.A.1 is the so–called "single–emitter plasma diode" or "Knudsen diode with surface ionization (KDSI)", which has been used extensively to describe both single–ended Q–machines (/75Mot/, /82Iizu/, /87Schritt/, and many others) and thermionic energy converters (/73Baksht/, /73Hat/, /91Bab/, and many others). The model is shown schematically in Fig. 3.B.1. One electrode (located at x = 0) represents the "emitter" or "hot plate", and the other (located at x = d) mimics the "collector" or "cold plate". The hot plate, whose temperature T typically exceeds 2000 K, produces electrons (via Richardson emission) and positive ions (via contact ionization of neutral alkali atoms hitting the surface), both species leaving the emitter surface at constant rates with half Maxwellian velocity distribution functions:

$$f_s(x=0,v_x>0) = 2n_{s0}^{+}\sqrt{\frac{m_s}{2\pi kT}}\exp\left[-\frac{m_s v_x^2}{2kT}\right] \qquad (s = e, i)\,, \tag{3.B.1}$$

where n_{s0}^{+} are the emission densities at the hot–plate surface. All particles are absorbed when hitting either electrode. Usually, the potential $\Phi(x,t)$ is gauged such that $\Phi(0,t) \equiv 0$, and the cold–plate potential is called Φ_C.

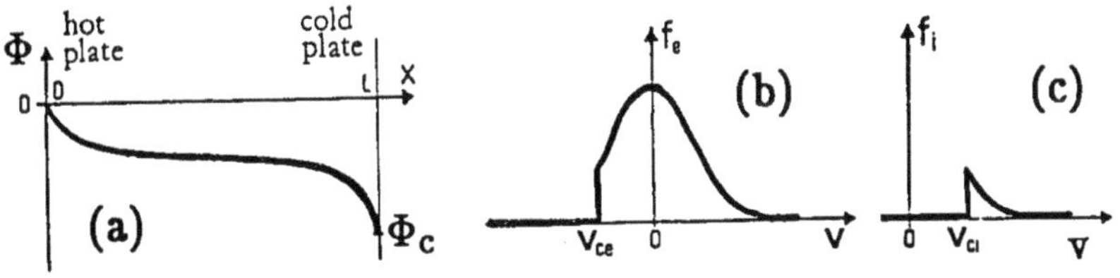

FIG. 3.B.1: (a) The single–emitter plasma diode (KDSI) model with a monotonically decreasing potential distribution, with related (b) electron and (c) ion velocity distributions (not to scale). Cutoff velocities are denoted by v_{ce} and v_{ci}.

Both the dynamic and dc states of this system are conveniently classified in terms of the three dimensionless parameters $\gamma \equiv n_{i0}^{+}/n_{e0}^{+}$ ("neutralization parameter"), $\eta_C \equiv e\Phi_C/kT$ (normalized cold–plate bias), and $\delta \equiv d/l_{e0}^{+}$ (normalized interelectrode distance), where k is Boltzmann's constant and $l_{e0}^{+} \equiv \sqrt{\{\epsilon_0 kT/(n_{e0}^{+}e^2)\}}$ is the "emission Debye length".

2. Previous results. With the method outlined in Sec. 3.A.2, the possible dc states of the KDSI model have been studied by many authors (/62McIn/, /67Ott/, /73Hat/, /81Kuhn/, /87Kuhn1/, and others) and are conveniently represented and classified by means of the (η_C,γ) parameter diagram. This diagram was given in its most general form in Fig. 3 of /67Ott/, and in a slightly simplified version in Fig. 1(a) of /81Kuhn/, Fig. 1 of /91Crys/, and elsewhere. The monotonically decreasing potential distribution shown in Fig. 3.B.1, which corresponds to parameter region 1' of the (η_C,γ) parameter diagram, provides the basis for the (1d,2v) collisional and (2d,3v) collisionless extensions outlined in Secs. 5 and 6, respectively.

Note that these dc states are only *formal dc solutions* of the basic equations, and whether they will actually be observable in practice depends on their stability. In fact, while the negative–bias distributions generally seem to be stable, the positive–bias ones are usually strongly unstable, so that in these parameter regions large–amplitude "potential relaxation oscillations" are observed, cf. Sec. 3.D.

3.C. Cylindrical and spherical (1d,1v), dc states

Following the basic principles outlined in Sec. 2.D, *Zhykharsky* has formalized and generalized the calculation of collisionless, dc, (1d,1v) bounded–plasma states (/91Zhy/, /92Zhy2/, /92Zhy3/) and given a systematic classification of such states in various cartesian, cylindrical and spherical geometries. A detailed application to a spherical probe problem was given in /92Zhy1/, and a few similar results can also be found in the previous literature (/63Park/, /64Prince/).

3.D. Cartesian (1d,1v), time–dependent systems (Ender–Kuznetsov method)

An analytic–numerical trajectory–simulation method for investigating nonlinear phenomena in KDSI's (Sec. 3.B) has been developed and applied by Ender and Kuznetsov (/77Kuz/, /78Kuz/, /79End/, /79Kuz1,2/, /80Kuz/, /81Kuz/, /83Kuz/, /86Kuz/, /88Kuz/). To the present author's knowledge, this "Ender–Kuznetsov method" is the only method besides particle simulation to date that can handle *strongly nonlinear time–dependent* collisionless plane–diode problems. More recently, it has also been applied to the Pierce diode (/92End/), in which only the electron dynamics are considered (Sec. 3.A.3).

In the KDSI case, the ion velocity distribution $f_i(x,v,t)$ is determined by tracing the time–*dependent* (!) ion trajectories back to the emitter (Sec. 2.B), while the electron distribution $f_e(x,v,t)$ is calculated quasistatically (Sec. 3.A.2) for the instantaneous ion distribution. In this manner, the strongly nonlinear KDSI oscillations ("potential relaxation oscillations") occurring at positive values of η_c have been treated in great detail, cf. the references given above. These oscillations generally involve phenomena on two different timescales, namely "slow stages", largely determined by the ion dynamics, and "fast stages", arising at certain moments and leading to a rapid and substantial rearrangement of both the electron and and the potential distributions. These fast stages are due to the fact that at certain points of the slow evolution the electron distribution becomes unstable to aperiodic Pierce–type instability (Sec. 3.A.3) which then evolves, on an electron time scale, into the nonlinear regime, thus setting the initial conditions for a new slow stage. The detailed time evolution of these fast stages is only now becoming the subject of detailed study (/92End/), whereas at the present stage of development of the Ender–Kuznetsov method it is still approximated by an *instantaneous* switching from one quasistatic electron distribution to another (/77Kuz/). [For a given ion distribution, there may be more than one quasistatic electron distribution, which can be classified by means of the so–called "(η,ϵ) diagram" (/77Kuz/).]

For more detailed reviews of the Ender–Kuznetsov method, the reader is referred to Refs. /87Kuhn1,2/.

4. ITERATIVE SCHEME FOR SELF–CONSISTENT KINETIC TRAJECTORY SIMULATION OF TIME–INDEPENDENT (DC) SYSTEMS

4.A. Introduction and outline

The scheme to be described here, which is basically the one used in the collisional (1d,2v) and collisionless (2d,3v) studies outlined in Secs. 5 and 6, respectively, represents a major new development in kinetic bounded–plasma modeling. It usually requires discretization of both configuration and velocity spaces, in which case it is understood that $[\mathbf{x},\mathbf{v}]$ is a generic *grid*point in phase space (e.g., $[\mathbf{x}_{ij}, \mathbf{v}_{pqr}]$), and that the relevant equations and boundary conditions (Sec. 2) are actually used in their appropriate discretized forms.

To initiate the calculation, we must prescribe an "initial" potential distribution $\Phi^{(0)}(\mathbf{x})$ and "initial" velocity distribution functions $f_s^{(0)}(\mathbf{x},\mathbf{v})$, with the latter correlated with $\Phi^{(0)}$ in a meaningful manner. These must be given all over the relevant configuration space including the boundaries, and should already be as good approximations as possible to their expected self–consistent counterparts. Since the $f^{(0)}$'s are only needed for the approximate evaluation of collision terms, we can always set $f_s^{(0)}(\mathbf{x},\mathbf{v}) \equiv 0$ if particle species s is a collisionless one.

Then, the "main iteration block" (Sec. 4.B) is entered for the first time: It takes $\Phi^{(0)}$ and $f_s^{(0)}$ as input and yields new distributions $\Phi^{(1)}(\mathbf{x})$ and $f_s^{(1)}(\mathbf{x},\mathbf{v})$, which under appropriate conditions will be better approximations to the self–consistent ones than the initial guesses. Next, the main iteration block is entered again, taking $\Phi^{(1)}$ and $f_s^{(1)}$ as input and yielding the next–better approximations $\Phi^{(2)}(\mathbf{x})$ and $f_s^{(2)}(\mathbf{x},\mathbf{v})$, etc. After each main iteration m, a convergence check is performed which decides on whether another main iteration m+1 is to be performed or whether $\Phi^{(m)}(\mathbf{x})$ and $f_s^{(m)}(\mathbf{x},\mathbf{v})$ already approximate the self–consistent distributions to within the prescribed accuracy.

4.B. The main iteration block

The *main iteration block* is a package of routines whose task consists in calculating, on the basis of given input potential and velocity distributions $\Phi^{(m-1)}(\mathbf{x})$ and $f_s^{(m-1)}(\mathbf{x},\mathbf{v})$ (with $m \geq 1$), improved output distributions $\Phi^{(m)}(\mathbf{x})$ and $f_s^{(m)}(\mathbf{x},\mathbf{v})$, respectively. Its basic way of operation can be subdivided into the following five steps, of which Nos. 2 and 3 are the ones requiring most of the computing time:

*1. **Get the input distributions*** $\Phi^{(m-1)}(\mathbf{x})$ and $f_s^{(m-1)}(\mathbf{x},\mathbf{v})$. For m = 1 (first main iteration), these are the initial guesses, whereas for m > 1 they are the result of the preceding main iteration.

*2. **Calculate the new velocity distribution functions*** $f_s^{(m)}(\mathbf{x},\mathbf{v})$ via trajectory integration (Sec. 2.B). The fields to be inserted into (2.B.3b) are the given magnetic field $\mathbf{B}(\mathbf{x})$ and the input electric field $\mathbf{E}^{(m-1)}(\mathbf{x}) \equiv -\nabla\Phi^{(m-1)}(\mathbf{x})$, and the velocity distribution functions to be inserted into the collisional trajectory integral (2.B.9) are the input ones, $f_s^{(m-1)}(\mathbf{x},\mathbf{v})$.

3. Calculate the new particle densities $n_s^{(m)}(\mathbf{x})$ by evaluating (2.A.3), with $f_s(\mathbf{x},\mathbf{v},t)$ replaced by $f_s^{(m)}(\mathbf{x},\mathbf{v})$.

4. Calculate the new electric charge density $\rho^{(m)}(\mathbf{x})$ by evaluating the second equality in (2.A.2), with $n_s(\mathbf{x},t)$ replaced by $n_s^{(m)}(\mathbf{x})$.

5. Calculate the new potential distribution $\Phi^{(m)}(\mathbf{x})$ by solving Poisson's equation (2.A.2) with $\rho(\mathbf{x},t)$ replaced by $\rho^{(m)}(\mathbf{x})$, for which purpose we use a standard relaxation scheme involving an appropriate number of sweeps across the spatial grid (/87Koon/).

4.C. Refinements

The "basic" iteration scheme outlined in Secs. 4.A and 4.B works in principle but is far from optimal as far as stability and computing time are concerned. Hence, one has to introduce refinements whose details will usually depend on the specific situation considered (/93Ped/).

One refinement which seems to be of general usefulness is the following: If $\rho^{(m)}(\mathbf{x})$ is not yet sufficiently close to the exact distribution, experience shows that in step 5 Poisson's equation must not be solved "too exactly" for $\Phi_s^{(m)}(\mathbf{x})$ because otherwise the system blows up numerically. In other words, in step 5 one should not use an exact Poisson solver but rather a "truncated Poisson solver", involving only a small number of sweeps across the spatial grid.

Much more effort is required to optimize the iterative scheme outlined above. For further details, the reader is referred to /90Ped2/, /92Mai1/ and /93Ped/.

5. A CARTESIAN (1d,2v) Q–MACHINE MODEL INCLUDING CHARGE–EXCHANGE COLLISIONS

In a first application of the iterative scheme of Sec. 4 to a *collisional* bounded plasma, Maier and Kuhn (/92Mai1,2/) have extended the (1d,1v), dc, *collisionless* single–emitter plasma diode model with a monotonically decreasing potential (Fig. 3.B.1) into a (1d,*2*v) *collisional* one. A *finite*, uniform magnetic field is applied along the x–direction, and the ion velocity distribution function is now of the form $f_i(x,v_x,v_n)$. Instead of the *1* v emission distribution functions (3.B.1) we now have the *2* v ones

$$f_s(x=0,v_x>0,v_n) \;=\; 2n_{s0}^{+}\left[\frac{m_s}{2\pi kT}\right]^{\frac{3}{2}}\exp\left[-\frac{m_s}{2kT}\,(v_x^2+v_n^2)\right] \quad (s = e, i)\,, \tag{5.1}$$

with v_n the velocity component normal to the magnetic field. The electrons are still collisionless, but the ions are now assumed to undergo symmetric charge–exchange collisions with the given neutral background gas, whose velocity distribution function $F_b(x,V_x,V_n)$ is chosen to be a 2v Maxwellian with Temperature T_b for $V_x > 0$ and zero for $V_x \lesssim 0$. Accordingly, the ions satisfy a linearized Boltzmann equation, namely (2.A.1) [or, equivalently, (2.B.1)] with $s = i$ and a collision term of the form $\mathscr{C}_i(f_i F_b | x, v_x, v_n)$. The differential and total cross sections have been obtained by fitting well–known analytic approximations (/67Fahr/) to relevant experimental data (/76Sin/).

In Figs. 5.1(a) and 5.1(b) we show some exemplary results (/92Mai1,2/) for a "Q–machine" (or thermionic diode) with length $d = 5$ mm at two different background (K) densities n_b ("low": $10^{18}\ m^{-3}$, marked by "1", and "intermediate": $10^{19}\ m^{-3}$, marked by "2"). The other parameters characterizing the system are: $\Phi_c = -2$ V, $T = 2100$ K, $T_b = 500$ K, electron emission density $n_{e0}^{+} = 1.5{\times}10^{16}\ m^{-3}$, and neutralization parameter $\gamma = 0.01$. This system is "short" in the sense that there is no significant quasineutral plasma region between the emitter and collector sheaths. Figure 5.1(a) shows the potential distributions, and the ion distribution function of Fig. 5.1(b) exhibits a distinct peak of slow ions resulting from charge–exchange collisions with the fairly cold background. Clearly, these charge–exchange collisions tend to lower the ion average velocity and, hence, to raise the ion density. Consequently, the region of positive space charge is widened, and the "plasma point" (defined by exact space–charge neutrality) is shifted to the left. These tendencies become stronger with increasing background density.

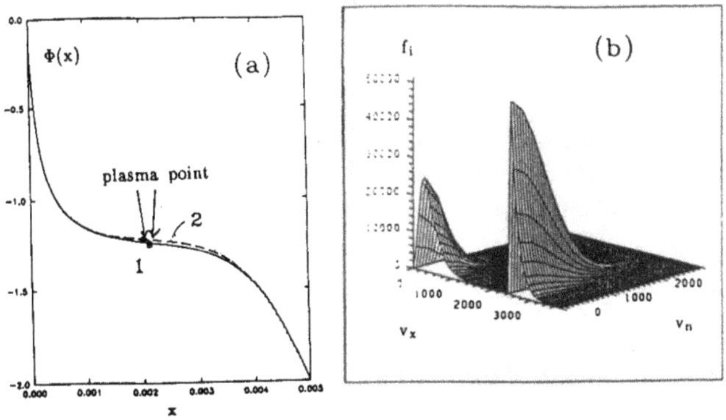

FIG. 5.1. (a) Potential distributions for (1) low and (2) intermediate background densities. (b) Ion distribution function near the plasma point for intermediate background density.

6. A CARTESIAN (2d,3v), COLLISIONLESS Q–MACHINE MODEL

In a first application of the iterative scheme of Sec. 4 to a *2d* bounded plasma, Pedit and Kuhn (/90Ped1,2/, /92Ped/, /93Ped/) have extended the (1d,1v), dc, collisionless single–emitter plasma diode model with a monotonically decreasing potential (Fig. 3.B.1) into a (2d,3v) collisionless slab model (Fig. 6.1). The potential distribution again decreases monotonically in the *axial* direction, i.e., the direction of the constant, finite applied magnetic field).

Although the model is still cartesian, the y–coordinate simulates to some extent the radial coordinate of the *cylindrical* Q–machine geometry. The system is specified by prescribing the following parameters: axial length d, emitter "radius" R_e, collector "radius" R_c, conducting–wall "radius" R_w; emitter potential $\Phi_e \equiv 0$, collector potential Φ_c, conducting–wall potential Φ_w; emitter temperature T, electron emission density n_{e0}^{+}, ion emission density n_{i0}^{+} (or, equivalently, the neutralization parameter γ). Along the *insulating* walls between the electrodes and the conducting wall, the potential is assumed to vary linearly. The charged–particle emission distribution functions, $f_s(x=0,|y|\leq R_e,v_x\geq 0,v_n\geq 0)$, are again of the form (5.1) All particles reaching the collector, the emitter or the surrounding wall are absorbed. While the ion trajectories are obtained by integrating the full equations of motion, the electron ones are calculated in a *guiding–center* approach.

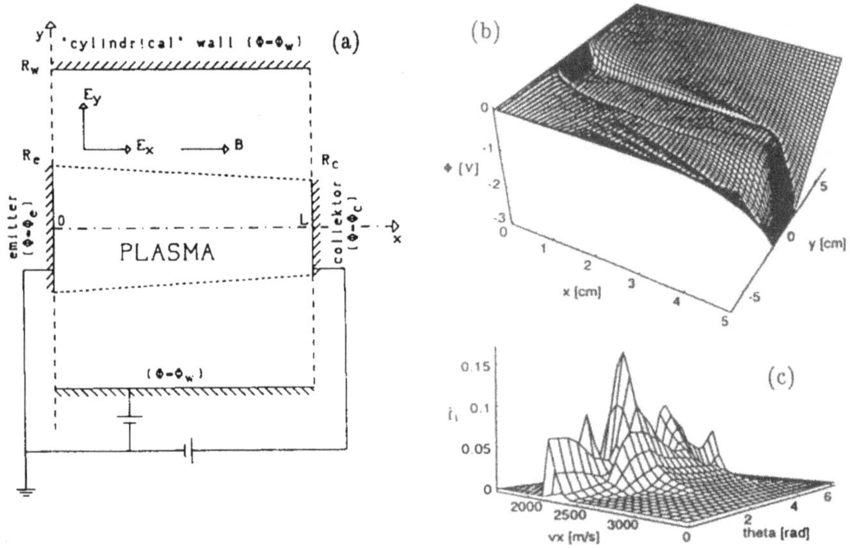

FIG. 6.1. (a) The 2d Q–machine model. (b) 2d potential distribution. (c) Example of ion velocity distribution function for v_n = const. at plasma–vacuum transition.

Figure 6.1(b) shows the 2d potential distribution for a system with the following parameters (/92Ped/): d = 5 cm. $R_e = R_c = 1.5$ cm, $R_w = 6.5$ cm, $\Phi_e = \Phi_w = 0$, $\Phi_c = -3$ V, T = 2000 K, $n_{e0}^{+} = 5.12\cdot 10^{14}$ m^{-3}, and $\gamma = 0.1$. The electrode sheath regions and the plasma– vacuum transition region have been calculated with enhanced spatial resolution. For the same system, Fig. 6.1(c) shows an ion velocity distribution function for a position lying within the plasma–vacuum transition region, which demonstrates the complexity of these functions and the high resolution the trajectory–simulation method provides in velocity space as well. To the present author's knowledge, this is the most detailed kinetic model to date of any bounded plasma system.

7. CONCLUSIONS AND PERSPECTIVES

In this paper we have reviewed the *"Analytic–Numerical Plasma Simulation (ANPS)"* approach to the kinetic modeling of bounded plasmas, which is becoming increasingly important in both fundamental plasma research and applied disciplines such as plasma technology and fusion research. This approach, which is based on a combination of analytic and numerical methods closely related to the *method of characteristics*, is becoming *at least complementary, if not a genuine alternative to particle simulation.* ANPS naturally allows for *realistic ion–electron mass ratios* and *high resolution* in both configuration and velocity spaces. It either yields analytic solutions or, where such are impossible, is being applied in the form of a semi–analytic method termed "trajectory simulation".

The work reviewed in Secs. 5 and 6 clearly demonstrates that the ANPS approach and, in particular, the iteration scheme outlined in Sec. 4 are indeed capable of kinetically modeling even nontrivial bounded plasma systems in great detail and with high resolution. However, although significant improvements have been achieved already, much more effort addressing the details of the algorithm will be needed to increase its efficiency and to thus make it suitable for the treatment of larger and physically more complex systems.

In the author's opinion, some problems which are of particular importance and, hence, should be tackled by means of ANPS in the near future are these: (i) Inclusion of more types of collisions (up to now, only charge–exchange collisions have been included, cf. Sec. 5); (ii) linear eigenmode and stability analysis of collisional and/or 2d dc states, taking into account external–circuit properties; (iii) small–amplitude nonlinear periodic oscillations of collisional and/or 2d systems, taking into account external–circuit properties.

The ultimate goal would be to develop ANPS to such a level of sophistication and efficiency that it could handle practically any bounded–plasma system of interest. Clearly, this would enable plasma physicists to, e.g., replace expensive laboratory experiment with much cheaper computer studies aiming at optimizing some bounded plasma system of interest, but despite his basically optimistic attitude, even the present author is enough of a realist to admit that we are still a long way from reaching that goal.

ACKNOWLEDGEMENTS. This work was partly supported by the Austrian Science Foundation (FWF) under contracts P7005 and P8405.

REFERENCES

/44Pier/ J.R. Pierce, J. Appl. Phys. **15**, 721–6 (Oct 1944).

/62Kuch/ R.Ya. Kucherov and L.E. Rikenglaz, Zh. Tekh. Fiz. **32**, 1275–84 (Oct 1962); Sov. Phys. Tech. Phys. **7**, 941–6 (Apr 1963).

/62McIn/ R.G. McIntyre, Adv. Energy Conv. **2**, 405–16 (1962).

/63Park/ J. Parker, Phys. Fluids **6**, 1657–8 (Nov 1963).

/64Baksht/ F.G. Baksht, Zh. Tekh. Fiz. 34, 926–39 (May 1964); Sov. Phys. Tech. Phys. **9**, 713–22 (Nov 1964).

/64Derf/ H. Derfler, Phys. Fluids **7** (10), 1625–37 (Oct 1964).

/64Kuch/ R.Ya. Kucherov and Yu.A. Shuander, Zh. Tekh. Fiz. **34**, 66–76 (Jan 1964); Sov. Phys. Tech. Phys. **9**, 50–7 (Jul 1964).

/64Prince/ D.C. Prince, Jr., and N.P. Jeffries, Final Rep., Contr. NAS 8–5214, Contr. No. TP 3–84060, GE, Cincinnati, OH (Dec 1964).

/67Fahr/ H.J. Fahr and K.G. Mueller, Zeitschr. Physik **200**, 343–65 (1967).

/67Ott/ W. Ott, Zeitschr. Naturf. **22A**, 1057–67 (Jul 1967).

/73Baksht/ F.G. Baksht et al., *Thermionic Converters and Low–Temperature Plasma*, edited by Moizhes and Pikus (Nauka, Moscow, 1973; ERDA, Washington, D.C., 1976).

/73Hat/ G.N. Hatsopoulos and E.P. Gyftopoulos, *Thermionic Energy Conversion. Vol. I: Processes and Devices, Vol. II: Theory, Technology, and Application* (MIT Press, Cambridge, MA, and London, 1973 and 1979).

/74Bab/ V.I. Babanin and A.Ya. Ender, Zh. Tekh. Fiz. **44**, 102–12 (Jan 1974); Sov. Phys. Tech. Phys. 19, 60–5 (Jul 1974).

/75Mot/ R.W. Motley, *Q Machines* (Academic Press, 1975).

/76Sin/ S. Sinha and J.N. Bardsley, Phys. Rev. A **14**, 104–13 (Jun 1975).

/77Kuz/ V.I. Kuznetsov and A.Ya. Ender, Zh. Tekh. Fiz. **47**, 2237–46 (Nov 1977); Sov. Phys. Tech. Phys. **22**, 1295–300 (Nov 1977).

/78Kuz/ V.I. Kuznetsov and A.Ya. Ender, Preprint 575 (in Russian), A.F. Ioffe Physico-tech. Inst., Leningrad (21 Mar 1978).

/79Dec/ V.K. Decyk and J.M. Dawson, J. Comput. Phys. **30** (3), 407–28 (Mar 1979).

/79End/ A.Ya. Ender and V.I. Kuznetsov, J. Phys. (Paris), Colloque C7, Suppl. au n° 7, **40** (7), C7–523 (Jul 1979).

/79Kuz1/ V.I. Kuznetsov and A.Ya. Ender, Preprint 604 (in Russian), A.F. Ioffe Physico-tech. Inst., Leningrad (7 Mar 1979).

/79Kuz2/ V.I. Kuznetsov and A.Ya. Ender, Zh. Tekh. Fiz. **49**, 2176–2179 (Oct 1979); Sov. Phys. Tech. Phys. **24**, 1199–201 (Oct 1979).

/80Kuhn/ S. Kuhn, Phys. Rev. A **22**, 2460–7 (Dec 1980).

/80Kuz/ V.I. Kuznetsov and A.Ya. Ender, Zh. Tekh. Fiz. **50**, 67–77 (Jan 1980); Sov. Phys. Tech. Phys. **25**, 37–42 (Jan 1980).

/81Kuhn/ S. Kuhn, Plasma Phys. **23**, 881–902 (Oct 1981).

/81Kuz/ V.I. Kuznetsov and A.Ya. Ender, Zh. Tekh. Fiz. **51**, 2250–59 (Nov 1981); Sov. Phys. Tech. Phys. **26**, 1326–31 (Nov 1981).

/82Iizu/ S. Iizuka, P. Michelsen, J.J. Rasmussen, R. Schrittwieser, R. Hatakeyama, K. Saeki, and N. Sato, Phys. Rev. Lett. **48**, 145–8 (18 Jan 1982).

/83Kuz/ V.I. Kuznetsov and A.Ya. Ender, Zh. Tekh. Fiz. **53**, 2329–2338 (Dec 1983); Sov. Phys. Tech. Phys. **28**, 1431–6 (Dec 1983).

/84Kol/ I.N. Kolyshkin, V.I. Kuznetsov, and A.Ya. Ender, Zh. Tekh. Fiz. **54**, 1512–20 (Aug 1984); Sov. Phys. Tech. Phys. **29**, 882–87 (Aug 1984).

/84Kuhn1/ S. Kuhn, Phys. Fluids **27**, 1821–33 (Jul 1984).

/84Kuhn2/ S. Kuhn, Phys. Fluids **27**, 1834–51 (Jul 1984).

/85Crys/ T.L. Crystal and S. Kuhn, Phys. Fluids **28**, 2116–24 (Jul 1985).

/85Law/ Wm.S. Lawson, J. Comput. Phys. **61**, 51–63 (Oct 1985).

/86Kuz/ V.I. Kuznetsov and A.Ya. Ender, Preprint 1061 (in Russian), A.F. Ioffe Physicotechnical Institute, Leningrad (4 Sep 1986).

/87Koon/ S.E. Koonin, *Physics on the Computer* (in German; R. Oldenbourg Verlag, Munich, 1987), Vol. 1, p. 48.

/87Kuhn1/ S. Kuhn, 1987 Int. Conf. on Plasma Phys., Kiev, 6–12 Apr 1987, Invited Papers, Vol. **2**, 954–1008.

/87Kuhn2/ S. Kuhn, 3rd Symp. on Plasma Double Layers and Related Topics, Bucharest, 14–15 Apr 1987. Analele Ştiinţifice ale Universităţii "Al. I. Cuza" DIN Iaşi, Vol. **34** – Supplement, 25–58.

/87Schritt/ R. Schrittwieser, ibidem, 9–23.

/88Kuz/ V.I. Kuznetsov and A.Ya. Ender, 23rd Intersociety Energy Conversion Engineering Conference (1988 IECEC), Denver, Colo., 31 Jul – 5 Aug 1988, 581–4.

/89Law/ W.S. Lawson, J. Comput. Phys. **80**(2), 253–76 (Feb 1989).

/90Hoer/ M. Hörhager and S. Kuhn, Phys. Fluids B **2** (11), 2741–63 (Nov 1990).

/90Ped1/ H. Pedit and S. Kuhn, 17th EPS Conference on Controlled Fusion and Plasma Heating, Amsterdam, 25–29 Jun 1990, Contributed Papers, Pt. **IV**, 1782–5.

/90Ped2/ H. Pedit, , Diploma Thesis (in German), Institute for Theoretical Physics, University of Innsbruck, Austria (Nov 1990).

/91Bab/ V.I. Babanin, I.N. Kolyshkin, V.I. Kuznetsov, A.S. Paschina, V.I. Sitnov, and A.Ya. Ender, 20th Conference on Phenomena in Ionized Gases, Pisa, Italy, 8–12 Jul 1991, Contributed Papers **1**, 180–1.

/91Crys/ T.L. Crystal, P.C. Gray, W.S. Lawson, C.K. Birdsall, and S. Kuhn, Phys. Fluids B **3** (1), 244–54 (Jan 1991).

/91Vahe/ V. Vahedi, M.A. Lieberman, M.V. Alves, J.P. Verboncoeur, C.K. Birdsall, J. Appl. Phys. **69**(4), 2008–14 (15 Feb 1991).

/91Zhy/ A.V. Zhykharsky, Phys. Scripta **44**(6), 606–16 (Dec 1991).

/92Bau/ F. Bauer and H. Schamel, Physica D **54**, 235–52 (1992).

/92Bul/ S.V. Bulanov, S. Kuhn, and A.S. Sakharov, Phys. Fluids B **4** (9), 2871–84 (Sep 1992).

/92End/ A.Ya. Ender, S.Kuhn, and V.I. Kuznetsov, these proceedings.

/92Mai1/ H. Maier, Diploma thesis (in German), Institute for Theoretical Physics, University of Innsbruck, Austria (Apr 1992).

/92Mai2/ H. Maier and S. Kuhn, 1992 International Conference on Plasma Physics, Innsbruck, Austria, 29 Jun – 3 Jul 1992, Contributed Papers **III**, 2001–4.

/92Ped/ H. Pedit and S. Kuhn, ibidem, 1777–80.

/92Zhy1/ A.V. Zhykharsky, Phys. Scripta **45** (2), 100–9 (Feb 1992).

/92Zhy2/ A. V. Zhykharsky, 1992 International Conference on Plasma Physics, Innsbruck, Austria, 29 Jun – 3 Jul 1992, Contributed Papers **II**, 1553–56.

/92Zhy3/ A. V. Zhykharsky and V.N. Shmykov, ibidem, 1557–60.

/92Zhy4/ A. V. Zhykharsky and V.N. Shmykov, these proceedings.

/93Ped/ H. Pedit and S. Kuhn, to be published.

PROPERTIES AND EVOLUTION OF NONSTATIONARY DOUBLE LAYERS IN NONEQUILIBRIUM PLASMA

V.I.Maslov, Kharkov Institute of Physics & Technology, 310108 Kharkov,Ukraine

It is shown in 1 that ion-acoustic turbulence in one-dimensional nonequilibrium current-carrying plasma is unstable relative to the modulation. This instability results in turbulence splitting into widely spaced short perturbations. The width of the latter is of the order of the wavelength. The energy exchange between particles of nonequilibrium plasma and these nonhomogeneous perturbations as well as nonlinear generation of higher harmonics can lead to the formation of a solitary perturbation, viz. the potential well. The possible formation of the correlated perturbation - the potential well - has been investigated with ion-acoustic instability development in 2-9. The selfconsistent evolution of a nonmonotonous ion-acoustic double layer (potential well φ with potential shock $\Delta\varphi$ in its vicinity) is considered in this paper in one-dimensional current-carrying bounded plasma, $0<x<L$. We choose the initial well of small amplitude (depth)- φ_0 ; of width Δx smaller than the system length L and moving with a velocity V_0. Initially electrons move with respect to ions with a current velocity V_d satisfying $\sqrt{T_e/m_i} = V_s<V_d<V_{th}= \sqrt{T_e/m_e}$. Here T_e is the temperature of electrons; m_e,m_i are the electron and ion masses. At boundaries, the injected electrons have a Maxwellian distribution of velocities (with $V_d\neq 0$), while the distribution of leaving electrons is determined selfconsistently.

Owing to the reflection from the potential well of resonant electrons, the velocity distribution of which is asymmetric relatively to V_0, the quasineutrality is broken near the well: the electrons are short in the first front of the well and they are surplus on the back one. The restoration of the latter is due to the acceleration or to the deceleration of electrons. This is caused by the selfconsistent formation of the potential shock $\Delta\varphi$ near the well. Hence the resonant region in electron velocity space is wider in front of the well, $\xi=x-V_c t>0$, by the value of potential shock, than behind the well, $\xi<0$.

The potential well is unstable in a current-carrying plasma. It reflects the resonant electrons (thus the current is decreasing), absorbing their energy, and the well amplitude is growing $\varphi_0(t)$. When the well amplitude increases up to certain value the slow evolution of well starts as compared with the dynamics of general part of electrons. It is true when the inverse time of interaction $\tau^{-1} = V_{tr}/\Delta X = (2e\varphi_0/m)^{1/2}/\Delta X$ of the general part of resonant electrons with the well exceeds the growth rate $\gamma(\varphi_0)=\partial \ln \varphi_0/\partial t$

of the well amplitude $V_{tr}(\varphi_0) > \gamma(\varphi_0)\,\Delta X(\varphi_0)$. In this case the changed velocity distribution of resonant electrons moves from the well with the relative velocity equal to V_{tr}. We describe the well evolution using the expanding of the small parameter $\alpha = \gamma \Delta X / V_{tr}$

We obtain from the Vlasov equation the expression for the perturbation of electron distribution function in the first order of $(V_d - V_0)/V_{th} = V_c < 1$ and in the second order of $e\varphi/T_e = \varphi < 1$. From the latter the expression for the electron density perturbation follows

$$\delta n = \varphi + \varphi^2/2 - \mathrm{sign}(\xi)\{2\Delta\varphi\sqrt{(\varphi_0+\varphi)/\pi} + V_c\sqrt{2/\pi}\,[\sqrt{\varphi_0(\varphi_0+\varphi)} + \varphi\ln(\sqrt{-\varphi}/(\sqrt{\varphi_0}+\sqrt{\varphi_0+\varphi}))]\} \tag{1}$$

Using the quasineutrality condition outside the well we obtain

$$\Delta\varphi/\varphi_0 = V_c\sqrt{2/\pi}\,(1+2\sqrt{\varphi_0/\pi}) \tag{2}$$

The potential shock is proportional to the amplitude φ_0 and the electron current velocity V_c.

From the Vlasov equation we obtain the expression for the perturbation of ion distribution function. From the latter the expression for the ion density perturbation follows

$$\delta n_i' = \varphi' + 2\dot{\varphi} - 2(V_e - 1)\varphi' + 3\varphi\cdot\varphi' \tag{3}$$

Here "." $= \partial/\partial t\omega_{pi}$, "," $= \partial/\partial(\xi/r_d)$, $V_e = V_0/V_s$, ω_{pi} is the ion plasma frequency, r_d is the Debye shielding distance. Substituting (1),(3) in the Poisson equation, we derive the evolution equation

$$\dot{\varphi} - (V_e-1)\varphi' + \varphi\cdot\varphi' + \varphi'''/2 + \mathrm{Sign}(\xi)(\Delta\varphi/2)\,\varphi'[(1-\sqrt{\varphi_0/\pi})/\sqrt{\varphi_0(\varphi_0+\varphi)} + \varphi_0^{-1}(1-2\sqrt{\varphi_0/\pi})\ln(\sqrt{-\varphi}/(\sqrt{\varphi_0}+\sqrt{\varphi_0+\varphi}))] = 0 \tag{4}$$

of KdV-type but with nonquadratic nonlinearity. The equation (4) describes the space distribution of nonhomogeneous potential, its deformation, the increase of $\varphi_0(t)$ in time and also the motion of the potential well with the velocity V_0. The last term 4,7,9 in (4) is determined by the resonant electrons.

From (4) we derive, in the zeroth order of α, the equation describing the space distribution of the potential:

$$(\varphi')^2/2 = (V_e-1)\varphi^2 - \varphi^3/3 - \Delta\varphi\{-(\varphi_0+\varphi) + (\varphi^2/2\varphi_0)(1-2\sqrt{\varphi_0/\pi})\times \ln(\sqrt{-\varphi}/(\sqrt{\varphi_0}+\sqrt{\varphi_0+\varphi})) + (2/3\sqrt{\pi})\varphi_0^{3/2} + \sqrt{1+\varphi/\varphi_0}\,[\varphi_0(1-(2/3)\sqrt{\varphi_0/\pi}) + (\varphi/2)(1+(2/3)\sqrt{\varphi_0/\pi})]\} \tag{5}$$

From (5) and $\varphi'|_{\varphi=-\varphi_0}=0$ we obtain the expression for V_e

$$V_e = 1 - \varphi_0/3 + (2/3\sqrt{\pi})(\Delta\varphi/\sqrt{\varphi_0}) \tag{6}$$

The profile of the well is characterized by its width. We determine roughly the width from (5):

$$\Delta X = \{V_c\sqrt{2/\pi}\,[1-3/2\sqrt{2}+0.25\ln(1+\sqrt{2})+(1-2\sqrt{2}/3)\sqrt{\varphi_0/\pi}\,]\}^{-1/2} \tag{7}$$

To describe the well evolution in time we determine from (4),(5) the well amplitude growth rate $\gamma = \partial \ln \varphi_0/\partial t$

$$\gamma = (2/\pi)^{3/4}(V_c^{3/2}/\sqrt{2})[1+(8/3)\sqrt{\varphi_0/\pi} - \sqrt{\pi/2}\,(\varphi_0/3V_c)]^{1/2} \tag{8}$$

There is a threshold in V_c for the formation of a large-amplitude well 3. Moreover, the stationary ion-acoustic perturbation in the form of the potential well doesn't exist in equilibrium plasma.

Hence, let us consider the properties and the evolution of a potential well at slow ion mode in current-carrying plasma, $V^i_{th}<V_d<V_{th}$. In other words, we describe the well, propagating with the velocity V_0 approximately equal to the ion thermal velocity, $V^i_{th}=(Ti/M)^{\frac{1}{2}}$. Taking into account that the resonant electrons are reflected from the well, we obtain from the Vlasov equation the expression for the velocity distribution function of electrons

$$f(t,\xi,u) = \begin{cases} f_0^e((u^2-2e(\varphi-\Delta\varphi)/m)^{1/2}+V_d), & u \le A(\varphi)\,\mathrm{sign}(\xi) \\ f_0^e((u^2-2e(\varphi+\Delta\varphi)/m)^{1/2}-V_d), & u > A(\varphi)\,\mathrm{sign}(\xi) \end{cases} \tag{9}$$

Here $A(\varphi)=(2e(\varphi_0+\varphi)/m)^{\frac{1}{2}}$, $u=V-V_0$. Integrating (9) over the velocities, we obtain the expression for the density perturbation of electrons in the first order of $(V_d-V_0)/V_{th}=V_c$ and in the second order of $e\varphi/T_e=\varphi$

$$\delta n_e = \varphi(1+\varphi/2) - \mathrm{sign}(\xi)\{2\Delta\varphi\sqrt{(\varphi_0+\varphi)/\pi} + \sqrt{2/\pi}\,V_c[\sqrt{\varphi_0(\varphi_0+\varphi)} + \\ + \varphi\ln(\sqrt{-\varphi}/(\sqrt{\varphi_0}+\sqrt{\varphi_0+\varphi}))]\} \tag{10}$$

From the Vlasov equation we obtain the expression for the perturbation of ion distribution function. From the latter the expression for the ion density perturbation can be derived.

$$\delta n_i = \delta n_1 + \delta n_2, \quad \delta n_2' = -\theta^{3/2}\dot{\varphi}\,[1-z^2-R(z)(1-2z^2)]/z\sqrt{2}$$
$$\delta n_1 = -\theta[\varphi - \Delta\varphi\,\mathrm{sign}(\xi)]R(z) + \theta^2(\varphi/2)^2[R(z)(3-2z^2)-1] \tag{11}$$
$$z = V_0/V^i_{th}\sqrt{2}, \quad \theta = T_e/T_i; \quad R(z) = 1+(z/\sqrt{\pi})\int_{-\infty}^{\infty}dt\,(t-z)^{-1}\exp(-t^2)$$

In the quasi-stationary approximation from (10),(11) at $\xi > \Delta x$ we obtain (2). Substituting (10),(11) in the Poisson equation, we derive the evolution equation

$$\dot{\varphi}\theta^{3/2}[1-z^2+(2z^2-1)R(z)]/z\sqrt{2}+\varphi'[1+\theta R(z)]+\varphi\cdot\varphi'\{1+[1+(2z^2-3)R(z)]\theta^2/2\}-\varphi'''+\sqrt{2/\pi}\,V_c|\varphi'|[(1+\sqrt{\varphi_0/\pi})/\sqrt{\varphi_0+\varphi}+\ln(\sqrt{-\varphi}/(\sqrt{\varphi_0}+\sqrt{\varphi_0+\varphi}))]=0 \qquad (12)$$

From (12) we derive the equation describing the space distribution of the potential:

$$(\varphi')^2=\varphi^2(1+\theta R(z))+(\varphi^3/6)\{2+\theta^2[1+(2z^2-3)R(z)]\}-\mathrm{Sign}(\xi)(\Delta\varphi/\varphi_0)\times\{\varphi^2(1-2\sqrt{\varphi_0/\pi})\ln(\sqrt{-\varphi}/(\sqrt{\varphi_0}+\sqrt{\varphi_0+\varphi}))+\sqrt{\varphi_0(\varphi_0+\varphi)}(2\varphi_0+\varphi)-2\varphi_0^2+2\varphi_0\varphi\theta R(z)+\varphi_0[5(\varphi_0+\varphi)^{3/2}-3\varphi_0\sqrt{\varphi_0+\varphi}-2\varphi_0^{3/2}]/3\sqrt{\pi}\} \qquad (13)$$

From (13) and $\varphi'|_{\varphi=-\varphi_0}=0$ we obtain (similarly to 10) the expression for V_0

$$z=z_0[1+1/\theta-(\varphi_0/6)(\theta+3-2z^2+2/\theta)],\quad z_0=0.924 \qquad (14)$$

The well velocity is close to V_{th}^{i} and decreases with increasing φ_0.
From (13),(14) it may be concluded that the potential well at the slow ion mode (in contrast to the ion-acoustic mode case) may exist in the current-free plasma.
We determine roughly the width of the well from (13):

$$(\Delta x)^{-2}=(\varphi_0/48)\{2+\theta^2[1+(2z^2-3)R(z)]\}+\mathrm{Sign}(\xi)(\Delta\varphi/\varphi_0)\{0.25(1-2\sqrt{\varphi_0/\pi})\ln(1+\sqrt{2})+2-3/2\sqrt{2}+(\varphi_0/9\pi)^{1/2}(3/2+3/\sqrt{2}-5/2\sqrt{2})\} \qquad (15)$$

The well is asymmetric: the first front is steeper than the back one. The well width decreases with increasing amplitude. The area of the "hole" $\Delta x\cdot\sqrt{\varphi_0}=\{48/[2+\theta(\theta+3-2z_o^2)]\}^{1/2}$ in the ion phase space is independent of φ_0.
Now we determine from (12),(13) the well amplitude growth rate

$$\gamma=\dot{\varphi}_0/\varphi_0=(z/\sqrt{3\theta})/[1-z^2+(2z^2-1)R(z)] \qquad (16)$$

There exists a threshold in θ. The growth rate is not zero for $\theta > \theta_{th}=1/2(1-z_o)$.
For large amplitudes, $\varphi_0 \geq 1$, we consider the properties of the well in the current-free plasma. In the latter case we have

$$n_e=n_0\exp(\varphi),\qquad n_i=(n_0/\sqrt{\pi})\int_{-\infty}^{\infty}dt\,\exp(-t^2)/[1-\theta\varphi/(z-t)^2]^{1/2} \qquad (17)$$

From (17) and the Poisson equation we obtain

$$[\varphi'(\varphi=-\varphi_0/2)]^2 = \sqrt{2} - 2 + 2\exp(-\varphi_0/2) - \sqrt{2}\exp(-\varphi_0) \qquad (18)$$

φ_0 cannot exceed the critical value φ_a determined by

$$1 - \sqrt{2} + \sqrt{2}\exp(-\varphi_a/2) - \exp(-\varphi_a) = 0 \qquad (19)$$

The width of the well increases with growing φ_0. Therefore we take into account the trapped ions. We assume their distribution function f^i_{tr}=const=$f^i_0(V_0)$ Then their density equals

$$n^i_{tr} = 2\sqrt{-\theta\varphi}\,(n_0/\sqrt{\pi})\exp(-z_0^2) \qquad (20)$$

From (17),(20) and Poisson equation we find, that the width and velocity of the well increase with growing φ_0 (in contrast to the case of small amplitude φ_0) Also the solitary perturbation can form in a magnetized, plasma-loaded waveguide 11-16. Instead of treating this potential well emergence we describe its evolution and properties when it is excited by the "hot" electron beam in a magnetized plasma-loaded waveguide. We consider the case of the strong external longitudinal magnetic field, $H_0 \to \infty$. Then the electron dynamics is one-dimensional.

We choose the initial perturbation in the form of three-dimensional potential well $\varphi = J_0(k_\perp r)\varphi_w(z,t)$ of small amplitude (depth)-φ_0 of width z smaller than the system length and moving with a velocity V_0, satisfying the inequalities $V_b - V^b_{th} < V_0 < V_b$. Here $J_0(k_\perp r)$ is the zeroth order Bessel function, $k_\perp R_0$ is its zero, R_0 is the radius of waveguide, V_b is the beam velocity, V_{th} is the thermal velocity of beam electrons. Now we describe the well evolution. Below we neglect the nonadiabatic dynamics of electrons with the velocities $/V-V_0/<V_{tr}=(2e\varphi_0/m)^{\frac{1}{2}}$ in the field of potential tails. The latter moves with the velocity V_{tr}. Consideration of the nonadiabatic dynamics leads to the additional terms of higher order as compared to the used ones. Taking into account that the resonant electrons are reflected from the well, we obtain from the Vlasov equation the expression for the velocity distribution function of beam electrons

$$f_b(t,\xi,u) = \begin{cases} f_b((u^2-2e\varphi/m)^{1/2}+V_0-V_b), & u > A(\varphi)\,\mathrm{sign}(\xi) \\ f_b((u^2-2e\varphi/m)^{1/2}-V_0+V_b), & u < A(\varphi)\,\mathrm{sign}(\xi) \end{cases} \qquad (21)$$

$$A(\varphi) = (2e(\varphi_{00}+\varphi)/m)^{1/2}, \quad \varphi_{00} = \varphi_0 J_0(k_\perp r)$$

In order to describe the slow evolution of a system we shall seek for solution of Vlasov and Poisson equations in the form (23) and the potential distribution in the form of a well with amplitude $\varphi_0(t)$ slowly varying in time. Integrating (21) over velocities, we obtain the expression for the density perturbation of beam electrons

$$\delta n_B(\xi) = (en_{B0}/T_B)\{\varphi(1+e\varphi/2T_B) - \\ -\mathrm{sign}(\xi)\sqrt{2/\pi}((V_B-V_0)/V_{th}^B)[\sqrt{\varphi_{00}(\varphi_{00}+\varphi)} + \varphi \ln(\sqrt{-\varphi}/(\sqrt{\varphi_{00}}+\sqrt{\varphi_{00}+\varphi}))]\} \tag{22}$$

Finding from the Vlasov equation the expression for the density perturbation of plasma electrons, and substituting it and (22) in the Poisson equation, we derive the evolution equation

$$2\dot{\varphi}_w/V_0^3+\omega_p^{-2}\varphi_w'''+\varphi_w'(V_0^{-2}-V_a^{-2}-V_m^{-2})-\varphi_w'\{(e/m)\beta\varphi_w(3V_0^{-4}+V_m^{-2}V_{Bth}^{-2})- \\ -\mathrm{sign}(\xi)\sqrt{2/\pi}((V_B-V_0)/V_{th}^B V_m^2)[(1+\varphi_w/\varphi_0)^{-1/2}+\ln(\sqrt{-\varphi_w}/(\sqrt{\varphi_0}+\sqrt{\varphi_0+\varphi_w}))]\}=0 \tag{23}$$

Here V_a is the maximum phase velocity of a wave in waveguide with no beam, $V_m=V_{th}^B(N_0/N_{b0})^{\frac{1}{2}}$ is the maximum phase velocity of electron-acoustic mode 17, $\omega_p^2 = 4\pi n_0 e^2/m$, $\beta = \int_0^{R_0} J_0^3(k_\perp r) r\,dr / \int_0^{R_0} J_0^2(k_\perp r) r\,dr$.

Let us obtain from (23) the expressions for properties of potential well. For this we choose symmetric $\Phi(\xi)$ and asymmetric $\delta\Phi(\xi)$ parts of potential distribution $\varphi_w(\xi) = \Phi(\xi)+\delta\Phi(\xi)$. $\delta\Phi(\xi)$ describes the asymmetric potential, added in presence of a beam, to the symmetric well. $\delta\Phi(\xi)$ changes from $-\Delta\varphi$ at $\xi<-\Delta z$ to $\Delta\varphi$ at $\xi>\Delta z$. Assuming $\Delta\varphi \ll \varphi_0$ we obtain from (23) the equation describing the space distribution of the potential:

$$(\Phi')^2 = \omega_p^2\Phi^2(V_a^{-2}+V_m^{-2}-V_0^{-2})+\Phi^3\beta(e/m)(\omega_p^2/V_0^4+d_B^{-2}V_{Bth}^{-2}/3), \\ d_B = V_m/\omega_p \tag{24}$$

From (24) and $\Phi'|_{\Phi=-\varphi_0}=0$ we obtain the expression for the well velocity V_0

$$V_0 = V_{00} + \Delta V(\varphi_0), \qquad V_{00} = V_m V_a/(V_m^2+V_a^2)^{1/2} \\ \Delta V(\varphi_0) = (e\varphi_0\beta/2T_B)(V_{00}^3/3V_m^2 + V_{Bth}^2/V_{00}) \tag{25}$$

V_0 increases with increasing φ_0 and is equal to V_m at $V_a \gg V_m$ and V_a at $V_m \gg V_a$. From (24),(25) it may be concluded that the well is determined by two modes with the linear dispersion relation: the mode in magnetized plasma-loaded waveguide and the electron-acoustic mode.

In the quasi-stationary approximation from (23) at $\xi>\Delta z$ we obtain the rough expression for the potential shock near the well

$$e\Delta\varphi/m = \sqrt{2/\pi}((V_B-V_0)/V_{th}^B)/\beta(V_{Bth}^{-2}/3 + V_m^2/V_{00}^4) \tag{26}$$

In the case being considered the shock is greater by a large factor $T_e/e\varphi_0$ in comparison with the case, when fronts of electrons reflected from the well reach the boundaries of the system 9.

We determine roughly the well width from (24):

$$\Delta z = \{(e\varphi_0\beta/8m)[(\omega_p/V_{\infty}^2)^2 + (d_B V_{th}^B)^{-2}/3] + \\ + \mathrm{Sign}(\xi)\sqrt{2/\pi}\, d_B^{-2}((V_B - V_0)/V_{th}^B)[1 - 3/2\sqrt{2} + 0.25\ln(1+\sqrt{2})]\}^{-1/2} \tag{27}$$

The well is asymmetric in the presence of the beam: the first front is steeper than the back one. The well width decreases with increasing amplitude. Now we determine from (23),(24) the well amplitude growth rate $\gamma = \partial \ln\varphi_0/\partial t$

$$\gamma = (1+\sqrt{2})((V_B - V_0)/V_{th}^B 2\sqrt{\pi})\sqrt{e\varphi_0\beta/m}\,(V_{\infty}^3/V_m^2)(\omega_p^2/V_{\infty}^4 + d_B^{-2} V_{Bth}^{-2}/3)^{1/2} \tag{28}$$

Nonlinear growth rate increases with increasing well amplitude.

In a megnetized, plasma-loaded waveguide two types of solitary structure may exist 12. One is above described potential well. The other type of the structure moves with a thermal velocity of plasma electrons, V_{th}. This note is concerned with the properties of the second-type potential solitary structure. In the case of a small soliton amplitude, φ_0 ($\ll T_e/e = mV_{th}^2/e$), from the Vlasov equation we obtain the expression for the velocity distribution function of electrons. Integrating the latter over velocities, we obtain the expression for the electron density perturbation in the second order of $\varphi = e\varphi/T_e$

$$\delta n' = \dot{\varphi}[y + (1-2y^2)(1-R(y))/y] + \varphi' R(y) + \varphi\cdot\varphi'[1 - y^2 + (3/2 - y^2)(R(y)-1)] \\ R(y) = 1 + (y/\sqrt{\pi})\int_{-\infty}^{\infty} dt\, \exp(-t^2)/(t-y), \quad y = V_0/V_{th}\sqrt{2} \tag{29}$$

V_0, φ are the velocity and potential of the soliton. Substituting (29) in the Poisson equation, we derive the KdV evolution equation

$$\dot{\varphi}[y + (1-2y^2)(1-R(y))/y] + \varphi' R(y) + \varphi\cdot\varphi'[1 - y^2 + (3/2 - y^2)(R(y)-1)] - \varphi''' = 0 \tag{30}$$

From (30) we derive the equation describing the space distribution of the potential:

$$(\varphi')^2 = \varphi^2 R(y) - [1 + (2y^2 - 3)R(y)]\varphi^3/6 \tag{31}$$

From (31) and $\varphi'|_{\varphi=\varphi_0} = 0$ we obtain (similarly to 10) the expression for V_0

$$V_0 = 1.32\, V_{th}(1 - e\varphi_0/6T_e) \tag{32}$$

V_0 is close to V_{th} and decreases with increasing φ_0.
We determine roughly the soliton width from (31):

$$\Delta X = \sqrt{48\, T_e / e\varphi_0} \tag{33}$$

The soliton width decreases with increasing amplitude. The soliton is the "hole" in the electron phase space, the area of which, $\Delta X \sqrt{e\varphi_0/m}$, is independent of φ_0.
In the case of large amplitudes, $e\varphi_0/T_e \gtrsim 1$, from the Vlasov equation we have the expression for the velocity distribution function of plasma electrons (without electrons trapped by a soliton field) $f = f_0[(u^2 - 2e\varphi/m)^{1/2} + V_0 \,\mathrm{Sign}(u)]$ for $|u| = |V - V_0| > \sqrt{2e\varphi/m}$. Here f_0 is the Maxwellian distribution function. Thus we derive the equation for the soliton shape

$$(\varphi')^2 = -\varphi + (2/\sqrt{\pi}) \int_{-\infty}^{\infty} dt\, (t-y)^2 \exp(-t^2) \{[1 + \varphi/(y-t)^2]^{1/2} - 1\} \tag{34}$$

From (34)

$$\Delta X = (2e\varphi_0 / T_e (\sqrt{2} - 1))^{1/2} \tag{35}$$

we conclude that the soliton width increases with φ_0. Hence, taking into account the trapped electrons is important. Assuming for their density distribution $n_{tr}(x) = n_2 \exp[e\varphi(x)/T_{tr}]$, we obtain similarly to (35) that the width and velocity of the soliton as well as of the potential well at slow ion mode increase with amplitude growing (in contrast to the case of small amplitude φ_0).
These properties of the soliton and its amplitude dependences were observed in experiments and numerical simulations 12.
Let us consider the possibility of double layer (DL) formation at electron beam injection into the plasma. When an electron beam of density, n_B, comparable with the plasma density, n_i, is injected from the isolated gun into the plasma; the plasma electrons are accelerated to the electron beam source in the field appearing between the source and the beam. In this case their density falls and they do not compensate a positive charge of the source. The beam returns to the source and neutralizes its charge. Then, the potential separates from the source and moves inside the plasma with a velocity V_h. Thus, the DL may be formed (i.e., a potential jump from the boundary value $\varphi_0 = \varphi(z=0)$ to zero in a small space interval δz).
We investigate the properties of such a DL in the case of relativistic beam injection into one-dimensional plasma.

The DL formation needs a charge separation. In the case of the monotonous DL formation by the electron-beam-plasma mode two groups of injected electrons are necessary. Let us consider the half-space plasma, $z>0$, into which a high energy beam, $V_b \gg V_{th}$, V_{th}^{b}, and a slowly moving beam $V_{sl}<V_{th}^{s}\ll V_b$, of densities n_b and n_s are being injected. $V_{th}^{b} = (T_b/m)^{1/2}$, $V_{th}^{s} = (T_s/m)^{1/2}$ are the thermal velocities of the electron beams. The velocity of slowly moving beam V_{sl} can be chosen equal to V_h.
The density of the slow-moving beam exponentially falls off

$$n_s(z) = n_s \exp((\varphi - \varphi_0)e/T_s), \tag{36}$$

resulting in positive charge at $\varphi_a < \varphi < \varphi_0$, while the density of both the high energy beam and plasma electrons

$$\begin{gathered} n_b(z) = n_b (1-\gamma_0^{-2})^{1/2} (1-(1+e\varphi/mc^2)^{-2})^{-1/2} \\ n_e(z) = n_e (V_h/c)[1-(1+(\gamma_0-1)\varphi/\varphi_0)^{-2}]^{-1/2}, \quad \gamma_0 = (1-(V_b+V_{th}^{b})^2/c^2)^{-1/2} \end{gathered} \tag{37}$$

grows in accordance with the power law. This leads to negative charge at $0<\varphi<\varphi_a = \varphi_0(\gamma_0-1)^{-1}[\sqrt{(\gamma_0+1)/2} - 1]$. At $\varphi_c = \varphi_0 (2V_{th}^{b}/c)\gamma_0^2$ the beam starts to be reflected. As a result, the quasineutrality is restored outside the DL. Since the nonresonant beam electrons propagate into the plasma, where they are decelerated, their density grows. Therefore, the quasineutrality condition in front of the DL requires $V_h<V_{th}$, and the density of the beam electrons passing through the DL should be small $n_b^0 \ll n_e$. Hence $\varphi_0 \simeq (mc^2/e)(\gamma_0-1)$.
Using the momentum- and energy-balance equations and also the quasineutrality condition at the boundary of beam injection, we obtain

$$\begin{gathered} V_h/c = (n_b^0/n_e)(1-\gamma_0^{-2})^{1/2} \ll 1 \\ n_b = (n_i/2 - n_s(T_s/2e\varphi_0))\gamma_0/(\gamma_0+1), \quad n_s = [1+(T_0/e\varphi_0)\gamma_0/(\gamma_0+1)]n_i/(\gamma_0+1) \end{gathered} \tag{38}$$

n_s is inversely proportional to γ_0, and n_b tends to $n_i/2$ if γ_0 is large. From the Poisson equation and (36),(37) we obtain

$$(\partial\varphi/\partial z)^2/8\pi e = n_i\sqrt{\varphi_0\varphi}\,\{[(\varphi/\varphi_0)(\gamma_0+1)+2]^{1/2}/\sqrt{\gamma_0+1} - \sqrt{\varphi/\varphi_0}\} \tag{39}$$

We determine roughly the DL width from (39)

$$\delta z = (c/4e)(2m/\pi n_i)^{1/2}(\gamma_0-1)/[1-\sqrt{2/(\gamma_0+1)}]^{1/2} \tag{40}$$

At $\gamma_0 >> 1$, $\delta z = (c/\sqrt{2}w_p)\gamma_0$, where $w_p = (4\pi n_i e^2/m)^{\frac{1}{2}}$, the DL width is proportional to γ_0 .

In the DL field only ions gain the momentum. So, the ions, passing through the DL, are accelerated to an energy $E_i = mc^2(\gamma_0 - 1)$, comparable with the beam energy.

During DL formation time the plasma electrons have no time for the response to the field initiation, and are trapped by the DL field. In this way the slow electron group is formed, which is necessary for the DL formation on injecting only the high energy beam. But now the DL is nonmonotonous: first, the potential slowly rises from φ_0 to φ_1 and then abruptly drops from φ_1 to zero. Since the DL slowly drifts downstream, the density of the trapped electrons n_s in the nonmonotonous DL decreases because of the growth of the region, where these electrons are localized. The stability analysis of the electron fluxes relative to the low frequency perturbations shows that if the density of the trapped electrons falls below the critical value

$$(V_{th}^s / V_b)^2 / \gamma_0^3 > n_s / n_i \tag{41}$$

the DL becomes unstable relatively to the perturbation with the phase velocity equal to V_h. The computer simulation 18 has shown that the DL, which has penetrated deeply into the plasma, decays with the generation of a single vortex structure in the electron phase space. At the same time, a new DL arises at the boundary.

Similar processes for the potential distribution and beam behaviour have been observed in a full-scale experiment 20 and numerucal simulations 18,21. The beam injection into the plasma leads in numerical simulations 21 in certain conditions to the DL formation and moving it inside the plasma at a thermal velocity of plasma electrons.

REFERENCES

1. A.S.Bakai, V I.Maslov, Plasma Physics, 13 (1987), 1019.
2. H.Schamel, Phys. Rep., 140 (1986), 163.
3. T.Sato, H.Okuda, Phys.Rev.Lett., 44 (1980), 740.
4. K.Nishihara et al., Riso-R-472 (1982), 41.
5. T.H.Dupree, Phys. Fluids, 29 (1986), 1820.
6. C.Chan, M.H.Cho, N.Hershkowitz, T.Intrator, Phys.Rev.Lett, 52(1984),1782.
7. J.E.Allen, Plasma Phys. and Contr. Fus. 27 (1985), 1343.
8. A.N.Sekar, Y.C.Saxena, Plasma Phys. and Contr. Fus., 27 (1985), 181.
9. V.I.Maslov, Plasma Physics, 16 (1990), 759.
10. H.Schamel, Plasma Physics, 14 (1972), 905.
11. S.M.Krivoruchko et al., JETP, 67 (1974), 2092.
12. I.P.Lynov et al., Physica Scripta, 20 (1979), 328.
13. H.Ikezi et al., Phys. Fluids, 14 (1971), 1997.
14. B.N.Rutkevich et al., JTP, 42 (1972), 493.
15. V.I.Kurilko, A.P.Tolstolushskij, JTP, 44 (1974), 1418.
16. V.K.Sayal, S.R.Sharma, Phys. Lett., 149 (1990), 155.
17. V.K.Jain, N.T.Tsintsadze, IC/89/150.
18. H.Okuda, R.Horton, M.Ono, M.Ashour-Abdalla, Phys.Fluids, 30(1987),200.
19. V.I.Maslov, Ukr. Fiz. Zh., 33 (1988), 1342.
20. T.Obayashi, N.Kawashima, K.Kuriki et al., Science, 225 (1984), 195.
21. N.Singh, R.W.Schunk, Plasma Phys. and Contr. Fus., 26(1984), 859.

IONIZATION-PRODUCED DOUBLE LAYERS IN MAGNETIZED AND UNMAGNETIZED PLASMAS

R.L. Merlino, Bin Song,[a] and N. D'Angelo
Department of Physics and Astronomy
University of Iowa, Iowa City, Iowa 52242 USA

Abstract: Recent experiments on anode double layers at the University of Iowa are reviewed. Two cases have been studied: (1) nearly spherical anode double layers, 'fireballs,' produced in an unmagnetized discharge plasma and (2) anode double layers formed in a nonuniform magnetic field geometry.

1. INTRODUCTION: Anode, or ionization double layers, were first discussed by Langmuir[1] in his classic paper on the interaction of electron and positive ion space charges. Consider a planar electrode immersed in an unmagnetized, weakly ionized, hot-filament discharge in argon at a pressure of $\sim 10^{-3}$ Torr. If a positive potential is applied to this electrode, an anode glow appears near the electrode as the voltage approaches the ionization potential of the gas. When the voltage exceeds the ionization potential, and at sufficiently high neutral gas pressure, the rate of positive ion production in the anode sheath becomes large enough to overcome the electron space charge causing an additional discharge to develop near the anode. In Langmuir's words "this second plasma usually takes the form of a *globular luminous protuberance* from the anode." The interior of this 'luminous protuberance' is a typical electrically neutral plasma and its boundary is a double layer with an inner positive space charge and an outer negative space charge. The voltage drop across this layer is usually on the order of the ionization potential of the gas.

There have been a number of experimental and theoretical studies of anode double layers. Most of the experimental work has dealt with anode double layers formed in plasma columns in uniform magnetic fields,[2–4] although usually only the electrons were well magnetized. The paper of Torvén and Andersson[2] contains the essential physical arguments indicating the role of ionization in converting the anode sheath into a double layer. There seems to be some evidence that it is more difficult to form stable anode double layers in plasma columns with relatively strong uniform magnetic fields.[4] However, stable anode double layers have been produced in strong nonuniform magnetic field geometries.[5,6] In this paper we will review our recent work[7] on the effect of a magnetic field gradient on anode double layers.

---

[a] Present address: Plasma Physics Laboratory, Columbia University, New York, NY.

Interestingly, relatively less work has been done recently on anode double layers in magnetic field-free plasmas. In the next section of this paper we will summarize some of our recent work in this area.[8] We have studied the formation of nearly spherical luminous objects ('fireballs') anchored to a positively biased disk electrode in argon, krypton, and xenon discharge plasmas. The evolution of these 'fireballs' from an unstable anode sheath was studied, and measurements were made of the fireball diameter as a function of neutral pressure.

The first attempt to find numerical double-layer solutions of Poisson's equation including ionization was presented by Andersson[3] and by Andersson and Sørensen.[9] These time-independent calculations showed that double-layer solutions were possible for a wide range of parameters when the ions in the high-potential region were created by ionization due to the accelerated electrons. More recently, Cook and Katz[10] performed a numerical simulation of the effect of volume ionization within an anode sheath. Their results clearly showed that at a sufficient value of the neutral density, the density of ions formed by ionization, begins to dominate in some regions of the sheath. At this point the sheath edge begins to move outward to include more electrons, and in so doing, increases the ion production proportionally, leading to further sheath expansion. Their calculations, however, were not carried out to sufficient times to observe the formation of stable anode double layers.

The balance of this paper is as follows. In Section 2 we present the experimental results and analysis pertaining to the case of anode double layers in an unmagnetized plasma. In Section 3 the experimental results on the effect of a magnetic field gradient on anode double layers will be presented and discussed. Final remarks and conclusions are collected in Section 4.

2. IONIZATION-PRODUCED DOUBLE LAYERS IN UNMAGNETIZED PLASMAS:

These experiments were performed in an unmagnetized plasma discharge chamber 60 cm in diameter and 100 cm in length. Typically, argon gas was used, although some observations were made in krypton or in xenon. The plasma was produced by biasing a set of tungsten filaments, hot enough to emit thermionic electrons, at -60 V relative to the grounded chamber walls. Typically, the plasma parameters were: plasma density $\sim 10^9$ cm^{-3}, $T_e \approx 2$ eV, $T_i \lesssim 0.2$ eV, and neutral pressures in the range of 10^{-4} Torr to 10^{-3} Torr. Plasma potentials were inferred from the floating potential of an emissive probe, heated to emission by a direct current. The emissive probes consisted of a 0.025 mm diameter tungsten wire 3 mm length and supported on two insulated shafts of 0.1 mm diameter. The use of miniature probes was necessary to minimize probe perturbations to the potential structures.

The anode double layers, or 'fireballs' as we called them, were produced by applying a positive potential of 40–50 V to a metal disk of 3 cm diameter coated on one side with a dielectric material. The disk was located in the center of the chamber in front of a large viewing port to allow visual observations and photographing of the 'fireballs.' To begin the presentation of the results, we show in Fig. 1 a current vs. voltage characteristic of the 3cm disk in an argon plasma. For this case the argon pressure was $P = 1.5 \times 10^{-3}$ Torr

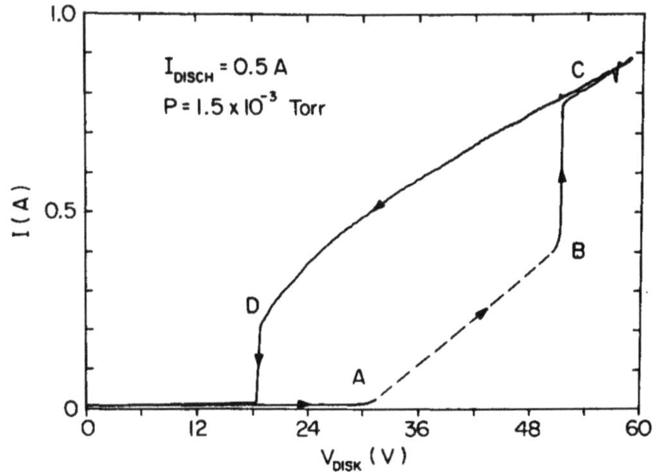

Fig. 1. $I-V$ characteristic of the 3 cm disk in an argon plasma. In the dashed portion from A to B the current consisted of regularly spaced 'spikes.'

and the plasma discharge current $I_{DISCH} = 0.5$ A corresponding to a plasma density $n_e \approx 3 \times 10^8$ cm^{-3}. As the disk voltage was increased from 0 V up to $\sim$ 30 V, the disk current increased somewhat up to a value of $\sim 10^{-2}$ A (at $V_{DISK} = 30$ V). Visually, at just above 30 V, a luminous plasma sheath was observed on the conducting side of the disk with a thickness of perhaps $2-3$ mm. In the portion of the characteristic from $A \rightarrow B$ (dashed line) the current was not steady, but rather consisted of regularly occurring spikes, the time between successive spikes decreasing steadily as the disk voltage increased. During this part of the characteristic, we observed a diffuse luminosity extending as much as $15-20$ cm from the disk. At $V_{DISK} \approx 50$ V a sudden jump in the current took place ($B \rightarrow C$). Further increasing V_{DISK} produced only a moderate increase in the current. On the upper portion of the characteristic ($C \to D$), the current was steady until about $V_{\mathrm{DISK}} \lesssim 22$ V, when the current fell back to a level of $\sim 10^{-2}$ A. After the current jump from $B \to C$, a nearly spherical intensely luminous region ('fireball') with a very strong boundary was observed to be attached to the disk. This 'fireball' persisted as V_{DISK} was lowered from $C \to D$, until at ~ 20 V, when it suddenly disappeared.

A well-developed 'fireball' could be studied by probing it with either a Langmuir probe or an emissive probe. Figure 2 shows a profile of the space potential across a fireball in argon. The potential drop across the boundary double layer is always just above the ionization potential of the particular gas used. Measurements within the fireball indicated that the thermal electron densities were several times larger than in the surrounding plasma.

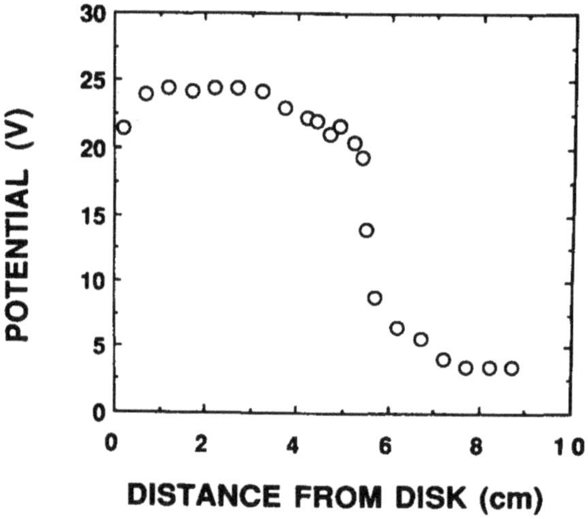

Fig. 2. Profile of the space potential across a 'fireball' in argon at $P = 1.2 \times 10^{-3}$ Torr and $V_{\mathrm{DISK}} = 40$ V.

The dependence of the fireball diameter on neutral pressure was studied for argon, krypton, and xenon gases and the results are shown in Fig. 3. As the neutral pressure is increased, the fireball diameter decreases, although their size was limited to no less than about 5 cm presumably due to the finite size of the disk (3 cm diameter) to which they were anchored. If the plasma density was varied, the fireball diameter remained

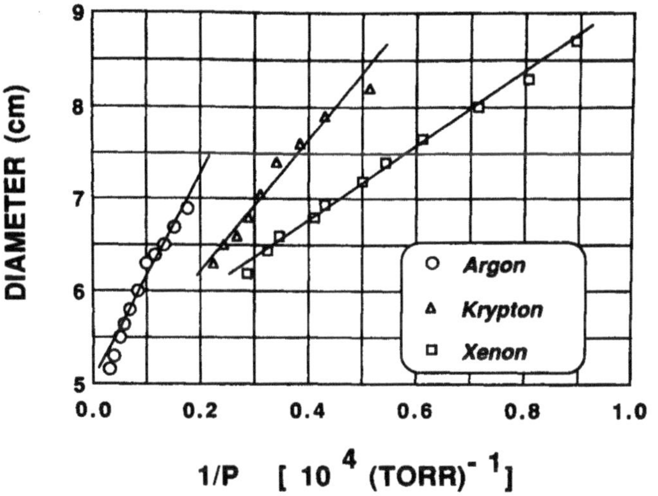

Fig. 3. The 'fireball' diameter vs. inverse pressure in Ar, Kr, and Xe plasmas.

nearly constant, although there was a pronounced increase in their luminosity when the plasma density was increased.

For a stable fireball, the rate of ion production within the fireball must equal the rate of ion loss into the surrounding plasma. If we assume a double layer which is nearly spherically symmetric (except for the small region where the fireball is attached to the anode), the total number of ions lost to the surrounding plasma per second is $4\pi(D/2)^2\varphi_i$, where D is the fireball diameter and φ_i is the ion flux, while the total number of ions produced per second within the fireball is $4\pi(D/2)^2\varphi_e D\sigma_i N$, where φ_e is the electron flux into the fireball, σ_i is the ionization cross section of the neutral gas corresponding to an energy slightly above (1 − 2 eV) the ionization energy, and N is the neutral gas density. The quantity $D\sigma_i N$ is, of course, the ratio between an ionizing electron path length within the fireball, D, and the ionization mean free path $1/\sigma_i N$. Balancing ion production and loss, we can then write

$$\varphi_e D\sigma_i N = \varphi_i \tag{1}$$

and since φ_e and φ_i are related by the usual Langmuir condition $\varphi_e = (m_i/m_e)^{1/2}\varphi_i$, we have

$$D\sigma_i N = \left(\frac{m_e}{m_i}\right)^{1/2} \tag{2}$$

Using parameters appropriate to argon gas and for typical pressures at which the fireballs are formed, Eq. (2) predicts $D \gtrsim 5$ cm, which is on the order of the observed fireball diameter. Also (2) predicts a fireball diameter which decreases with increasing pressure (or N), in agreement with the data of Fig. 3. The variation of the slopes of the D vs. $1/P$ lines in Fig. 3 for the three gases is also in agreement with (2). Finally, Eq. (2) does not contain the density of the plasma surrounding the fireball, in agreement with the observation that the fireball diameter is independent of plasma density.

Reference 8 contains more details of the properties of the fireballs including time-resolved potential measurements on the unstable fireballs. The stability of the spherical fireballs is discussed in a separate paper.[11]

3. ANODE DOUBLE LAYERS IN A NONUNIFORM MAGNETIC FIELD: Previous experiments on anode double layers in nonuniform magnetic fields have revealed that the magnetic field inhomogeneity seems to stabilize the position of the double layers.[5,6] For relatively strong magnetic fields, the double layer only seems to be formed if the anode is located in a region of nonuniform magnetic field. In the experiments of Fujita et al.,[4] performed in a uniform magnetic field, it was noted that lower magnetic fields were more conducive to stable double-layer formation. When the magnetic field was increased, the double layer expanded toward the plasma source. We have carried out experiments investigating the effect of a magnetic field gradient on anode double layers. Full details of this work are given in Ref. 7.

The experiments were performed in a single-ended Q-machine in which potassium atoms were contact ionized on a 6 cm diameter tantalum hot plate. Typical plasma parameters are electron densities $n_e \sim 10^9 - 10^{10}$ cm^{-3}, and electron and ion temperatures $T_e \approx T_i \approx 0.2$ eV. The axial magnetic field for the radial confinement of the K^+ ions and electrons is produced by a set of 10 magnet coils in two 5-coil sets (west and east coils) which could be energized with different currents to produce a variable magnetic field geometry. Magnetic field profiles for various combinations of west and east coil currents are shown in Fig. 4 with the hot plate plasma source located at $z = 65$ cm. The double layers are produced by biasing a 6 cm diameter cold plate (located at $z = 0$ in Fig. 4) to voltages above the ionization potential of argon gas which is leaked into the system up to pressures of $\sim 10^{-3}$ Torr. The anode double layers evolve, in the usual manner, from an anode sheath if sufficient ion production occurs within the sheath due to the accelerated electrons. The Ar^+ anode plasma that is formed is separated from the ambient K^+ plasma by a stable, stationary double layer. The anode plasma appears as

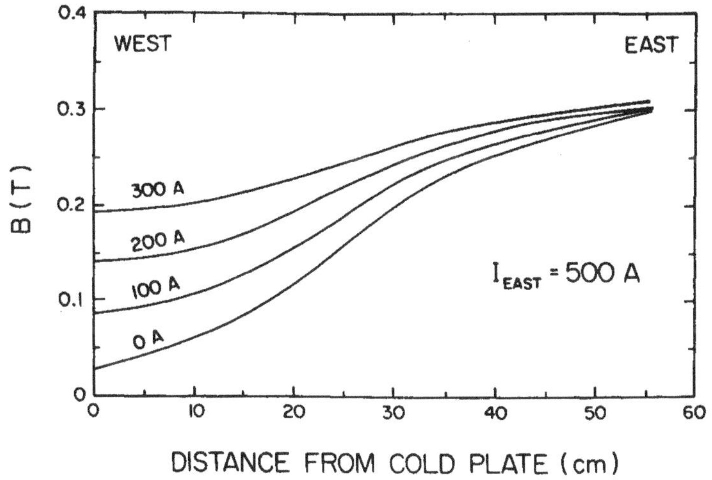

Fig. 4. Axial magnetic field profiles for $I_E = 500$ A and $I_W = 300$ A, 200 A, 100 A, and 0 A.

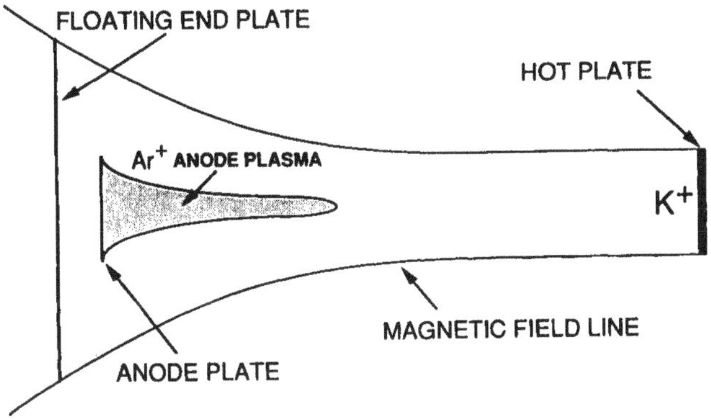

Fig. 5. Schematic diagram showing the formation of the Ar^+ anode plasma in the nonuniform magnetic field.

a luminous axisymmetric structure, elongated along the magnetic field, out to distances of 30–40 cm from the anode plate, as shown schematically in Fig. 5.

Axial potential profiles of the double layers for different magnetic field geometries are shown in Fig. 6. The magnetic field profiles for Fig. 6 correspond to those in Fig. 4 when the east coil current is fixed at 500 A and the west coil current is varied from 0 A to 300 A. The double layers always form in the region of strong magnetic field nonuniformity and

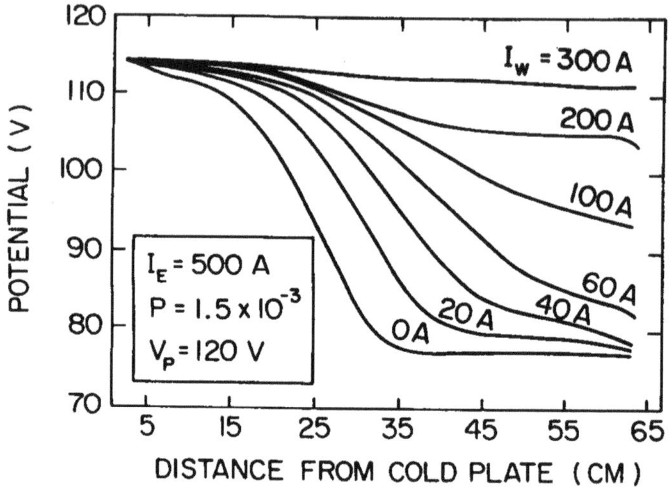

Fig. 6. Axial profile of plasma potential for $I_E = 500$ A and various west coil currents.

their position shifts away from the anode plate as the west coil current is increased. If the magnetic field gradient is reduced below some critical value (i.e., for $I_W > 300$ A), the double layer suddenly shifts all the way to the hot plate, so that the entire central portion of the plasma column is a region of high-potential argon discharge plasma.

We also investigated the effect of varying the magnitude of the magnetic field everywhere but keeping the magnetic field geometry fixed. This is accomplished by setting the west coil current $I_W = 0$ and varying the east coil current. Varying the magnitude of B with a fixed geometry seems to have a much smaller effect on the position of the double layers. On the other hand, if we reverse the direction of the magnetic field gradient (i.e., when the magnetic field increases in the direction toward the cold anode plate), there was no stable position for the double layers except adjacent to the hot plate.

In contrast to the unmagnetized anode double layers discussed in Section 2, we now find that increasing the neutral pressure causes the double layer to move *away* from the cold plate, as shown in Fig. 7. This behavior, which was also observed in our previous experiments,[5] obviously cannot be explained solely on the basis of the analysis leading to Eq. (2). However, this type of analysis can be extended to the present case if we take into account the effect of the magnetic field gradient on the ions.

Again let φ_e represent the flux of electrons entering the double layer. In passing through the double layer the electrons are energized and are then capable of ionizing the argon

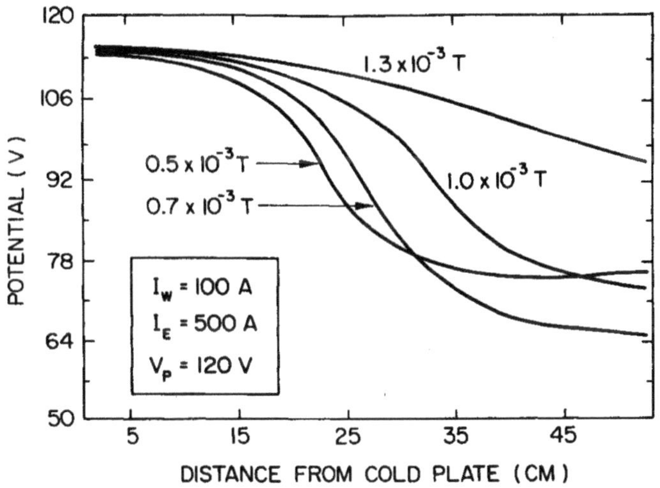

Fig. 7. Axial profile of plasma potential for fixed magnetic field geometry and various neutral argon pressures.

gas. If the magnetic field were uniform and if there were no radial losses, all of the ions produced by volume ionization would pass down the double layer to the low-potential region. However, in the actual converging magnetic field geometry, some fraction of the ions would be reflected by the magnetic mirror force. If ϵ represents the fraction of ions transmitted through the magnetic mirror, then the ion flux φ_i, through the double layer, in this case, is given by

$$\varphi_i = \epsilon(N\sigma_i z_D)\varphi_e \tag{3}$$

where z_D is the axial position of the double layer from the anode plate and $(N\sigma_i)^{-1}$ is the ionization mean free path. Since the fluxes must obey the Langmuir condition, Eq. (3) becomes

$$\epsilon N\sigma_i z_D = \sqrt{\frac{m_e}{m_i}} \tag{4}$$

Equation (4) determines the position of the double layer z_D. From an isotopic ion distribution ϵ is given by $B_0/B(z_D)$ where B_0 is the magnetic field strength in the region near the anode plate and $B(z_D)$ is the magnetic field strength at the position of the double layer. From (4) we can now see the effect of the magnetic geometry on the position of the double layer. For example, for a fixed plate voltage and neutral pressure, if the west coil current is increased, as in Fig. 6, the double layer moves away from the plate to a larger z_D where $B(z_D)$ is also larger so as to maintain the balance suggested by Eq. (4). However, there will be some critical value of I_W above which Eq. (4) cannot be satisfied, so no stable double-layer position will exist in the nonuniform field region. If the neutral pressure is increased for a fixed magnetic geometry, the double layer again shifts to a larger z_D (larger $B(z_D)$ and smaller ϵ), so that a greater fraction of ions will be reflected by the magnetic mirror.

This simple physical model seems, at least qualitatively, to account for most of the observations and to provide some insight into the role of the magnetic field gradient in stabilizing the double-layer position. A more quantitative analysis, however, would probably require the inclusion of the effect of radial ion losses which could be important for this type of double layer.

4. DISCUSSION AND CONCLUSIONS: Measurements of the properties of anode (ionization) double layers in both unmagnetized and magnetized plasmas have been presented. Many of the observations have been accounted for in terms of a simple model balancing ionization production and loss, and requiring that the Langmuir condition be satisfied. For the unmagnetized case this simple model predicts 'fireball' diameters close to those observed and also correctly predicts the dependence of their diameter on neutral pressure. This model can be applied also to the magnetized case if we take into account the effect of ion reflection by the converging magnetic field.

In closing, it is important to point out that the formation of anode double layers can greatly increase the current collection capability of an electrode (see, e.g., Fig. 1). This result is of some interest from the point of view of the so-called 'plasma contactors' to be used on spacecraft to provide a low-impedance path between a conductor and the dilute ionospheric plasma.[12] There has been one observation reported where anode double layers have been produced around a positively charged rocket during a neutral gas release.[13]

ACKNOWLEDGEMENTS: This work was supported by ONR and NASA.

REFERENCES:

1. I. Langmuir, *Phys. Rev.* **33** (1929), 954.
2. S. Torvén and D. Andersson, *J. Phys. D.: Appl. Phys.* **12** (1979), 717.
3. D. Andersson, *J. Phys. D.: Appl. Phys.* **14** (1981), 1403.
4. H. Fujita, S. Yagura, and K. Matsuo, *Phys. Fluids* **27** (1984), 2948.
5. M.J. Alport, S.L. Cartier, R.L. Merlino, and G. Knorr, *Proceedings of the Second Symposium on Plasma Double Layers and Related Topics*, edited by R. Schrittwieser and G. Eder, University of Innsbruck, Austria, 1985, p. 224; S.L. Cartier and R.L. Merlino, *Phys. Fluids* **30** (1987), 2549; M.J. Alport, S.L. Cartier, and R.L. Merlino, *J. Geophys. Res.* **91** (1986), 1599.
6. R. Plamondon, J. Teichmann, and S. Torvén, *J. Phys. D.: Appl. Phys.* **21** (1988), 286.
7. B. Song, R.L. Merlino, and N. D'Angelo, *Phys. Scr.* **45** (1992), 395.
8. B. Song, N. D'Angelo, and R.L. Merlino, *J. Phys. D.: Appl. Phys.* **24** (1991), 1789.
9. D. Andersson and J. Sørensen, *J. Phys. D.: Appl. Phys.* **16** (1983), 601.
10. D. L. Cook and I. Katz, *J. Spacecr. Rockets* **25** (1988), 132.
11. B. Song, N. D'Angelo, and R.L. Merlino, *J. Phys. D.: Appl. Phys.* in press (1992).
12. P.J. Wilbur and T.G. Laupa, *Adv. Space Res.* **8** (1988), 221.
13. B.E. Gilchrist, P.H. Bandes, T. Neubert, P.R. Williamson, N.B. Meyers, W.J. Raitt, and S. Sasaki, *J. Geophys. Res.* **95** (1990), 2469.

COHERENT STRUCTURES IN PLASMA TURBULENCE

H.L. Pécseli, Institute of Physics, University of Oslo, Boks 1048 Blindern, N-0316 Oslo 3, Norway

ABSTRACT: Basic properties of ion phase space vortices are discussed with particular attention to their possible role in the development of small amplitude double layers. The possible importance of an externally imposed magnetic field for the stability of spatially compact ion vortices is pointed out. The statistical aspects of the problem are addressed and illustrated by a numerical simulation, where the distributions of basic parameters as amplitude and velocity of ion beam generated phase space vortices can be obtained by simple methods.

1. INTRODUCTION: Large scale coherent structures play an important role in many aspects of fluid as well as of plasma turbulence. Such structures have gained interest relatively recently. The spectral representations of turbulence emphasizes the inertial subranges which can have universal properties. These subranges are however usually associated with small spatial scales, which are of little importance for instance for convective transport which is a phenomenon associated with large structures or eddies. The properties of these are often dominated by the boundary conditions of the actual system and the generation mechanism of the turbulence. Examples are electrostatic vortical flute type structures in magnetized plasmas. These phenomena can adequately be described in two spatial dimensions, i.e. in a plane perpendicular to the magnetic field[1]. In this limit the basic or lowest level equations describing the problem are identical to those for two dimensional neutral flows. In other cases, however, a plasma can support coherent nonlinear structures which are associated with specific properties of collisionless plasmas, the socalled BGK equilibria. The existence of these phenomena relies on a delicate balance between free and trapped particles in plasma phase space[2,3]. Plasma double layers and phase space vortices (or phase space holes) represent the two best known examples for these structures. They are extensively studied in the laboratory and numerically as well as theoretically.

Ion phase space vortices in particular are considered important for the development of small double layers. Being associated with a local density depletion and a corresponding potential well, they act as a barrier for an electron flow. Slow electrons are reflected by the barrier, and in case of a net electron flow a small potential difference builds up due to the resulting imbalance in electron charge at the two flanks of the ion vortex. It has been argued on the basis of theoretical as well as numerical studies carried out in one spatial dimension[4,5] that the momentum exchange between reflected electrons and an ion phase space vortex gives rise to a deceleration and, in particular, an *amplification* of the vortex so the process can be initiated by very small vortices[5].

Laboratory experiments where the conditions for these processes were fulfilled[6] did not, however, show any sign of small double layers developing, although a conditional sampling technique clearly demonstrated that spatially localized ion phase space vortices *were* present (they were in that case directly excited by the ion-ion beam instability[7]). The analysis of the experimental data clearly demonstrated that the average velocity of the vortices was constant, independent of the imposed conditions, in variance with the theoretical expectations[5] based on analysis in one spatial dimension. In another experiment[7] a single, essentially one dimensional, relatively long lived ion phase space vortex was excited externally. A hole in the ion energy distribution was in this case observed directly by an ion energy analyzer. In both experiments the ion vortices were found to decay on a time scale which was larger than the ion bounce period (but not necessarily *much* larger[6]). Of course one may speculate on the importance of boundary conditions in the experiments, collisional losses etc., but it may be worth pointing out that the one dimensional arguments suffer in one important aspect; they only consider momentum exchange between the perturbation in potential and slow, reflected, particles. The other particles are assumed to have transient effects only. Indeed this is so in one spatial dimension, but as the simple sketch in Fig. 1b indicates, in two or three dimensions particles can exchange momentum with *localized* potential structures[6] of both polarities.

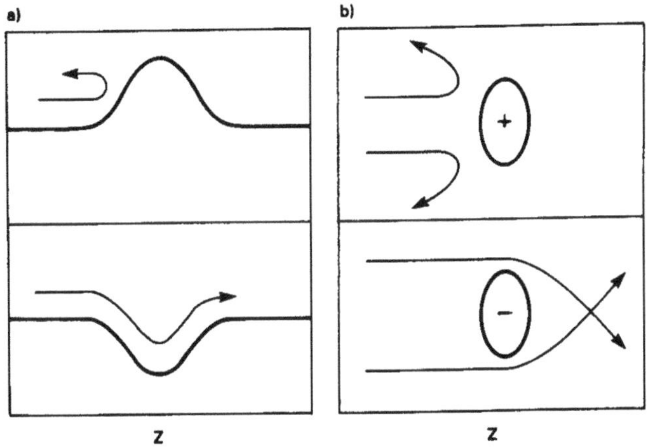

Fig. 1. Schematic diagrams for discussing the difference between particles interacting with localized potential variations in one and in higher spatial dimensions.

It is not evident how these effects are to be incorporated in the original analysis. In a strong homogeneous magnetic field of course the particle dynamics become essentially one dimensional as in Fig. 1a and it is in these conditions one might expect to recover long lived, spatially compact small amplitude ion phase space vortices. It is actually worth mentioning that in this case nonlinear fully *three* dimensional BGK-equilibria can be constructed trivially from the Vlasov equation in its appropriate form, i.e. the drift kinetic equation

$$\partial_t f + \nabla_\perp \cdot (\mathbf{v}_\perp f) + v_z \partial_z f - \frac{e}{M} \partial_z \Phi \, \partial_{v_z} f = 0 \tag{1}$$

with $\mathbf{v}_\perp = [-\nabla_\perp \Phi \times \hat{b} - \Omega_{ci}^{-1} \nabla_\perp (\partial_t \Phi + v_z \partial_z \Phi)]/B_o$, where (1) is to be completed with Poisson's equation and equations for the electron dynamics. The second term in $\mathbf{v}_\perp$ is the polarization drift which originates from the time variation of the $\mathbf{E}\times\mathbf{B}_o/B_o^2$-velocity experienced in the particle frame of reference. This effective acceleration, $\mathbf{g}$, gives rise to a slow particle drift in the $\mathbf{g} \times \mathbf{B}_o$-direction in or out of the potential well, resulting in a violation of the assumed time stationarity of the particle distribution. In the limit of large $\mathbf{B}$-fields, i.e. large Ω_{ci}, the polarization contribution to $\mathbf{v}_\perp$ is negligible and a steady state solution to (1) is $f = F[\frac{1}{2}Mv_z^2 + e\Phi(\mathbf{r}_\perp, z)]$ with arbitrary F, where then the rest of the $\mathbf{v}_\perp \cdot \nabla_\perp f$ term vanishes identically by $(\nabla\Phi \times \hat{\mathbf{b}}) \cdot \nabla\Phi = 0$. Insertion of the expression for f into Poisson's equation gives

$$\partial_{r_\perp}^2 \Phi + \partial_z^2 \Phi = \frac{e}{\epsilon_o} \left(n_e - \int_{-\infty}^{\infty} F \left[\frac{1}{2} M v_z^2 + e\Phi(\mathbf{r}_\perp, z) \right] dv_z \right) . \tag{2}$$

A corresponding expression for $n_e = \int_{-\infty}^{\infty} f_e \, dv_z$ can be obtained similarly where the polarization drift can be ignored from the outset, but in some problems it might safely be assumed that the electrons are isothermally Boltzmann distributed at all times which simplifies the relation between n_e and Φ. If $\Phi = \Phi(\mathbf{r}_\perp, z)$ is prescribed then (2) can be considered as an equation for determining F with standard methods available[8], where F is assumed to be composed of free and of trapped particle contributions. The $\mathbf{r}_\perp$ dependence of Φ imposes then the $\mathbf{r}_\perp$ variation of F. The solutions are not restricted to cylindrical symmetry. Nothing is known however about the stability of these solutions with respect to two or three dimensional perturbations.

Although the arguments given above explicitly assumed large magnetic fields, it might be expected that they are applicable also for intermediate field strengths provided the polarization term in $\mathbf{v}_\perp$ is small, i.e. $v_{T_i} \nabla_\perp \partial_z \Phi / \Omega_{ci}$ is negligible, taking v_{T_i}, the ion thermal velocity, as representative for v_z. In the extreme case with $\partial_z \Phi = 0$ a cylindrical BGK-equilibrium can be obtained, although this limit is not so interesting.

Ionospheric data obtained by the VIKING-satellite[9,10] show frequent occurence of small localized density depletions with an associated double-spiked electric field pointing *into* the density minimum. Since the velocity of these structures is often of the same order of magnitude as v_{T_i}, it seems reasonable to interpret than as ion phase space vortices. The **B**-transverse scale is at present rather uncertain, but is evidently an important quantity for the detailed understanding of these structures. Also the origin of them seems to be somewhat uncertain. Ion bursts have been suggested[11] and also more complicated nonlinear electron streaming instabilities[10]. The data seems to require an extensive statistical analysis to resolve these questions. This study is in progress[12].

2. A STATISTICAL ANALYSIS: Even in the case where the dynamics of *one* ion phase space vortex, and its role in small amplitude double layer formation, is understood, there remains a formidable problem in accounting for the full potential drop along e.g. the auroral field lines. With an individual potential drop of typically 1 V, many thousands of such structures are needed to be present simultaneously. In essence this becomes a statistical problem. Central questions will be the amplitude and width distributions of ion holes, their spatial density and velocity distributions. This type of information is available from e.g. numerical simulations.

Here the results of such a numerical simulation[11,13] will be discussed. These studies were carried out in one spatial dimension, which can be assumed to be adequate for the particle dynamics in strongly magnetized plasmas (although, of course, the full spatial dimensionality will generally be retained in Poisson's equation). The code assumes the electrons to be isothermally Boltzmann distributed at all times, i.e. $n_e = n_o \exp(e\Phi/T_e)$. The electron dynamics are then considerably simplified on the expense of a nonlinear Poisson equation. This assumption has the indirect advantage that ion hole formation can unambigously be attributed to the *ion* dynamics since electron streaming instabilities are absent. The problem is solved as a boundary value problem with an ion beam being injected into a stationary background. A typical phase space plot with approximately 60000 simulation particles and spatial potential variations at selected times are shown in Fig. 2. The fluctuating potential is also recorded as a function of time in a number of fixed (typically 30) probing positions.

Visual inspection of data as those shown in Fig. 2b is rather discouraging, i.e. some structures might be discriminated (best in the phase space plot if the paper is tilted and seen from below), but it seems unlikely that any statistically significant information can be obtained. It turns out however, that a careful analysis can provide a quite detailed insight into the distributions of relevant parameters. Various methods were employed. A standard correlation and tripple correlation analysis demonstrated a significantly non-Gaussian component of the fluctuations. Without this information it may not be worthwhile to pursue the analysis, i.e. for a strictly Gaussian random process all the available statistical information is exhausted by the correlation function. A conditional analysis[6,11,13] clearly demonstrated the propagation of coherent structures, but this method provided only information about *average* properties (which should be interpreted with care[1]) and not about the statistical distribution of characteristic parameters.

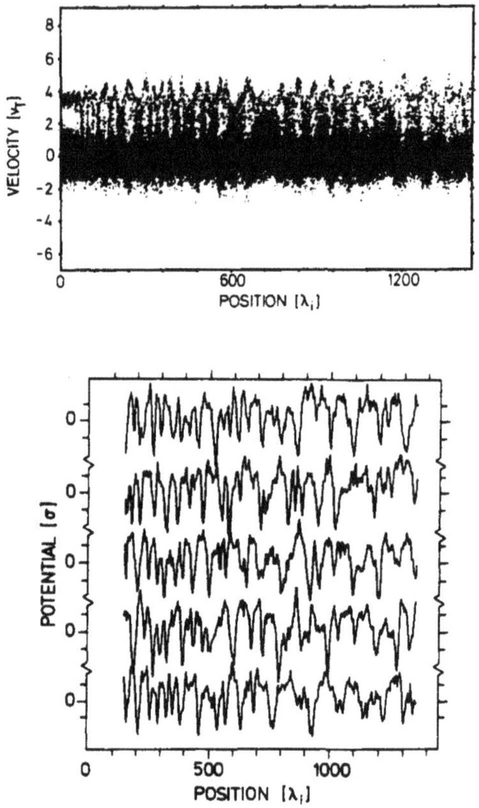

Fig. 2. Ion phase space at $t = 300\omega_{pi}^{-1}$ *and spatial potential variations at selected times with 250* ω_{pi}^{-1} *interval. Parameters are* $V_b = 3.5v_T$ *and* $n_b/n_o = 0.2$ *with* $T_e = 20T_{io}$.

Using a simple method it is however possible to follow individual structures and estimate their basic characteristics. A detailed description of the method is given by Kofoed-Hansen et al.[11]; a short summary will suffice here. Basically the idea relies on the possibility of identifying an *average* shape of a typical structure which can be characterized by a modest number of parameters such as amplitude, width and velocity. This average shape could in our case be determined equally well by a conditional analysis or by fitting correlation and triple correlation functions to simplified model.

A typical, average temporal shape is shown in Fig. 3, as it will be seen by an observer at rest while the structure propagates with an average characteristic velocity.

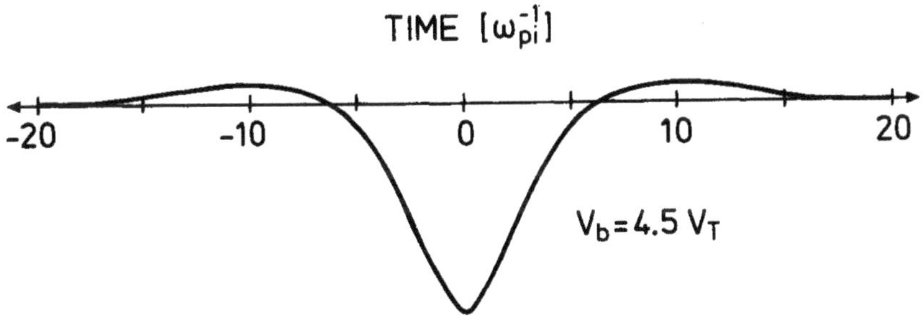

Fig. 3. Temporal variation at a fixed position of a moving average structure.

By a combination of a filtering and "filling-in" procedure[11] it is possible to remove essentially all structures from the signal by a straight forward numerical procedure. In this method an amplitude resolution in units of the rms-level, σ, is chosen, typically as $\delta a = \sigma/20$. Then each of the probing records are searched for the deepest local minimum. (In some cases it was advantageous to remove a high frequency noise first by filtering). In this position a fraction δa of a standard pulse is filled in. The process is then repeated many times until the signal level is everywhere smaller than a prescribed value. The entire procedure is ultimately repeated for varying basic parameters of the elementary pulse. Analyzing the operations used to fill in the potential dips of the ion holes it is then possible to estimate parameter distributions. The method was first calibrated against a Monte-Carlo simulation and applied to various particle simulations, also some where the results could be checked by simple inspection.

In Fig. 4 we have examples for an amplitude probability density and velocity distributions for different parameters v_b which is the velocity of the incoming ion beam in units of the background ion thermal velocity. Also the spatial density of vortices can be obtained as approximately $\mu = 7.8 \cdot 10^{-3}\lambda_{Di}^{-1}$ for $V_b = 4.5\ v_T$ and $\mu = 29.1 \cdot 10^{-3}\lambda_{Di}^{-1}$ for $V_b = 3.5\ v_T$ with $\pm 20\%$ uncertainty on μ. Here λ_{Di} is the ion Debye length for the background ion component. Average amplitudes are correspondingly $a = 0.1\ T_e/e$ and $a = 0.046\ T_e/e$ with an approximately $\pm 20\%$ standard deviation. Pulse widths had only a moderate spread of approximately $\pm$ 20% around that characterizing the standard pulse. With trajectories in an $x-t$ diagram available, it was possible to

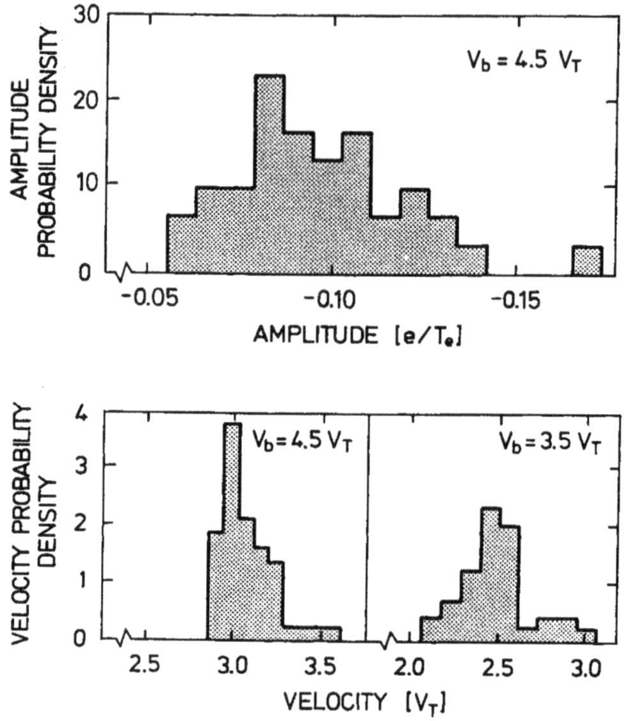

Fig. 4. Amplitude and velocity probability densities of ion phase space vortices.

obtain indications of clustering of groups of phase space vortices, although individual structures belonging to a cluster performed intricate orbits along the average trajectory in such a way that two structures never occupied the same region in phase space (some overlap could occur, though). A "filling-in" procedure as outlined here can of course be applied to *any* signal, with apparent success if the resolution in amplitude is taken sufficiently small. However, only in those cases where identifiable coherent structures are propagating, do we obtain clear trajectories in an $x-t$ diagram as shown in Fig. 5. Here, black dots indicate positions where more than two δa-fractions of an elementary pulse were filled in.

The methods outlined here, as described in more detail elsewhere[11], allow a systematic study of signals where individual coherent structures can be identified only with difficulty by other methods. It is thus plausible that for instance VIKING data could be analyzed with greater depth by these methods.

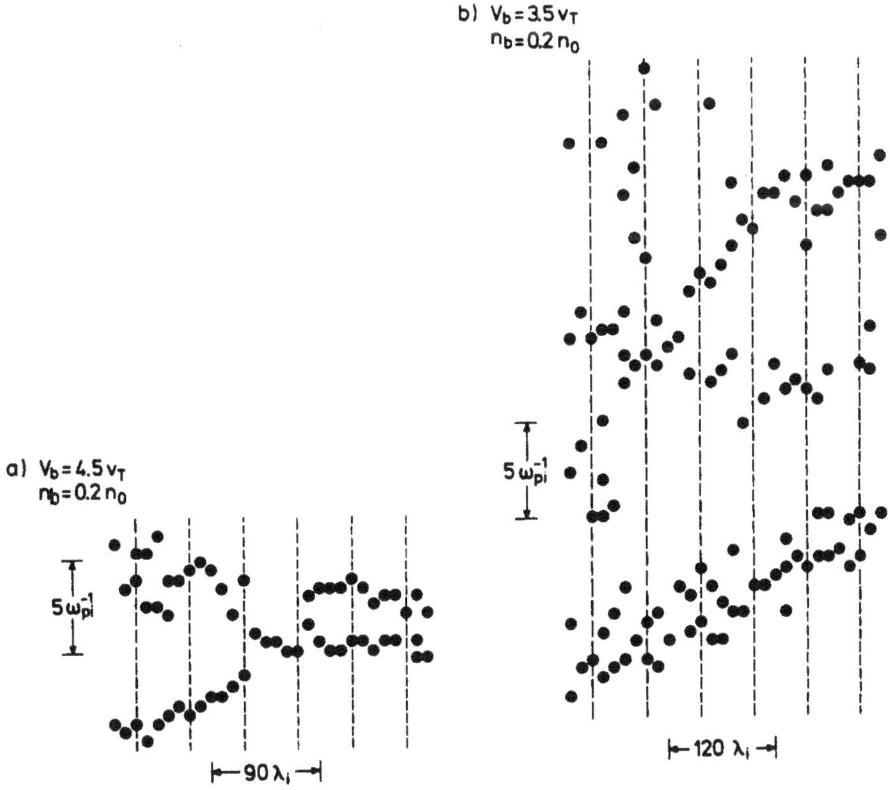

Fig. 5. Samples of vortex trajectories in an x-t diagram with time increasing from top to bottom. The figures are shown in a rest frame moving with an average vortex velocity.

3. CONCLUSION: In this summary we addressed some questions concerning ion phase space vortices. It was argued that the presence of an external magnetic field might be important in determining their lifetime. *Without* such a field, numerical simulations in two and three dimensions demonstrated that an *ensemble* of phase space vortices rapidly disintegrated. *Isolated* vortices are presumably more stable. Incidentally the very first experimental observation of a phase space vortex (electron vortex) was carried out in a strongly magnetized plasma[14]. An estimate of the polarization drift for *intermediate* magnetic fields is presumably an important element for assessing in particular the stability of spatially localized phase space vortices in such conditions.

Also the statistical aspect of the problem was addressed here. In particular it was demonstrated by an example how it is possible to devise methods which can provide the basic statistical information by a straight forward analysis of the signals. An interesting point in such studies is to investigate a possible clustering of phase space vortices, which might occur when their density exceeds a certain level. Such a clustering might give rise to localization of large scale potential drops when electron streams are present.

Several mechanisms for generation of ion phase space vortices can be envisaged. Here we emphasized ion-ion two stream instabilities or localized ion bursts. These mechanisms have proved effective in numerical simulations as well as laboratory experiments. In particular it might be emphasized that also local enhancements of plasma density can in some cases give rise to formation of ion vortices because of the ion acceleration caused by the electric fields at the front and back flanges of the density enhancement. (This mechanism is most effective when these local plasma density gradients are large.) The density depletions observed by the VIKING satellite can apparently be interpreted as ion phase space vortices, but their origin is yet uncertain.

REFERENCES:

1. T. Huld, A.H. Nielsen, H.L. Pécseli and J. Juul Rasmussen, *Phys. Fluids* **B3** (1991) 1609.
2. H. Schamel, *Phys. Reports* **140** (1986) 161.
3. H.L. Pécseli, in *Proceedings from the Second Symposium on Double Layers and Related Topics*, Innsbruck, Austria, 1984, ed. R. Schrittwieser and G. Eder (Studia, Innsbruck, 1984) p. 81.
4. T.H. Dupree, *Phys. Fluids* **26** (1983) 2460.
5. R.H. Berman, D.J. Tetreault, T.H. Dupree and T. Boutros-Ghali, *Phys. Rev. Lett.* **48** (1982) 1249.
6. H. Johnsen, H.L. Pécseli and J. Trulsen, *Phys. Fluids* **30** (1987) 2239.
7. H.L. Pécseli, J. Trulsen and R. Armstrong, *Phys. Scripta* **29** (1984) 241.
8. I.B. Bernstein, J.M. Greene and M.D. Kruskal, *Phys. Rev.* **108** (1957) 50.
9. R. Boström, B. Holback, G. Holmgren and H. Koskinen, *Phys. Scripta* **39** (1989) 782.
10. A. Mälkki, H. Koskinen, R. Boström and B. Holback, *Phys. Scripta* **39** (1989) 787.
11. O. Kofoed-Hansen, H.L. Pécseli and J. Trulsen, *Phys. Scripta* **40** (1989) 280.
12. A.I. Eriksson, P.-O. Dovner, A. Mälkki, R. Boström and G. Holmgren, Poster presented at VIKING workshop, Tammsvik, Sweden, May 17-18, 1992.
13. H.L. Pécseli and J. Trulsen, *Phys. Fluids* **B1** (1989) 1616.
14. K. Saeki, P. Michelsen, H.L. Pécseli, J. Juul Rasmussen, *Phys. Rev. Lett.* **42** (1979) 501. J.P. Lynov, P. Michelsen, H.L. Pécseli, J. Juul Rasmussen, K. Saeki and V.A. Turikov, *Phys. Scripta* **20** (1979) 328.

EXPERIMENTAL RESULTS RELATED WHICH PROCESSES INVOLVED IN THE PHENOMENOLOGY OF DOUBLE LAYERS

M.Sanduloviciu and G.Popa
Department of Plasma Physics of the "Al.I.Cuza" University
6600 Iasi, Romania

INTRODUCTION. Commonly, double sheaths, now known as double layers (DLs), are observed in laboratory as cathode and anode double sheaths /1-3/, as double sheaths developed in a region where the vessel containing a current carrying plasma has a reduction of the cross section area /4-5/ as well as in a constriction region of the plasma column diameter produced by an external imposed magnetic field /6/.

The above specified different regions of a gaseous conductor where a DL can be observed justify the assertion /7-9/ that the premise for the DL generation is the presence of a local acceleration of electrons. Consequently, in a collisional plasma a direct relationship of the local electrical space charge arrangement as that corresponding to a DL and the different regions in which the accelerating electrons achive kinetic energies for which the excitation and ionization cross section increase suddenly can be established /10/.

At the beginning we will point out the necessity to make a distinction between the today most accepted DL model /11/ which considers its generation as a passive answer of a collisionless plasma submitted to external locally imposed causes (external electric field which determine the space charge arrangement) and the collisional DL model /8-10/ which considers its genesis as a result of a self-organization process during of which internal causes determine in a current carrying plasma the resulting space charge arrangement.

In order to resolve the controversy /12/ wheter the inelastic collisions are essential in the DL phenomenology we firstly consider necessary to work out a new definition of the DL. This definition must start, in our opinion, with the experimentally proved fact that the DL represents an autonomous body /12/ with a propre identity, whose "viability" is assured by processes taking place inside it.

It is interesting to specify that very probably the processes taking place during the genesis of DL in plasmas are the same as those which determine, in general, the transition from a disordered state to a spatially and temporally ordered state of matter /13/.

An important advantage of the investigations performed in gaseous conductors (plasma) is the fact that the geometrical dimensions of the DL is relatively large and, more important, the fact that there exist a very large set of investigation possibilities which allow to specify the processes implied in their physical scenario. Considering the role played by the DL in the appearance of instabilities /8/ it is sure that the undestanding of the physical scenario of the DL generation and dynamics is one of the most important question of the plasma physics /8,14/.

In the following we present experimental results obtained in different plasma devices which evidence many essential similarities between DL formed under collisional plasma conditions and those formed under collisionless presumed plasma conditions.

SPATIAL AND TEMPORAL ORDERED STRUCTURES APPEARED IN THE VICINITY OF A POSITIVE ELECTRODE IMMERSED IN A GASEOUS CONDUCTOR. A direct connection between certain well localized space charge structures and the appearance of ordered or disordered temporal structures (oscillations, pseudo-waves, noise, chaos) frequently observed in gaseous conductor was proved only in the last time /3,7,8,13/. Thus, investigation performed on the space charge structures known as anode spots have proved their DL structure and also the fact that the anode spot acts as an energy reservoir able to stimulate different time-dependent processes /3,8,9/. Since the anode spot appears in an usual low pressure d.c. discharge in its final state simultaneously with the ignition of the discharge for investigation of the DL forming process an experimental device /7/ as shown in Fig.1 was used. Between a plane cathode K and a hole anode HA placed in a glass vessel filled with Argon at pressure around 10^{-1} Torr a d.c. discharge was produced. Through the hole H a negative glow plasma containing a great number of fast electrons is formed between K and a plane electrode E. In the plasma laterally diffusing from the negative glow, a circular plate anode A, having a central hole through which a very fine movable heated probe can be introduced, allows to investigate the different phases of the anode spot formed in front of A. In Fig.2 the axial potential profile of the electrical space charge structure formed for different positive bias of A is presented /7/. The result proves that, simultaneously with the increase of the potential applied at A, in front of it, a double layer is formed (Fig.2). This anode DL is stable only as long as its potential drop V_{DL} is lower than the ionization potential V_i of the used gas (in our case Ar V_i = 15,7 V). The visual aspect of a stable anode DL is presented in Fig.3.

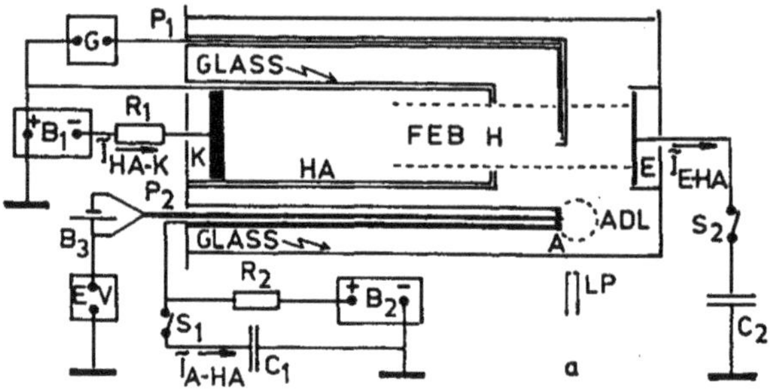

Fig.1. Experimental device

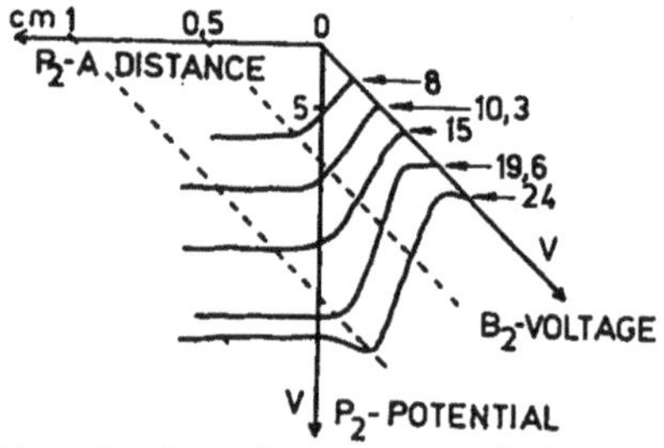

Fig.2. Potential variation in front of A plotted as a function of the P_2-A distance for different B_2-voltage. Gas Ar, $p=1.3\times10^{-1}$ torr, V_{K-HA} = 1949 V, I_{K-HA}=7.8 mA

A very important experimental fact is the observation that the stable anode DL transit into an unstable one in the moment when $V_{DL} = V_i$. In this moment the anode DL is shelling off from the anode surface and moves (expands) as a independent space charge structure toward the surrounding plasma. In Fig.4 one shows the axial space-time dynamics of the unstable DL.

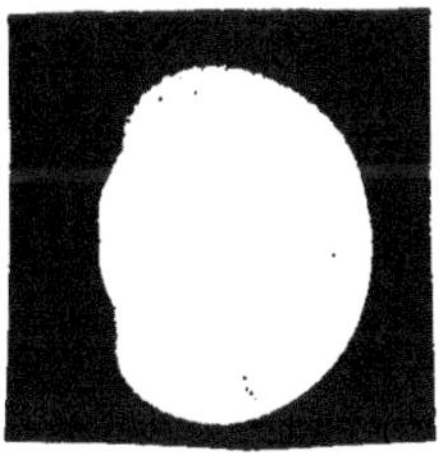

Fig.3.Anode glow aspects

In recent published papers Bin Song at al. /15/ presents results which prove that also for gas pressures much lower than those existent in a usual d.c. gas discharge anode spots with a DL boundary are formed in the vicinity of a positively biased electrode. The transition to the moving phase of the anode DL is considered as a exploding process which takes place when $V_{DL}=V_i$.

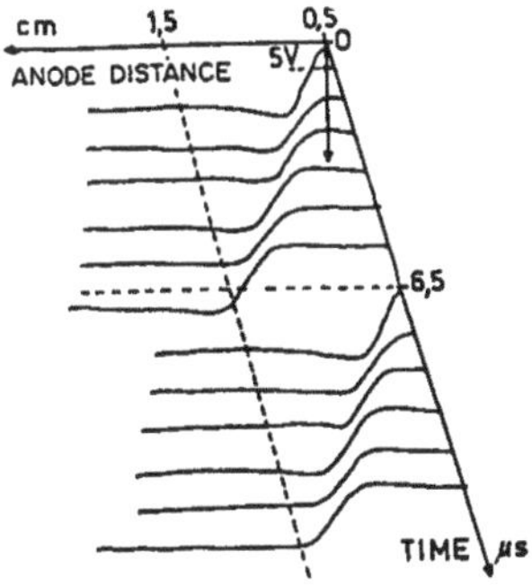

Fig.4. Space time (stroboscopic) diagram to emphase the evolution process of a movable anode double layer. The successive positions of the double layer at different time intervals are obtained using the movable heated probe P_2

A comparison of the above described experimental results proves very strong similarities between the two investigated anode DL obtained, one for low d.c. current intensities and relative high gas pressures and the other for low gas pressures and relatively high current intensities. These results seem to prove that essential for the DL phenomenology is the presence of a sufficiently great number of inelastic collision processes. As is known, this number depends on the neutral concentration (gas pressure) as well as of the number of the accelerated electrons (current intensity) which achive kinetic energies for which the cross sections of neutral excitation and ionization increase suddenly. The strong luminous phenomenon observed as a fireball prove irrefutably the presence of a very high excitation (and desexcitation)·rate. Therefore, in our opinion, a correct theoretical description /15/ of the above phenomena must consider, beside the ionization processes, also the presence of excitation processes.

Additional detailed experimental results which give essential information on the DL dynamics are presented in a paper published in 1987 /16/. The used device is showed in Fig.5.

Fig.5. Experiment device

Visual aspect of the stable (V_{DL} < V_i) and unstable (V_{DL} = V_i) DL. Ar, p = 0.04 Torr

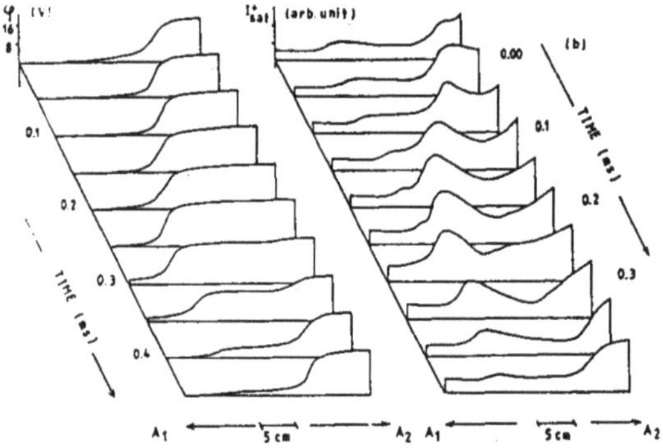

Space time evolution of the local potential (V) and of the ion saturation current I_{sat}

In this experiment the DL is formed in the contact region between two independent working d.c. gas discharges. The discharge was produced between the plate cathodes K_1 respectively K_2 and the ring anodes A_1 respectively A_2. The K_1A_1 respectively K_2A_2 distances were chosen so that the ends of the two negative glows are placed in the A_1 respectively A_2 regions. The experiment was performed in Ar and He. The DL appears in the region between A_1 and A_2 (Fig.5) when one of the anode is positively biased in respect to the other. One observe that the DL behaves stable only as long as the potential drop developed over them does not achieve the ionization potential of the used gases (15,7 V for Ar and 24,6 V for He). In the moment when V_{DL} = V_i the DL become unstable. During its instability the DL moves from its initial position toward the anode with the lower potential. After covering a certain distance the DL disrupts. The spatio-temporal evolution of an unstable DL is presented in Fig.5. From this figure it results that during the DL motion in the region where the DL was initially formed, a new DL appear. The motion of this new DL starts again after a very short time interval in respect to the moment when the previously formed DL is disrupted. This result proves the presence of an internal feedback mechanism which triggers the dynamics of the successively formed DLs /16/.

As is known wandering DLs was observed also in Q-machines where usually the plasma is considered collisionless /6/. A correct answher to the problem wheter between the last mentioned phenomena observed in collisional plasma and those above described qualitative differences exist is possible, only after investigations performed in the next vicinity of the hot plate. It is, for example, possible that here the gas pressure is comparable with to existent in an usual gas discharge so that the unstable DL is formed in the vicinity of the hot plate. In this case a similar dynamic as that described in /16/, limited only in the proximity of the hote plate, determine in the plasma column phenomena known as relaxation instabilities /17/. Additionaly, the self-organization mechanism considered at the origin of the DL generation /10,18/ in an usual gas discharge implied also a self-confi-nation process which locally produces a matter concentration much different than that existent in the region where usually the gas pressure is meassured /18/. Consequently, it is very probable that also in collisionless plasma device, in the region where the DL is formed and maintened, the local neutral concentration corresponds to that existent in a collisional plasma /18/.

ON THE STABILITY OF SPHERICAL DOUBE LAYERS. Under certain conditions in low potential arcs, as well as in thermionic converters, stable globular space charge structures without visible connection to the anode surface appears /19,20/. Related to different anode-voltage the visual aspect of the discharge is presented in Fig.6 /21/.

Fig.6

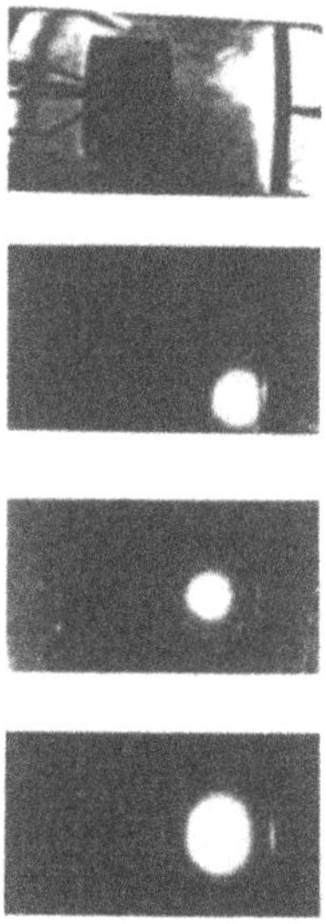

Anode sheath

Formation of the globular anode spot

Stable globular DL (Electrode-free)

Unstable globular DL

A phenomenological explanation of the above mentioned phenomenon /20/ starts with the experimentally observed fact that the electrode-free globular space charge structure develops from a planar DL initially formed near the anode surface (see Fig.6) By increasing the anode potential, a luminous protuberance in a certain well limitated region of the anode surface appears. This protuberance has a positive "nucleus" which acts as a virtual anode /20/. Taking into account the strong electrostatic forces which act between the two adjacent layers with opposite electric sign (which form the boundary of the DL) this spontaneously transits toward a spherical shape characterized by a minimal value of its potential energy. The phenomenon is similar to the formation of liquid drops where the acting force is the surface tension. The resulting globular space charge structure appears as a very luminous globus, usually known as ball of fire /19/. The nucleus of the ball of fire is a plasma enriched with positive ions. This nucleus is delimited from the surrounding electron rich plasma by a DL. Considering the described physical scenario we believe that only ionizations are not able to assure the stability of the spherical DL /15/.

The most interesting observation related to the appearance of the ball of fire is the fact that the main potential inside it is, in certain condition, higher than the potential of the anode. This intriguing fact was explained /20/ considering the electrical space charge separation conditioned by diffusion (difference between of the mobilities of the electrons and the positive ions). Such a space charge separation appears in the regions of the anode surface where the current is concentrated so that a strongly plasma heating process takes place.

As it was shown /20/ the ball of fire remains in a stable state only as long as the potential drop developed over the spherical DL is lower than the ionization potential. As well as the nearly unidimensional DL already described, the ball of fire transits into an unstable state when $V_{DL} = V_i$. During the unstable state the DL at the boundary of the ball of fire disrupts after the expandion and it is periodically reformed.

Similar globular space charges known as plasmoides appear also in HF gas discharges, A detailed description of them is given in /22/.

TEMPORAL STRUCTURES RELATED TO UNSTABLE DOUBLE LAYERS. The transition to an unstable state of a DL determins, in the electrical circuit at which the gaseous conductor is connected, variations of the current intensities. Depending on the circuit elements, this variation can be ordered (oscillations, pseudo-waves) or disordered (noise, chaos). The first mentioned case appears when an external or internal feedback mechanism assures a periodical triggering of the DL disruptions /7,16/. The ability of a DL to act as an energy reservoir for time depending phenomena is related to an energy accumulation process during of which kinetic energy of accelerated electrons is transformed, through the agency of quantum processes, in electrical field energy stored in the resulting space charge structure.

One of the most investigated temporal structure observed in usual low pressure gas discharges and up today not well understood /23/ is the as known anode instability phenomenon. A correct explanation of this phenomenon is possible, in our opinion, only by a circuit theory /9,24/ in which the DL acts as a negative resistance. The essential experimental fact not considered up today is the presence of an important capacitive alternative current which flows from the tube wall to the surrounding. This current is equal to the anode current and much greater than the current flowing from the DC power supply toward the cathode /24/. This current appears since the secondary electron emission with a coefficient greather than 1 produced by fast electrons, accelerated in the cathode fall, attribute to the inner surface of the dielectric (glass) tube wall the property of one plate of a capacitor whose other plate is the surrounding /17/. Together with an inductor placed between the inner surface of the tube and the anode, this capacitor form a parallel oscillatory circuit whose reactive elements determin mainly the frequency of the described oscillations /7/. In this oscillatory circuit the oscillations are stimmulate by the unstable anode DL whose dynamics is triggered by the secondary electrons originated at the inner surface of the wall and collected by the anode. The periodical variation of potential of the inner surface of the tube determined by the oscillations assures the changing of the secondary electron flux collected by the anode. Consequently, a periodical reaching of the ionization potential assures the DL energy releasing process (DL disruption). Besides the above described periodical phenomena, which represent genuine oscillation taking place in an oscillator circuit connected to a negative resistance (the DL), in d.c. gas discharge with a developed positive column appears also the well known moving striations which determin a periodical variation of the d.c. current. As it was shown /9 / such phenomena really represent moving DLs whose periodical formation and disruption is assured by an internal triggering mechanism 16/.

Using a capacitive probe it was possible to evidence beside the known as ionization wave phenomenon also the internal feedback mechanism (echo phenomenon /25/) which assures the periodical triggering of the disruption (and reformation) of the successively formed moving DLs incorectly considered as a wave phenomenon.

When the positive ions production rate at the high potential side of the DL exceeds a certain value, the DL disruption process is not started and, consequently, the phenomenon known as soliton propagate through the plasma column /9/.

ANOMALOUS RESISTANCE, BISTABILITY AND HISTERESIS PHENOMENA RELATED TO THE DOUBLE LAYERS DYNAMICS. The processes implied in the DL generation mechanism and their dynamics also determin the profile of the I-V characteristics. Thus, for a d.c. glow discharge, this characteristic contains different branches /10/. During the DL forming process the I-V characteristic shows a branch corresponding to a negative resistance related to the generation of the double space charge (Fig.7). The charges separation taking place during the forming phase of the DL determins a decrease of the d.c. current intensity althugh the external imposed potential difference increases. The transition toward a nearly spherical unstable DL determines an increase of the main current intensity. This current increase is related to the production of new free charges during the periodical disruptions and reformation of the DL. The histeresis phenomena also evidenced in fig.8 prove the tendency of the pulsating nearly globular anode spot to maintain its individuality when the potential is decreased. The histeresis phenomenon is related to the energy accumulated during the DL generation process /26/.

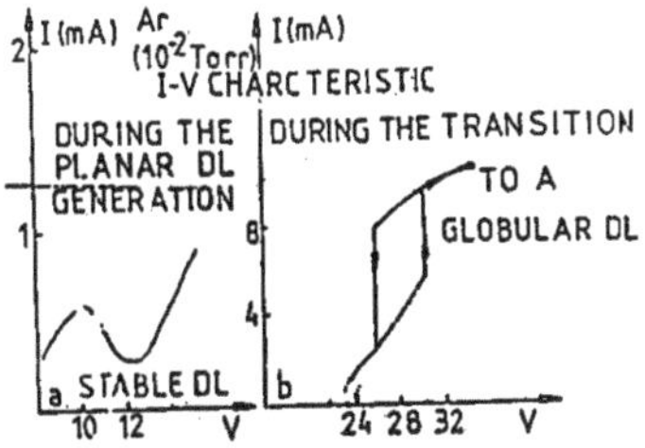

Fig.7

Similar profile of the I-V characteristics was obtained for very low gas pressures also in other devices/15/ and also in Q machines /28/. This fact seems to prove that the physical scenario involved in the DL generation and dynamics is the same in all plasma devices.

QUANTUM PROCESSES INVOLVED IN THE DOUBLE LAYER PHENOMENOLOGY. Spectral investigations performed in the region where a stable DL is formed confirm that inelastic collisions as excitations and ionizations produced by accelerated electrons are at the origin of the opposite electrical cha-rges separation as those existent in a DL. In Fig.8 one present such results /28/. It is evident that the profile of the DL potential drop is related to the two adjacent regions where the cross sections corresponding to different neutral excitations energy levels and to the ionizations energy level begin to increase suddenly /10/. This fact confirmed by the position of the spectral lines in respect to the electron and positive ion concentration show that in current carrying gaseous conductors at the origin of the DL potential profile there is an electrical charges separation sustained by internal causes (quantum processes). The global effects (opposite space charges separation) related to these processes assure themselves the potential drop which produces the local electron acceleration necessary, at its part, to make the DL maintenance possible /10/. An argument that the described selforganization process is at the origin of the DL generation, in all plasma devices are the well localized visible DLs images, also present in devices where the plasma is usually considered collisionless /20/.

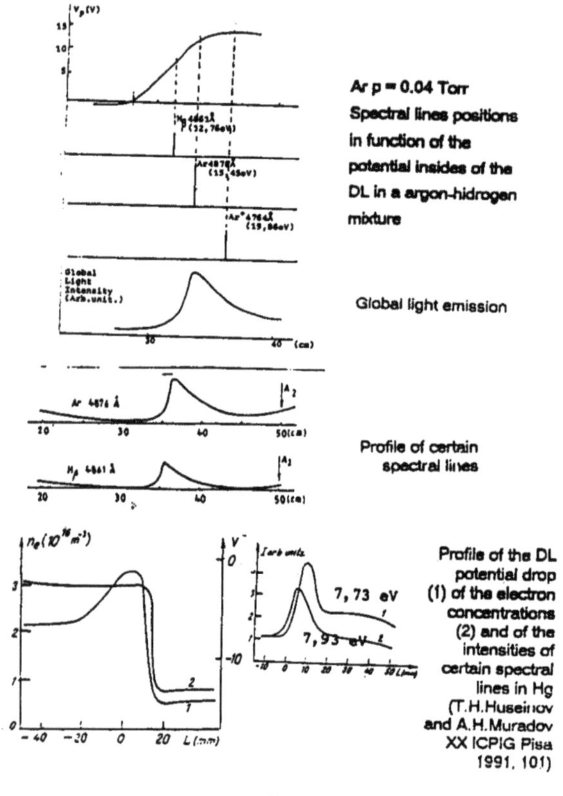

Fig.8

REFERENCES

1. I.Langmuir 33 (1929) Phys. Rev. 955
2. F.W.Crawford and A.B.Cannara 36 (1965) J.Appl.Phys. 3135
3. M. Sanduloviciu, Symp.on Plasma Double Layers, Riso, Denmark (1982) p.231
4. F.W.Crawford and I.L.Freeston (1963) ICPIG, Serma, p.461
5. I.E.Allen 27 (1985) Plasma Phys. Contr. Fus., 1343
6. N.Sato (1984) ICPP, Lausanne. Proc.Inv. Papers p.555
7. M. Sanduloviciu, E. Lozneanu, 28 (1986) Plasma Phys. Contr. Fus., 585
8. M. Sanduloviciu, ICPP,(1987) Kiev, Proc. Invited Papers, 293
9. M. Sanduloviciu, G.Popa, (1987) Third Int.Symp.on Plasma Double Layers and Related Topics, Bucharest 34 (1988) Suppl. An. St. Univ. Iasi, 83
10. M.Sanduloviciu (1992) Romanian Journ.of Phys. in print
11. H.Schamel 3 (1986) Phys. Rep. Rev.Phys. Lett. 161
12. R.Schrittwieser (1992) Symp.on Atomic and Surface Phys. Italy
13. M.Sanduloviciu (1991) Int. Workshop on Plasma Phys. Pichl, Austria
14. H.Alfven (1986) IEEE Trans. on Plasma Sci. (dec.1986)
15. Bin Song, N.D'Angelo and R.Merlino (1991) J. Phys. D. Appl. Phys., 24 1789 and 25 938
16. D. Alexandroaei, M. Sanduloviciu,A122 (1987) Phys. Lett.173
17. S.Iizuka, P.Michelsen, J.J.Rasmusen, R.Schrittwieser, R.Hatakeyama, K.Saeky and N.Sato, 48 (1982) Phys. Rev. Lett.145
18. M.Sanduloviciu (1992) ICPP, Innsbruck 1771
19. G.Musa 36 (1988) An. St. Univ. Iasi, 49
20. M.Sanduloviciu (1991) XX ICPIG, Pisa, 265
21. S.Klagge (1971) Doctor Thesis Univ. Greifswald, Germany
22. M.Sanduloviciu 29 (1987) Plasma Phys. Contr.Fus.1687
23. K.G.Emeleus 52 (1982) Int. J.Electr.407
24. M.Sandulovciu and E.Lozneanu,(1989) XIX ICPIG, Belgrad, p.794
25. G.Popa, K.Ohe, N.Dumitrascu 22 (1989) J.Phys.D.Appl.Phys. 1327
26. M.Toma and M.Sanduloviciu (1991) XX ICPIG, Pisa. 271
27. G. Popa, R. Schrittwieser, M.Sanduloviciu (1991) XX ICPIG, Pisa, Italy,
28. D.Alexandroaei and M.Sanduloviciu (1985) XVII ICPIG, Budapest, p.52
29. S.I.Zweben and S.S.Medley B1 (1989) Phys.Fluids 2058

THE DYNAMICS AND STABILITY OF WEAK DOUBLE LAYERS

M.A. Raadu, Department of Plasma Physics, Alfvén Laboratory, Royal Institute of Technology, S-100 44 Stockholm, Sweden

ABSTRACT: Steady state double layers (DL's) have been extensively studied as time independent solutions of the coupled Vlasov and Poisson equations, and their structure is well understood. However the accessibility of these double layers from general initial plasma states and their stability to perturbations raise many questions. The requirement of exact time independence is to some extent an overrestriction on the models and existence criteria. In general the most straightforward way to treat time dependent DL's is by computer simulation; but such studies are restricted to one case at a time, making it difficult to obtain general results unless they can be complemented with analytical work. Weak DL's, with amplitudes comparable to or less than the plasma thermal energy, have been classified in terms of steady state solutions and in the case of time dependence are amenable to both analytical and computational treatment. The dynamics of a whole class of weak ion-acoustic DL's has been shown to be described by a modified Korteweg-de Vries equation. Inverse scattering theory has led to the discovery of the necessary and sufficient conditions for the formation of such weak DL's from arbitrary initial conditions. The description may be generalised to two dimensions in the case of magnetised (modified Zakharov-Kuznetsov equation) and unmagnetised (modified Kadomtsev-Petviashvili equation) plasmas. Perturbation analysis shows stability in the magnetised and instability in the unmagnetised case. The effects of Landau damping may be added to the mKdV description, but the full consequences of this remain to be analysed. More detailed comparisons with experiments are needed.

1. INTRODUCTION: Double layers (DL's) may occur in current carrying plasmas where they can withstand potential differences with local electric fields in excess of those needed to maintain a uniform steady state. Thus, if a current is driven by an applied potential difference to overcome ohmic dissipation, the appearance of DL's within the plasma will increase the required applied potential so that viewed as a whole the plasma will appear to have an anomalously large resistivity. The excess dissipated energy goes first of all to local particle acceleration within the DL's. The emerging particle beams may be dissipated by interaction with the ambient plasma over distances depending on the plasma density and temperature as well as the beam properties, and in this case the net effect, enhanced plasma heating, may be

indistinguishable from that of anomalous resistivity. Such a situation has been found in particle simulations of long systems, i.e. for a simulated plasma of several hundred Debye lengths, where several ion-acoustic DL's form separated by distances sufficient to dissipate the accelerated particle beams[1]. Experiments have also been reported corresponding to such a scenario[2]. These results may also be relevant to space observations[3], although in this case it is believed that some of the particles may pass through many weak ion-acoustic DL's to reach high energies, behaving as a run-away population. An alternative state found in many experiments[4] is one in which a single strong DL is formed accelerating ions and electrons to energies much larger than the thermal energy of the surrounding plasma. The electron beam usually interacts strongly with the plasma producing high frequency fluctuations on the high potential side of the DL.

From the point of view of energy balance a DL acts as an electrical load dissipating energy at a rate $(I_e + I_i)\,\phi_{DL}$ where I_e and I_i are the electron and ion currents and ϕ_{DL} is the net potential difference sustained over the DL. The force exerted by the electric field on particles within the DL is balanced by their inertia, in contrast to an ohmic situation where particle collisions provide the balance. It is however clear from this that the potential over the DL must be applied by external means which are also the ultimate source of the energies gained by the particles, in close analogy to the case of a potential over a region with ohmic or anomalous resistivity.

2. EQUILIBRIUM: Steady state double layers have been extensively studied as time independent solutions of the coupled Vlasov and Poisson equations, and their structure is well understood. (See for example a recent presentation[5].) In one dimension[6] such solutions are particular examples of BGK states[7], and are a straightforward generalisation of laminar electrostatic shocks[8]. A first integral of the equations gives the condition of stress balance point by point through the DL (the only term arising from the Maxwell stress tensor is that due to the electric field E):

$$\int_{-\infty}^{+\infty} m_i v^2 f_i(v,\phi)\,dv + \int_{-\infty}^{+\infty} m_e v^2 f_e(v,\phi)\,dv - \frac{1}{2}\varepsilon_0 E^2 = const \tag{1}$$

The first two terms are the total (i.e. the sum of random and dynamic) particle pressures for ions and electrons with distribution functions $f_{i,e}(v,\phi)$. To satisfy the time independent Vlasov equations it is sufficient that these are chosen to be functions of an integral of motion, in this case the total energy $1/2\, m_{i,e}v^2 \pm e\phi$. The integrals over velocity v lead to a function of the potential ϕ only which may be denoted $-S(\phi)$, where $S(\phi)$ is referred to as the Sagdeev[9] or classical potential. Since the electric field $E = -d\phi/dx$, equation (1) may be integrated directly to give $x(\phi)$ implicitly giving the potential as a function of position x. At the edges of the DL, $E = 0$ and equation (1) implies stress balance between the particles on either side of the DL, $S(0) = S(\phi_{DL})$, a generalisation of the Langmuir criterion[10] which must be satisfied for DL solutions to be possible. In the case of strong DL's this gives directly the ratio between the ion and electron currents in the rest frame of the DL. Further conditions follow from requiring charge neutrality at the edges (implying $S'(0) = 0$ and $S'(\phi_{DL}) = 0$) and the correct sign for the total net charge just inside i.e. the Bohm criteria ($-S''(0)>0$ and $-S''(\phi_{DL})>0$ so that $E^2>0$).

DL's may be crudely classified as "strong", "intermediate" or "weak". A DL is "strong" if the energies of the accelerated ions and electrons are very large compared to the energies of incoming and reflected particles. The internal structure is essentially determined by the accelerated particles and is that of a Langmuir[10] DL. In the case of "weak" DL's the potential energies, $O[e\ \phi_{DL}]$, are comparable to or less than particle kinetic energies. The parameter range between "strong" and "weak" corresponds to the remaining class of "intermediate" DL's. The "analytical"[11] DL's, based on truncated and shifted Maxwellian particle distributions satisfying continuity conditions, fall into all three classes. Two types[12,13,14] of DL's related respectively to electron and to ion holes have a "weak" DL limit, and from the terminology of linear electrostatic modes they are called slow electron acoustic DL's (SEADL's) and slow ion-acoustic DL's (SIADL's) respectively. Non-monotonic DL's in the form of asymmetric electron or ion holes are also possible[15,16,17]. The asymmetric ion holes have a potential minimum on the low potential side and are related to ion-acoustic DL's. The ion-acoustic DL's have been studied in plasma simulations[18] and analytically[19].

"Cold ion" DL's, in which all the ions form a simple cold fluid or beam distribution and the electrons have a non-Maxwellian distribution, are of particular interest since in the weak DL limit they may be described using the mKdV$^-$ equation[20] (see below). For such distributions both large amplitude and weak solitary waves may be found, and depending on the relative proportions of the hot and cold electron components even negative solutions, density dips, may be found[21,22]. At the transition between negative and positive solitary wave solutions the Sagdeev potential takes the correct form for a DL solution[22]. The properties of both the large[23] and small[24] amplitude solutions for these DL's have been discussed.

Steady state weak DL models are derived by expanding the Sagdeev potential as a power series. The particle distribution functions may have discontinuities of varying degree at the separatrix between particles originating from distinct sources, and it is found that the power series may then require fractional powers of the potential[25]. This accounts for the forms of the Sagdeev potentials found for weak SEADL's and SIADL's[14], which are dependent on discontinuities in the derivatives of the distribution functions at the separatrices.

3. EVOLUTION OF DOUBLE LAYERS: The derivation of double layer solutions of the steady state Vlasov and Poisson equations, still leaves as an open question the accessibility of such states from given initial conditions in a plasma. The Bohm criteria may in reality be overrestrictions on the DL solutions, since experimentally there are nearly always strong time dependent fluctuations at the edges of a DL where they should apply. Dynamical presheaths are expected allowing a transition between the DL and the distant plasma[26] analogously to the case of plasma wall sheaths[27]. This point of view is supported by computer simulation results, which allow the time dependent nonlinear evolution of DL's to be followed. Weak DL's can also have the character of solitons and it has been possible to find various evolution equations of mKdV type.

Computer simulation provides a direct means of studying time dependent DL's. The growth of ion-acoustic instabilities has been shown to lead to the formation of weak ion-acoustic DL's. Their formation is associated with a growing potential minimum which reflects electrons, leading to an accompanying evacuation process[28]. In this case the development can also be described by a modified form of the Korteweg-de Vries equation including the effects of electron reflection at the potential minimum[28]. There is an overpopulation of those electrons reflected by the potential minimum that are moving in the direction of the mean electron drift velocity. This leads to the growth of the potential, and comparisons between the analytical theory and the numerical simulation show good agreement. Simulations show that although large amplitude electron holes may form on the high potential side of the ion-acoustic DL, these propagate away from it leaving the DL relatively undisturbed[5]. The combination of an analytical treatment with simulations enables general conclusions to be argued. Such a combined approach has been applied to the dynamical processes in plasma two-stream and diode systems[29], where holes, sheaths and DL's appear as non-dissipative structures. Using numerical studies alone it is difficult to obtain general results for a whole parameter range, and the presence of purely numerical effects is often an unresolved question. For this reason the results obtained by the application of analytical methods for weakly nonlinear DL's are of special significance.

4. WEAK DOUBLE LAYER DYNAMICS: The dynamics of a whole class of weak ion-acoustic double layers has been shown to be described by a modified Korteweg-de Vries equation (mKdV$^-$) with generalisations to three dimensions. For small but finite perturbations weakly nonlinear ion-acoustic waves have been described by the ordinary Korteweg-de Vries (KdV) equation[30] which in standard form may be written:

$$\frac{\partial \phi}{\partial t} \pm 6\,\phi \frac{\partial \phi}{\partial x} + \frac{\partial^3 \phi}{\partial x^3} = 0 \qquad (2)$$

The two forms of the equation are essentially the same as may be seen by the substitution $\phi \rightarrow -\phi$. The modified Korteweg-de Vries (mKdV) equation may be applied to a variety of plasma physical problems. It can take two distinct forms:

$$\frac{\partial \phi}{\partial t} \pm 6\,\phi^2 \frac{\partial \phi}{\partial x} + \frac{\partial^3 \phi}{\partial x^3} = 0 \qquad (3)$$

denoted by mKdV$^\pm$ according to the sign of the nonlinear term. (These are unchanged by the transformation $\phi \rightarrow -\phi$.) The mKdV$^+$ equation supports both positive and negative solitons and applies, for example, to a plasma with a negative ion component[31]. The mKdV$^-$ equation may be applied to a plasma with cold ions and a non-Maxwellian electron distribution[20,32]. (The mKdV$^-$ equation has even been found in a case with finite ion temperature[33].) For the mKdV$^-$ equation there are both soliton and DL (shocklike) solutions. The DL solutions are true solitons[34,35] and are characteristic for the mKdV$^-$ equation (the mKdV$^+$ equation has no such solutions). They appear as limiting forms of the general soliton solutions and take the form:

$$\phi = \pm\phi_0\, tanh\{\phi_0\, x + 2\phi_0^3\, t + const\,\} \qquad (4)$$

For a double Maxwellian electron distribution the mKdV$^-$ equation with DL solitons applies for a specific range of the parameters $T_{eh,c}$ and $n_{eh,c}$, the temperatures and densities of the <u>h</u>ot and <u>c</u>old electron populations[36,37]. Such a situation is possible when distinct plasmas, e.g. ionospheric and magnetospheric plasma in the Earth's magnetosphere, interpenetrate. Another possibility is for experiments where primary electrons ionise a neutral gas to produce a population of lower energy electrons.

To determine the conditions under which a DL will form for given initial conditions an inverse scattering method (ISM) has been developed[36,37]. After a transformation this incorporates the ISM for the KdV equation[38] in a generalised form[39] allowing different initial asymptotic values for $\phi(x,0)$ as $x \rightarrow \pm\infty$. A DL will appear if and only if the initial potential $\phi(x,0)$ changes sign between $\pm\infty$. The potential $\phi(x,0)$ is related to the actual physical potential by the linear transformation introduced to reduce the mKdV$^-$ equation to standard form[36,37]. Expressed directly: the DL potential is determined by the plasma parameters and a DL will appear if and only if at least half of this potential is applied over the plasma.

An important generalisation of the mKdV analysis[25] includes the effects at the separatrix in phase space between electrons arriving from the physically distinct sources at $\pm\infty$. Normally a discontinuity may be expected at the separatrix, and this modifies the perturbation analysis. An evolution equation of modified KdV type is then found:

$$2\frac{\partial\phi}{\partial t} - \frac{\partial}{\partial x}\left(\frac{3}{5}\phi^{1/2} + \phi^{3/2}\right) - \frac{\partial^3\phi}{\partial x^3} = 0 \qquad (5)$$

Essentially the same result has recently been rederived[40], but with a higher order discontinuity so that the first nonlinear term does not appear. (In the same paper[40] the mKdV$^-$ equation[36,37] is also rederived in the case of a double Maxwellian distribution, now including two ion species and the effects of a uniform magnetic field.) Equation (5) has a steady state DL solution[25], but it has not been shown whether or not this behaves as a soliton.

The mKdV analysis of DL's may also be generalised by extending the description to three dimensions. More of the physics may be added by including the effects of Landau damping.

5. STABILITY AND EXTENSIONS TO THREE DIMENSIONS: The local stability of a one-dimensional DL model has been treated[41] by applying the Penrose criterion[42]. A similar analysis has also been made asymptotically for laminar shock waves[43]. At the edges of the DL the conditions for stability are in fact identical to the Bohm criteria[5], given that there is a minimum at zero velocity in the combined distribution function at the edges of a DL. (However, unstable distributions of injected particles are often required by the Bohm criteria, associated with further minima in the combined distribution at non-zero velocities.) In contrast at the centre of a DL the Penrose criterion implies local instability[5]. This contradiction shows the necessity for a fully non-local stability analysis. A stability analysis for weak slow ion-acoustic and slow electron acoustic DL's (SIADL's and SEADL's) has established linear stability with respect to one-dimensional perturbations and instability for

two-dimensional electrostatic perturbations in both cases[14]. For a full stability analysis of weak ion-acoustic DL's an extension of the mKdV$^-$ equation to three dimensions is need. The form of the generalised equation depends on whether or not a magnetic field is present. In the case of an unmagnetised plasma the mKdV equation has been generalised to three dimensions where it takes the form of a Kadomtsev-Petviashvili[44] (KP) equation with a modified nonlinear term corresponding to the mKdV equation:

$$\frac{\partial}{\partial x}\{\frac{\partial \phi}{\partial t} \pm 6\,\phi^2 \frac{\partial \phi}{\partial x} + \frac{\partial^3 \phi}{\partial x^3}\} = \kappa\,\{\frac{\partial^2 \phi}{\partial y^2} + \frac{\partial^2 \phi}{\partial z^2}\} \qquad (6)$$

where $\kappa = \pm 1$ according to whether the dispersion is positive or negative. An mKP$^+$ equation describes ion-acoustic waves (for which $\kappa = -1$) in a plasma with negative ions[45]. For a double Maxwellian electron distribution the mKdV$^-$ equation generalises to three dimensions as the mKP$^-$ equation with negative dispersion[46,47]. In the case of a magnetised plasma the mKdV$^-$ equation generalises to the Zakharov-Kuznetsov[48] (ZK) equation in a modified form[49]:

$$\frac{\partial \phi}{\partial t} - 6\,\phi^2 \frac{\partial \phi}{\partial x} + \frac{\partial^3 \phi}{\partial x^3} + \frac{\partial}{\partial x}\{\frac{\partial^2 \phi}{\partial y^2} + \frac{\partial^2 \phi}{\partial z^2}\} = 0 \qquad (7)$$

corresponding in the linear approximation to ion sound waves, as seen in a frame of reference moving with the effective ion-acoustic velocity parallel to the magnetic field. (These have a stronger negative dispersion perpendicular to the magnetic field, so that the length scale that is chosen in that direction must be shorter to give the form of the dimensionless equation shown here.) The one-dimensional DL solutions of the mKdV$^-$ equation are also solutions of the mKP$^-$ and mZK$^-$ equations, and using these it is possible to follow the response of the DL to three-dimensional perturbations. A linearised perturbation analysis has been made to determine the stability properties of the ion-acoustic DL in magnetised and unmagnetised plasmas[46,47]. In the magnetised case stability to transverse perturbations was established both by solving the spectral problem and by using an "energy principle". In the unmagnetised case, governed by the mKP$^-$ equation, instability was found for long wavelength transverse perturbations.

6. LANDAU DAMPING EFFECTS: As referred to above the growth of ion-acoustic DL's has been described using a modified form of the Korteweg-de Vries equation including kinetic effects[28]. Similar equations have been derived for other situations[50,51]. Landau

damping effects can be included in the mKdV$^-$ equation for the potential by adding a term proportional to the Hilbert transform of the electric field[32]:

$$\frac{\partial\phi}{\partial t} - 6\,\phi^2\frac{\partial\phi}{\partial x} + \frac{\partial^3\phi}{\partial x^3} - \beta H\,[\frac{\partial\phi}{\partial x}] = 0 \tag{8}$$

A linearised analysis is useful to see how the extra term arises in the mKdV$^-$ and related equations. A full weakly nonlinear perturbation analysis would be necessary to properly establish the validity of the resulting equations with the nonlinear term restored. However in what follows only the linear analysis will be presented. Starting from a standard linearised kinetic analysis for ion-acoustic waves, including two Maxwellian electron distributions, one "hot" and one "cold", with temperatures, densities etc. denoted by $T_{eh,c}$, $n_{eh,c}$ etc. and possibly several cold ion components, an approximate dispersion relation for long wavelengths (i.e. for inverse wave number $k^{-1} << \lambda_{D,eff}$, an effective Debye length) may be found:

$$\frac{\omega}{\omega_{pi,t}} \approx \lambda_{D,eff}k\,\{1 - \frac{1}{2}(\lambda_{D,eff}k)^2\} + \frac{i\pi}{2}\lambda^3_{D,eff}k\,|k|\sum_{j=h,c}\omega^2_{pj}\frac{df_{e,j}(\omega/k)}{dv} \tag{9}$$

Here the positively propagating branch has been chosen, the electron distribution functions $f_{e,j}(v)$ are normalised to unit density, and the following definitions are assumed:

$$\lambda^{-2}_{D,eff} \equiv \sum_{j=h,c}\frac{\omega^2_{pe,j}}{c^2_{e,j}} = \sum_{j=h,c}\frac{n_{e,j}e^2}{\varepsilon_0 KT_{e,j}}\ ; \qquad \omega^2_{pi,t} \equiv \sum_{j=1,2...}\omega^2_{pi,j} \tag{10}$$

The total ion plasma frequency $\omega_{pi,t}$ is defined through a summation over ion species. Waves propagate with an effective ion-acoustic velocity $c_{s,eff} = \lambda_{D,eff}\,\omega_{pi,t}$ and are modified by a negative dispersion and Landau damping. The dispersion relation, equation (9), is for waves proportional to $exp(i[kx-\omega t])$, so that multiplying by the appropriate Fourier transforms in space and time of, for example, the potential ϕ and taking the Fourier inversions the corresponding differential equation in the rest frame of the ions is:

$$\frac{\partial\phi}{\partial t} + \frac{\partial}{\partial x}\{1 + \frac{1}{2}\frac{\partial^2}{\partial x^2}\}\,\phi - \frac{\pi}{2}\,\alpha\,H\,[\frac{\partial\phi}{\partial x}] = 0 \tag{11}$$

where time and space are normalised by $\omega_{pi,t}^{-1}$ and $\lambda_{D,eff}$ (so that $c_{s,eff} \Rightarrow 1$). $H[]$ is the

Hilbert transform and α the Landau damping parameter defined by:

$$H[E](x) \equiv -\frac{1}{\pi} P \int_{-\infty}^{+\infty} \frac{E(s)}{s-x} ds \ ; \quad \alpha \equiv \sum_{j=h,c} (\lambda_{D,eff}\omega_{pe,j})^2 \frac{df_{e,j}(c_{s,eff})}{dv} \tag{12}$$

Taking equation (11) in a frame moving with the ion-acoustic velocity $c_{s,eff}$ and adding the appropriate nonlinear term gives equation (8) above. The dispersion relation, equation (9), may be generalised to three dimensions, introducing the wave vector (k_x, k_y, k_z) and assuming $k_y, k_z << k_x$, by expanding k as a power series:

$$\frac{\omega}{\omega_{pi,t}} \approx \lambda_{D,eff}k_x \{1 + \frac{1}{2}\frac{k_y^2+k_z^2}{k_x^2} - \frac{1}{2}(\lambda_{D,eff}k_x)^2\} + \ldots \tag{13}$$

It is assumed that $\lambda_{D,eff}k_{y,z} = O[(\lambda_{D,eff}k_x)^2]$ so that the only modification apart from replacing k by k_x is the additional term shown, which originates from the ion-acoustic velocity term $\lambda_{D,eff}k$ in equation (9). To find a corresponding differential equation as for the one-dimensional case the dispersion relation must first be multiplied by $\lambda_{D,eff}k_x$. Fourier inversion then gives in the ion rest frame:

$$\frac{\partial}{\partial x}\{\frac{\partial \phi}{\partial t} + \frac{\partial}{\partial x}\{1 + \frac{1}{2}\frac{\partial^2}{\partial x^2}\}\phi - \frac{\pi}{2}\alpha H[\frac{\partial \phi}{\partial x}]\} = -\frac{1}{2}\{\frac{\partial^2 \phi}{\partial y^2} + \frac{\partial^2 \phi}{\partial z^2}\} \tag{14}$$

and the Hilbert transform is applied over the x-coordinate. In the ion-acoustic velocity frame adding the nonlinear term gives the mKP equation (6) modified to include Landau damping effects.

To estimate the importance of Landau damping effects, consider the case of Maxwellian distributions for the electrons drifting with velocities $u_{d,h}$ and $u_{d,c}$ and assuming $c_{s,eff} << c_{e,h}$, $c_{e,c}$ so that:

$$\alpha \approx \sum_{j=h,c} \left[\frac{n_{e,j}/c_{e,j}^2}{\sum_{i=h,c} n_{e,i}/c_{e,i}^2}\right] \frac{1}{\sqrt{2\pi}} \frac{(u_{d,j} - c_{s,eff})}{c_{e,j}} \tag{15}$$

For zero drift velocities $\alpha<0$ corresponding to damping, and the weighting factors (square bracket in equation (15)) sum to unity. In this case α is given by:

$$\alpha \approx -\frac{1}{\sqrt{2\pi}}\left[\{\sum_{j=h,c}\bar{n}_{e,j}/T_{e,j}^{3/2}\}\,/\,\{\sum_{j=h,c}\bar{n}_{e,j}/T_{e,j}\}^{3/2}\right]\sqrt{\frac{m_e}{\langle m_i\rangle}} \tag{16}$$

where the electron densities are normalised (indicated by $\bar{n}_{e,j}$) to sum to unity and the average ion mass $<m_i>$ is defined to correspond to the definition of $\omega_{pi,t}$. It follows that for zero drift velocities $\alpha = O(\{m_e/<m_i>\}^{1/2})$. For sufficiently high drift velocities α changes sign and ion-acoustic instabilities set in. In the marginal case with $\alpha \approx 0$ the Landau term may be arbitrarily small. If the Landau term is to be of the same order as the dispersion term this restricts the length scale L of perturbations, so that $L \approx 1/\alpha^{1/2} \approx \{<m_i>/m_e\}^{1/4}$ ≈ 6.5 for a single ion component hydrogen plasma or ≈ 16.5 in the case of argon (assuming $u_{d,h}=u_{d,c}=0$). This makes it difficult to apply the equations to experiments, since if the length scales are made large ($L >> 1$) as required for validity of the weakly nonlinear analysis the Landau damping term will dominate. Alternatively from equation (8) if the nonlinear term is to be significant then $\phi \approx \alpha^{1/2}$ and the amplitude is not very small making it difficult to justify the expansion procedure used in the nonlinear analysis. A way round this difficulty would be to raise the plasma current so that the plasma is close to the marginal condition for ion-acoustic instabilities. The consequences of Landau damping require further analysis, and further experiments are needed for proper confirmation of the theory.

ACKNOWLEDGEMENTS: This work was supported by the Swedish National Science Research Council.

REFERENCES:

1. Sato, T. and Okuda, H., J. Geophys. Res. 86 (1981), 3357.
2. Chan, C. and Hershkowitz, N., *Phys. Fluids* **25** (1982), 2135.
3. Boström, R., Gustafsson, G., Holback, B., Holmgren, G. and Koskinen, H., *Phys. Rev. Lett.* **61** (1988), 82.
4. Torvén, S., *Wave Instabilities in Space Plasmas*, P.J. Palmadesso and K. Papadopoulos (eds), D. Reidel Publ. Co., Dordrecht, Holland (1979), p.109.
5. Raadu, M.A., *Phys Reports* **178** (1989), 25.
6. Andrews, J.G. and Allen, J.E.: *Proc. Roy. Soc. Lond.* **A320** (1971), 459.
7. Bernstein, I.B., Greene, J.M. and Kruskal, M.D., *Phys. Rev.* **108** (1957), 546.
8. Montgomery, D. and Joyce, G., *J. Plasma Phys.* **3** (1969), 1.
9. Sagdeev, R.Z., *Reviews of Plasma Physics* **4** (1966), edt M.A. Leontovich, Consultants Bureau, New York, p.23.

10. Langmuir, I., *Phys. Rev.* **33** (1929), 954.
11. Schamel, H. and Bujarbarua, S., *Phys. Fluids* **26** (1983), 190.
12. Hudson, M.K., Lotko, W., Roth, I. and Witt, E., *J. Geophys. Res.* **88** (1983), 916.
13. Kim, K.Y., *Phys. Lett.* **97A** (1983), 45.
14. Schamel, H., *Z. Naturforsch.* **38a** (1983), 1170.
15. Schamel, H., *Symposium on Plasma Double Layers*, P. Michelsen and J.J. Rasmussen (eds.), Risø, June 16-18, 1982, Risø-R-472, (1982), p. 13.
16. Schamel, H., *Physica Scripta* **T2/1** (1982), 228.
17. Schamel, H., *Phys. Reports*, **140** (1986), 161.
18. Sato, T. and Okuda, H., *Phys. Rev. Lett.* **44** (1980), 740.
19. Hasegawa, A. and Sato, T., *Phys. Fluids* **25** (1982), 632.
20. Torvén, S., *Phys. Rev. Lett.* **47** (1981), 1053.
21. Buti, B., *Physics Lett.* **76A** (1980), 251.
22. Nishihara, K. and Tajiri, M., *J. Phys. Soc. Japan* **50** (1981), 4047.
23. Bharuthram, R. and Shukla, P.K., *Phys. Fluids* **29** (1986), 3214.
24. Goswami, K.S. and Bujarbarua, S., *Phys. Lett.* **108A** (1985), 149.
25. Torvén, S., *Physica Scripta* **33** (1986), 262.
26. Raadu, M.A. and Rasmussen, J.J., *Astrophys. Space Sci.* **144** (1988), 43.
27. Cipolla, J.W. and Silevitch, M.B., *J. Plasma Phys.* **25** (1981), 373.
28. Chanteur, G., Adam, J.C., Pellat, R. and Volokhitin, A.S., *Phys. Fluids* **26** (1983), 1584.
29. Bauer, F. and Schamel, H., *Physica D* **54** (1992), 235.
30. Washimi, H. and Taniuti, T., *Phys. Rev. Lett.* **17** (1966), 996.
31. Watanabe, S., *J. Phys. Soc. Jpn.* **53** (1984), 950.
32. Tajiri, M., and Nishihara, K., *J. Phys. Soc. Jpn.* **54** (1985), 572.
33. Baboolal. S., Bharuthram, R.and Hellberg, M.A., *J. Plasma Phys.* **40** (1988), 163.
34. Perelman, T.L., Fridman, A.Kh. and El'yashevich, M.M., *Sov. Phys. JETP* **39** (1974), 643.
35. Perelman, T.L., Fridman, A.Kh. and El'yashevich, M.M., *Phys. Lett.* **A 47** (1974), 321.
36. Raadu, M.A. and Chanteur, G., *Physica Scripta* **33** (1986), 240.
37. Chanteur, G. and Raadu, M.A., *Phys. Fluids* **30** (1987), 2708.
38. Gardner, C.S., Greene, J.M., Kruskal, M.D. and Miura, R.M., *Phys. Rev. Lett.* **19** (1967), 1095.
39. Khruslov, E. Ya., *JETP Lett.* **21** (1975), 217.
40. Reddy, R.V. and Lakhina, G.S., *Planet. Space Sci.* **39** (1991), 1343.
41. Knorr, G. and Goertz, C.K., *Astrophys. Space Sci.* **31** (1974), 209.

42. Penrose, O., *Phys. Fluids* **3** (1960), 258.

43. Biskamp, D., *J. Plasma Phys.* **3** (1969), 411.

44. Kadomtsev, B.B. and Petviashvili, V.I., *Dokl. Akad. Nauk SSSR* **192** (1970), 753 [*Sov. Phys. Dokl.* **15** (1970), 539].

45. Matsukawa, M. and Watanabe, S., *J. Phys. Soc. Japan* **57** (1988), 2725.

46. Turitsyn, S.K., Rasmussen, J.J. and Raadu, M.A., *"Nonlinear World" Proc. IV Intern. Workshop on Nonlinear and Turbulent Processes in Physics*, Kiev, USSR 9-22 Oct 1989 (eds A.G. Sitenko, V.E. Zakharov and V.M. Chernousenko) (Naukova Dumka, Kiev) **1** (1989), p.187.

47. Turitsyn, S.K., Rasmussen, J.J. and Raadu, M.A., *TRITA -EPP* **-91-01**, Dept of Plasma Phys., Alfvén Lab., Royal Inst. Technology, Stockholm, Sweden (1991).

48. Zakharov, V.E. and Kuznetsov, E.A., *Zh. Eksp. Theor. Fiz.* **66** (1974), 594 [Sov. Phys. JETP **39** (1974), 285].

49. Bharuthram, R. and Shukla, P.K., *Physica Scripta* **34** (1986), 732.

50. Karpman, V.I., Lynov, J.P., Michelsen, P., Pécseli, H., Rasmussen, J.J. and Turikov, V.A., *Phys. Rev. Lett.* **43** (1979), 210.

51. Nishihara, K., Sakagami, H., Taniuti, T. and Hasegawa, A., *Symposium on Plasma Double Layers*, P. Michelsen and J.J. Rasmussen (eds.), Risø National Laboratory, Roskilde, Denmark, June 16-18 1982, (Risø-R-472) (1982), p41.

FORMATION OF ELECTROSTATIC POTENTIALS IN COLLISIONLESS PLASMAS FLOWING ALONG MAGNETIC FIELD LINES

Noriyoshi Sato

Department of Electronic Engineering,
Tohoku University, Sendai 980, Japan

Abstract : Measurements are presented of plasma potentials formed by modifying local plasma structures in collisionless plasmas under strong magnetic fields. After a short survey of electric double layers under mirror configurations of magnetic field, two experiments are described on field-aligned electric field and related electron acceleration. One of them is concerned with potential jumps formed in high-speed plasma flows injected into a region of converging magnetic field lines. The other is on potential formation due to localized RF electric fields (electron plasma waves) generated externally under a uniform magnetic field.

1. INTRODUCTION

Formation of plasma potentials along magnetic field lines is of crucial importance in conjunction with particle acceleration in space plasmas[1] and plasma confinement in fusion-oriented open-ended devices.[2] From these points of view, there has been much interest in field-aligned potential differences in magnetized plasmas.

Potential formation is closely related to spatial variations of plasma structure. As is well known, in discharge tubes with local change of diameter, an electric double layer is formed, being accompanied by electron acceleration, in a region of the wall constriction.[3] Generally speaking, there are various methods for generating local structure changes in plasmas. In this paper, measurements are presented of field-aligned potential formation and related electron acceleration, which are provided by modifying local plasma structures, in magnetized collisionless plasmas. Here plasma structures are locally modified by spatial changes of magnetic field or localized RF electric fields.

Section 2 makes a short survey of electric double layers under mirror configurations of magnetic field. Section 3 presents field-aligned potential formation in high-speed plasma flows injected into a region of converging magnetic field lines. Potential formation due to localized RF electric fields is presented in Section 4. Section 5 contains conclusions.

2. DOUBLE LAYERS UNDER MAGNETIC MIRROR CONFIGURATIONS

A nonuniform magnetic field has a big influence on local structures in plasmas. A lot of measurements have been made on effects of nonuniform magnetic fields on electric double layers in collisionless plasmas at Tohoku University. Here typical results are presented to demonstrate clear effects of mirror-type magnetic fields on double layers. In the experiments, collisionless plasma columns are produced by surface ionization at hot plates (plasma sources: S_1, S_2) of 3.5 cm in diameter at both ends under strong magnetic fields. Double layers are generated by applying potential differences between S_1 and S_2.[4] The double layer position is set by controlling the plasma densities supplied by S_1 and S_2.

Under a uniform magnetic field, the double layers generated always fluctuate along the magnetic field. The fluctuations are due to dynamics of a small potential dip formed on the low-potential tail of the double layers. A periodic change of formation, movement, and disruption of the potential dip gives rise to a "back and forth" motion of the low-potential tail along the magnetic field lines, although a percentage of the fluctuations becomes small with an increase in the potential differences across the double layers. Thus, time-averaged double-layer widths are much larger than those for stationary double layers.

A magnetic mirror (bump) with mirror ratio R_m in the range 1.0-1.8 is provided somewhere midway between S_1 and S_2.[5] When the double layer position is around the magnetic bump, an effect of the magnetic bump on the double layers is quite drastic, as found in Fig. 1(a), where axial profiles of potential ϕ and magnetic field B are shown with R_m as a parameter. With an increase in R_m, the double-layer width L becomes small, approaching the value for stationary double layers with no fluctution [see Fig. 1(b)]. Even for $R_m \simeq 1.02$, we can find a clear change of the double-layer potential shape. Being consistent with this decrease of the width, the potential fluctuations decrease on the low-potential tail, as shown in Fig. 2. The fluctuations are found to be localized in the narrow region near the mirror point. This is because the "back and forth" motion of the low-potential tail is restricted by the magnetic mirror.

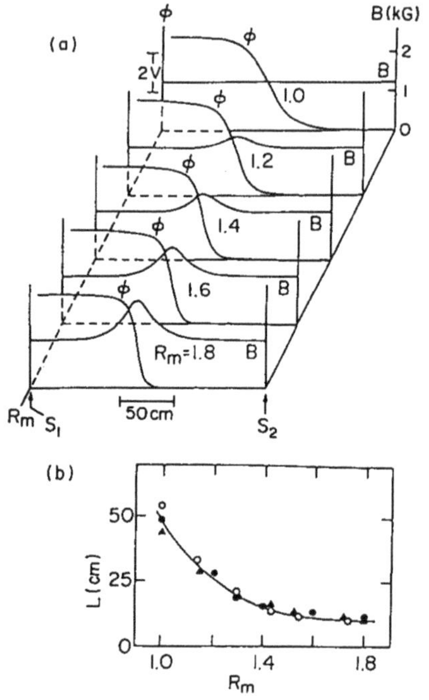

Fig. 1. A dependence of double-layer width on mirror Ratio R_m of the magnetic bump.

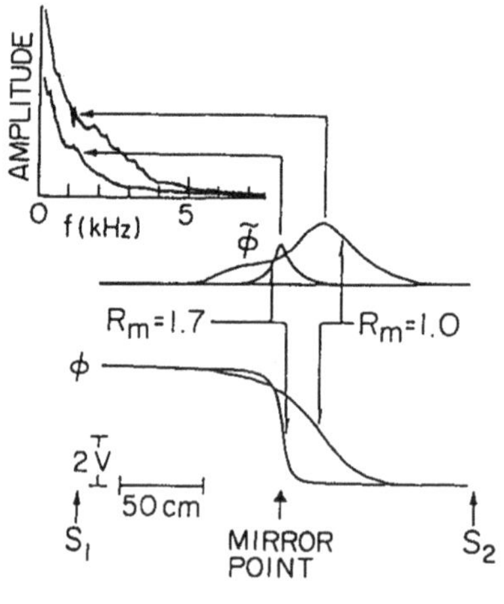

Fig. 2. A dependence of double-layer fluctuation on mirror ratio R_m of the magnetic bump.

As mentioned above, the fluctuations are closely related to the potential dip formed on the low-potential tail, which limits electric current passing through the double layers. When the potential dip is disrupted, electrons can immediately flow without this current limitation toward the high-potential side. This fast electron motion results in a penetration of high potential into the low-potential region, yielding an increase in the double-layer width, although the process is followed by formation of a new potential dip on the low-potential side. In the presence of the magnetic bump, however, this electron motion is restricted by the magnetic mirror. This is the reason why the fluctuations and thus the widths are reduced by the magnetic bump.

When the magnetic bump is axially shifted, the double-layer position is found to shift together with the bump. But, if the double layer is far from the bump position, such a clear influence of the magnetic bump is not recognized.

Measurements have also been made on V(or U)-shaped double layers under a strong magnetic field.[6] In this case, the diameter of S_2, 0.8 cm, is smaller than that of S_2, 3.5 cm. Therefore, the plasma is a mixture of two plasmas with different potentials and diameters. Typical structures of the V(or U)-shaped double layers are demonstrated by equipotential lines in Fig. 3, where the magnetic field lines (dotted lines) are converged toward the high-potential side of the double layers. The V(or U)-shaped double layers have a large radial electric field in their low-potential region. As found in this figure, an axial shift of the converging region of magnetic field lines is followed by the double layers, as in case of the one-dimensional double layers mentioned above.

In the presence of the V(or U)-shaped double layers, there appears another type of electrostatic fluctuations, in addition to the axial fluctuations of the low-potential tail. The frequencies are a little above the harmonics of ion cyclotron frequency. The fluctuations have a peak amplitude at the radial position where the radial potential profiles give the maximum slope on the low-potential side of the double layers. They are observed to propagate with the azimuthal mode number m=1, 2, 3,···· in the direction of $E_r \times B$ plasma rotation, suggesting a mechanism of E×B driven instability.

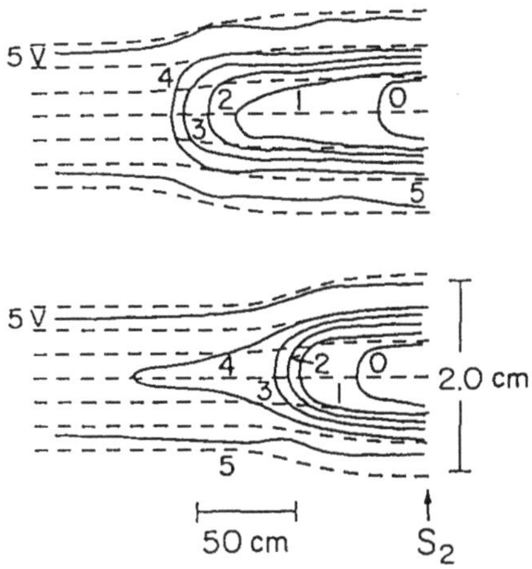

Fig. 3. Equipotential lines of V(or U)-shaped double layers along converging magnetic field lines.

3. POTENTIAL FORMATION IN HIGH-SPEED PLASMA FLOWS ALONG CONVERGING MAGNETIC FIELD LINES

As predicted theoretically,[7] a field-aligned potential difference is generated by different pitch angle anisotropies of electrons and ions in plasmas flowing into a region of converging magnetic field lines.

In order to demonstrate this mechanism, measurements have been made on high-speed plasma flows produced in a modified single-ended Q machine under a mirror configuration of magnetic field.[8] The plasma flow speed V_o is varied in the range V_o/C_s =5-26 [$C_s=(T_e/M)^{1/2}$: ion acoustic speed]. The plasma flow is introduced into a region of converging magnetic field lines, which is located in the central part of the machine. The mirror ratio R_m is in the range 1.0-7.5.

A typical evolution of potential formation is shown for V_o/C_s=20 and R_m=7.0 in Fig. 4. Here the endplate is biased at ϕ_T=-20 V with respect to the plasma source which is electrically grounded. Since V_o is so large that the ion pitch-angle distribution is highly shifted in the direction toward the endplate, most of the ions pass through the $\nabla B>0$, being finally absorbed by the endplate. A fraction of the electrons are reflected by $\nabla B>0$. The electrons passed through toward the endplate are reflected by ϕ_T. The plasma, which is injected at t(time)=0, flows along the magnetic field, being accompanied by potential profiles presented at subsequent times in Fig. 4. Before the plasma reaches the region of $\nabla B>0$ ($t \lesssim 80 \mu s$), the potential ϕ is almost constant spatially up to the plasma front region, where ϕ decreases monotonically. In the plasma passing through the region of $\nabla B>0$, however, a spatial increase in ϕ is observed. After the plasma front

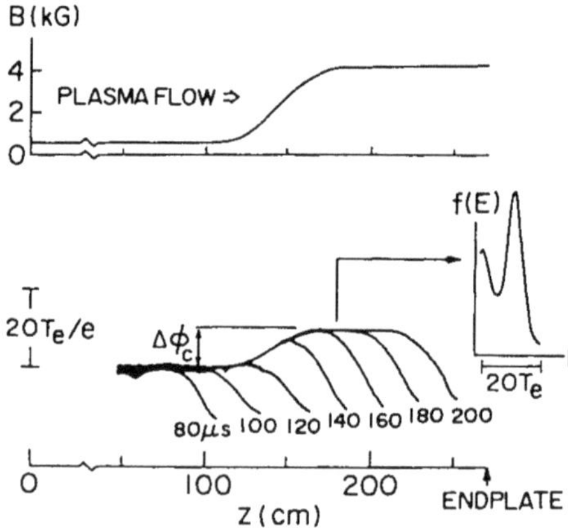

Fig. 4. Potential formation in a plasma flow along converging magnetic field lines.

passes through this region ($t \gtrsim 160\mu s$), the potential increase $\Delta\phi_c$ generated there is about $10.7T_e/e$ and ϕ is again constant beyond the region $\nabla B>0$ up to the plasma front. These features and the $\Delta\phi_c$ measured do not depend on $\phi_T(<0)$. For $\phi_T>0$, the potential structure in the region of the plasma front up to the endplate is modified by a spatial increase of the vacuum potential toward the endplate. But, the ϕ profiles up to the plasma front region at various times and $\Delta\phi_c$ in the region of $\nabla B>0$ are independent of ϕ_T, being the same as in case of $\phi_T<0$ unless the plasma comes in contact with the endplate.

The electron energy distribution f(E) parallel to the magnetic field is obtained from a derivative of the Langmuir-probe characteristics. The result is also presented in Fig. 4, where the measured position is on the side of higher magnetic field just beyond the region of $\nabla B>0$ at $t=180\mu s$. There are two peaks in f(E). The separation between them corresponds to the observed $e\Delta\phi_c$. This means that the peak with higher energy is due to electron acceleration by $\Delta\phi_c$, although most of the electrons accelerated are reflected by $\nabla\phi(<0)$ in the plasma front region. The other peak is observed around the plasma potential, indicating an existence of stationary electrons with finite energy spread.

According to the measurements, with an increase in R_m, $\Delta\phi_c$ increases initially, being followed by a gradual saturation. The value of R_m at which this saturation starts increases as V_o increases. At a fixed value of R_m, for example, at $R_m=7.0$, $\Delta\phi_c$ is observed to increase with an increase in V_o, but $e\Delta\phi_c/T_e \simeq 11$, independent of V_o for $V_o/C_s \gtrsim 10$. This saturation appears at a smaller value of V_o when R_m is smaller.

An effect of ϕ_T on the ϕ profiles is observed after the plasma front comes in contact with the endplate. In Fig. 5, the ϕ profiles for $\phi_T=15$ and -20 V are presented at t=220 and 300μs, together with those in a final state ($t=\infty$) under the same conditions as in Fig. 4. The front tail of the plasma flow reaches the endplate at $t \simeq 220$ s. For $\phi_T=15$ V, the plasma front is strongly modified by ϕ_T even at t=220μs. After the plasma comes in contact with the endplate, there appears a drastic increase of the potential difference in the region of $\nabla B>0$, in addition to an increase of ϕ in the whole plasma region along the magnetic field. In the final state ($t \gtrsim 500\mu s$), the potential difference is again localized in the region of $\nabla B>0$, although ϕ decreases spatially beyond the region of $\nabla B>0$ toward the endplate. The final difference $\Delta\phi_{co}$ in the region of $\nabla B>0$ is about $30T_e/e$ ($\simeq 2.8\Delta\phi_c$). On the other hand, for $\phi_T=-20$ V, there appears a gradual decrease of the potential difference in the region of $\nabla B>0$ after the plasma reaches the endplate. For $\phi_T<0$, $e\Delta\phi_{co}/T_e=1.0-2.0$, independent of ϕ_T. For $\phi_T>0$, however, $\Delta\phi_{co}$ increases with an increase in ϕ_T. It is to be noted that $\Delta\phi_{co}$ is, roughly speaking, proportional to the electric current passing through the endplate which increases as ϕ_T is increased.

In the presence of the potential jump, there appear high-frequency fluctuations in the high-potential region, as shown for $V_o/C_s=26$ and $R_m=7.0$ with ϕ_T as a parameter in Fig. 6. The electron beam produced as a result of electron acceleration is responsible for the fluctuations observed. It must be mentioned that the fluctuations disturb electron orbits, resulting in generation of trapped electrons in the high-potential region of the plasma flows.

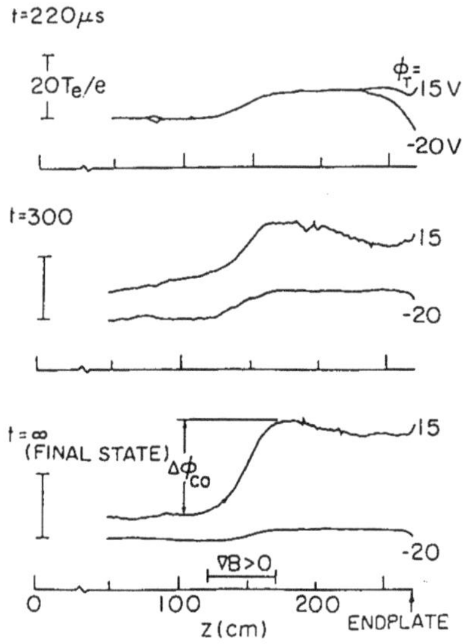

Fig. 5. An effect of endplate bias on potential formation in a plasma flow along converging magnetic field lines.

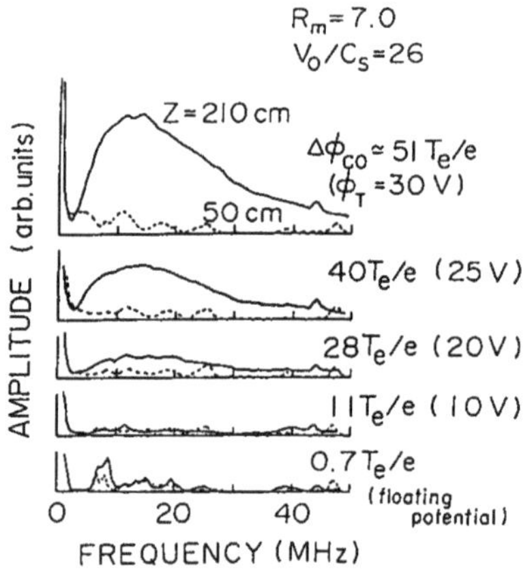

Fig. 6. High-frequency fluctuations in the high-potential region for various values of endplate bias.

4. POTENTIAL FORMATION DUE TO LOCALIZED RF ELECTRIC FIELDS IN PLASMA FLOWS ALONG A UNIFORM MAGNETIC FIELD

Here we are interested in potential formation due to localized RF electric fields in plasma flows along a uniform magnetic field. In this experiment,[9] the localized RF electric fields are provided by extremely large-amplitude electron plasma waves generated externally. There appears a clear potential change in the region around the wave generation, where the waves are localized because of enhanced wave damping.

A plasma, produced also by surface ionization at the hot plate of 5.2 cm in diameter, flows along a strong uniform magnetic field. Electron plasma waves are excited by applying external RF potentials V_{RF} ($\lesssim$50 V peak to peak) with frequency $\omega/2\pi$ (=10-100 MHz) to a ring exciter surrounding the plasma column axisymmetrically at a distance of 100 cm (z=0) from the hot plate. The plasma flows downstream without being disturbed by the exciter at V_{RF}=0 V even when a dc (negative or positive) potential $\lesssim$ 50 V is applied to the exciter.

When the RF potentials applied to the exciter are so small as $V_{RF} \lesssim 50$ mV ($\ll T_{eo}/e$), the electron plasma waves propagate from the exciter in the upstream (z<0) and downstream (z>0) directions along the plasma column, satisfying the linear dispersion relation (Trivelpiece-Gould mode) with the Landau damping.[10] With an increase in V_{RF}, however, there appear gradually nonlinear phenomena such as enhanced initial damping and amplitude oscillation.[11, 12] For a further increase in V_{RF} (>3 V), the waves damp even more strongly, yielding big changes of plasma parameters. Such nonlinear behaviors appear clearly under the condition $\omega/\omega_{pe}<1$ ($\omega_{pe}/2\pi$: electron plasma frequency), where the Landau damping of linear waves is small.

Typical examples of the results are demonstrated in Fig. 7, where $\omega/2\pi$=45 MHz and V_{RF}=8 V. As a result of the heavy damping, the electron plasma wave is strongly localized around z=0. In spite of almost no change of ϕ in the upstream region, there appears a decrease in ϕ, with a peak generation at z$\simeq$0, in the downstream region, resulting in a large-scaled potential drop, $\Delta\phi=20T_{eo}/e$. We can also find an increase in electron temperature T_e, $\Delta T_e=30T_{eo}$ and a decrease in plasma density n, $\Delta n=0.2n_o$. Here a subscript "o" stands for plasma parameters in the absence of the large-amplitude waves. The characteristic length of $\Delta\phi$ is several ten times the Debye length, being nearly equal to the width of stationary double layer.

A gradual change of the axial electron energy distribution function f(E) with an increase in V_{RF} is presented in Fig. 8, where f(E) (E=-eV_c) is measured at z=50 cm and V_c is the collector bias of the analyzer with respect to the hot plate. As V_{RF} is increased, high-energy electrons are produced around the wave phase velocity, forming a broad bump on tail

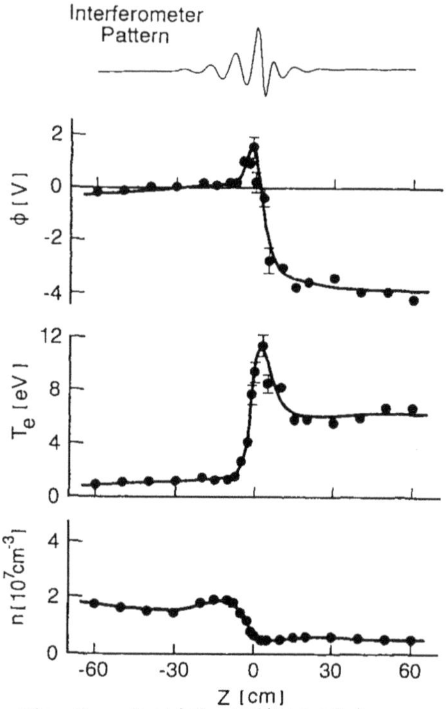

Fig. 7. Spatial profiles of ϕ, T_e, and n in the presence of localed large-amplitude electron plasma wave with $\omega/2\pi$=45 MHz (ω/ω_{pe}=0.71) and V_{RF}=8 V.

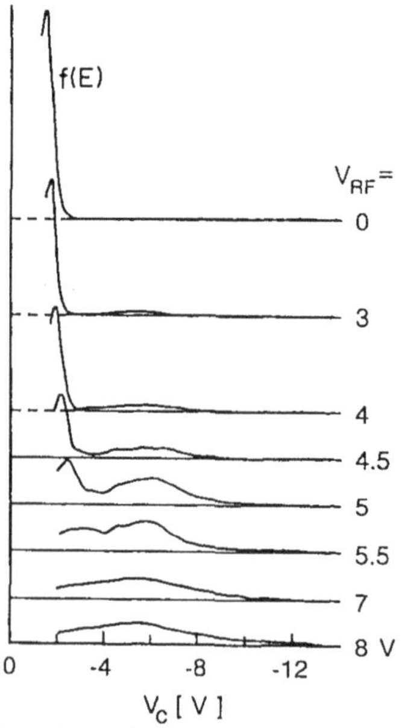

Fig. 8. Electron energy distribution f(E) at z=50 cm with V_{RF} as a parameter at $\omega/2\pi$=20 MHz (ω/ω_{pe} =0.91). V_c(=-E/e) is retarding potential of the analyzer.

(V_{RF}=3-4 V). For much larger values of V_{RF}, where the large-amplitude wave is clearly localized around the exciter, the plasma consists of two almost Maxwellian energy distributions with low and high temperatures (two-temperature plasma). The low-energy peak of f(E) is found to shift gradually toward a large value of $-V_c$ (V_{RF}=4.0-5.5 V). For $V_{RF}\geq 6$ V, a nearly one-component Maxwellian distribution with high temperature is formed with its peak shifted toward a much larger value of $-V_c$. Since ϕ at z<0 is almost independent of V_{RF}, this peak also indicates the $\Delta\phi$ generation. According to time-resolved measurements of f(E), this high-temperature Maxwellian distribution yielding ΔT_e and also Δn is formed within 100-200 μs well before the $\Delta\phi$ generation.

Now we consider a mechanism of the potential formation observed. In our experiment, the extremely large-amplitude electron plasma waves are generated in a magnetized collisionless plasma flowing toward the endplate kept floating or biased negatively. The waves are strongly damped spatially, resulting in the RF localization near the exciter. The localized RF field provides a ponderomotive force repelling electrons.[13] Thus, there should appear a potential increase in this region of the localized RF field. A part of the ions, although they are accelerated by the electron sheath (potential drop≃1.5 V) in front of the hot plate, are reflected by this local increase in ϕ. As a result, n decreases in the downstream region. On the other hand, the enhanced wave damping gives rise to a number of detrapped electrons,[14] generating a bump on tail in the energy distribution f(E). With an increase in V_{RF}, this bump is deformed into generation of appreciable group of electrons with high energy. The result is clearly observed in the downstream region, where all electrons passed through z=0 move in the direction of the wave propagation. In the upstream region, however, there are many electrons flowing from the hot plate in the direction opposite to the wave propagation. The electron-energy increase implies a density decrease in the downstream region because the electron flux is conserved. The potential drop is generated in the downstream region in a way to accelerate the ions passed through the local potential increase at z≃0, providing a necessary decrease of the ion density for charge neutrality. A theoretical analysis for this model gives us a reasonable agreement with the experimental results.[15]

In this experiment, the localized RF fields give rise to the electron acceleration which is accompanied by the potential drop in the plasmas flowing along the magnetic field.

5. CONCLUSIONS

The measurements have been presented of potential formation induced by local changes of plasma structure in magnetized collisionless plasmas. After a short survey of double layers observed in plasmas under mirror configurations of magnetic field, two experiments have been presented of potential formation due to local changes of plasma structure in plasma flows along strong magnetic fields.

One of them is concerned with potential jumps generated in high-speed plasma flows injected into a region of converging magnetic field lines, which result in field-aligned electron acceleration. The results are qualitatively consistent with the predictions based on different pitch angle anisotropies of electrons and ions along the magnetic field. A drastic endplate effect on the generated potential jumps is observed when the plasmas come in contact with the endplate, showing an importance of boundary conditions for the potential formation.

The other experiment has been carried out on potential formation due to localized RF electric fields. Large potential drops are formed along plasma flows in the presence of localized large-amplitude electron plasma waves under a uniform magnetic field. The phenomenon is attributed to ponderomotive force on electrons and nonlinear electron acceleration (or heating) due to enhanced wave damping.

Two experiments show different features of relation between potential formation and electron acceleration in plasmas flowing along strong magnetic fields.

ACKNOWLEDGEMENTS

I would like to thank the members of the plasma physics group at Tohoku University, who have done the experiments presented here. The works were supported in part by Grant-in-Aid for Scientific Research from the Ministry of Education, Science, and Culture of Japan.

REFERENCES

1. H. Alfvén and P. Carlqvist, Solar Phys. 1(1967)220.

2. M. Inutake et al., Phys. Rev. Lett. 55(1985)939.

3. M. J. Schönhuber, Z. Angew. Phys. 15(1963)454.

4. N. Sato et al., Phys. Rev. Lett. 46(1981)1330; N. Sato et al., J. Phys. Soc. Jpn. 52(1983)875; Iizuka et al., J. Phys. Soc. Jpn. 54(1985)2516.

5. N. Sato, Proceedings of the Symposium on Plasma Double Layers, Roskilde, 1982 (Risø National Laboratory, Roskilde, 1982), p.116; N. Sato, Proceedings (Invited Papers) of the 1984 International Conference on Plasma Physics, Lausanne, 1984 (Center de Recherches en Physique des plasmas, Ecole Polytechnique Federale de Lausanne, 1984), Vol.II, p.555.

6. N. Sato et al., Phys. Rev. Lett. 57(1986)1227.

7. H. Alfvén and C.-G. Fälthammar, Cosmical Electrodynamics (Oxford Univ. Press, London, 1963), p.162.

8. N. Sato et al., Phys. Rev. Lett. 61(1988)1615.

9. N. Y. Sato et al., Proceedings of the 1992 International Conference on Plasma Physics, Innsbruck, 1992.

10. J. H. Malmberg and C. B. Wharton, Phys. Rev. Lett. 17(1966) 175.

11. T. O'Neil, Phys. Fluids 8(1965)2255; R. Sugihara and K. Yamanaka, Phys. Fluids 18(1975)114.

12. J. H. Malmberg and C. B. Wharton, Phys. Rev. Lett. 19(1967) 775; R. N. Franklin et al., Phys. Rev. Lett. 28(1972)1114.

13. H. Motz and C. J. H. Watson, Advances in Electronics and Electron Physics (Academic Press, New York, 1967), Vol.23, p.153.

14. T. P. Starke and J. H. Malmberg, Phys. Rev. Lett. 37(1976) 505; N. Sato et al., Phys. Rev. Lett. 37(1976)1983.

15. N. Y. Sato et al., Proceedings of the Fourth Symposium on Double Layers and Other Nonlinear Potential Structures in Plasmas, Innsbruck, 1992.

THE DYNAMIC RESPONSE OF A COLLISIONLESS PLASMA TO THE APPLICATION OF A VOLTAGE DROP

S. Torvén, M. Bohm and I. Axnäs, Alfvén Laboratory, Royal Institute of Technology, 10044 Stockholm, Sweden

1. INTRODUCTION: Double layers form in laboratory plasmas as the response to an applied or induced voltage drop. In plasmas maintained by local ionization of the background gas due to electron impacts, a double layer evolves from an electrode sheath when a sufficiently large voltage is applied to an electrode in the plasma. The significant mechanism for the double layer formation is in this case the generation of positive ions in the electrode sheath. Ionization may be neglected in a stationary anode sheath only if $\lambda_i \gg d(m_i/m_e)^{1/2}$.[1,2,3,4] Here λ_i is the ionization mean free path of the electrons, d the thickness of the anode sheath, m_e the electron mass, and m_i the ion mass. The voltage drop over these "ionization double layers" is limited to relatively small values which are related to the ionization potential of the background gas.[5] To avoid local ionization caused by the double layer, one may work with voltage drops less than the ionization potential of the background gas[6] or work at such low neutral particle densities that the plasma becomes collisionless. "Collisionless double layers", with very large potential drops, have been observed in a Q-machine[7] and in a triple plasma machine.[8] In the latter device, ionization was suppressed in the central chamber by differential pumping.

In this paper we report some new results on double layer formation when a step voltage is applied to a collisionless plasma. Experiments with the magnetized plasma columns of a "double-ended" Q-machine[9] and a differentially pumped triple plasma machine[10] are described in Section 2. They demonstrate the different responses obtained for an axially homogeneous and an axially inhomogeneous plasma column. In the homogeneous case, the plasma potential drop first concentrates in a "cathode sheath". Then this drop begins to propagate into the plasma as a double layer on the time scale of the ion motion. When the initial plasma has an ion density cavity (a local ion density minimum with a width of many Debye lengths), it may support large potential drops, extended over the cavity, for several electron transit times. A double layer forms during the subsequent ion motion by a steepening of this profile.

In the homogeneous case, computer simulations[11] give similar results as the experiments. Computer simulations of the inhomogeneous case,[12] with periodic boundary conditions, have demonstrated the importance of periodic ion density cavities for the voltage supporting capability of a collisionfree plasma, and the results have been used to interprete electric fields measured above the aurora.[13] Periodic density cavities can also strongly modify the nonlinear evolution of plasma oscillations.[14] These investigations cover only the electron dynamics. The ion dynamics in a density cavity have also been studied both theoretically and by computer simulations,[15] but the the formation of a double layer was not studied.

In more recent computer simulations,[16,17] the boundary conditions were non-periodic and similar to those in the experiments. The electron dynamics was studied, and a particle-in-cell code[18] (PDW1) was used. These simulations are presented in Section 3. They show that an ion density cavity in a plasma diode can support large voltage drops for several electron transit times in agreement with the experiments. However, when the applied potential drop exceeds a certain critical value, which depends on the cavity depth, the drop instead concentrates in a "cathode sheath" which also is the response obtained for homogeneous initial plasma. The existence regions for these two different responses in the appropriate parameter plane are found.

In Section 3 we also present a simulation with an initial ion density profile consisting of several cavities. A weak double layer is formed at each density minimum. Such results are of interest to predict the voltage supporting capability of a plasma, in which ion density cavities are produced by nonlinear ion waves or by the ponderomotive force from high frequency wave packets.

A new analytical model is presented in Section 4. The model gives explicitely the potential profiles over an ion density cavity in a plasma diode. The existence regions for the two different types of responses, observed in the simulations, also agree with the existence regions predicted theoretically.

Simulations with a longer system giving beam-plasma interaction are presented in Section 5. In Section 6 we discuss the results and present the conclusions.

2. EXPERIMENTAL RESULTS: The time resolution of potential measurements by electron emitting probes has been improved by using a low resistance in the circuit connecting the probe to the reference potential.[19] A strong improvement of the time resolution has also been obtained by minimizing the distributed probe capacitance to ground, and step voltage responses with rise and fall times less than 0.5 μs have been obtained in this way in a triple plasma machine.[20] The ion motion may be neglected during such a short time. Potential changes caused by essentially the electron motion can accordingly be detected, and the dynamic phase leading to double layer formation can be studied in detail.

Fig. 1 shows the step voltage response of the axially homogeneous plasma column in the Q-machine experiment.[9] The machine was operated "double-ended", that is, both end plates were emitting ions and electrons, and the particle injection was "electron rich". The potential drop first concentrates at the low potential end of the plasma column. Then it propagates into the plasma and a double layer forms. This is not stationary, but tends to move back and forth with amplitudes of the order of the thickness of the double layer.

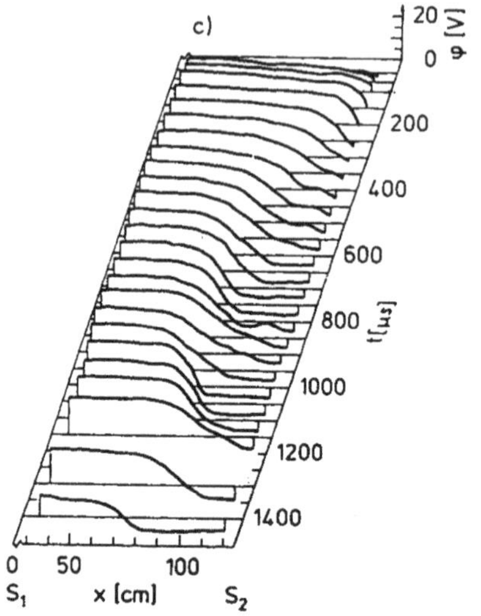

Fig.1. Axial potential profiles in a "double ended" and "electron rich" Q-machine, measured at different times after the application of a step voltage between the emitting plates.[9] Most of the potential drop first concentrates in a narrow region at the low potential plate. Then this drop begins to propagate into the plasma column and a double layer forms. This tends to oscillate back and forth with amplitudes of the order of the thickness of the double layer.

As shown in Fig. 2 (left), the same type of response can be obtained in the weakly magnetized plasma column in the triple plasma machine.[10] The plasma column is maintained in the central chamber by a plasma source on each side of the chamber. In the sources, plasmas are produced by discharges between filaments and the source chamber walls. The sources are connected to the chamber by circular apertures which determine the diameter of the plasma column. At the apertures, the electron number density drops sharply from the value N_s in the sources to N_1 at the ends of the column. To reproduce the result from the Q-machine experiment, the density ratio N_s/N_1 at the low potential aperture must not be too large ($N_s/N_1 \approx 3$ in the case shown). The plasma column was also kept as axially homogeneous as possible and the maximum density difference along the plasma column was less than 10 %. The boundary conditions at the apertures are similar to the boundary conditions at the emitting plates in the Q-machine experiment, but the electron temperature is higher (10 eV).

A quite different type of response was obtained in the triple plasma machine when the plasma column was kept axially inhomogeneous. Due to radial ion losses, the density may drop along the plasma column, from the value N_1 at the apertures to a density minimum with the value N_1-ΔN. Then the whole plasma column forms a ion density cavity with a width of the order of the length of the column. For symmetric operation of the machine, the minimum forms half-way along the plasma column. The cavity can be characterized by the relative cavity depth $\Delta N/N_1$, which can be varied ($0.1<\Delta N/N_1<0.7$) by varying the magnetic field,

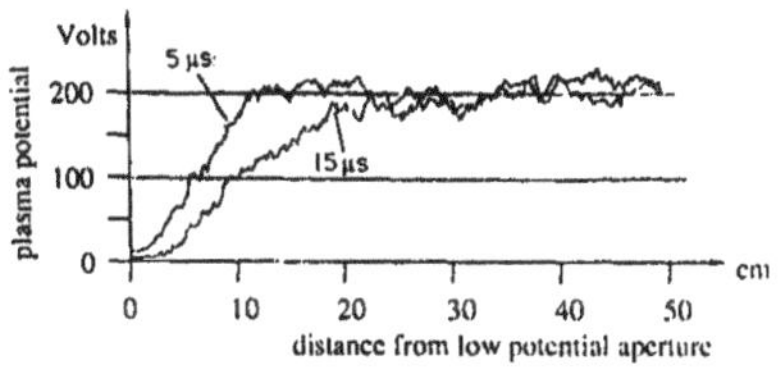

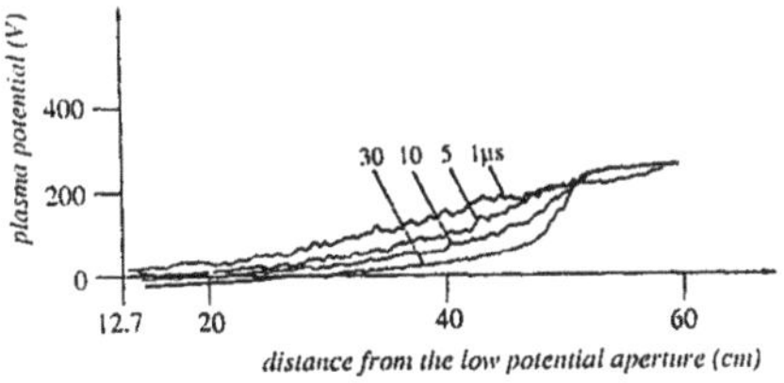

Fig.2. Axial potential profiles in the magnetized plasma column of the triple plasma machine, measured at different times after the application of a step voltage between the plasma sources. (left) Axially homogeneous plasma column. The potential drop is first supported by a narrow region near the low potential aperture, and the potential is approximately constant along the plasma column. Then this drop begins to propagate into the plasma on the time scale of the ion motion. (right) Plasma column with an initial ion density cavity. The potential drop is first (t=1µs) supported by an almost linearly varying potential along the magnetic field. A double layer forms on the time scale of the ion motion by a steepening of this profile.

and the electric potential of the end plates of the central chamber.

The potential response for $N_S/N_1 \approx 10$, $\Delta N/N_1 \approx 0.7$ is shown in Fig. 2 (right). The plasma potential drop is now extended over the cavity, and the profiles have local minima close to the low potential aperture (not shown in the diagrams). A double layer forms on the time scale of the ion motion by steepening of this profile. The different initial conditions accordingly do not only lead to different electron responses during the first few electron transit times but also to quite different ion motions. The computer simulations and the analytical model presented below give the result that the critical voltage drop for the transition between the two states depends on only $N_0/(N_1-\Delta N)$ and the applied voltage drop, normalized to $k_B T_e/e$. Here N_0 is the density of the electrons incident at the boundaries.

3. SIMULATIONS: A plasma diode in the region $0 < x < L$ is considered. Electrons and singly charged ions are injected symmetrically at the two boundaries with time-independent, half-Maxwellian velocity distributions. The number densities of the particles injected is N_0. The initial electrons and ions have Maxwellian velocity distributions with the same temperatures as the injected particles, and $T_i=T_e$. The natural ion (argon) to electron mass ratio is used to obtain the correct division between the electron and ion motion. This limited the simulation time to about 3 electron transit times between the boundaries, corresponding to about 1 µs in the experiment.[10] For the simulation times used, test runs using different initial ion velocity distributions gave the same results as when a Maxwellian distribution was used. This indicates that the ion motion is of little significance during the first few electron transit times. In contrast to this, the initial ion density distribution is of great importance since an ion density gradient gives rise to an electric field when the electrons are displaced. As a suitable test profile for the initial ion density cavity we use

$$n_{ip}(x) = N_1 - \Delta N \sin^2(\pi x/L) \qquad (1)$$

The profile describes a symmetric cavity with an ion number density N_1 at the boundaries and a relative cavity depth $\Delta N/N_1$. Only the case $N_0 > N_1$ is considered here. This ensures that the electron number density at the boundaries always is larger than N_1. Electron rich conditions are then obtained, as in the experimental plasmas, and electron reflecting sheaths form at the boundaries for any value of the applied voltage. The time step was close to $(2\omega_{pe})^{-1}$ and the grid size $\lambda_D/4$. Here $\lambda_D = (\varepsilon_0 k_B T_e/N_1 e^2)^{1/2}$ and $\omega_{pe} = (N_1 e^2/m_e \varepsilon_0)^{1/2}$. The applied potential drop Φ_a is normalized, $\phi_a = e\Phi_a/k_B T_e$, and ϕ_a is varied from small values up to about 10.

Fig. 3 shows examples of potential profiles obtained. The initial potential profile was obtained by a pre-simulation of a short-circuited diode ($\phi_a=0$) during $0 < t < 25\omega_{pe}^{-1}$. For t=0 we put $n_e(x) = n_i(x)$ so that the electric potential vanished everywhere. After about $20\omega_{pe}^{-1}$, a symmetric, quasi-steady state was obtained with a potential profile approximately determined by the Boltzmann relation and with sheath potential drops approximately given by $(k_B T_e/e)\ln(2N_0/N_1)$. This is shown in Fig. 3(a) for t=19. A potential drop was applied instantaneously between the boundaries at $t=25\omega_{pe}^{-1}$. Fig. 3(a) also shows profiles at later times, averaged over $50\omega_{pe}^{-1}$. Fig.3(b) shows instantaneous profiles for the same parameter values. The average profiles remain approximately stationary during the whole simulation time after an initial transient lasting for about one electron transit time. Fig. 3(c) shows the steepening of the profile for large cavity depths. In the case shown in 3(d), a new state is obtained as further discussed below. The parameters are the same as in (a), but ϕ_a has been increased to 5. Most of the potential drop is now concentrated in a cathode sheath.

In the experimental plasmas there are also ion density drops in the sheaths. We have tested the significance of such drops by using the following initial profile in $0<x<L$.

$$n_i(x) = n_{ip}(x) + (2N_0 - N_1)\{2 - \tanh(x/d) - \tanh((L-x)/d)\} \qquad (2)$$

The profile is symmetric around L/2 and has a sharp density increase from N_1 to the value $2N_0$ over a small distance at the boundaries, determined by the parameter d. Simulations for $d=5\lambda_D$ have been presented elsewhere[17]. They show that the potential profiles within the plasma region do not depend on N_0 and d, and the profiles agree with those obtained when (1) is used. The main difference is that (2) gives sheath thicknesses of the order of $10\lambda_D$, compared with about $3\lambda_D$ when (1) is used. For some parameter values, the fluctuation level is also higher,[17] but profiles averaged over 10 to 50 ω_{pe}^{-1} remain constant and agree with

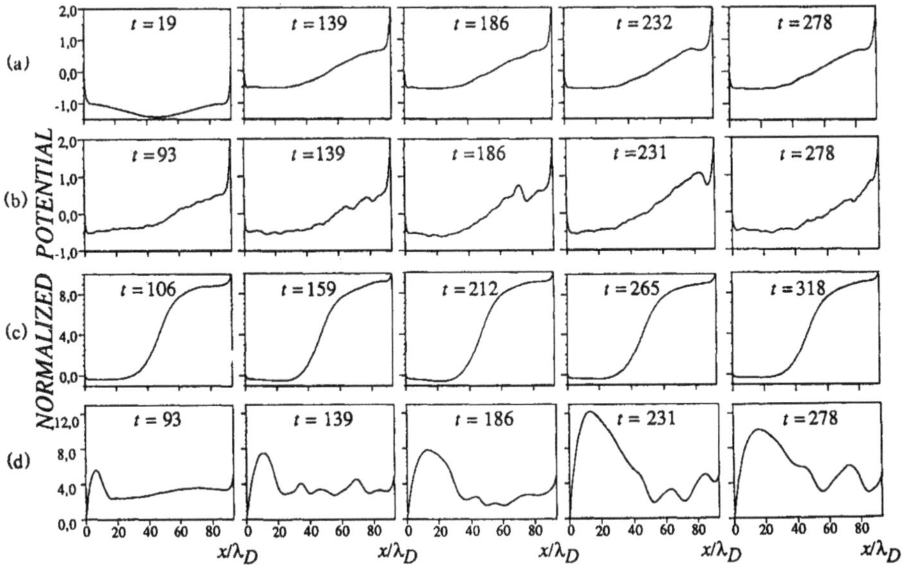

Fig.3. Simulated potential profiles at different times given in ω_{pe}^{-1}. The initial ion density profile is given by (1). The zero level for the potential is at the left boundary. A voltage drop was applied between the boundaries at $t \approx 25\ \omega_{pe}^{-1}$. (a) The initial profile ($t=19\ \omega_{pe}^{-1}$) and profiles at later times which are averaged over 50 ω_{pe}^{-1} ($\phi_a=2$, $\Delta N/N_1=0.3$, $N_0/N_1=1.36$, and $N_0/(N_1-\Delta N)=1.94$). (b) Instantaneous profiles for the same parameters. (c) The steepening of the profile for large cavity depths ($\phi_a=10$, $\Delta N/N_1=0.83$, $N_0/N_1=1.04$, and $N_0/(N_1-\Delta N)=6.12$). In (d), ϕ_a exceeds the critical voltage drops shown in Fig.4 below ($\phi_a=5$, $\Delta N/N_1=0.3$, $N_0/N_1=1.36$, and $N_0/(N_1-\Delta N)=1.94$). The profiles in (c) and (d) are averages over 10 ω_{pe}^{-1}.

those determined by (1). As seen from Fig. 3 (a), (b) and (c), a potential minimum is present in the cavity potential profile.

A number of simulations showed that there is a critical value of the applied voltage, $\phi_a = \phi_{al}$, such that most of the applied voltage is taken up by a cathode sheath for $\phi_a > \phi_{al}$ and by the cavity for $\phi_a < \phi_{al}$. In the latter case the average potential profile always has a local minimum within the diode. Simulations for different values of $\Delta N/N_1$, N_0/N_1, ϕ_a, and d indicate that ϕ_{al} is a function of only $N_0/(N_1-\Delta N)$. From the analytical model presented in Section 4, the maximum voltage drops consistent with the existence of a potential minimum within the diode have been obtained as a function of the same quantity. A comparison is made in Fig. 4. As is seen from the diagram, the critical values of ϕ_a observed in the simulations agree well with the theoretical curve separating the two states.

Additional potential profiles are shown in Section 4 where they are compared with analytical profiles.

Finally we give an example of the potential profiles obtained when the initial ion density distribution has several periodic cavities (Fig. 5). By generalizing (1) to $n_{ip}(x) = N_1 - \Delta N \sin^2(q\pi x/L)$, where q is an integer, we obtain an initial ion density profile with q minima

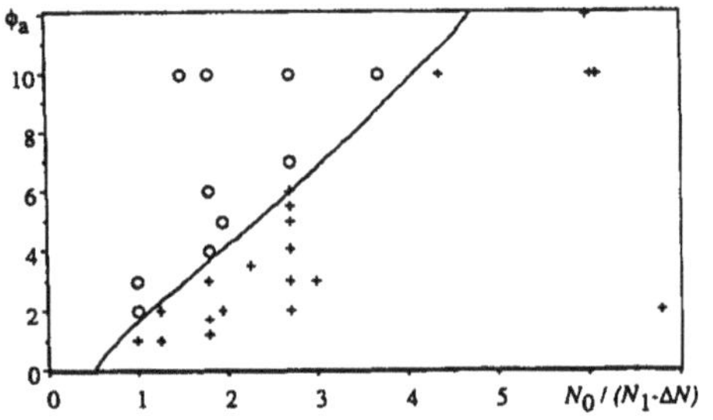

Fig.4. The circles denote simulations giving voltage drops concentrated in a cathode sheath and the crosses those giving a potential drop over the initial cavity. The line gives the maximum value of the applied voltage ϕ_a consistent with the existence of a potential minimum according to the theory presented in the next Section (Eq. (7) for ϕ_c=0).

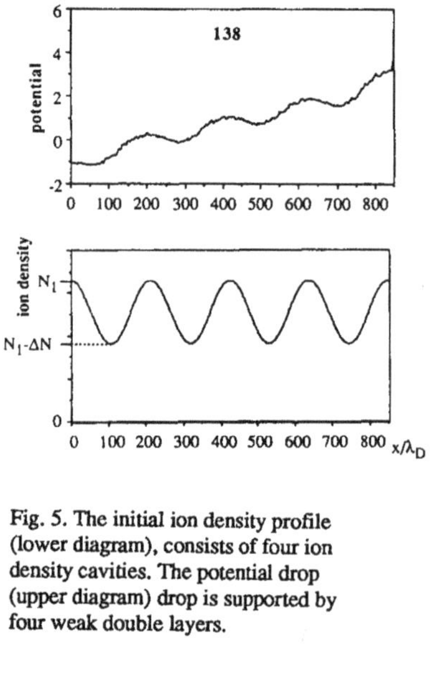

Fig. 5. The initial ion density profile (lower diagram), consists of four ion density cavities. The potential drop (upper diagram) drop is supported by four weak double layers.

and maximum at each boundary. The lowest diagram in Fig. 5 shows the initial ion density profile for q=4. The upper diagram shows an example of the potential profiles at t= 138 ω_{pe}^{-1} (ϕ_a= 5, N_0/N_1=2.3, and $\Delta N/N_1$ = 0.45). After an initial transient, the potential profiles remained approximately constant during the simulation time, 200 ω_{pe}^{-1}. As an interesting effect, we see that the applied voltage drop is shared between the cavities during the simulation time so that four weak double layers, "coupled in series", support the applied potential drop.

4. AN ANALYTICAL MODEL FOR THE RESPONSE OF A DIODE WITH A DENSITY CAVITY: The ion motion is assumed to be negligible during the first few electron transit times so that the ion density profile is given by the initial profile (1). The electrons present initially have left the diode after about an electron transit time, and here only the stationary

solution of the Vlasov equation will be considered. As observed in the simulations, we assume that the normalized potential ϕ decreases monotonically from the value ϕ_c at x=0 to a minimum at $x=x_0$, where we put $\phi=0$ (in the diagrams $\phi-\phi_c$ has been plotted), and then increases monotonically to $\phi_c+\phi_a$ at x=L. Here ϕ_a is the voltage drop applied over the diode. The electron number density can then be evaluated as a function of ϕ, and the quasi-neutrality condition is used to calculate $\phi(x)$. For very deep cavities ($\Delta N/N_1 \rightarrow 1$), further investigations, including Poisson's equation, are necessary.

The electron distribution function f_e is given by

$$f_e = N_0 \exp(-\phi_c+\phi)\, F_e(v), \qquad \mathrm{sgn}(x-x_0)v_e\sqrt{\phi} < v < \infty \tag{3}$$

$$f_e = N_0 \exp(-\phi_a-\phi_c+\phi) F_e(v), \quad -\infty < v < \mathrm{sgn}(x-x_0)v_e\sqrt{\phi} \tag{4}$$

Here $F_e(v)$ is a Maxwellian velocity distribution, normalized to unity when integrated over the interval $(0,\infty)$, and $v_e=(2k_BT_e/m_e)^{1/2}$. By integrating (3) and (4), the density becomes

$$n(\phi) = N_0 \exp(-\phi_c+\phi)(1-\exp(-\phi_a))(\coth(\phi_a/2) \pm \mathrm{erf}\sqrt{\phi}) \tag{5}$$

The upper sign gives the density $n_1(\phi)$ for $x<x_0$ and the lower sign the density $n_2(\phi)$ for $x>x_0$. erf is the error function normalized so that it approaches unity for large values of its argument. $n_1(\phi)$ increases monotonically. By differentiation it can be shown that $n_2(\phi)$ decreases from n(0) to a minimum for $\phi=\phi_m$ and increases monotonically for $\phi>\phi_m$. ϕ_m is determined by only ϕ_a through the relation

$$\coth(\phi_a/2) = \mathrm{erf}\sqrt{\phi_m} + \exp(-\phi_m)/(\pi\phi_m)^{1/2} \tag{6}$$

(6) defines ϕ_m as a monotonically increasing function of ϕ_a. Here we consider ϕ_a as given, and ϕ_m is accordingly known. Equation (6) is used to eliminate $\coth(\phi_a/2)$ in (5). Then, as a first condition, we require that ϕ_c must assume such a value that quasi-neutrality prevails at the density minimum, $\phi=\phi_m$ and x=L/2. This gives

$$N_1-\Delta N = N_0\{1-\exp(-\phi_a)\}\exp(-\phi_c)/(\pi\phi_m)^{1/2} \tag{7}$$

Eq. (7) determines ϕ_c, which can be evaluated by specifying values of $N_0/(N_1-\Delta N)$ and ϕ_a. As has been shown elsewhere[17], (7) defines ϕ_c as a monotonically <u>decreasing</u> function of ϕ_a for any fixed value of $N_0/(N_1-\Delta N)$, and ϕ_c vanishes for a critical value $\phi_a = \phi_{al}$. For larger values of ϕ_a, there are no solutions with a potential minimum in the diode. ϕ_{al} is a function of only $N_0/(N_1-\Delta N)$ which already has been shown in Fig. 4.

We can now normalize n_1 and n_2 to the minimum density given by the right hand side of (7). This gives the normalized electron number densities F_1 and F_2, given by

$$F(\phi) = \exp(\phi-\phi_m) + (\pi\phi_m)^{1/2}\exp(\phi)(\mathrm{erf}\sqrt{\phi_m} \pm \mathrm{erf}\sqrt{\phi}\,) \qquad (8)$$

Here the upper sign gives F_1 and the lower sign F_2. The electron number density at any potential level is now known, and the quasi-neutrality condition becomes

$$(N_1-\Delta N)F(\phi) = N_1 - \Delta N \sin^2(\pi\, x/L) \qquad (9)$$

Eq. (9) gives x explicitly as a function of ϕ when values of $\Delta N/N_1$ and ϕ_a are specified. n_i (x) varies little over the sheath regions, and the density $N_{i1} \approx N_{e1}$ at the sheath-plasma boundaries can be approximated by N_1. Then we obtain the potential levels ϕ_1 and ϕ_2 at the sheath-plasma boundaries from $F_1(\phi_1) = F_2(\phi_2) = N_1/(N_1-\Delta N)$, and the sheath potential drops, $\phi_c - \phi_1$ and $\phi_c + \phi_a - \phi_2$, can be evaluated.

Fig. 6a and 6b demonstrate the good agreement between the analytical profiles and the profiles simulated. These are averaged over 10 ω_{pe}^{-1}. The sheath potential drops evaluated theoretically are indicated in the theoretical diagrams by straight lines, and they agree well with the simulations. The shape of the profiles are independent of N_0 according to the analytical model. A variation of N_0 only shifts the potential level relative to the boundaries by a change of ϕ_c and thereby the sheath potential drops. This agrees with the simulation results as shown in Fig.6b.

5. SIMULATIONS OF A LONGER SYSTEM WITH ELECTRON-BEAM PLASMA INTERACTION: The electron distribution functions simulated show good agreement with those given by (3) and (4) for the diode lengths considered so far. An example of this is given in Fig. 7. The parameters are the same as those used in Fig. 6b (N_0/N_1=1.04), and the velocity distribution is evaluated at $x=87\lambda_D$. There the distribution function is unstable giving beam-plasma interaction. To get a first evidence of modifications of the cavity potential profile due to this interaction, a longer system was considered. The initial ion number density $n_{ip}(x)$ was given by (1) in the interval $0 < x/\lambda_D < 100$ and by a constant value, $n_{ip}(x)=N_1$ in $100 < x/\lambda_D < 200$. In this way an interaction region of 100 λ_D was introduced. Examples of profiles from such simulations are shown in Fig. 8. Apart from the system length, the parameters are again the same as in Fig. 6b (N_0/N_1=1.04). Fluctuations propagating in the direction of the beam

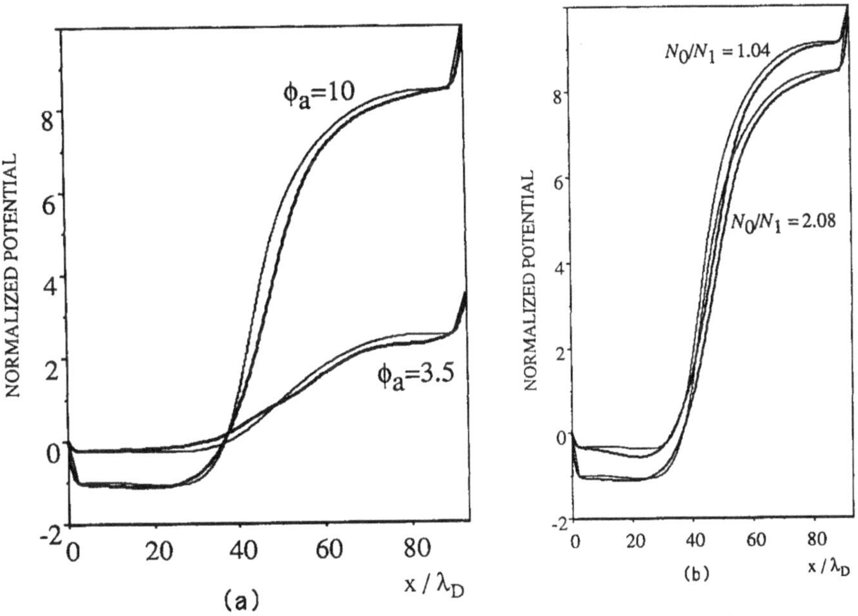

Fig. 6. Comparison between potential profiles evaluated theoretically (thin lines) and profiles simulated (heavy lines). The profiles simulated are averaged over 10 ω_{pe}^{-1}, and the average profiles remained constant during the simulation time (about 200 ω_{pe}^{-1}), after the initial transient. The sheath potential drops, evaluated by determining ϕ_1, ϕ_2 and ϕ_c, are indicated by straight lines in the theoretical profiles. In Fig. 6a, $\Delta N/N_1$=0.54, N_0/N_1=1.04, $N_0/(N_1-\Delta N)$=2.26 (ϕ_a=3.5) and $\Delta N/N_1$=0.83, N_0/N_1=2.08, $N_0/(N_1-\Delta N)$=12.2 (ϕ_a=10). Fig. 6b demonstrates that the cavity profiles are essentially unchanged when N_0 is varied, and only the sheath potential drops changes (ϕ_a=10, $\Delta N/N_1$=0.83).

are now excited. However, as shown by the two lower diagrams in Fig. 8, the potential profile over the cavity, averaged over $10\omega_{pe}^{-1}$, agrees with the one in the shorter system. The theory presented is accordingly still valid in the cavity region. From plots of profiles at different times, the propagation velocity was estimated to $0.7v_b$. Here v_b is the velocity that an electron, starting at rest, would gain in the cavity potential drop. Due to the large phase velocity, the thermal electrons, that are injected at the high potential boundary and move towards the cavity, are not trapped by the waves. Hence the potential profile over the cavity remains the same as in the shorter system. Studies of phase space diagrams show the formation of phase space vortices. The velocity distribution has become stable at $x=200\lambda_D$, and the beam velocity distribution has been replaced by a plateau. The electrostatic energy density shows a sharp maximum for x/λ_D between 100 and 150.

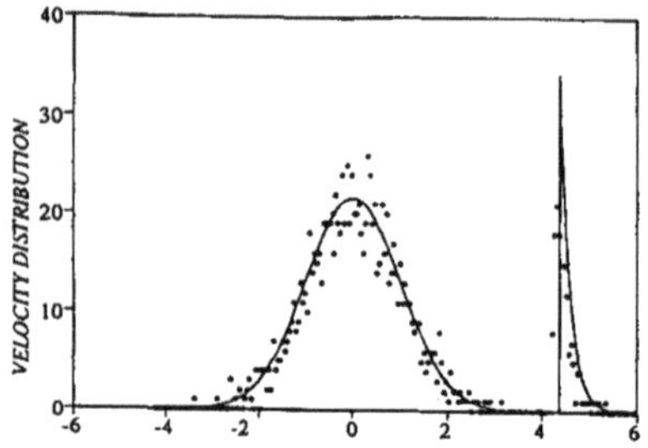

Fig.7. Comparison between the theoretical electron velocity distribution (full lines) and the simulated one (dots). The parameters are the same as in Fig. 6b (N_0/N_1=1.04), and the position is x=87λ_D. The velocity is given in units of $v_t = (k_B T_e/m_e)^{1/2}$. The sampled velocity interval is 0.05 v_t and the space interval 2λ_D.

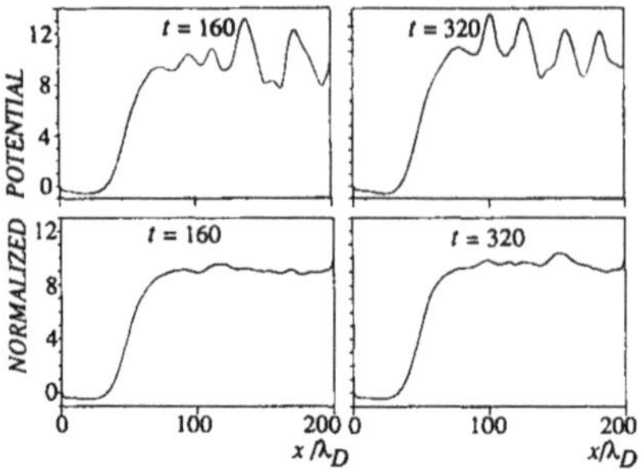

Fig. 8. The parameters are the same as in Fig.6b (N_0/N_1=1.04) but an interaction region of 100λ_D with a constant ion number density is added. The upper diagrams show instantaneous potential profiles and the lower ones, averages over $10\omega_{pe}^{-1}$. The time-averaged cavity potential drop remains the same as for the shorter system. For x=200λ_D, a plateau has formed in the velocity distribution.

6. DISCUSSION AND CONCLUSIONS: The cavity potential profiles obtained from the simulations and the analytical model are similar to those observed experimentally, and the limiting values of the potential, ϕ_{al} shown in Fig. 4, are consistent with the experimental results.[9,10] The normalized values of the applied potential drops in the experiments are comparable since the electron temperature is much lower in the Q-machine than in the triple plasma machine. However, new experiments are required to get conclusive experimental evidence of the voltage supporting capacity of an ion density cavity.

The good agreement between the potential profiles simulated and those calculated shows that the theoretical approach gives an appropriate description of the potential drops over a cavity. These do not depend on the particular shape of the ion density profile in the sheaths but only on the cavity depth, and they are also independent of the sheath potential drops. The basic assumption in the theoretical model is the existence of a potential minimum. This introduces the depth of the minimum as a free parameter which is determined so that quasi-neutrality can be maintained. As another interesting consequence of the theory we have found a condition for the existence of a potential minimum in the diode and shown that this coincides with the condition for the existence of a cavity potential drop over the plasma. When a minimum exists, all the field lines from the surface charge on the diode plates end on charges in the plasma. The plates are "screened" from each other. When the position of the potential minimum reaches the cathode plate, the negative charge distributed in the diode exactly compensates for the positive charge on the anode plate. For larger voltage drops applied, the field from the the anode plate begins to penetrate the plasma diode, and this leads to a new state as shown by the simulations.

ACKNOWLEDGMENT: Most of the work presented has been financed by the Swedish Natural Science Research Council. We wish to thank Dr P. Gray for providing us with a version of the PDW1 code and Dr N. Brenning, Dr A. Peratt, Dr M. A. Raadu, and Dr N. Singh for useful discussions. We also thank the National Supercomputer Centre, Linköping, Sweden.

REFERENCES

1. S. Torvén and D. Andersson, *J. Phys. D: Appl. Phys.* (1979), **12**, p. 717; S. Torvén and L. Lindberg, *J. Phys. D: Appl. Phys.* (1980), **13**, p. 2285
2. D. Andersson and J. Sörensen, *J. Phys. D: Appl. Phys.* (1983), **16**, p. 601.
3. D. L. Cooke and I. Katz, *Ionization - induced instability in an electron collecting sheath, S-Cubed* (1985) (unclassified report, Maxwell Laboratories, PO Box 1620, LaJolla, CA 92093, USA).
4. H. Fujita, K .Matsuo and S. Yagura, *Plasma Phys. Contr. Fusion* (1984), **26**, p. 1039; H. Fujita, S. Yagura and K. Matsuo, *Phys. Fluids* (1984), **27**, p. 2948.
5. S. Iizuka, P. Michelsen, J. Juul Rasmussen, R. Schrittwieser, R. Hatakeyama, K. Saeki and N. Sato, *Proc. XVth Int. Conf. on Phenomena in Ionized Gases (Minsk)* (1981), Contributed Papers, p. 1020
6. N. Hershkowitz, G. L. Payne, C. Chan and J. R. DeKock, *Plasma Physics* (1981), **23**, p.903.
7. N. Sato, R. Hatakeyama, S. Iizuka, T. Mieno, K. Saeki, J. Juul Rasmussen and P. Michelsen, *Phys. Rev. Lett.* (1981), **46**, p. 1330.
8. S. Torvén, *J. Phys. D: Appl. Phys.* (1982), **15**, p. 1943.
9. S. Iizuka, P. Michelsen, J. Juul Rasmussen, R. Schrittwieser, R. Hatakeyama, K. Saeki and N. Sato, *J. Phys. Soc. Japan* (1985), **54**, p. 2516.
10. I. Axnäs, M. Bohm and S. Torvén, *1992 Int. Conf. Plasma Physics*, Innsbruck, Contributed Papers
11. N. Singh, *Plasma Phys.* (1982), **24**, p. 639.
12. H. L. Rowland and P. J. Palmadesso, *J. Geophys. Res.* (1983), **88**, p. 7997; H. L. Rowland and P. J. Palmadesso, *J. Geophys. Res.* (1987), **92**, p. 299.
13. M. Temerin, K. Cerny, W. Lotko, and F. S. Mozer, *Phys. Rev. Lett.*. (1982), **48**, p.1175; R. Boström, G. Gustafsson, B. Holback, G. Holmgren, H. Koskinen and P. Kintner, *Phys. Rev. Lett.* (1988), **61**, p. 82.
14. E. Infeld, G. Rowlands and S. Torvén, *Phys. Rev. Lett.* (1989), **62**, p. 2269.
15. N. G. Belova, A. A. Galeev, R. Z. Sagdeev and Y. S. Sigov, *JETP Lett.* (1980), **31**, p. 518; M. A. Raadu and P. Carlqvist, Astrophys. Space Sci. (1981), **74**, p. 189; A. A. Galeev, R. Z. Sagdeev, V. D. Shapiro and V. I. Shevchenko, *Sovj. Phys. - JETP* (1981), **54**, p. 306.
16. M. Bohm and S. Torvén, *Phys. Rev. Lett.* (1991), **66**, p. 604.
17. M. Bohm and S. Torvén, *IEEE Transact. Plasma Sci.*, in the press
18. C. K. Birdsall and A. B. Langdon, *Plasma Physics via Computer Simulation* (1985), New York: McGraw-Hill; W. S. Lawson, *J. of Computational Phys.* (1989), **80**, p. 253.
19. S. Iizuka, P. Michelsen, J. Juul Rasmussen, R. Schrittwieser, R. Hatakeyama, K. Saeki and N. Sato, *J. Phys. E: Sci. Instrum.* , **14**, 1291.
20. M. Bohm, *Report TRITA-EPP-91-04* (1991), Dept. of Plasma Physics, Alfvén Laboratory, Royal Inst. of Technology, S-10044 Stockholm, Sweden.

DOUBLE LAYERS AND SOLITONS IN DUSTY PLASMAS

Frank Verheest, Instituut voor theoretische mechanika, Universiteit Gent, Krijgslaan 281, B–9000 Gent, Belgium

Nonlinear double layers and solitons are investigated in magnetized dusty plasmas. The dust dynamics is modelled by a number of cold, highly negatively charged and very massive fluids, in the presence of electrons and protons. Besides the usual modes several new low–frequency motions occur which are typical for the dust components. Some of the nonlinear modes are described by canonical model equations such as (modified) Korteweg–de Vries equations for electrostatic waves and the derivative nonlinear Schrödinger equation for electromagnetic waves. One can include in the formalism equilibrium drifts and even charge density waves, where the grain charges fluctuate with the perturbations in the plasma, typically at average dust plasma frequencies. Applications are given to dusty plasmas with one kind of negative grains, in the presence of hot and cold protons, all electrons having been accreted onto the dust grains. Most of the preceding conclusions are relevant for different kinds of astrophysical and heliospheric (cometary and magnetospheric) plasmas.

1. INTRODUCTION: In recent years there has been growing interest in the study of dusty plasmas. Dust is quite common throughout the universe and represents most of the solid matter in it. On the other hand, the gaseous component of matter is often fully or at least partially ionized, and it is the coexistence of dust and plasma which is called a dusty plasma. Moreover, in a dusty plasma the dust particles or grains are in general charged and will be affected by electromagnetic phenomena in many ways. So we will refer to dusty plasmas as plasmas which contain enough charged dust grains to make the distinction with ordinary plasmas meaningful. One can readily think of space applications such as planetary rings, asteroid zones and cometary comae and tails, and also in Earth's lower magnetosphere. These have recently been reviewed by Goertz[1] and Northrop.[2]

The typically micron–sized grains can have very high negative charges, up to $10^4 \sim 10^6$ times the electron charge, and even higher masses, of the order of $10^6 \sim 10^8$ times the proton mass. The grains are mostly negative because the electron flux onto the dust is usually much larger than the positive ion flux. The characteristic frequencies of the dust components of dusty plasmas are thus considerably smaller than the corresponding electron or ion quantities. Also, due to the high charges of the dust grains and in order to maintain charge neutrality in equilibrium, the number density of electrons is much smaller than that of the positive ions, the balance in negative charge being made up by the dust grains. Another characteristic of such dusty plasmas is that the charge–to–mass ratio of the dust grains is not fixed at all, but one encounters a great many possibilities.

Depending on their concentration, one can either focus attention on the dust grains in a plasma from a particle dynamics point of view, or one can look at the collective effects of the charged dust on the plasma. Processes such as radial mass transport and angular momentum redistribution in planetary ring systems, arising due to the non-Keplerian orbits of the charged dust grains, seem to be governed by collective processes like waves and instabilities in dusty plasmas. In a first approximation the charged dust is then described as one or more fluids, besides the classic electron and positive ion fluids. Many-fluid descriptions for collective modes in microparticle plasmas have been first given by James and Vermeulen,[3] elaborating on our earlier theoretical work for linear multispecies waves.[4] It is only very recently that attention has returned to such plasmas, following the increasingly successful space missions to comets and planets. For low-frequency and long-wavelength modes the dust grains can be described essentially as negative ions, albeit with charge-to-mass ratios which can vastly differ from what one is accustomed to.

Ion- and dust-acoustic modes in dusty plasmas have recently been treated by a number of authors,[5–13] usually on the basis of Korteweg-de Vries (KdV) equations and modified versions thereof. We will come back to their conclusions later on. Since the charges, sizes and masses of the dust grains are usually widely distributed, we will allow in the theoretical framework for a multispecies description of the different dust grain species. Also, equilibrium drifts for the dust grains can be included, so that the dust fluids could move with respect to the electron and positive ion constituents of the plasma. The possibility to distinguish between hot and cold ions is of crucial importance in allowing for dust-acoustic double layers. Where the equations will be given in a nondimensional form, the effective temperature takes into account all isothermal species, electrons as well as hot and cold positive ions. A typical dust-acoustic velocity then incorporates both the effective temperature and a typical dust grain mass. In the limit one could even restrict the plasma to just one kind of positive ions and one class of dust grains. The electrons would then all have been accreted onto the dust grains, as is thought to be the case *e.g.* in the F-ring of Saturn.[1,2] We will focus on this case in some of our applications.

On the other hand, the charged dust grains also modify and affect the electromagnetic modes, in first instance through the substantial lowering of the Alfvén speed.[14] For the nonlinear evolution of low-frequency Alfvén modes the characteristic equation is the derivative nonlinear Schrödinger (DNLS) equation, and for this purpose recent multi-species descriptions can be used.[15–17]

We would also like to mention the self-similar expansion of dusty plasmas in a vacuum, given both through a fluid description[18] and from a kinetic theory point of view.[19] The expansion is governed by the mass of the charged dust particles and by the pressure of the usual plasma particles. Finally, in non-uniform dusty plasmas, several phenomena have been treated such as the Kelvin-Helmholtz instability,[20] linear and nonlinear dust drift waves[21] and the appearance of dipolar vortices,[22] all with particular emphasis on the specific influence of the charged dust grains.

2. MODEL AND FORMALISM: We start from the well known set of basic fluid equations per species, namely the continuity equations

$$\frac{\partial}{\partial t}\sigma_\alpha + \nabla \cdot (\sigma_\alpha \mathbf{u}_\alpha) = 0, \tag{1}$$

and the equations of motion

$$\left(\frac{\partial}{\partial t} + \mathbf{u}_\alpha \cdot \nabla\right)\mathbf{u}_\alpha + \frac{1}{\rho_\alpha}\nabla p_\alpha = \frac{\sigma_\alpha}{\rho_\alpha}(\mathbf{E} + \mathbf{u}_\alpha \times \mathbf{B}). \tag{2}$$

In these equations σ_α, ρ_α, $\mathbf{u}_\alpha$ and p_α respectively refer to the charge density, mass density, fluid velocity and scalar pressure per species, labelled with the generic subscript α. When needed, the subscripts e, i and d will refer to the electrons, the positive ions (protons usually) and the negative dust grains. $\mathbf{E}$ and $\mathbf{B}$ are the electric and magnetic fields, respectively, for the evolution of which we also need Maxwell's equations

$$\nabla \times \mathbf{E} + \frac{\partial}{\partial t}\mathbf{B} = 0, \tag{3}$$

$$c^2 \nabla \times \mathbf{B} = \frac{\partial}{\partial t}\mathbf{E} + \frac{1}{\varepsilon_0}\sum_\alpha \sigma_\alpha \mathbf{u}_\alpha, \tag{4}$$

$$\nabla \cdot \mathbf{E} = \frac{1}{\varepsilon_0}\sum_\alpha \sigma_\alpha. \tag{5}$$

In equilibrium the plasma is globally charge and current neutral, if we admit (in a de Hoffmann–Teller frame) possible drifts U_α parallel to the static magnetic field $\mathbf{B}_0$,

$$\sum_\alpha \sigma_{\alpha 0} = 0, \qquad \sum_\alpha \sigma_{\alpha 0} U_\alpha = 0. \tag{6}$$

The subscript 0 is used to denote the corresponding equilibrium quantities. The reason why charge densities σ_α are used here, rather than the customary expressions $n_\alpha q_\alpha$ in function of the number densities n_α and the charges q_α, is that for the dust grains the charge q_α is not really a fixed quantity. The implication is that, whereas for ordinary electron or ion oscillations charge density fluctuations are really number density fluctuations,

$$\delta\sigma_e = -e\delta n_e, \tag{7}$$

for the dust one could have fluctuations in the charge itself due to the wave,

$$\delta\sigma_d = -Z_{d0} e \delta n_e - e n_{d0} \delta Z_d. \tag{8}$$

Here Z_d is the number of unit charges for a given dust species, and Z_{d0} its average value. We will return to this question later.

3. ELECTROSTATIC MODES, DOUBLE LAYERS AND SOLITONS: We cannot deal with all aspects of electrostatic modes, but will emphasize the dust–acoustic modes most typical for the dust constituents. We consider isothermal electrons and hot and cold protons, in the presence of a number of cold negatively charged dust grain species and assume that the characteristic frequency is much smaller than the dust–cyclotron frequencies. Since the linear wave velocity for dust–acoustic modes is much smaller than the electron and proton thermal velocities, we neglect electron and proton inertia and use Boltzmann distributions for their densities.[1,8,9] The nonlinear behaviour of dust–acoustic electrostatic modes propagating parallel to the z–direction is then described by the following set of fluid equations for each dust species,

$$\frac{\partial \rho_d}{\partial t} + \frac{\partial}{\partial z}(\rho_d u_d) = 0, \tag{9}$$

$$\frac{\partial u_d}{\partial t} + u_d \frac{\partial u_d}{\partial z} = Q_d \frac{\partial \phi}{\partial z}, \tag{10}$$

together with Poisson's equation for the plasma as a whole,

$$\frac{\partial^2 \phi}{\partial z^2} = N_e e^{\beta_e \phi} - N_{ic} e^{-\beta_{ic}\phi} - N_{ih} e^{-\beta_{ih}\phi} + \sum_d Q_d \rho_d. \tag{11}$$

The subscripts ic and ih now refer to the cold and hot isothermal proton species. In the above equations lengths and time have been expressed in units of a typical Debye length $(\varepsilon_0 \kappa_B T_{\text{eff}}/n_0 e^2)^{\frac{1}{2}}$ and plasma period $\omega_{\text{ref}}^{-1} = (\varepsilon_0 m_{\text{ref}}/n_0 e^2)^{\frac{1}{2}}$, where κ_B is Boltzmann's constant, $n_0 = n_{ic0} + n_{ih0}$ the equilibrium number density of all the isothermal protons together and m_{ref} a typical dust grain mass. We have normalized mass densities ρ_d by $n_0 m_{\text{ref}}$, fluid velocities u_d by a reference dust–acoustic velocity $c_S = (\kappa_B T_{\text{eff}}/m_{\text{ref}})^{\frac{1}{2}}$, the absolute values of the charge-to-mass ratios Q_d by e/m_{ref} and finally the electrostatic potential ϕ by $\kappa_B T_{\text{eff}}/e$. The effective temperature, which is taken as the appropriate mean for all isothermal species

$$T_{\text{eff}} = \frac{T_e T_{ic} T_{ih}}{(N_{ic} T_{ih} + N_{ih} T_{ic}) T_e + N_e T_{ic} T_{ih}}, \tag{12}$$

generalizes what is used in the treatment of ordinary ion–acoustic modes, and the different β's are given by $\beta_\alpha = T_{\text{eff}}/T_\alpha$.

We now look at wave phenomena travelling in the z–direction with a dimensionless velocity (or Mach number) M, so that all quantities will depend upon $\xi = z - Mt$. The above equations may then be integrated with the one-sided boundary conditions

$$\phi,\ \frac{d\phi}{d\xi} \to 0, \qquad \rho_d \to \rho_{d0}, \qquad u_d \to U_d \qquad \text{as} \qquad \xi \to +\infty. \tag{13}$$

Since the electron and proton densities are described by Boltzmann distributions, this implies that we are working in a reference frame where the usual plasma is essentially at rest. Charge neutrality in equilibrium means that

$$N_e + \sum_d Q_d \rho_{d0} = N_{ic} + N_{ih} = 1, \tag{14}$$

and we find from the fluid equations (9)–(11) that

$$\rho_d = \rho_{d0}\left(1 + \frac{2Q_d\phi}{(M-U_d)^2}\right)^{-\frac{1}{2}}, \tag{15}$$

$$u_d = M - (M - U_d)\left(1 + \frac{2Q_d\phi}{(M-U_d)^2}\right)^{\frac{1}{2}}. \tag{16}$$

This allows us to integrate Poisson's equation (11) in the form

$$\frac{1}{2}\left(\frac{\mathrm{d}\phi}{\mathrm{d}\xi}\right)^2 + V(\phi, M) = 0, \tag{17}$$

with the Sagdeev potential $V(\phi, M)$ given by

$$V(\phi, M) = \frac{N_e}{\beta_e}\left(1 - e^{\beta_e\phi}\right) + \frac{N_{ic}}{\beta_{ic}}\left(1 - e^{-\beta_{ic}\phi}\right) + \frac{N_{ih}}{\beta_{ih}}\left(1 - e^{-\beta_{ih}\phi}\right)$$
$$+ \sum_d \rho_{d0}(M - U_d)^2\left\{1 - \left(1 + \frac{2Q_d\phi}{(M-U_d)^2}\right)^{\frac{1}{2}}\right\}. \tag{18}$$

Equation (17) represents the energy integral for a classical particle of unit mass moving with velocity $\mathrm{d}\phi/\mathrm{d}\xi$ in a potential $V(\phi, M)$, M appearing as a parameter. In order that mass densities and velocities remain real, there is a limitation on the $\phi < 0$ range but solutions with $\phi > 0$ are *a priori* unrestricted. A note of caution has to be sounded here concerning the use of the labels *compressive* and *rarefactive* to characterize the soliton or double layer solutions. For positive ϕ both the positive ion and the charged dust densities *decrease*, whereas the electron density increases. Since in the dust–acoustic part of the spectrum the most important role is *not* played by the electrons, but by the charged dust grains and the protons, it would seem logical to label modes with $\phi > 0$ rarefactive and those with $\phi < 0$ compressive. This is the reverse of what occurs for ordinary ion–acoustic modes, where compressive or rarefactive refers to the behaviour of the electron density, and this has to be kept in mind when comparing the results of different authors.

For small but finite–amplitude dust–acoustic modes, double layers or solitons, we expand (18) up to fourth order in ϕ. This yields

$$\frac{1}{2}\left(\frac{\mathrm{d}\phi}{\mathrm{d}\xi}\right)^2 = \frac{A}{2}\phi^2 + \frac{B}{3}\phi^3 - \frac{C}{4}\phi^4, \tag{19}$$

with

$$A = 1 - \sum_d \frac{Q_d^2 \rho_{d0}}{(M - U_d)^2}, \tag{20}$$

$$B = \frac{3}{2} \sum_d \frac{Q_d^3 \rho_{d0}}{(M - U_d)^4} + \frac{1}{2}(N_e \beta_e^2 - N_{ic} \beta_{ic}^2 - N_{ih} \beta_{ih}^2), \tag{21}$$

$$C = \frac{5}{2} \sum_d \frac{Q_d^4 \rho_{d0}}{(M - U_d)^6} - \frac{1}{6}(N_e \beta_e^3 + N_{ic} \beta_{ic}^3 + N_{ih} \beta_{ih}^3). \tag{22}$$

The soliton solutions to (19) are[6,8,11]

$$\phi_{\mp}(\xi) = -\frac{6A \operatorname{sech}^2(\kappa\xi)}{a_{\pm} - a_{\mp} \tanh^2(\kappa\xi)}, \tag{23}$$

where the integration constant in ξ has been put equal to zero for simplicity, and

$$\kappa = \frac{\sqrt{A}}{2}, \tag{24}$$

$$a_{\pm} = 2B \pm \sqrt{4B^2 + 18AC}. \tag{25}$$

In order that we have localized solutions, A has to be positive, which leads to supersonic modes with $M > c_{ph}$. The phase velocity c_{ph} itself obeys the linear dispersion law

$$\sum_d \frac{Q_d^2 \rho_{d0}}{(c_{ph} - U_d)^2} = \sum_d \frac{\omega_{pd}^2}{(c_{ph} - U_d)^2} = 1. \tag{26}$$

Here ω_{pd} refers to the dimensionless plasma frequencies associated with the charged dust grains. For one dust species without equilibrium streaming the phase velocity is of the order of the dust–acoustic velocity in dimensional units, usually much lower than in ordinary plasmas because the average mass of a charged dust grain is so big.

In order to see what the solutions (23) imply, we specialize the above expressions to a dusty plasma consisting only of one kind of protons and one kind of charged dust grains. This is admittedly a very extreme occurrence, but such a situation could model in first approximation what one encounters in the F–ring of Saturn, where there are few free electrons left, as almost all electrons have been accreted onto the dust grains. Then the coefficients (20)–(22) become

$$A = 1 - \frac{1}{M^2}, \tag{27}$$

$$B = \frac{3}{2M^4} - \frac{1}{2}, \tag{28}$$

$$C = \frac{5}{2M^6} - \frac{1}{6}, \tag{29}$$

while (26) shows that $c_{ph} = 1$. We then use for weakly supersonic solutions that

$$M \simeq 1 + \mu^2, \tag{30}$$

with μ small, and find that up to terms of $\mathcal{O}(\mu^2)$ (27)–(29) reduce to

$$A \simeq 2\mu^2, \tag{31}$$

$$B \simeq 1 - 6\mu^2, \tag{32}$$

$$C \simeq \frac{7}{3} - 15\mu^2. \tag{33}$$

The solution for the compressive soliton becomes

$$\phi_-(\xi) = -\frac{12\mu^2 \mathrm{sech}^2(\mu\xi/\sqrt{2})}{4 - 3\mu^2 + 21\mu^2 \tanh^2(\mu\xi/\sqrt{2})}, \tag{34}$$

whereas the rarefactive soliton would be described by

$$\phi_+(\xi) = \frac{12\mu^2 \mathrm{sech}^2(\mu\xi/\sqrt{2})}{21\mu^2 + (4 - 3\mu^2)\tanh^2(\mu\xi/\sqrt{2})}. \tag{35}$$

While the maximum amplitude of the compressive soliton,

$$|\phi_-|_{\max} = \frac{12\mu^2}{4 - 3\mu^2} \simeq 3\mu^2, \tag{36}$$

is fully within the approximations used so far, the rarefactive soliton amplitude,

$$|\phi_+|_{\max} = \frac{4}{7} = 0.571, \tag{37}$$

really is outside the acceptable limit for the series expansion used in deriving (19), as the series converges then much too slowly. This can easily be checked by going to the next terms in the expansions. So only small–amplitude compressive solitons are possible, whereas rarefactive dust–acoustic solitons have to be described by the full Sagdeev potential (18). Ordinary ion–acoustic solitons only have positive localized humps in the potential, and thus would be rarefactive in our terminology!

Similar difficulties will be encountered when looking for double–layer solutions. Here the requirement is that $V(\phi, M)$ possesses a double root not only in $\phi = 0$, as was already the case for soliton solutions, but also in $\phi = \phi_{\max}$, so that the potential ϕ changes from 0 to $\phi_{\max}$. In between $V(\phi, M)$ has to be negative to guarantee real solutions. In the language of classical mechanics this ensures that a particle with coordinate ϕ transits just once from 0 to $\phi_{\max}$ or *vice versa*. Equation (19) is rewritten in the form

$$\frac{1}{2}\left(\frac{d\phi}{d\xi}\right)^2 = -\frac{C}{4}\phi^2(\phi - \phi_{\max})^2, \tag{38}$$

and the conditions for *weak* double layers thus follow as

$$\phi_{\max} = \frac{2B}{3C}, \qquad 2B^2 + 9AC = 0, \tag{39}$$

with $A > 0$, since it is obvious that C has to be negative. The double layer itself is compressive when $B > 0$, and rarefactive for $B < 0$, with typical solution

$$\phi = \frac{B}{3C}(1 - \tanh(\kappa\xi)), \tag{40}$$

where κ has been defined in (24). However, the requirements (39) imply that

$$B = -\frac{3A}{\phi_{\max}}, \qquad C = -\frac{2A}{\phi^2_{\max}}, \tag{41}$$

showing that the coefficients in the expansion tend to increase rapidly if $\phi_{\max}$ is small. So in general, the expansion will converge too slowly in order to describe even weak double layers by (19) or (38), as one will neglect then terms comparable to the ones retained,[23] and for double layers one has to return to the full Sagdeev potential (18). Unfortunately, this allows in most cases a numerical investigation only. The additional requirements for double layers on the Sagdeev potential (18) are that

$$\begin{aligned} V(\phi_{\max}, M) = {} & \frac{N_e}{\beta_e}\left(1 - e^{\beta_e\phi_{\max}}\right) + \frac{N_{ic}}{\beta_{ic}}\left(1 - e^{-\beta_{ic}\phi_{\max}}\right) + \frac{N_{ih}}{\beta_{ih}}\left(1 - e^{-\beta_{ih}\phi_{\max}}\right) \\ & + \sum_d \rho_{d0}(M - U_d)^2\left\{1 - \left(1 + \frac{2Q_d\phi_{\max}}{(M - U_d)^2}\right)^{\frac{1}{2}}\right\} = 0 \end{aligned} \tag{42}$$

and

$$\begin{aligned} \frac{\partial V}{\partial\phi}(\phi_{\max}, M) = {} & -N_e e^{\beta_e\phi_{\max}} + N_{ic}e^{-\beta_{ic}\phi_{\max}} + N_{ih}e^{-\beta_{ih}\phi_{\max}} \\ & - \sum_d Q_d\rho_{d0}\left(1 + \frac{2Q_d\phi_{\max}}{(M - U_d)^2}\right)^{-\frac{1}{2}} = 0. \end{aligned} \tag{43}$$

Moreover, the second derivatives have to be negative in both 0 and $\phi_{\max}$ in order to ensure the desired type of solution

$$\frac{\partial^2 V}{\partial\phi^2}(0, M) = -1 + \sum_d \frac{Q_d^2\rho_{d0}}{(M - U_d)^2} < 0, \tag{44}$$

$$\begin{aligned} \frac{\partial^2 V}{\partial\phi^2}(\phi_{\max}, M) = {} & -N_e\beta_e e^{\beta_e\phi_{\max}} - N_{ic}\beta_{ic}e^{-\beta_{ic}\phi_{\max}} - N_{ih}\beta_{ih}e^{-\beta_{ih}\phi_{\max}} \\ & + \sum_d \frac{Q_d^2\rho_{d0}}{(M - U_d)^2}\left(1 + \frac{2Q_d\phi_{\max}}{(M - U_d)^2}\right)^{-\frac{3}{2}} < 0. \end{aligned} \tag{45}$$

The easiest case to discuss is when all electrons have been accreted onto the dust grains, as typical for the F–ring of Saturn, and the double layers are maintained by both ion species. For the values $N_{ih}/N_{ic} = 10$ and $T_{ih}/T_{ic} = 20$ and restricting for simplicity the description to one kind of charged dust grains, one finds $\phi_{\max} = 1.11$, hence a rarefactive double layer, propagating at a Mach number of 1.16.

If one admits a sprinkling of electrons, in the sense that the electrons have densities comparable to the cold proton density, $N_e \simeq N_{ic}$, but at the temperature of the hot protons, $T_e \simeq T_{ih}$, then the foregoing conclusions are qualitatively unaltered, except that both the maximum amplitude and the Mach number are lower, $\phi_{max} = 0.69$ and $M = 1.02$. For further details we refer to the recent paper by Bharuthram and Shukla.[9] This has to be contrasted with ordinary ion–acoustic double layers maintained by two electron populations, and which tend to be destroyed in the presence of negative ions.[23–28] So one sees that the influence of the massive dust fluid changes the picture quite radically!

4. CHARGE DENSITY FLUCTUATIONS: As remarked already, charge density fluctuations are possible because the charge on the dust grains is not really a fixed quantity. We therefore recall (9)–(11), but rewritten in terms of the charge and mass densities, rather than using mass densities alone and rather than assuming that Q_d is constant,

$$\frac{\partial \sigma_d}{\partial t} + \frac{\partial}{\partial z}(\sigma_d u_d) = 0, \tag{46}$$

$$\frac{\partial u_d}{\partial t} + u_d \frac{\partial u_d}{\partial z} = \frac{\sigma_d}{\rho_d}\frac{\partial \phi}{\partial z}, \tag{47}$$

$$\frac{\partial^2 \phi}{\partial z^2} = N_e e^{\beta_e \phi} - N_{ic} e^{-\beta_{ic}\phi} - N_{ih} e^{-\beta_{ih}\phi} + \sum_d \sigma_d. \tag{48}$$

Without equilibrium streaming this yields the linear dispersion law

$$\omega^2 = \frac{k^2}{1+k^2} \sum_d \frac{\sigma_{d0}^2}{\rho_{d0}} = \frac{k^2}{1+k^2} \sum_d \omega_{pd}^2. \tag{49}$$

We thus see that charge density fluctuations tend to occur at frequencies wholly determined by the dust components of the plasma. More generally, one can describe the dust by an average over a random distribution of charged dust particles.[29]

5. ELECTROMAGNETIC MODES AND SOLITONS: For right– or left–circularly polarized, parallel electromagnetic modes the dispersion law is

$$\omega^2 = c^2 k^2 + \sum_\alpha \frac{\omega \omega_{p\alpha}^2}{\omega \pm \Omega_\alpha}. \tag{50}$$

Here $\omega_{p\alpha}$ is the plasma frequency per species, defined in dimensional units through $\omega_{p\alpha}^2 = \sigma_{\alpha 0}^2/\varepsilon_0 \rho_{\alpha 0}$, and $\Omega_\alpha = \sigma_{\alpha 0} B_0/\rho_{\alpha 0}$ the signed gyrofrequency of each species. Equation (50) includes the electron, proton and dust contributions. If we deal with low–frequency modes such that $|\Omega_d| \ll \omega \ll \Omega_i \ll |\Omega_e|$, we can approximate (50) as

$$\omega^2 \left(1 + \frac{c^2(\rho_{\text{total}} - \rho_{\text{dust}})}{V_A^2 \rho_{\text{total}}}\right) \pm \omega \sum_d \frac{\omega_{pd}^2}{\Omega_d} - c^2 k^2 - \sum_d \omega_{pd}^2 = 0. \tag{51}$$

Here $\rho_{\rm total}$ and $\rho_{\rm dust}$ refer to the total mass density of the plasma as a whole and to the mass density of all the dust grains, respectively. Furthermore, V_A is the Alfvén velocity of the whole plasma, including the charged dust. Since the dust grains typically have a much lower charge–to–mass ratio than ordinary protons, the Alfvén velocity V_A is usually much smaller than the usual one, $V_{A,u}$. So (51) admits modes with frequencies

$$\omega = \frac{1}{2}\omega_D \pm \frac{1}{2}\sqrt{\omega_D^2 + 4\omega_a^2}, \tag{52}$$

where

$$\omega_D = \mp \sum_d \frac{\omega_{pd}^2}{\Omega_d} \Big/ \left(1 + \frac{c^2}{V_{A,u}^2}\right), \quad \omega_a^2 = \left(c^2k^2 + \sum_d \omega_{pd}^2\right) \Big/ \left(1 + \frac{c^2}{V_{A,u}^2}\right), \tag{53}$$

since clearly

$$V_A^2 = \frac{B_0^2}{\mu_0 \rho_{\rm total}} = V_{A,u}^2 \frac{\rho_{\rm total} - \rho_{\rm dust}}{\rho_{\rm total}}. \tag{54}$$

At $k = 0$, the mode is fully controlled by the dust grains.[11] At even lower frequencies, such that $\omega \ll |\Omega_d|, \Omega_i, |\Omega_e|$, one sees that the modes (50) become purely Alfvénic,[30]

$$\omega^2 = k^2 V_A^2 \Big/ \left(1 + \frac{V_A^2}{c^2}\right), \tag{55}$$

albeit at a much lower Alfvén velocity than customary. So this is yet another indication that the dust modes have frequencies and phase velocities much lower than usual.

The nonlinear evolution of Alfvén waves in multispecies plasmas is typically described by a DNLS equation, obtained by reductive perturbation analysis and written in its dimensionless vector form as

$$F\frac{\partial \mathbf{B}_{\rm wave}}{\partial \tau} + \frac{G}{4}\frac{\partial}{\partial \xi}\left(|B_{\rm wave}|^2 \mathbf{B}_{\rm wave}\right) - H \mathbf{e}_B \times \frac{\partial^2 \mathbf{B}_{\rm wave}}{\partial \xi^2} = 0. \tag{56}$$

The coefficients F, G and H depend on the precise form of the pressure closure,[15–17] but are given by

$$F = \sum_d \rho_{d0}(c_{\rm ph} - U_d), \tag{57}$$

$$G = 1, \tag{58}$$

$$H = \sum_d \frac{\rho_{d0}(c_{\rm ph} - U_d)^3}{2Q_d}, \tag{59}$$

in the simplest case of a cold–plasma approach for the dust constituents and by neglecting electron and ion inertia. The linear phase velocity $c_{\rm ph}$ is then defined through

$$\sum_d \rho_{d0}(c_{\rm ph} - U_d)^2 = 1. \tag{60}$$

The dimensionless variables in the above expressions are similar to what was used for the electrostatic modes, except that the characteristic velocity is now the global Alfvén velocity rather than a typical dust- acoustic speed. For one dust species in the absence of free electrons, the normalization can be chosen such that $c_{\rm ph} = F = 1$ and $H = 1/2$. Again, it has to be remembered that the characteristic frequencies and velocities are much lower than in ordinary plasmas in the absence of charged dust. The DNLS equation (56) and related nonlinear equations do not admit double layer solutions, only various kinds of envelope solitons.[31] The paradigm is one of the vanishing solitons, given by[15]

$$|B_{\rm wave}|^2 = \frac{8FV}{\sqrt{2}\cosh\left(\frac{FV}{H}(\xi - V\tau)\right) - 1}, \tag{61}$$

where V refers to the envelope speed measured in a frame moving with $c_{\rm ph}$.

6. CONCLUSIONS: In this review we have been mostly concerned with low–frequency electrostatic and electromagnetic parallel waves in dusty plasmas in the presence of an external magnetic field. It is found that the presence of charged dust grains can significantly modify the existing wave spectra, and in particular that the dust dynamics introduce novel eigenmodes at the very low–frequency end. This is because the charged grains are much more massive than ordinary (negative) ions, and hence typical dust–acoustic and Alfvén velocities are so much lower than what we are accustomed to. Electrostatic dust–acoustic modes can propagate nonlinearly in the form of supersonic solitons of either positive or negative electrostatic potential. In the case of weak compressive solitons a modified KdV equation can be used. Also, if hot and cold positive ions are present, dust–acoustic double layers can be obtained, even in the total absence of free electrons, supported by the positive ion non- isothermality. These double layers, however, are only of the rarefactive type. On the other hand, the nonlinear evolution of parallel electromagnetic Alfvén modes is governed by a vector DNLS equation, which only admits soliton solutions.

While it is obvious that the above described modes can be of some help in the interpretation of low–frequency electrostatic and electromagnetic noise in many space and astrophysical plasmas, it has to be admitted that the observational data on dusty plasmas in space are at present too infrequent to permit a ready validation of the theoretical ideas. Nevertheless, since double layers can accelerate charged particles in space plasmas such as on the auroral field lines in planetary magnetospheres,[32] they could produce dust beams as evidenced in cometary environments. To conclude, dusty plasmas admit new and typical modes, although many of these have been left out of this review, in particular those propagating in oblique and perpendicular directions.

ACKNOWLEDGMENTS: It is a pleasure to thank W. Hereman, G.S. Lakhina, W. Malfliet, P.K. Shukla and W. Vermeir for enlightening discussions.

REFERENCES:

1. C.K. Goertz, *Rev. Geophys.* **27** (1989) 271–292
2. T.G. Northrop, *Physica Scripta* **45** (1992) 475–490
3. C.R. James and F. Vermeulen, *Canad. J. Phys.* **46** (1968) 855–863
4. F. Verheest, *Physica* **34** (1967) 17–35
5. U. de Angelis, V. Formisano and M. Giordano, *J. Plasma Phys.* **40** (1988) 399–406
6. N.N. Rao, P.K. Shukla and M.Y. Yu, *Planet. Space Sci.* **38** (1990) 543–546
7. N. d'Angelo, *Planet. Space Sci.* **38** (1990) 1143–1146
8. F. Verheest, *Planet. Space Sci.* **40** (1992) 1–6
9. R. Bharuthram and P.K. Shukla, *Planet. Space Sci.* **40** (1992) 465–471
10. M. Salimullah, M.H.A. Hassan and A. Sen, *Phys. Rev. A* **45** (1992) 5929–5934
11. P.K. Shukla, *Physica Scripta* **45** (1992) 504–507
12. P.K. Shukla and V.P. Silin, *Physica Scripta* **45** (1992) 508
13. R. Bharuthram, H. Saleem and P.K. Shukla, *Physica Scripta* **45** (1992) 512–514
14. P.K. Shukla and L. Stenflo, *Astrophys. Space Sci.* **190** (1992) 23–32
15. F. Verheest, *Icarus* **86** (1990) 273–282
16. F. Verheest and B. Buti, *J. Plasma Phys.* **47** (1992) 15–24
17. F. Verheest, *J. Plasma Phys.* **47** (1992) 25–37
18. K.E. Lonngren, *Planet. Space Sci.* **38** (1990) 1457–1459
19. H. Luo and M.Y. Yu, *Phys. Fluids B* **5** (1992) 1122–1125
20. N. d'Angelo and B. Song, *Planet. Space Sci.* **38** (1990) 1577–1579
21. P.K. Shukla, M.Y. Yu and R. Bharuthram, *J. Geophys. Res.* **96** (1991) 21343–21346
22. R. Bharuthram and P.K. Shukla, *Planet. Space Sci.* **40** (1992) 647–654
23. S. Baboolal, R. Bharuthram and M.A. Hellberg, *J. Plasma Phys.* **46** (1991) 247–254
24. R. Bharuthram and P.K. Shukla, *Phys. Fluids* **29** (1986) 3214–3218
25. S. Baboolal, R. Bharuthram and M.A. Hellberg, *J. Plasma Phys.* **40** (1988) 163–178
26. S. Baboolal, R. Bharuthram and M.A. Hellberg, *J. Plasma Phys.* **44** (1990) 1–23
27. R.V. Reddy and G.S. Lakhina, *Planet. Space Sci.* **39** (1991) 1343–1350
28. R.V. Reddy, G.S. Lakhina and F. Verheest, *Planet. Space Sci.* **40** (1992) to appear
29. A. Forlani, U. de Angelis and V.N. Tsytovich, *Physica Scripta* **45** (1992) 509–511
30. N.N. Rao, Phys. Res. Lab. (Ahmedabad, India) preprint PRL–TH/92–20 (1992)
31. T. Hada, C.F. Kennel and B. Buti, *J. Geophys. Res.* **94** (1989) 65–77
32. M.A. Raadu, *Phys. Reports* **178** (1989) 25

CONTRIBUTED PAPERS

EXPERIMENTS

COMPARATIVE STUDY OF THE CHAOTIC AND NONCHAOTIC DYNAMICS OF THE DOUBLE LAYERS

D.Alexandroaei, M.Sanduloviciu
Department of Plasma Physics, "Al.I.Cuza" University
6600 Iasi, Rumania

INTRODUCTION. Although in the last years a great number of papers have been devoted to the plasma double layers (DLs), the interest about this question has remained. Acting as storing and releasing energy elements /1 / DL can sustain periodical or non periodical variations of the current or other global parameters of the plasma in which DLs appear. The characteristics of these variations are given by the proper DL dynamics which is in connection with elementary processes taking place inside of the DL and the elements of the circuit which include DL-containing plasma /2/. By relating different time sequences (formation and disruption) of the evolution of an unstable DL with experimental parameters involving elementary process, it was possible to find the internal feedback and triggering mechanism for the maintaining of periodical processes in the plasma devices /3/.

Some previous experimental results /4/ show that the formation of multiple DL /5/ is correlated with the chaotic phenomena taking place in the plasma devices. In the present paper some experimental results concerning the dynamics of single and multiple DL are presented. These results prove that the global evolution of the plasma system containing such multiple DL is the result of the mutual influence of two or more DLs coupled by the current flowing through them. The chaotic regime (observed as chaotic variations of the current and other plasma parameters) appears when the correlation between the proper DLs dynamics does not exist.

EXPERIMENTAL PROCEDURE AND RESULTS. The experiments have been carried out using the arrangement shown in figure 1. It was extensively described in /3/. For certain experimental conditions (discharges currents, gas pressure) in the contact region of the two independent glow discharges a DL may appear when a biasing potential is applied between the ring anodes A_1 and A_2. In figure 2(a) a time-averaged $\bar{I}_{12}$ - $\bar{U}_{12}$ current voltage characteristic of the A_1A_2 discharge interval is presented. On this characteristic different behaviours of the plasma device are pointed out (AB-region of double layer formation (no oscillation); BC-single-period oscillations; CD-appearance of the second DL - single-period oscillations; DE - period-doubling oscillations; EF -chaotic oscillations). In figure 2(b) the main frequencies of the I_{12} current variations versus the biasing U_{12} potential are presented. The length of the vertical line segments is proportional to the amplitude of the I_{12} oscillations.

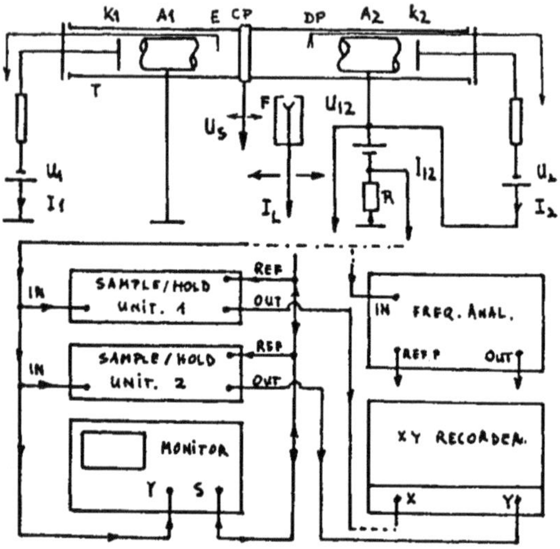

Figure 1

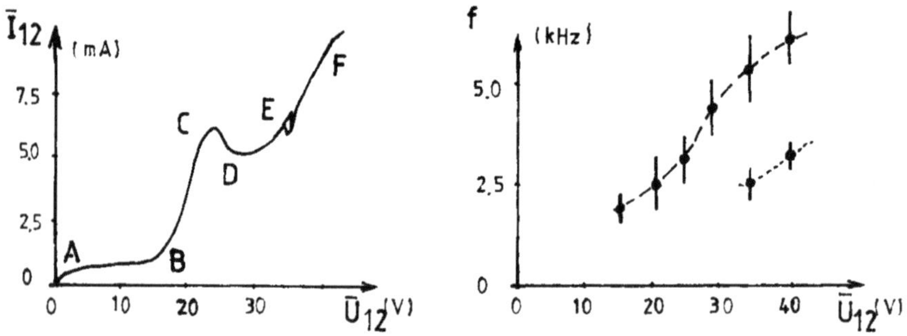

Figure 2. a) A time-averaged $\bar{I}_{12}$ - $\bar{U}_{12}$ current voltage characteristic of the A_1A_2 discharge interval; b) main frequencies of I_{12} current variations versus the biasing $\bar{U}_{12}$ potential (Argon, P=0.09 Torr, I_1=5mA, I_2=6mA)

The DL structure and its characteristics were tested using corresponding diagnostic procedures - both emissive and cold probe as well as photomultipliers. With a sample/ hold electronics, time-resolved measurements of some plasma parameters during DL dynamics were performed.

Different time sequences (during DL dynamics) were related with experimental parameters able to be correlated with elementary processes taking place in plasma in which DL is formed. It is observed that when the potential drop φ_{DL} is approximatively equal to the ionization potential of the used gas the DL becomes unstable. This fact is emphasized by the appearance of coherent variations of I_{12} current and U_{12} potential. These oscillations are correlated with the space-temporal variations of the plasma potential φ , light intensity I_L and other parameters of the plasma in the A_1A_2 region. For example, in figure 3, correlated with the time

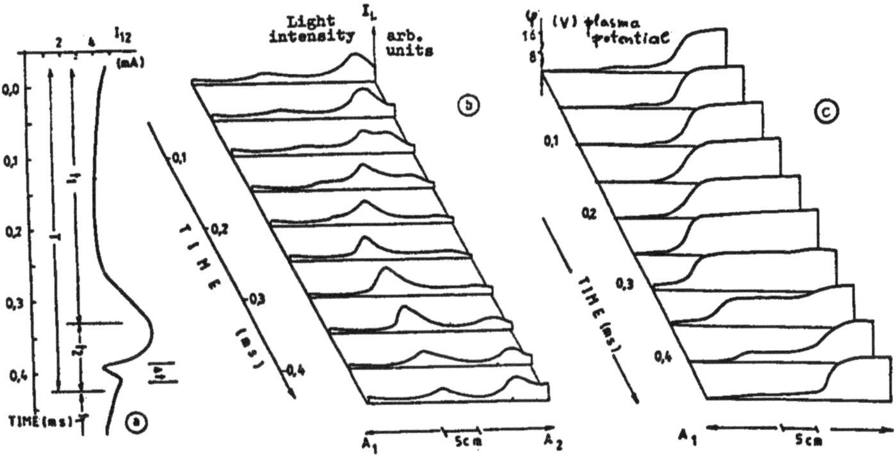

Figure 3.a) Time dependence of I_{12} current flowing between A_1 and A_2 in BC regime; b)time resolved I_L light intensity, c)φplasma potential distributions (P = 0.06 Torr, I_1 = 9 mA, I_2 = 8 mA)

dependence of I_{12} current, time-resolved plasma potential φ (c) and light intensity I_L (b) distributions along A_1A_2 region in the BC - regime [see figure 2(a)] are presented.

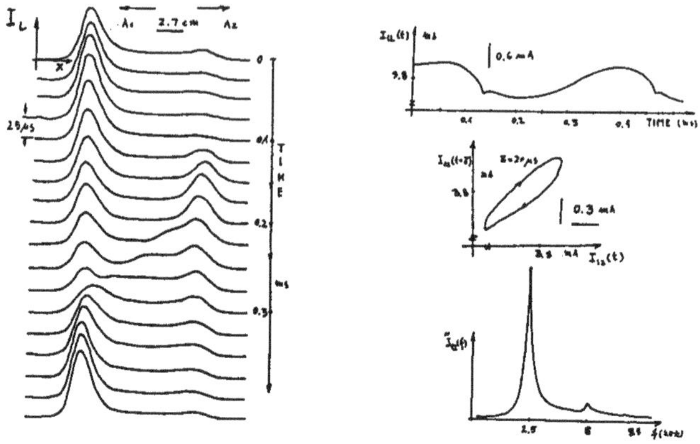

Fig.4 (a)

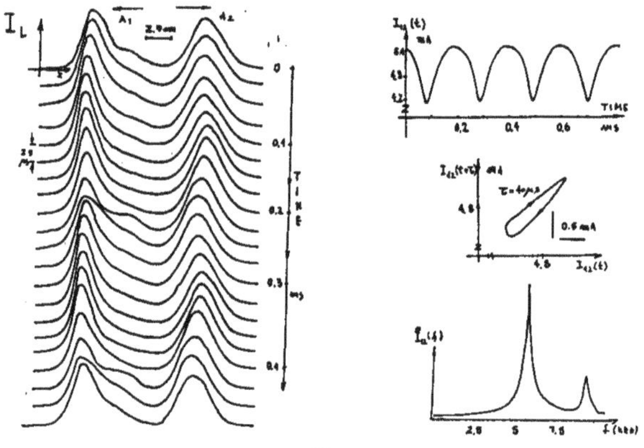

Fig. 4 (b)

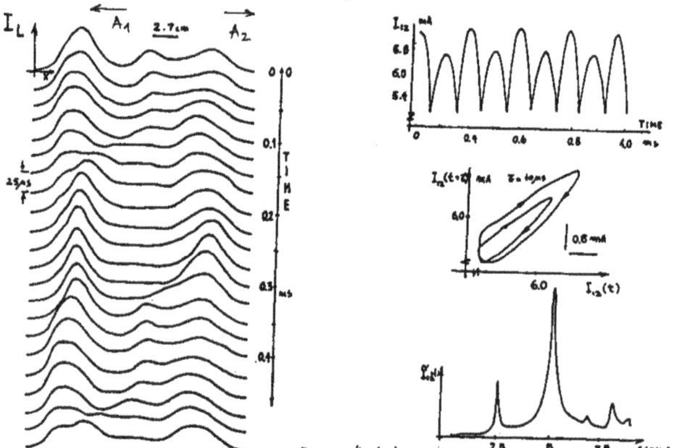

Fig. 4 (c)

Figure 4

For a comparative study of the periodical and pre-chaotic behaviours (period-doubling phenomenon - PDP) in figure 4(a,b,c) three kinds of space-time evolutions of the light intensity (correlated with plasma potential as in figure 3) in A_1A_2 region, which are correlated with the corresponding temporal variations of the I_{12} current, are presented.

The three presented situations correspond to the BC, CD and DE regions on the current voltage characteristic. The corresponding phase-plane plots $I_{12}(t)$ - $I_{12}(t + \tau)$ and the frequency spectra of I_{12} oscillations are also presented nearby in order to foresee the possibility of transition to chaos. As it is known PDP is a route for the transition to the chaotic behaviour of the plasma device /6/.

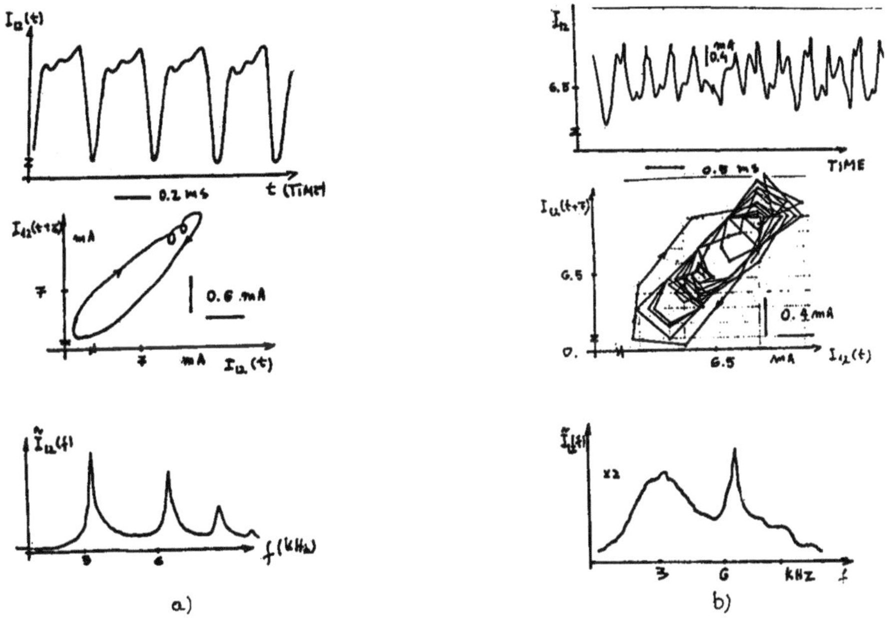

Figure 5

From the time-resolved light intensity distributions it is observed that the above mentioned PDP is connected with the appearance of the second DL (for corresponding $\bar{U}_{12}$

biasing potentials) at the positive side of the first formed one. Its proper dynamics is coupled to the dynamics of the first DL.

Other complicated behaviours together with the chaotic one (b) are presented in figure 5. In this last situation the simultaneous signals from capacitive probes /7/ placed between the A_1 and A_2 anodes show that the local oscillations of the plasma parameters are not in phase-constant relation on all length of the region of forming multiple DL.

DISCUSSION AND CONCLUSIONS. In the paper essential informations about the physical scenario involved in the appearance of the chaotic behaviour of a plasma device and its connection with the presence of the space charge structures - DLs - formed as result of a self-organization process are given.

The deterministic chaos appears after the formation of DL structures and the transition to chaos, via a period doubling route takes place under the conditions that the two coupled-by-current DLs present evolution which are not mutual correlated. That means that the two DL dynamics have different time constants of evolution, namely, the electrons lost after the disruption of a DL can assure the starting triggering mechanism /3/ for itself and/or for the next separately formed DL.

REFERENCES

1. H.Alfven, **PS-14** (1986) Trans.Plasma Phys., p.669
2. M.Sanduloviciu (1987) Int.Conf. on Plasma Phys., Kiev, Proc. Inv.Paper Vol., p.253
3. D.Alexandroaei, M.Sanduloviciu **122** (1987) Phys.Lett.A, p.173
4. D.Alexandroaei, I.Grosu, M.Sanduloviciu (1991) XX ICPIG, Pisa, Contr.Paper Vol.p.176
5. C.Hollenstein, M.Guiot, E.S.Weibel **45** (1980) Phys.Rev. Lett., 2010
6. P.Y.Cheung, A.Y.Wong **59** (1987) Phys.Rev.Lett., 551
7. M.Sanduloviciu, D.Alexandroaei, J.Talasman, B.Ianopol, (1992) Int.Conf.of Plasma Phys., Innsbruck, Contr.Pap.Vol.

DOUBLE LAYERS IN A STRONG COLLISIONAL HELIUM PLASMA

L. Biborosch and M. Sanduloviciu

Department of Plasma Physics, Al.I.Cuza- University
R-6600 Iasi, Romania

In recent papers /1/ the phenomenon known as anode spots was investigated in the very low gas pressure conditions ($\sim 10^{-3}$ Torr). Usually such anode spots appear in the d.c. discharges at much higher gas pressures of about 10^{-1} Torr/2/. A consequence of our data obtained in this last pressure range seems to be that the generation mechanism of these anode spots is the same in all these ranges of the neutral gas pressure.

This mechanism assumes an electrical charge separation with a potential profile which evolves to a double layer (DL) configuration, related with the variation of the excitation and ionization cross sections for the electron - neutral collisions /3/.

We investigate the appearance of luminous, sharply defined, spherical anode spots (balls of fire) of several millimetres in diameter, which are formed on cylindrical electrodes of 0,1 mm in diameter and 3,0 mm in length, biased at a positive voltage of 80-90 V above the anode (ground) potential and immersed in a helium plasma produced by a large hollow cathode at a gas pressure of 3.10^{-1} Torr.

EXPERIMENTAL SET-UP. The investigations were made in the d.c. discharge geometry shown in figure 1. The discharge plasma was produced by biasing the hollow cathode C at - 200 V relative to the grounded mesh-anode A, covering the whole inner surface of the glass discharge tube wall. Measurements with Langmuir probes indicate that the floating potential of about - 0,8 V and the plasma density remain practically constant over the large anode axis. This is also true if a weak (200 Gs) and narrow ($\sim$ 10 mm) transverse magnetic field $B_{\perp}$ is applied at the open end of the large hollow cathode in order to remove the fast electrons, which impeded the appearance of certain

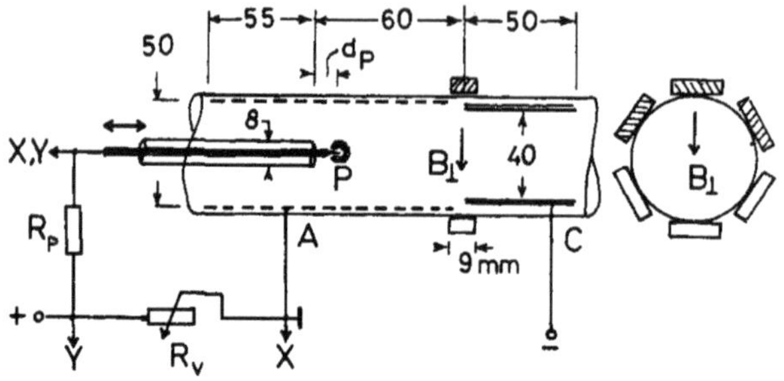

Fig. 1

features of the studied anode spots. A cylindrical electrode P made of tantalum wire, 0,1 mm in diameter and 3,0 mm in length was used to make possible (by biasing at a positive potential of 80-90 V in respect to the large hollow anode) the appearance of these spherical anode spots. Because the biasing of the electrode P with an additional power supply can also change the discharge current I, we have used a simple "self-biasing" method /4/. This consist in the d.c. electrical connection between the large mesh-anode A and the very thin cylindrical electrode P through a variable resistor R_v of about 100 kΩ.

EXPERIMENTAL RESULTS. By changing the value of this variable resistor R_v, the current-voltage (I_P-U_{PA}) characteristics /5/ of the thin electrode P can be plotted for the different axial positions d_P (see also figure 1). In the figure 2 it is seen that as the voltage U_{PA} on the electrode P is increased from 0 to 80 V, the electrode current I_P increases approx. linearly up to 25 % from the discharge current I and a thin cylindrical "anode" sheath is developed on the electrode surface for the currents I_P above the values corresponding to the broken lines of the I_P-U_{PA} characteristics.

The further increase of the voltage U_{PA} above a certain critical value, which increase from 80 to 90 V with the approaching of the electrode P nearer the open section of the cathode, induce a more pronounced and finally a sudden jump in the electrode current I_P, indicated by the vertical

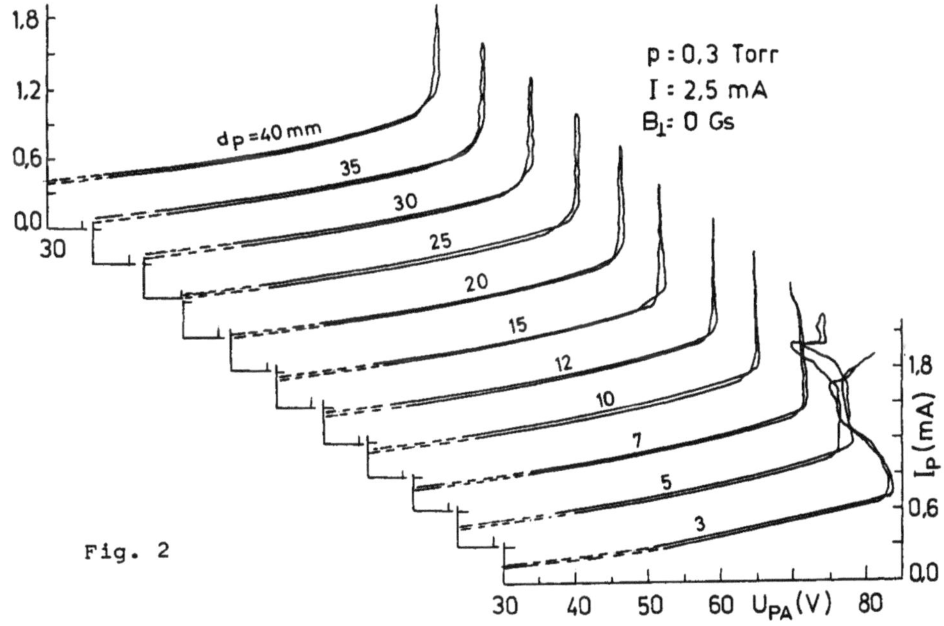

Fig. 2

branches of the same current-voltage (I_p-U_{PA}) characteristics showing in figure 2.

At the same time, the visual observations firstly revealed a diffuse luminosity expanded from the cylindrical anode sheath, which then evolves ("explodes") towards a nearly spherical DL (ball of fire) accomplished by the above mentioned sudden jump of the electrode current I_p. These balls of fire bordered by the DL remains anchored on the thin cylindrical electrode P and its spherical symmetry is also conserved for somewhat longer cylindrical but thin electrodes.

Only a small hysteresis can be seen on the same I_p-U_{PA} characteristics from the figure 2 (with the exception of the very small distances of the electrode P from both the axial boundaries of the large anode space). For these last positions d_p of the electrode P a well-developed negative branch can be seen on the corresponding I_p-U_{PA} characteristics.

Such negative branches of the I-U characteristic are known to occur as a general feature for the gaseous conductors which contain a DL and this is true also for our conditions. Thus, the negative branches of the I_P-U_{PA} characteristics become more evidently with the increasing strength of the transverse magnetic field $B_\perp$, kept fixed at the open end of the large hollow cathode (denoted by $d_{B_\perp} = 0$ in the figures 3 and 4). The same influence was found by bringing the thin electrode P nearer the location of the transverse magnetic field $B_\perp$, maintained in the same position ($d_{B_\perp} = 0$) but with a constant strength of $B_\perp = 175$ Gs (figure 4). In spite of these last somewhat more complicate discharge conditions, the I_P-U_{PA} characteristics are ended with the same values of the currents $I_P \hat{=} 1,8$ mA and voltages $U_{PA} \hat{=} 90$ V, respectively.

Independently of the action of this magnetic field, the well-developed spherical anode spots and DL are generally unstable. Therefore, even without any magnetic field, the electrode current I_P and the bias potential U_{PA} were not steady but consisted of spikes, firstly irregular which, by increasing the current I_P, become regular. The spike frequency f_s is increasing with the electrode distance d_P

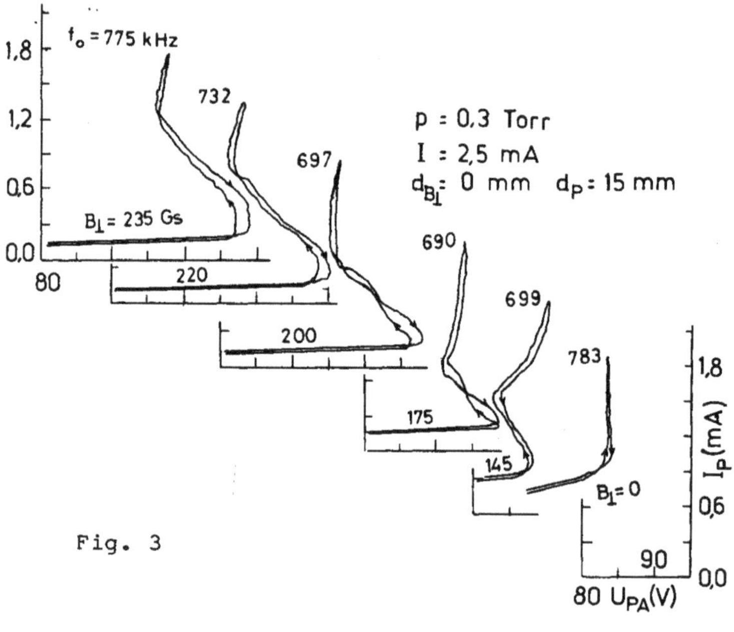

Fig. 3

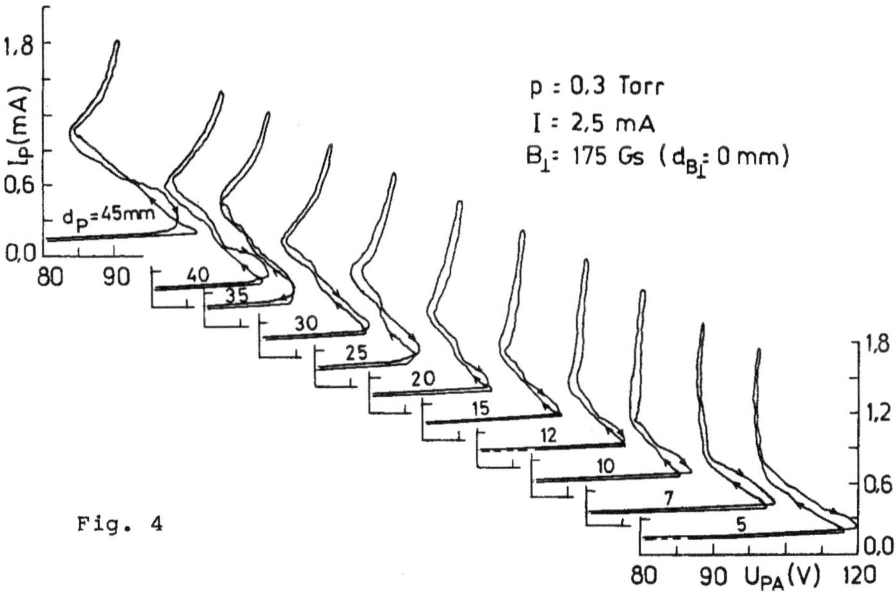

Fig. 4

from its initial location (see the broken curve in figure 5). At their peak values the spikes are amplitude modulated with a two order of magnitude higher frequency f_o, which more than likely corresponds to the so-called anode oscillations. These higher frequency f_o, on the contrary, show a steadily decrease with increasing distance d_p (see the solid curve in the same figure 5). But, both the two frequency dependencies approximately follow parabolical rules in respect to the axial distance d_p between the electrode P positions and its

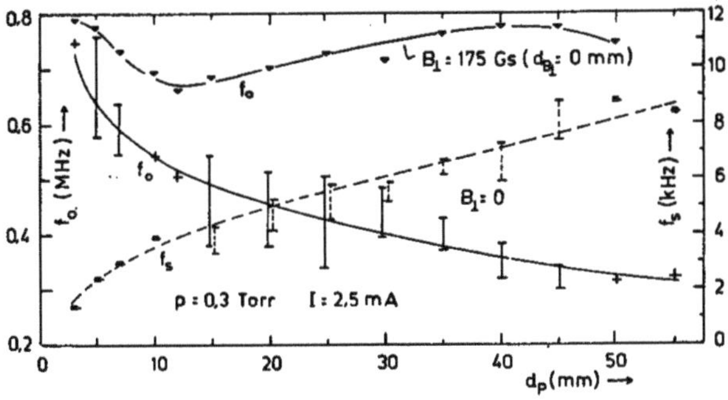

Fig. 5

initial location near the centre of the large anode A. It is seen that the above magnetic field has a certain compensating effect on the high frequency change with the axial distance d_p, whilst the current and potential spikes always dissapeared. These two effects of the magnetic field can be related to the elimination of the fast electrons by the action of this weak and narrow transverse magnetic field, particularly when it is located near the open edge of the large hollow cathode.

CONCLUSIONS. We reported some new data about the observation of the so-called anode spots with spherical symmetry of many millimetres in diameter, which developes on a cylindrical thin electrode.

The anode-wall geometry used in our d.c. discharge is, in principle, similar to those of the plasma device used in /1/ for the same investigations but at much lower gas pressure.

The sudden increase of the electrode current (during the current spikes) is likely induced by increase of the electron collecting area of the well-developed spherical anode spots, whereas the high frequency oscillations are appearing when the potential drop over the spherical DL reaches the ionization potential value of the used gas /3/.

REFERENCES

1. B. Song, N.D`Angelo and R.L.Merlino, 24 (1991) J.Phys.D; Appl.Phys.,1789; 25 (1992) J.Phys.D; Appl.Phys.,938

2. K. G. Emeleus, 52 (1982) Int.J. Electronics, 407; 56(1984) Int. J. Electronics, 441

3. M. Sanduloviciu and E. Lozneanu, 28 (1986) Plasma Phys. and Contr. Fusion, 585

4. L. Biborosch and M. Sanduloviciu, Pisa(1991) XXth ICPIG,460

5. L. Biborosch, M. Toma and M. Sanduloviciu, Innsbruck(1992) ICPP, 2029

PARAMETRIC INSTABILITY IN PLASMA SHEATH

L. T. Carneiro, C. da C. Rapozo and K. H. Tsui
Instituto de Física, Universidade Federal Fluminense
24.020 - Niterói, RJ, Brazil

In this paper we present some experimental results about parametric instability in plasma-sheath obtained in the linear mirror machine LISA. The helium plasma was created in the metallic ctlindrical vessel of the LISA device. The weakly ionized plasma was excited by the RF source built at Plasma Physics Laboratory at Physics Institute of Universidade Federal Fluminense(UFF-Niterói, RJ). Using the non-linear sheath capacitance, we can mention the influence of this property in the parametric instability that appear in this work. This effect was set up by Mathieu equation applied in the RLC circuit for a plasma sheath model where the sheath capacitance is a time dependence function.

1. INTRODUCTION

In plasma physics research, mainly in plasma diagnostics with resonant probes attention is forward on the mechanism that are important for the alternating current (AC) signals in the plasma sheath. In this work we consider the results obtained from the electrical model that we derive for plasma sheath region where the non-linear characteristic was used for the parametric amplification in a low density plasma sheath of radio frequency 5 MHz signal amplification in a low density plasma sheath [1] and the results obtained from the sheath region of negative conductance [2,3] to study the parametric instability which appears in our experiment. Parametric instabilities are not unique to plasmas, as they relate to any oscillatory system where our of the parameters (hence parametric is modulated at appropriate frequency). A rather extensive review of parametric instabilities in plasmas, including magnetized and unmagnetized plasmas, inhomogeneous and uniform plasmas, for linear and nonlinear electrostatic and electromagnetic waves is given by [4]. The experiment was carried out in the linear mirror machine LISA designed and constructed at the Max-Planck Institut für Plasmaphysik (Garching, Germany)(Fig. 1). The helium plasma is produced by a RF source in region A in a cylindrical stainless-steel vessel (diameter 17 cm, length 256 cm).

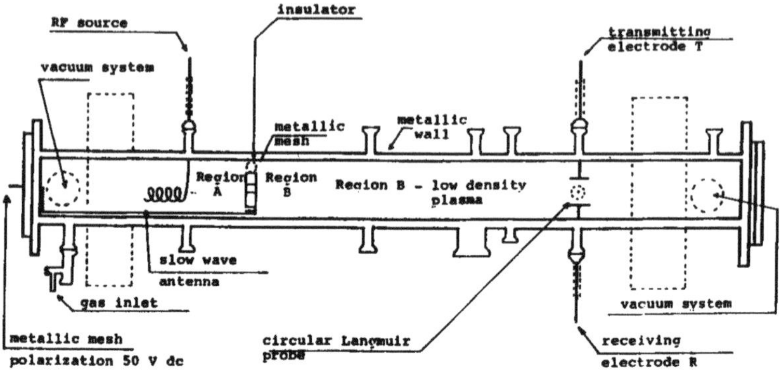

Fig. 1 - The linear mirror machine LISA.

2. EXPERIMENTAL RESULTS AND ANALYSIS

In this work the main goal is to study the instabilities that appear when the applied V_{RF} in the transmitting electrode T is increased. The helium plasma was created in the metallic cylindrical vessel of the linear mirror device LISA. More details about the device LISA can be seen elsewhere [5,6]. The normalized admittance profile versus applied pumping frequency Ω for the constant V_{RF} = 3.0 V(RMS) is shown in ref.[5]. This value of the RF voltage do not disturb sensibly the sheath capacitance.

Ω (MHz)	V_{RF} (V rms)
7.9	> 8.0
15.1	> 17.3
20.5	> 12.1
29.3	> 9.3
33.8	> 15.0

Table I - Values for Ω pumping instabilities and critical V_{RF}.

In the Table I we can see the values of the pumping frequency of Ω for which we obtained the sharp variations in the amplitude when applied V_{RF}(RMS) was varied up to 16.0 V(RMS). The plasma circuit, sheath and electrode can be interpreted like an RLC circuit [2]. The sheath capacitance [2] varies with Ω by the expression $C_s = C_{so}$ $(1 - m \cos\Omega t)$, where C_s is the parametric sheath capacitance, $m = 3/4$ (V_{RF}/V_{fo}), V_{RF} is the applied RF voltage, Ω is the pumping frequency, V_{fo} is the floating potential not disturbed, R is the plasma resistance and L is the plasma inductance. How the output V_{RF} pumping signal given by the amplifier system has a little impedance (Z ≅ 5 ohms) we can consider the circuit like an RLC(t) ensemble with excitation in the capacitance C(t) by the way for the C(t), we have

$$\frac{d^2}{dt^2} q + 2\alpha \frac{d}{dt} q + \omega_o^2 (1 + m \cos\Omega t) q = 0 \quad , \qquad (1)$$

where $\omega_o^2 = 1/LC_{so}$ is the resonant circuit for m = 0, and $\alpha = R/2L$ is the inverse time characteristic of the circuit. If we consider $q = y \exp(-\alpha t)$, eq.(1) becomes

$$\frac{d^2}{dt^2} y + (-\alpha^2 + \omega_o^2 + m\omega_o^2 \cos\Omega t) y = 0 \quad . \qquad (2)$$

If we make $\omega_\ell^2 = \omega_o^2 - \alpha^2$, $\tau = \Omega t/2$, $\delta = 4\omega_\ell^2/\Omega^2$, and $\gamma = 4m\omega_o^2/\Omega^2$, from eq.(2) we have the following expression

$$y'' + [\delta + \gamma \cos(2\tau)] \quad y \quad = 0 \qquad (3)$$

that is the Mathieu equation. We have many values of the δ for each value of the γ parameter. Consequently the solution of eq. (3) is a periodic Mathieu function of 2π period in relation to the variable τ [5]. For $\gamma = 0$, eq.(3) can be rewritten $y''(\tau) + \delta(0)y(\tau) = 0$, whose solution is $\cos\left[\sqrt{\delta(0)}\tau\right]$ and $\mathrm{sen}\left[\sqrt{\delta(0)}\tau\right]$. For that the above equation can be a periodical Mathieu function with 2π period in relation to the τ variable is necessary: $\delta(0) = n^2$ is a entire for the arbitrary value of the σ and δ, eq.(3) has solution as $y = A \exp(\mu\tau)\phi(\tau) + B \exp(-\mu\tau)\phi(-\tau)$, where A and B are constant, $\phi(\tau)$ is the periodical Mathieu function that depends on the τ and δ parameters. When V = 3.0 V(RMS), $\omega_o = \omega_s$ = 8.5 MHz, Ω = 13 MHz and V_{fo} = 17.0 V we have $m = (3/4)(V_{RF}/V_{fo}) = 0.13$, where ω_s is the sheath resonance. Then $\gamma = 4m\omega_o^2/\Omega^2 << 1$ the solution for eq.(4) will be written as $y \cong A \exp(\mu\tau) \cos(n\tau + \zeta) + B \exp(-\mu\tau) \cos(n\tau - \zeta) = y_1(\tau) + y_2((\tau)$, where ζ is a phase.

For a good understanding of the analysis we will be study μ in function of γ and for the δ parameter in the region quite near of the unit for the δ value, that is n = 1. In this case we make $y = y_1(\tau)$, that is particularly solution of eq.(3) which gives us

$$\mu^2 = -(\delta + 1) \pm 2\sqrt{\delta} \left[1 + \frac{1}{8\delta}\left[\frac{\gamma}{2}\right]\right] \qquad (4)$$

For μ real eq.(4) is instable and shown that for determined relation between a modulation deep m of the capacitance and the relative frequency of this modulation for some initial disturbed in the oscillations of the growth amplitudes can be appear.

How the solutions of the Mathieu equation are unstable for a depth modulation quite near to zero ($\gamma \to 0$) in the case that we have the value $\delta = 1, 4, 9$, it implies in $\Omega = 2\omega_1$; $\Omega = \omega_1$; $\Omega = 2/3\ (\omega_1)$ and so on. For small values of γ, eq.(4) will be rewritten as follows

$$\mu = \pm \sqrt{\frac{1}{\sqrt{\delta}} \left(\frac{\gamma}{4}\right)^2 - \left[\sqrt{\delta} - 1\right]^2} \tag{5}$$

and $q(t) = A \exp\{[(\Omega/2)\mu - \alpha]t\} \cos[(\Omega/2) + \xi] + B\exp\{[-(\Omega/2)\mu + \alpha]t\} \cos[(\Omega/2)t - \xi)$ that is a solution of eq.(1). If μ is real and larger than the relation $2\alpha/\Omega$ we have an instability in the system, that is $\mu_c = 2\alpha/\Omega$ where μ_c is a critical parameter and this value gives us a new expression for the above equation that can be rewritten as

$$\mu_c = \frac{\Omega^2}{2\omega_o^2\delta^{1/4}} \left[\frac{4\alpha^2}{\Omega^2} \left[\sqrt{\delta} - 1\right]^2\right]^{1/2} , \tag{6}$$

From fig. 4 of ref.[1] near $\omega_o^2 \approx \omega_s^2$ gives us the value of the merit factor Q = 2.43 that permits us to obtain the α value, that is $\alpha = \omega_o/2Q = 1.85 \times 10^6$ ohm/henry in the point $\omega_o = \omega_s$. Fig. 2 shows the μ_c profile versus pumping frequency Ω for $\delta = 1$; the vertical lines show the values of the frequency where instabilities occur with their V_{RF} values.

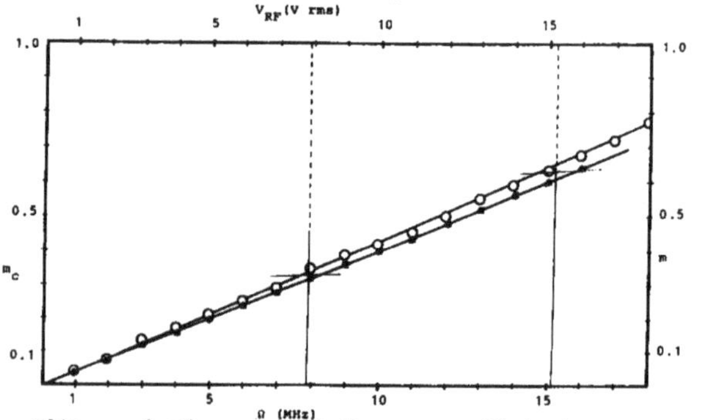

Fig. 2 - Skin depth modulation coefficient (m_c) profile versus pumping frequency Ω and applied RF voltage.

In first analysis we can be verified that for the values $\delta = 1$ and $n = 2$ there exists a instability point in $\Omega = \omega_1 = \sqrt{\omega_s^2 - \alpha^2} \cong 8.3$ MHz and the other point near to $2\omega_\ell \cong 16.6$ MHz. This value is quite near to the experimental value measured for the pumping frequency Ω = 15.1 MHz, and Ω = 7.9 MHz.

4. PHASE ANALYSIS

If we consider $\delta = n = 1$ in eq.(6), we obtain $y_1(\tau) = Ae^{u\tau}\cos(\tau + \xi)$. If we neglected the sum in $\cos3\tau$ and $\sin3\tau$ for the phase ξ we get $tg\xi = [(\gamma/2) + (u^2 + \delta - 1)]/2u$. Other particular case occurs when $|u| = \varepsilon/4$ and $\delta = 1$ and the above equation can be rewritten as $tg\xi = [(\gamma/2) + (\gamma/4)^2]/(\alpha\gamma/4) \cong 1 + (\gamma/4)^2(2/\gamma)$, $tg\xi = 1 + \gamma/2 \cong 1$, showing that the $\xi = 45^{\circ}$. In this case, the general solution for q(t) is $q(t) = A\exp([(\Omega/2)u - \alpha]t)\cos[(\Omega/2)t + 45^{\circ}] + B\exp(-[(\Omega/2)u + \alpha]t)\cos[(\Omega/2)t - 45^{\circ}] = q_1(t) + q_2(t)$.

Fig. 3a shows the profile of $C(t) = C_o/(1 + \gamma\cos\Omega t)$, $\cos[(\Omega/2) + 45^{\circ}]$ and $\cos[(\Omega/2) - 45^{\circ}]$, in the case $\Omega/2 = \omega_{\xi}$. From this figure we can see that the sheath capacitance has a good phase in relation to $q_1(t)$ and the pumping frequency gives a place to the growth of the amplitude in agreement with the law $\exp(-[(u\Omega/2) + \alpha]t)$. This analysis carried us to establish the following source board of the growth of the oscillations in the parametric circuit.

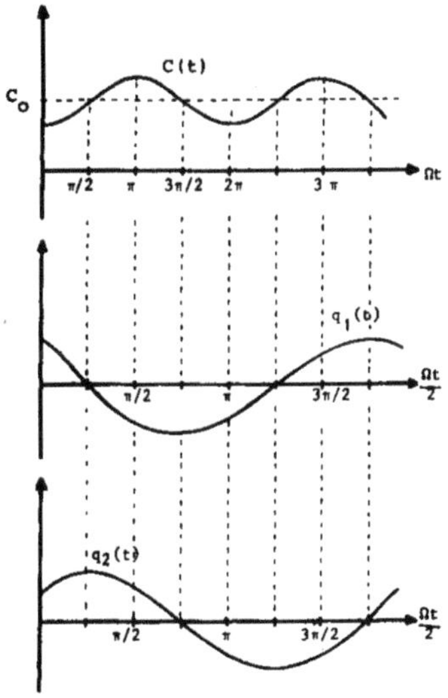

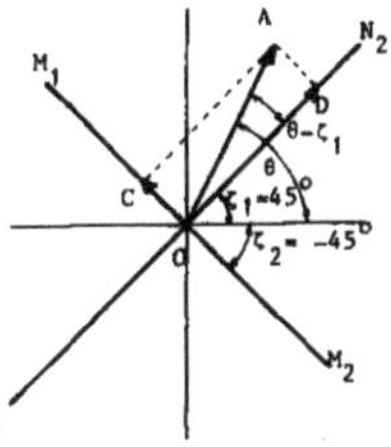

Fig. 3 – (a) Capacitance profile as a function of Ωt and $\Omega t/2$, for the cases $C(t)=C_o/(1+\gamma\cos\Omega t)$, $\cos((\Omega t/2)+45^{\circ})$ and $\cos((\Omega t/2)-45^{\circ})$. (b) Graphic analysis of the phase behaviour.

When we turn on the pumping oscillator, some disorded

oscillation, caused, e.g., by charged particles (thermal noise), will be in frequency $\Omega/2$ with randomics phase and amplitude. In a previous moment the analysed component that is important for us has an amplitude OA and phase θ (Fig. 3b). The vector OD coincident with the straight line N_1N_2 and equal to OA $\cos(\theta - \xi)$ represents the oscillation in phase with the pumping frequency. The vector OC coincident with the straight line M_1M_2 represents the oscillation that decreases by the effect of the variation of the capacitance. Because of this statements we concluded that the initial conditions for the parametric circuit are determined by the components whose phase agree with the pumping phase.

Thus, we can see that $\xi = 45^\circ$ and $\xi = 225^\circ$. If the vector OA coincides with the straight line M_1M_2 its projection above the straight line N_1N_2 is positive and it is negative in the opositive case. So, if we have a pre-fixed pumping phase, the oscillation phase of the parametric circuit can assume values different from 180°.

5. CONCLUSION

In conclusion we have shown that in LISA device in the two points observed where the analysis of the Mathieu equation predict the existence of the instabilities for the small value for γ the μ_c for given values Ω = 7.9 and 15.1 MHz is approximated 0.3 and 0.76, respectively and the index μ of the sheath capacitance modulation assumes a value $\mu > \mu_c$ in the points where $V_{RF} >$ 8.0 V(RMS) and $V_{RF} >$ 17.0 V(RMS), respectively agree with values of the applied V_{RF} for which we have an instability in Ω.

The experimental results show that the analysis of the plasma sheath system looks like a linear circuit that is a good way to study the sheath instabilities where the energy transfer processes can be seen as well as the Llewelyn how the modulation of the sheath thickness. In the intermediary region of the parameters δ where γ goes to Ω and Ω is quite small as well as we desire its necessary to solve the Mathieu equation for the experimental parameters when in sometime we need numerical method. We can be also seen that in the instability conditions the growth rate of the oscillations begins with a fluctuation of the thermal noise of the charged particle which has harmonics and randonics phase that have the same time for the instability criteria.

Acknowledgements: The authors are grateful to the opinions of Dr. A. S. de Assis and the contributions of Eng. H. Teixeira and M. G. Abreu. This work was supported by CNPq, FINEP, FAPERJ, IBM do Brasil Ltda. and PROPP/UFF

[1] L. T. Carneiro, C. da C. Rapozo, Physical Review **A43**, 6 (1991).

[2] L. T. Carneiro, C. da C. Rapozo, Phys. Scripta **44**, 484 (1991)

[3] R. J. Rosa, J. Phys. **A4**, 670 (1971).

[4] P. K. Kaw, W. L. Kruer, C. S. Liu, K. Nishikawa, "Advanced in Plasma Physics", ed. A. Simon and W. B. Thompson (Interscience NY, 1976) vol. 6, p. 1-269.

[5] C. da C. Rapozo et al, PlasmaPhysics Controll. Fusion 9, 1187(1988).

[6] C. da C. Rapozo et al, IEEE Transactions on Plasma Science 8, 6, 1038 (1990)

[7] C. da C. Rapozo et al., Physica Scripta **42**, 616 (1990).

[8] D. G. Swanson, Academic Press 335 (1989).

STRONG TURBULENCE AT A DOUBLE LAYER

R.T.Carpenter[a] and S. Torvén

Alfvén Laboratory, Royal Institute of Technology, S-100 44, Stockholm, Sweden

a) On leave from the University of Iowa

INTRODUCTION: The existence of high frequency (hf) fluctuations on the high potential side of a double layer (DL) has been known for some time. Investigations of these fluctuations have shown that the frequency spectrum has a peak at about the plasma frequency and that the phase velocity of the waves in a broad frequency interval is about constant and somewhat less than velocity of the electrons that have been accelerated by the DL potential drop.[1] The time averaged voltage level of the fluctuations has been found to increase by typically two orders of magnitudes, from the background noise level on the low potential side of the DL, up to a maximum on the high potential side, a few centimeters from the DL, and evidence of modifications of density profiles due to the ponderomotive force has been given.[2,3] This phenomenon has been interpreted as plasma oscillations excited by the beam-plasma interaction due to the "bump on tail" or two stream instability. This interaction may in general be nonlinear involving the formation of high frequency wave packets and associated density cavities.[4,5]

The present work gives the first experimental evidence of this nonlinear interaction at a DL using an experimental geometry designed to enhance the fluctuation level and to simplify the observation of the resulting oscillations by stabilization of the DL position through a weak magnetic mirror field. We find the essential new result that the waves are not continuous but rather occur in bursts or wave packets at the DL. These bursts occur randomly in time at a given point in space.The changes in the time averaged intensity of the fluctuations mentioned above occur not only because the amplitude of the bursts changes but most significantly because the burst rate changes. This is true whether the change results from moving the probe axially or radially or from changes in the plasma parameters. Thus the systematic investigation of high frequency oscillations in a double layer consists of a study of the properties of these wave packets as discussed more completely below.

The occurance of bursts of high frequency fluctuations is characteristic of strong turbulence[4,5] as has been demonstrated in experiments using low intensity beams of monoenergetic electrons. A crucial parameter in these experiments is the ratio η of the electron beam density to the background plasma density. At a DL, the beam-plasma parameters can approximately be expressed in terms of the DL parameters. Let n_1 and n_2 be the electron number densities on the low and high potential side, respectively. Assume that the electrons are incident at the low potential side of the DL with a half-Maxwellian velocity distribution with the temperature T_e. We introduce the normalized potential $\phi_{DL}=eU_{DL}/k_BT_e$ where U_{DL} is the potential drop over the double layer and restrict the discussion to strong DL, $\phi_{DL}>>1$. The contribution to n_1 from electrons on the high potential side is then negligible, and n_1 is the density of the incident electrons on the low potential side. Their density n_b on the high potential side is given by $n_1\exp(\phi_{DL})\{1-\mathrm{erf}(\phi_{DL})\}$ which reduces to $n_1/(\pi\phi_{DL})^{1/2}$ for $\phi_{DL}>>1$. The relative beam density is accordingly given by $n_b/n_2 \approx (n_1/n_2)(\pi\phi_{DL})^{-1/2}$. The beam velocity v_b may be defined as $(2eU_{DL}/m_e)^{1/2}$. It is conveniant to introduce $v_{th}=(2k_B T_e/m_e)^{1/2}$, and obtain the relative beam velocity as $v_b/v_{th} \approx (\phi_{DL})^{1/2}$. The half-width Δv_b of the beam distribution is given by $\Delta v_b/v_b \approx (\ln 2/\phi_{DL})^{1/2}$. Typical values in the experiment reported below are $\phi_{DL}\approx 10$, $n_1/n_2\approx 0.2$, $v_b/v_{th} \approx 7$, $\Delta v_b/v_b \approx 0.1$ and $\eta = n_b/n_2 \approx 0.04$. For comparison, it was found[2], using monoenergetic electron beams, that the threshold for the onset of strong turbulance was $\eta >0.001$ and for the creation of randomly propogating multiple bursts was $\eta>0.006$. It would seem, then, that strong turbulence should be a natural consequence of the experimental situation employed.

Besides theoretical models[6], several computer simulations of strong turbulence in 1, 2 and 3 dimensions, have appeared.[7] However, none seems to have dealt explicitly with the very large values of η found in a double layer or with variations of the burst rate due to variations in various plasma parameters. A discussion of the "nucleation time" for bursts has been given[8] with the result that the burst rate should be proportional to the spatially averaged electrostatic energy density, a condition that is difficult to compare with our results because we observe a strong spatial variation of the burst rate.

Nonlinear electron-beam plasma interaction has also been used to explain phenomena observed in astrophysical and space plasmas.[9]

EXPERIMENTAL DEVICE: The experiments were carried out in the "Green Tank" of the Alfvén Laboratory, Royal Institute of Technology, Stockholm. The device is described elsewhere.[2,3] It consists of a cylindrical stainless steel vacuum chamber of diameter 120 cm and length 140 cm. Argon is fed into the continously pumped tank, and a background pressure of 0.2 mTorr is used. At the right end of the tank there is a plane anode of diameter 10 cm. An "ionization DL" forms in front of the anode by drawing a sufficiently large current to it. A magnetic mirror field is used to stabilize the DL position. A coil at the anode position produces the field which has maximum strength of 50 mT in continous operation. At the left end of the chamber, where the magnetic field is about twenty times weaker, a ring of filaments with a diameter of about 100 cm is placed at right angles to the symmetry axis. A source of plasma is obtained by a discharge between the ring filaments and a ring anode which is concentric with the anode. The ring anode is insulated from the anode and has an outer diameter of 15 cm. The source discharge occurs essentially along the magnetic field, and the typical voltage drop is 20V. With this arrangement, only thermal electrons, with an energy that is small compared with eU_{DL}, are incident at the low potential side of the double layer, and these electrons constitute the beam electrons on the high potential side. This gives a high level of hf fluctuations and also a stationary density trough along the path of the electron beam which was attributed to the ponderomotive force.[3] A similar density trough has been observed in experiments where hf bursts were excited by an electron beam.[4]

EXPERIMENTAL RESULTS: Double probes[10] allowing simultaneous measurements of the relative hf power level are used. The difference signal, obtained in a hybrid tee, is fed to a spectrum analyzer or fed through a high pass filter to a fast oscilloscope or to an envelope detector giving the envelope of the rectified hf signal. All cables are matched to 50 Ω at the recieving ends.

Fig. 1 shows the variation of the time averaged hf power level measured on the central axis . In this measurement, the spectrum analyzer was used as a reciever, measuring the voltage level at the maximum in the frequency spectrum (300 MHz) with a bandwith of 3 MHz. The DL profile and the position the anode are also shown in Fig. 1. The hf voltage level increases by about two orders of magnitudes, from the background noise level on the low potential side of the DL, up to a maximum on the high potential side, at about 10 cm from the DL. Then it drops by about an order of magnitude up to the anode.

The time averaged frequency spectra at different positions along the axis show a single maximum at about 300 MHz, approximately independent of position, and the full width at half maximum is about 100 MHz. The phase velocity of the hf waves was estimated by direct observations of the time displacement between the signals from two double probes on a fast dual-beam oscilloscope (bandwidth 400 MHz). A propagation velocity of $3\cdot 10^6$ m/s was estimated which should be compared with $v_b=(2eU_{DL}/m_e)^{1/2} \approx 3.3\cdot 10^6$ m/s for U_{DL} =50 V. The phase velocity obtained gives a wave length of $\lambda_b \approx 1$ cm for 300 MHz.

The hf signal is strongly modulated, and it appears as well separated bursts in a region close to the DL. The envelope detector was used to investigate the bursts. Fig. 2 (upper traces) shows oscillograms of the hf signal obtained with a digitizing oscilloscope (sampling frequency 1 GHz). The lower traces show the corresponding outputs of the envelope detector. The wave envelope is accurately given by this method.

Fig. 3 shows the envelopes of the hf bursts taken at some different probe positions z (cf. Fig. 1) along the central axis. It shows that the increase of the time averaged hf power shown in Fig. 1 is due to an increase in the burst rate, with no large change in amplitude (z=32.6 cm and z=29 cm. For larger distances from the DL, the gaps between the bursts tend to disappear (z=21 cm). Finally there are no gaps between the bursts, and then the average amplitude also decreases (z=13 cm). The amplification for this trace is twice the amplification for the other traces.

Two hf probes with envelope detectors were used to estimate spatial correlation between the bursts. Fig. 4 shows superimposed traces of the two probes for several different probe separations along the central axis. As the probe separation increases, the similarity between the traces decreases. Inspection of a large number of such oscillograms gives an indication of the average axial extension which was estimated to 3 cm. However, individual pulses may occasionally show a good correlation even up 5 cm probe separation. Further investigations may be based on cross-correlation functions, evaluated for different probe separations d. Such functions were evaluted for several pairs of traces for each value of d. Variations from pair to pair were obtained, probably due to the limited memory in the digitizing oscilloscope. In spite of the statistical variations, the maximum of the correlation function was always obtained for $\tau=0$ when the probes are close ($d \approx 0$). Here τ is time shift between the two traces. For larger values of d, the maximum correlation was occasionally assumed for finite values of τ. Interpreting this as a propagation, a velocity along the magnetic field comparable with the ion sound speed was obtained.

To obtain an evidence of correlation between the hf bursts and density dips, a double probe allowing simultaneous measurements of the relative hf power level and the low frequency current at the average plasma potential was used. It consisted of hf probe described above in which the low frequency current was tapped from the high freqency circuit through two isolating resistances of 1 kΩ, and the sum of the low frequency currents to the probe tips was measured. The probe could be biased at different dc potential levels by a power supply. The wave packets were found to be correlated to probe current dips. However, due to the ambipolar field associated with a density dip, the plasma potential should also increase as a positive pulse over the dip, and evidence of such potential pulses was obtained. At a fixed probe potential bias, the probe current may accordingly dip also due to the plasma potential increase. Another probe construction was used to give further evidence. It consisted of the double probe just described and an emissive probe mounted on the same shaft with the hot wire parallel to the hf-probe wires and separated from them by a few electron gyro radii. The hot probe keeps the cold probe at a potential which is close to plasma potential.[10] The electron current collected by the cold probe is fed to the hot probe through a small signal resistor permitting current measurements by an isolation amplifier. The time resolution of this probe is determined by its stray capacitance to ground and its resistance to ground over plasma. Fig. 4 shows some examples of the correlation between between the hf bursts and probe current dips. These are typically between 10 and 20% of the total probe current at plasma potential. However, the actual probe current dips may be deeper since the probe current fluctuations are expected to be damped due to insufficient time resolution.

CONCLUSIONS: We have shown that the hf field excited at a DL is strongly modulated and appears as wave packets. These are localized in space with an extension of a few cm. The increase of the time averaged value of the hf amplitude to a maximum (Fig. 1) is mainly the result of increasing burst rates although increasing burst amplitudes also contrinbute close to the double layer. Measurements show that the wave packets are correlated to probe current dips, but further investigations are required to get conclusive evidence of the correlation to density dips.

ACKNOWLEDGEMENTS: This work was supported by the Swedish Natural Scince Research Council. We wish to thank Mr M. Wohlin, Dr I. Axnäs and M. Raadu for help with the data acquisition and the graphics and Mr J. Wistedt for the construction of probes and probe circuits.

REFERENCES

1. S. Torvén and L. Lindberg, *J. Phys. D: Appl. Phys.* (1980), 13, p. 2285; P. Leung, A. Y. Wong and B. H. Quon, *Phys. Fluids* (1980) 23, p. 992; D. Andersson, *J. Phys. D* (1981) **14** , p. 1403.
2. R. Schrittwieser, I. Axnäs, R. T. Carpenter and S. Torvén, *IEEE Transact. Plasma Sci.* (Proc. of the Alpbach

conference), in the press.
3. R. T. Carpenter et al., *Proc. Fourth Symp.on Double Layers*, Innsbruck(1992), Contributed Papers.
4. A. Y. Wong and B. H. Quon, *Phys. Rev. Lett.* (1975), **34**, p. 1499; H. Ikhezi, R. P. H. Chang and R. A. Stern, *Phys. Rev. Lett.* (1976), **36**, p. 1047; P. Y. Cheung and A. Y. Wong, *Phys. Fluids* (1985), **28**, p. 1538.
5. T. Intrator, C. Chan, N. Hershkowitz and D. Diebold, *Phys. Rev. Lett.* (1984), **53**, p. 1233.
6. V. E. Zakharov, *Sov. Phys. JETP* (1972), **35**, p. 908; L. M. Degtyarev, V. E. Zakharov and L. I. Rudakov, *Sov. J. Plasma Phys.* (1977), **2**, p. 240; B. Hafizi, J. C. Weatherwell, M. V. Goldman and D. R. Nicholson, *Phys. Fluids* (1982), **25**, p. 392.
7. S. I. Animosov, M. A. Berezovskii, M. F. Ivanov, I. V. Petrov, A. M. Rubenschick and V. E. Zakharov, *Phys. Lett.* (1982), A **92**, p. 32; H. L. Rowland, J. G. Lyons and K. Papadopoulos, Phys. Rev. Lett. (1981), **46**, p. 346; D. DuBois, H. A. Rose and D. Russell, *Physica Scripta* (1990), **T 30**, 137; J. P. Sheerin, *Physica Scripta* (1990), **T 30**, 181.
8. P. A. Robinson and D. L. Newman, *Phys. Fluids B* (1990), **2**, 2999.
9. S. Bardwell and M.V. Goldman, *Astrophys. J.* (1976), **209**, 912; K. Papadopolous, M. L. Goldstein and R. A. Smith, *Astrophys. J.* (1974), **190**, 175; D. A. Gurnett, J. E. Maggs, D. L. Gallagher, W. S. Kurth and F. L. Scarf, *J. Geophys. Res.* (1981), **86**, 8633; P. Y. Cheung, A. Y. Wong, T. Tanikawa, J. Santoru, D. F. DuBois, H. A. Rose and D. Russell, *Phys. Rev. Letters* (1989), **62**, 2676.
10. L. Lindberg, *J. Phys. E: Sci. Instruments* (1985), **18**, 214.

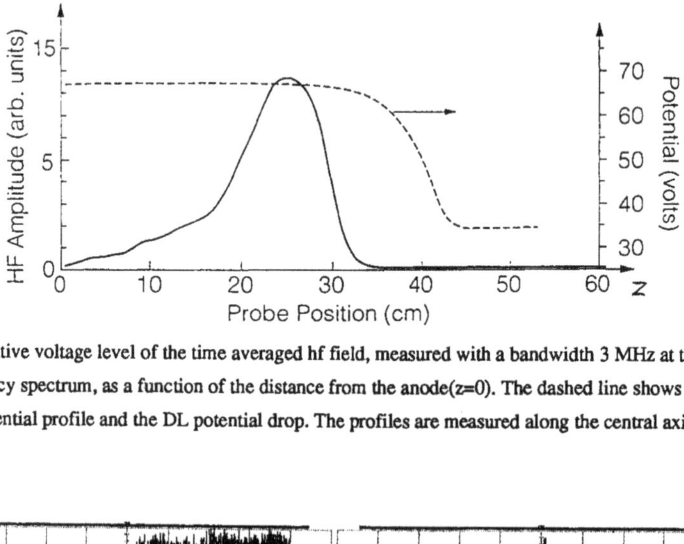

Fig. 1. Relative voltage level of the time averaged hf field, measured with a bandwidth 3 MHz at the maximum of the frequency spectrum, as a function of the distance from the anode(z=0). The dashed line shows the corresponding plasma potential profile and the DL potential drop. The profiles are measured along the central axis.

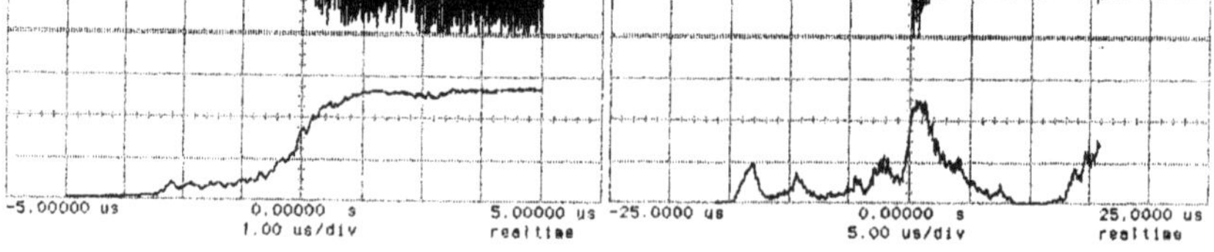

Fig. 2. Examples of hf bursts (upper traces) and the output from the envelope detector (lower traces)

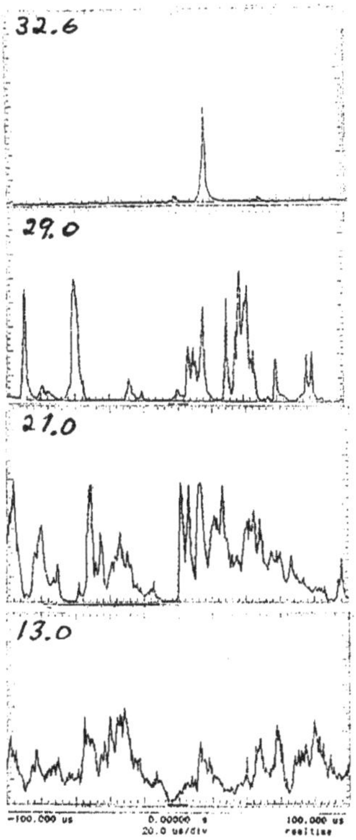

Fig. 3. Burst envelopes measured at the central axis and at different distances from the DL. The coordinates (cf. Fig.1) are given in cm at the left upper corner. With increasing distance from the DL, the burst rate first increases, and then individual bursts disappear and the signal becomes continous.

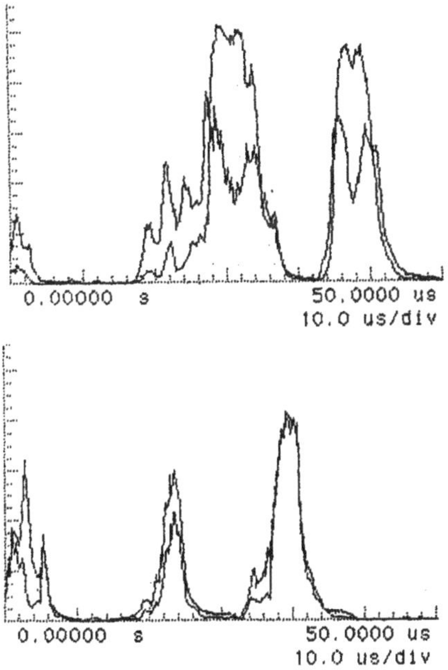

Fig.4. Burst envelopes from two probes superimposed, for a probe separation of 0.3 cm (lower diagram) and a probe separation of 1.1 cm, (upper diagram) along the central axis. From several such oscillograms, an average extension of 3 cm was estimated for the wave packet. However, some bursts showed a good correlation even up to a probe separation of 5 cm

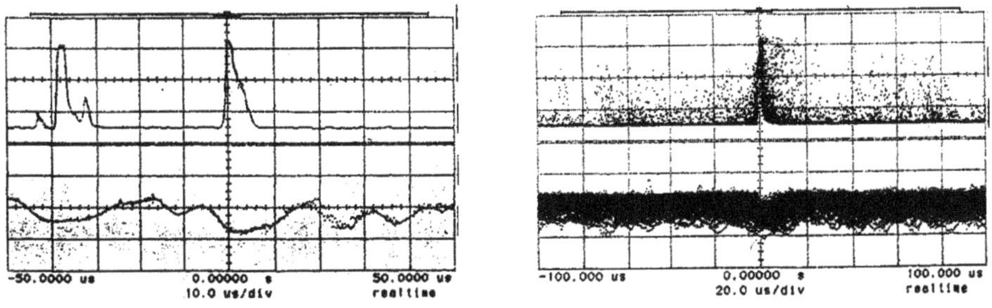

Fig. 5. (left) Oscillogram showing the correlation between the burst envelopes (upper trace) and probe current dips (lower trace). (right) A large number of overlapping traces showing the dip.

EVIDENCE OF PONDEROMOTIVE FORCE EFFECTS IN IONIZATION DOUBLE LAYERS

R.T.Carpenter, R. Schrittwieser and S. Torvén

Alfven Laboratory, Royal Institute of Technology, S-100 44, Stockholm, Sweden

1. INTRODUCTION: The occurence of high frequency (hf) fluctuations in the anode region of ionization double layers, with frequencies near the electron plasma frequency, has been observed in a few experiments.[1] Recently some evidence has been found indicating that these fluctuations are intense enough to produce modifications of the plasma density on the high potential side of the double layer,[2] presumably through the ponderomotive force, as reported in other investigations of the electron-beam plasma interaction.[3,4]

In the earlier work [2] the evidence of density depletion on the top of the double layer consisted of a flattening of the radial density contours in regions where intense hf fluctuations were detected. Plasma from a source was connected with the chamber through a 10 cm diameter hole, and it was guided to the anode by a converging, axisymmetric magnetic field forming a visible plasma column of about 3 cm diameter at the anode. The magnetic field stabilized the position of the double layer [5] so detailed measurements could be made in a simple way. The double layer was formed by increasing the anode voltage to a sufficently high value above the potential of the surrounding plasma. Electrons accelerated by the voltage drop in the anode sheath then caused additional ionization in the sheath, and an ionization double layer was formed.

In the present work we have rearranged the experimental configuration to obtain a much higher hf power level. This has been achieved by removing the tail of high energy electrons, that was present in the source plasma used in the earlier work and in general is present in double and triple plasma devices. Since only thermal electrons are now accelerated up the double layer, the "bump-on-tail" in the electron distribution function in the anode plasma is much more prominent. The ratio of the electron-beam density to background electron density on the high potential side of the double layer has also been increased. Both of these effects increase the growth rate of the instability. In this work we present profiles of plasma potential, density and relative hf voltage level for a much stronger hf field than obtained earlier, and radial density profiles now show clear minima along the path of the beam, much like those observed in other work on beam-plasma interactions[3] which were ascribed to caviton formation. The new experimental arrangement and some properties of the double layers formed are also dicussed. As will be shown in a subsequent paper,[6] the hf fluctuations appear as wave packets which are concentrated in small regions of space and associated with short-lived, local density depressions, as expected in nonlinear electron-beam plasma interaction. It is proposed that the steady density trough observed is an effect of these depressions.

2. EXPERIMENTAL SETUP: The experiment was performed in a large (1.5 m^3) cylinderical stainless steel vacuum chamber shown schematically in Fig. 1. As an important feature, an axially symmetric magnetic mirror field is used to stabilize the position of the double layer. Some test experiments guided the present construction of the experimental arrangement. Using the the same plasma source as earlier[2] with a high energy tail in the electron velocity distribution, the hf power level was found to increase strongly when the high energy electrons simply were removed by modifying the magnetic field so that the fast electrons were deflected and missed the double layer which formed in front of the anode. Measurements showed that the potential drop of the double layer increased, and the radial density scans showed minima in the region of intense hf fluctuations.

In the construction used here, care is taken to shield the anode region so that only thermal electrons are incident on the low potential side of the double layer. The maxium field strength, variable between 1 and 14 mT, occurs at the right end of the chamber. A ring of filaments, at right angles to the symmetry axis, is placed near the other end of the chamber and near the outer mantle surface where the magnetic field is about 20 times weaker. A source of plasma is obtained by a discharge between the filament ring and an anode ring with a smaller diameter, placed near the position of the maximum field, so the discharge occurs essentially along the magnetic field lines. The discharge is best characterized as an arc discharge, and the typical voltage drop is 20 V. The anode ring is concentric with, but insulated from, a plane anode. In front of this, the double layer forms by applying a sufficiently large voltage drop between the anode and the surrounding plasma. The plasma density could be increased additionally by another gas discharge plasma source, placed at the symmetry axis in the left end (Fig. 1) of the chamber and connected to the chamber through an aperture with a diameter of 4 cm. Some electrons from this source have high energies, and a metal plate at floating potential was used to prevent any direct tranfer of electrons between the source and the anode. Field lines through the central anode are all intercepted on the low potential side by the floating shielding plate. Thus when the central anode is biased positive, electrons are drawn to it from a region where the electron temperature is relatively low. Then only thermal electrons are incident at the low potential side of the double layer and these constitute the beam electrons on the high potential side. The beam density can be increased by increasing the source densities, and the density of thermal electrons on the high potential side of the double layer can be decreased by lowering the argon pressure which reduces the ionization rate.

Potentials in the plasma are measured by emissive probes using the floating potential method and high frequency waves are measured using an especially developed double probe with the inner conductors of two coaxial cables protruding a few mm into the plasma.[7] The high frequency signals from these wires are capacitively coupled to a hybrid tee, and the difference signal is connected to a spectrum analyzer or an oscilloscope with a bandwidth of 400 MHz. Each wire is connected through a 10 kohm isolating resistor to a dc bias voltage. The hf signal varies with the dc bias which usually was kept at or near the plasma potential. The plasma density is measured with a newly developed multiple probe which permits rapid scanning, even in the presence of potential variations. The probe is essentially an ordinary cold collecting wire which is connected, through a signal resistor, to an emissive probe located a few gyro radii from the cold probe. Thus the probe is always biased at a potential close to the local plasma potential, and the probe current is taken as a measure of the plasma density. A detailed discussion of this probe, including a critical evaluation of its performance, is given elsewhere.[8]

Data were recorded by a variety of means, including an X-Y plotter, an Analogic/Data Precision waveform analyzer (Data 6000), single traces on the fast oscilloscopes and GPIB links to a Macintosh Quadra 900. Standard data acquisition methods were used in all cases.

3. EXPERIMENTAL RESULTS AND DISCUSSION: Fig. 2 shows axial scans along the central axis of (a) the potential profile and (b) the hf voltage level at the maximum in the frequency spectrum (bandwidth 3 MHz). The frequency spectrum has a maximum at about 300 MHz and a full half-width of about 100 MHz. (c) shows probe current at plasma potential which is a measure of the electron number density. The profile of the hf voltage level assumes a maximum about 10 cm from the DL, and the voltage level there is about two orders of magnitudes larger than the back ground noise level on the low potential side of the double layer. The strength of the hf field depends strongly on the plasma parameters. The hf intensity increases when ring anode current I_r and the magnetic field strength B is increased and the background pressure p is decreased because these variations tend to increase the DL voltage drop and η, the ratio between the beam density and the total density in the anode plasma. Theoretically, these two parameters are expected to increase both the growth rate and the saturated amplitude of the hf field. Fig. 3 shows two extreme cases. The high intensity is obtained for $I_r \approx 3$ A, $B \approx 5$ mT, $p \approx 0.2$ mTorr and $I_a \approx 1$ A. I_a must not be too large because the DL voltage drop in general decreases with increasing I_a. η increases when the background pressure is decreased for constant I_a, because a decrease in pressure will cause a decrease in background plasma density independent of the beam density. A variation in the column current causes both electron beam and electron background density to vary and by the same factor. The low intensity case shown in Fig. 3 was obtained by combining the parameters to give a small potential drop over the DL and a low value of η.

Fig. 4 shows radial density profiles where the probe current at plasma potential is used as a measure of the density. Fig. 4a shows profiles when the level of the hf field is small. Fig. 4b shows profiles when the maximum voltage level of the time averaged hf field is about 12 times larger. On the high potential side of the double layer, there is now a density trough along the accelerated electron beam. Note that well defined density minima first show up at the point where hf power level begins to increase from the background noise. We propose that the density trough is caused by the ponderomotive force. Similar troughs have been observed in experiments where an electron beam is injected.[3] As seen in Fig. 2, there is no obvious variation of the axial density profile that can be correlated to the gradient of the hf profile.

This may be due to the fact that the force in terms of the gradient of the time averaged hf power is not in a simple way connected with the gradient of the time averaged hf level shown in these figures. As will be shown in a later paper, the hf field is concentrated in small regions of space, and the scale length of the ponderomotive force refer to these variations. A second thing to be noted in these data is that the minima extend all the way to the anode, even though the hf intensity has decreased considerably there. We believe this is due to a kind of wake effect. The density depletion is created in the vicinity of the maximum in the time averaged hf intensity and, as electrons flow to the anode along field lines of an increasing magnetic field, there is insufficient time for the minimum to be filled by diffusion across field lines.

4. CONCLUSIONS: We have shown the existence of depletions in plasma density on the tops of DL's which are correlated in position with regions of intense hf fluctuations. Since similar hf fluctuations have been observed in other DL experiments, it is reasonable to suppose that significant density depletions occur in a wide variety of DL's. In the present investigation we have shown a way to create the DL's that gives an enhanced hf intensity and increases the ease with which the density depletions can be studied.

5. ACKNOWLEDGEMENTS: This work was supported by the Swedish Natural Science Research Council. The many probe designs and design changes, changes of the plasma chamber and modifications and repair of the electronic and electrical equipment used in this work were performed with skill and enthusiasm by Mr. Jan Wistedt to whom the authors are grateful. The considerable assistance of Mr M. Wohlin and Dr. Ingvar Axnäs in the construction of programs to ease the data acquisition is gratefully acknowledged. One of us (RTC) was priveleged to spend a year's developmental assignment at KTH during the course of this investigation and would like to thank Professor Carl-Gunne Fälthammar for support and for his hospitality during the year.

REFERENCES

1. S. Torvén and L. Lindberg, *J. Phys. D* **13** (1980), 2285; P. Leung, A. Y. Wong and B. H. Quon, *Phys. Fluids* **23** (1980), 992; D. Andersson, *J. Phys. D* **14** (1981), 1403.

2. R. Schrittwieser, I. Axnäs, R. T. Carpenter and S. Torvén, *IEEE Transact. Plasma Sci.* (Proc. of the Alpbach conference), in the press.

3. P.Y. Cheung and A.Y. Wong, *Phys. Fluids* **28** (1985), 1538

4. A.Y. Wong and P.Y. Cheung, Phys. Rev. Lett.. **52** (1984), 1222; H.C. Kim, R.L. Stenzel

and A.Y. Wong, Phys. Rev. Lett. **33** (1974), **886**; T. Intrator, C. Chan, N. Hershkowitz and D. Diebold, *Phys. Rev. Lett.* (1984), **53**, p. 1233.

5. B. Song, R. L. Merlino and N. D`Angelo, *Physica Scripta* **45** (1992), 395; S. L. Cartier and R. L. Merlino, *Phys. Fluids* **30** (1987), 2549; R. Plamondon, J. Teichmann and S. Torvén, *J. Phys. D: Appl. Phys.* **21** (1988), 286.

6. R. T. Carpenter and S. Torvén, *Proc. Fourth Int. Symp. on Double Layers, Contributed Papers* (1992), Innsbruck

7. L. Lindberg, *J. Phys.E* **18** (1985), 214

8. I. Axnäs and S. Torvén, *Proc. 1992 Int. Conf. Plasma Phys., Contributed Papers*, Innsbruck.

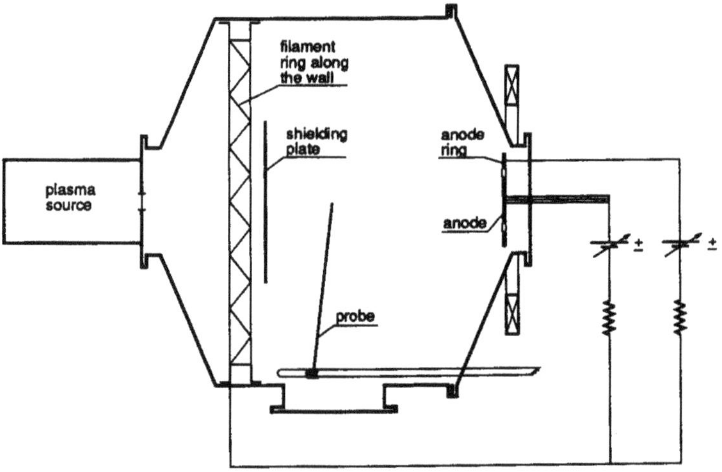

FIG. 1 Schematic of the experimental device. A magnetic mirror field, which is strongest at the anode, stabilizes the position of the DL. The main plasma source is a discharge along the magnetic field, between the filament ring at the left end and an anode ring concentric with, but insulated from, the anode.

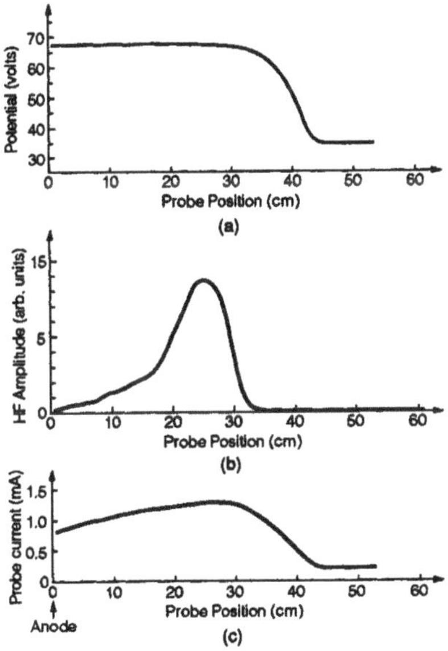

Fig. 2. Axial variation of (a) plasma potential, (b) hf intensity and (c) density along the central axis. The probe position zero is the anode position.

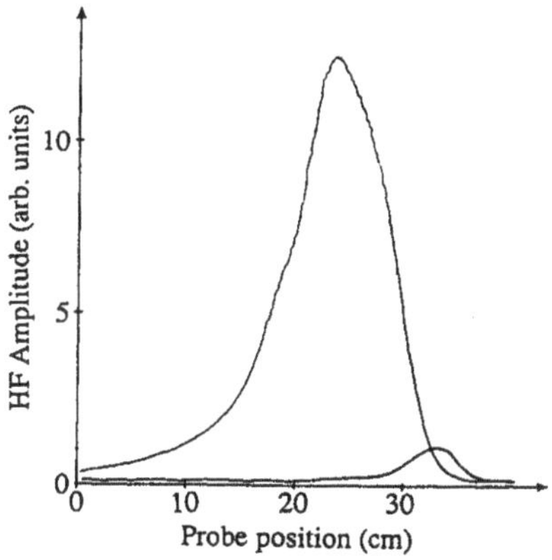

Fig. 3. Axial profiles of the hf voltage level for two different sets of plasma parameters. The high level is obtained when the parameters are chosen to give a high DL potential drop and a high value of the ratio of electron beam density to background density.

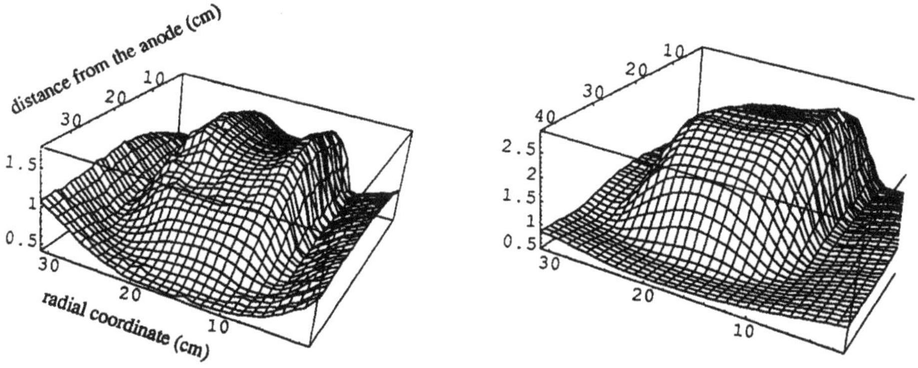

Fig. 4. Radial density profiles at different distances from the anode for the same DL position as shown in Fig.2. The vertical scales give the time averaged density in terms of probe current at plasma potential (mA). Profiles are given for a high (left) and a low (right) level of the time averaged hf field (cf. Fig. 3). For a high hf level a clear density trough is observed. The trough appears at a position approximately coinciding with the maximum of the time averaged hf field.

HYSTERESIS, SELFSPIKING AND CHAOS IN THERMIONIC DIODES WITH FILAMENT CATHODES

F. Greiner, T. Klinger, A. Piel, R. Timm
Institut für Experimentalphysik, Universität Kiel, 2300 Kiel, Germany

1. INTRODUCTION: Discharges with filament cathodes and magnetic box confinement [1, 2, 3] at pressures of about 0.01-1 Pa are in widespread use in double-plasma devices [5], as ion sources [4], or for low-pressure plasma processing. This kind of discharge has also attracted much attention because of its hysteresis in the I(U) characteristics [6, 7, 8, 9] and its low-frequency oscillations [6, 10]. Hysteresis and oscillations have also been invoked from the viewpoint of nonlinear dynamics and chaos. For periodic modulation mode-locking of the self-oscillations, Arnol'd tongues and quasiperiodic transitions to chaos were observed [11], whereas an intermittency route to chaos was found, when the selfoscillations were suppressed by a short anode-cathode separation [12]. In case of larger electrode distance, but with damped self-oscillations, a period-doubling route to chaos was observed [13]. In further experiments [14] the internal control parameter was identified as the number of newly generated ions in the anode sheath. The measured return map immediately showed a simple quadratic shape which suggests to model the dynamics by a 'noisy' logistic map. The dimensionality at the onset of chaos had a fractal value ($D_2 = (0.8 \pm 0.2)$). Modulation experiments [15] in the presence of self-oscillations showed mode-locking phenomena that resemble the dissipative circle-map. The sequence of mode-locked states forms a complete Farey-sequence. In contrast to earlier observations [11] the Arnol'd-tongues tend to overlap already at low modulation degree $m < 10\%$. In this contribution we present a comparison of new laboratory experiments and particle-in-cell simulations for hysteresis, self-oscillations and driven oscillations in thermionic discharges.

2. EXPERIMENTAL ARRANGEMENT AND HYSTERESIS: For an immediate comparison of thermionic discharges with the predictions of 1-D particle-in-cell simulations, we have operated a diode, consisting of a single straight tungsten filament (0.2mm dia. × 60mm) and a tantalum anode plate in a homogeneous axial magnetic field. Filling gas was argon at $p = 0.1$ Pa. The anode was either at 16cm or 85cm distance from the cathode. A flat sheet-like plasma is formed between filament and anode, which is characterized by the typical hysteresis effects (Fig. 1a).

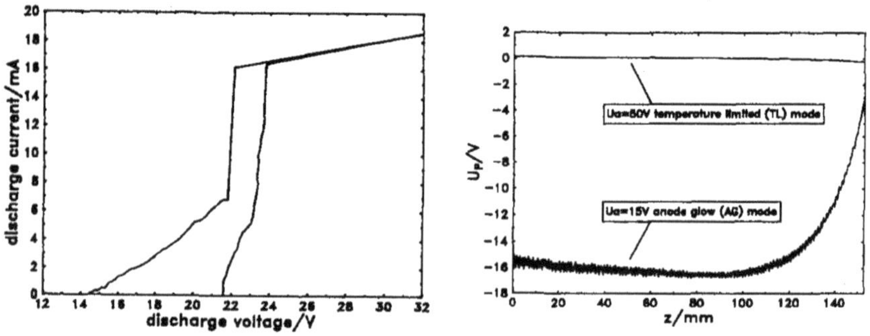

Figure 1: (a) Hysteresis curve of the 16cm long diode, (b) space potential distribution for the AGM and TLM (see text).

The space potential is slightly below cathode potential in the lower branch, and close to anode potential in the upper branch (Fig. 1b). Although the plasma is essentially collisionless, the two states resemble the 'anode-glow mode' (AGM) and the 'temperature-limited mode' (TLM), which are well known for thermionic discharges at pressures in the mbar regime [16, 17, 18, 19].

3. SELFSPIKING: In the AGM of this '1-D diode', close to the right hysteresis point, self-oscillations of the discharge with typical frequencies between 50-200 Hz (long diode) and 500-2000 Hz (short diode) appear above a minimum gas pressure. A typical pulse train of discharge current and plasma potential is shown in Fig. 2. During the current pulse the space potential rises by 1-2 V, typically. The pulse train is most regular close to the hysteresis point. The average pulse frequency decreases towards lower discharge voltage, but increases with gas pressure.

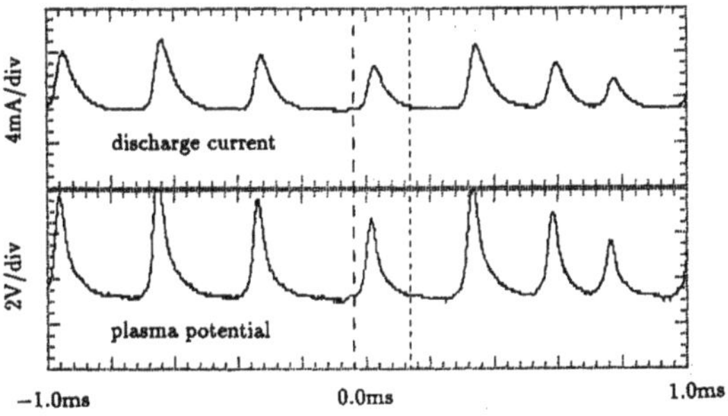

Figure 2: Current and plasma potential pulses in the selfspiking of the short diode.

4. PIC-SIMULATION OF HYSTERESIS AND SELFSPIKING: We use a modified version of the well-known PDP1-code [22]. Most of the simulation parameters were chosen as realistic as possible. The diode length was 15cm, the gas pressure was 0.1 Pa, and the collision cross-sections were matched to argon, except for a ten times increased ionisation cross-section, which models the confinement of primary electrons by the magnetic-box. The ion to electron mass ratio was chosen as 10 or 100, to study the influence of ion transit times. A spatial grid of 100 cells and a few hundred super-particles were used.

For operating voltages that slightly exceed the ionisation potential, the mechanism of formation for the AGM became evident: Ionisation is energetically allowed only in the anode sheath, and newly generated ions perform a nearly free fall towards the cathode. Because the mean free path for ion-neutral collisions is ten times larger than the system dimensions, only a small fraction of those ions is stopped by charge-exchange with slow neutrals. However, once an ion is stopped in the potential well of the virtual cathode, it becomes trapped there, and forms a cathodic plasma, which expands by the thermal motion of the ions. Although the filling mechanism differs from the diffusive transport at higher pressure, the final state is similar to the AGM, as can be seen from a typical phase-space and potential distribution in Fig. 3a. We have introduced a finite ion-lifetime to balance the continuous inflow of ions into the cathodic plasma. It represents surface losses towards non-emitting parts of the cathode or its supporting structure. Otherwise the cathodic plasma will always attain a selfspiking final state because of the unlimited increase in electron density.

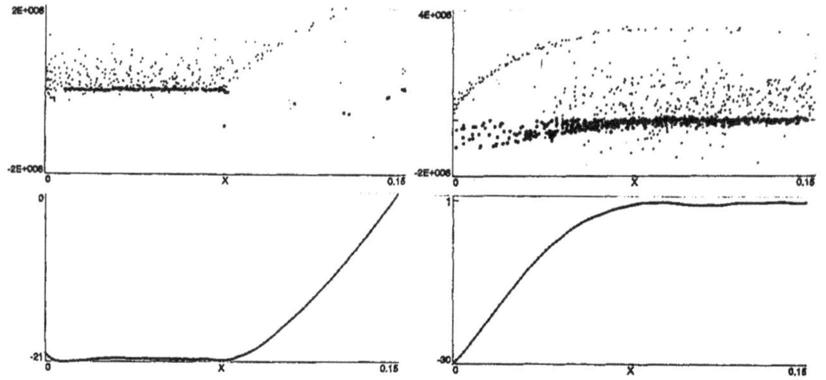

Figure 3: (a) $x - v_x$ phase space of electrons and ions in the AGM and potential distribution. (b) The same for the TLM

The TLM is attained if the ion production rate in the anode sheath is larger than the losses. In that case an anodic plasma forms, which expands towards the cathode until most of the discharge volume attains positive potential. The virtual cathode vanishes by the abundant ion space charge (Fig. 3b). By the plasma expansion the ionisation volume is increased. Hence, the fully developed TLM can still be maintained at a lower discharge voltage, which explains the hysteresis of the upper branch. The TLM is destroyed by the re-establishment of the virtual cathode by violation of the Langmuir-condition in the cathodic double layer. A typical hysteresis curve as obtained from the final states of such PIC-simulations is shown in Fig. 4.

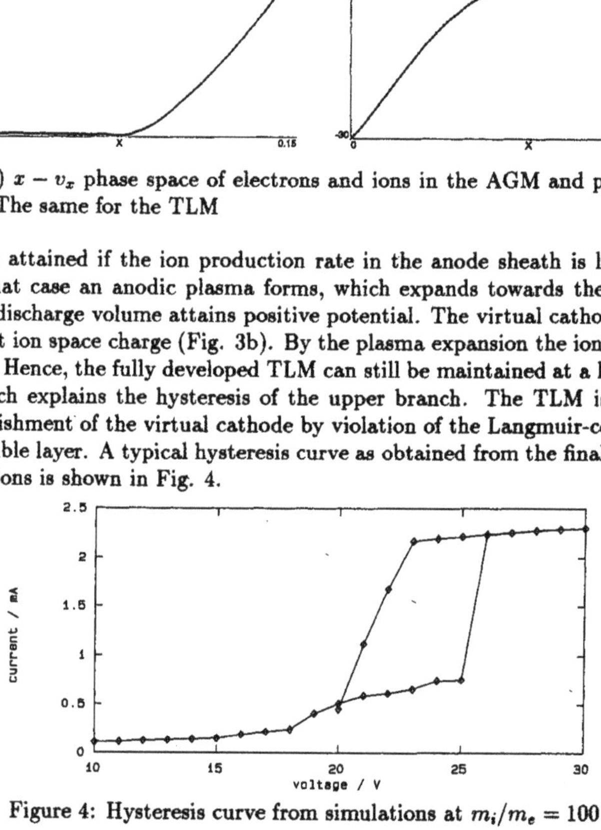

Figure 4: Hysteresis curve from simulations at $m_i/m_e = 100$.

Close to the point of instability of the AGM, selfspiking is observed. The sequence of elementary processes can be taken from Fig. 5: The cathodic plasma gradually expands by trapping ions from the source in the anode sheath. Once a critical length *and* density is reached, the condition for Buneman-instability is fulfilled: $\omega_{pe} \approx (\pi/L)v_b$, where L is the plasma length and v_b is the electron beam velocity. The formation of an electron hole [23] within the cathodic plasma starts, and transforms into a double-layer. This double-layer now accelerates part of the ions towards the cathode where they reduce the negative space charge, and enable the increase of the electron current. Note that the current spike is accompanied by a positive excursion of the mid-potential which represents the potential step of the double layer. The double-layer is destroyed by exhausting the ions on the high potential side of the double layer. The process starts again by refilling of ions into the cathodic plasma and subsequent expansion.

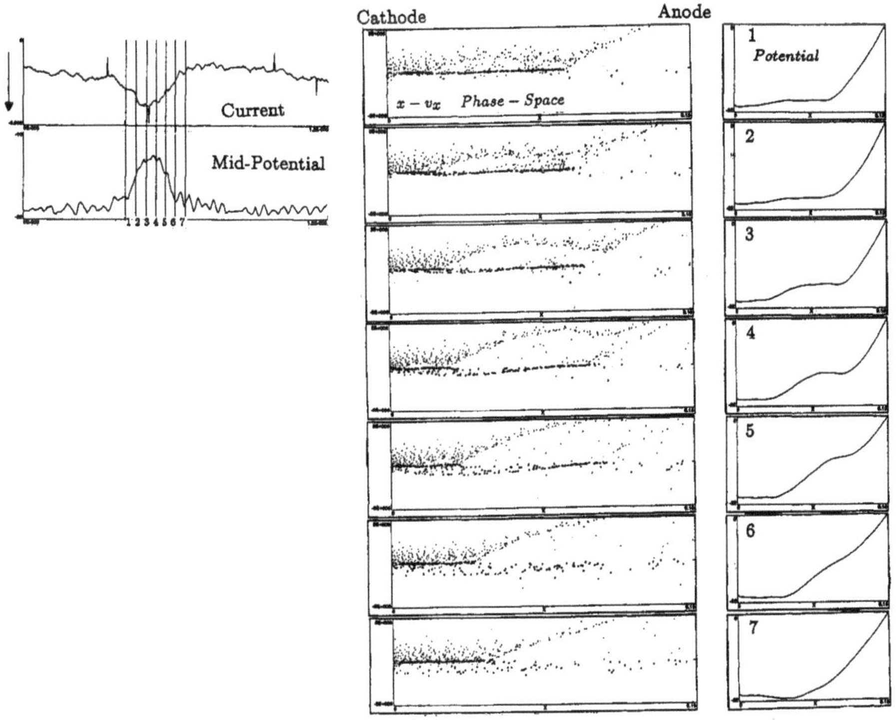

Figure 5: (a) A single current spike and corresponding mid-potential of the diode.
(b) $x - v_x$ phase pace and potential distribution for the indicated times in (a).

A similar sequence of events was observed in PIC-simulations for positively biased Q-machine conditions by Bauer and Schamel [24]. The essential difference between the models is the absence of neutrals and ion production at the cathode in [24].

<u>5. EXPERIMENTAL DETECTION OF UNSTABLE PLASMA WAVES</u>: We have tried to verify the role of the plasma instabilities in the onset of a current spike by detecting the expected hf-waves. A high-frequency probe in the cathodic plasma shows a characteristic burst which antecedes the current spike (Fig. 6a). The frequency spectrum is peaked at about 140 MHz, and shifts towards 300 MHz during the current pulse (Fig.6b). These frequencies agree with estimates of the electron plasma frequency before and during the current pulse. The spatial distribution of the signals peaks within the cathodic plasma. This finding hints at a two-stream instability rather than a Buneman-instability. The latter should have a frequency spectrum around $\omega \approx 2^{-4/3}(m_e/m_i)^{1/3}\omega_{pe}$, which is a factor of 100 smaller than the electron plasma frequency. We also observe lower frequency oscillati-

ons around 250 kHz, which may be associated with Buneman-type instabilities, but they appear not at a fixed phase relative to the current spike, but are randomly distributed bursts. Time resolved probe measurements indicate a double-humped electron distribution function during the spike in the cathodic plasma, which might be the origin of the observed hf- instability. This issue, however, needs further experimental investigations.

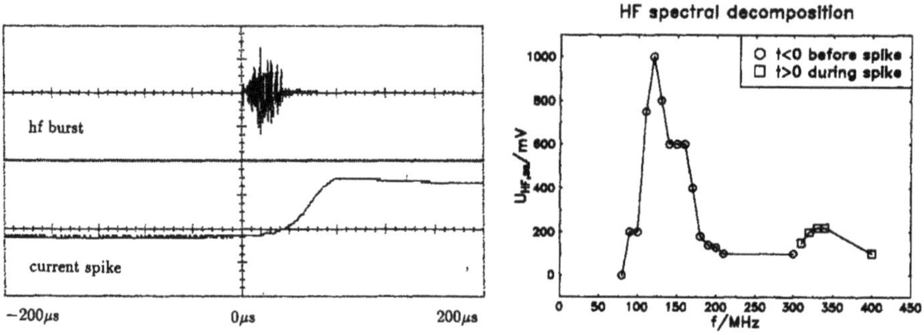

Figure 6: (a) Comparison of hf-burst position and current spike, (b) frequency spectrum of the hf-burst

6. PIC-SIMULATION OF THE MODULATED DIODE: The central question in the period doubling sequence reads: why is a strong current pulse strictly followed by a weak one? Some insight can be gained by studying the case of mode-locked states, where the driver frequency is twice the natural spiking frequency. Fig. 7 shows a simulation with a natural spiking frequency of 350 kHz and a modulation frequency of 750 kHz. A detailed inspection of the sequence of events demonstrates that each strong current pulse leads to a dramatic ion depletion in the cathodic plasma. At the subsequent (negative) peak of the cathode driver voltage, the cathodic plasma has not yet regained the necessary density for Buneman-type instability. The corresponding weak current pulse only reflects the nonlinear increase of discharge current with applied voltage.

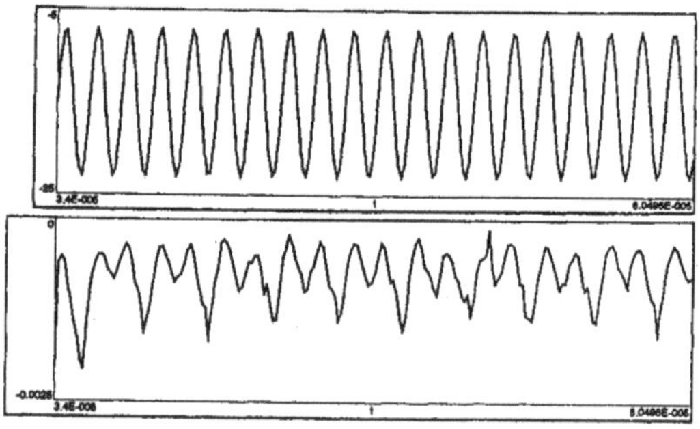

Figure 7: Simulation of a mode-lock-2 state. top: cathode voltage, bottom: discharge current. The discharge current attains a period-doubled state

7. SUMMARY AND CONCLUSIONS: The different parts of the hysteresis curve are re-interpreted in terms of the AGM and TLM of higher pressure thermionic discharges. Diffusive transport is replaced by free-fall and stopping by charge-exchange collisions as filling mechanism of the cathodic plasma. The instability of the AGM depends on the particle balance in the anode sheath, whereas the TLM becomes unstable by violating the Langmuir condition of the cathodic double layer. The PIC simulations reveal the sequence of events in the selfspiking process. The criterion for Buneman instability explains how selfspiking can be suppressed by short plasma length. The spike frequency is determined by the filling time of the cathode plasma. HF-bursts at the plasma frequency were found on the rising edge of the current pulse. PIC-simulations of a mode-lock-2 state demonstrate that a simple parabolic return map in the dynamics may be explained by the unability of the cathodic plasma to satisfy the Buneman-instability condition after ion depletion in a strong current pulse.

8. ACKNOWLEDGEMENTS: This work was supported by DFG grant Pi185/2-1.

References

[1] R. Limpaecher, K.R. MacKenzie, Rev. Sci. Instrum. **44**, 726 (1973)
[2] A.Y. Wong, Y. Nakamura, B.H. Quon, J.M. Dawson, Phys. Rev. Lett. **35**, 1156 (1975)
[3] K.N. Leung, G.R. Taylor, J.M. Barrick, S.L. Paul, R.E. Kribel, Phys. Lett. **57A** 145 (1976)
[4] K.W. Ehlers, K.N. Leung, Rev. Sci. Instrum. **50**, 1353 (1979)
[5] N. Hershkowitz, K.N. Leung, T. Romesser, Phys. Rev. Lett. **35**, 277 (1975)
[6] R.L. Merlino, S.L. Cartier, Appl. Phys. Lett. **44**, 33 (1984)
[7] G. Knorr, Plasma Phys. **26**, 949 (1984)
[8] R.A. Bosch, R.L. Merlino, Contrib. Plasma Phys. **26**, 1 (1986)
[9] R.A. Bosch, R.L. Merlino, Contrib. Plasma Phys. **26**, 13 (1986)
[10] Y. Jiang, H. Wang, C.X. Yu, Chin. Phys. Lett. **5**, 489 (1988)
[11] P.Y. Cheung, S. Donovan, A.Y. Wong, Int. Conf. Plasma Phys., New-Delhi, 1990, p.
[12] P.Y. Cheung, S. Donovan, A.Y. Wong, Phys. Rev. Lett. **61**, 1360 (1988)
[13] R. Timm, A. Piel, XXth ICPIG, Pisa, p. 259 (1991)
[14] R. Timm, A. Piel, Contrib. Plasma Phys. (tbp)
[15] T. Klinger, R. Timm, A. Piel, Int. Conf. Plasma Phys. 1992, Innsbruck
[16] L. Malter, E.O. Johnson, W.M. Webster, RCA Review, **XII**, 415 (1951)
[17] W.M. Webster, E.O. Johnson, L. Malter, RCA Review, **XIII**, 163 (1952)
[18] E.O. Johnson, W.M. Webster, RCA Review, **XVI**, 82 (1955)
[19] E.O. Johnson, RCA Review, **XVI**, 498 (1955)
[20] Bin Song, N. D'Angelo, R.L. Merlino, J. Phys. D **24**, 1789 (1991)
[21] C. Wilke, R. Leven, H. Deutsch, Phys. Lett. A**136**, 114 (1989)
[22] PDP1 reference manual V2.0, C.K. Birdsall, Plasma theory and simulation group, Univ. Cal. Berkeley (1990)
[23] H. Schamel Phys. Reports **140**, 161 (1986)
[24] F. Bauer, H. Schamel Physica D**54**, 235 (1992)

LOCAL MAGNETIC-MIRROR EFFECTS ON ELECTROSTATIC POTENTIAL IN A PLASMA FLOW WITH NON-MAXWELLIAN ELECTRON DISTRIBUTION

R. Hatakeyama, B. Kuramori, K. Aoyama, and N. Sato

Department of Electronic Engineering, Tohoku University, Sendai 980, Japan

Measurements are performed on potential profile along a DC-discharge plasma flow in regions of magnetic mirrors. When the plasma of a non-Maxwellian electron energy distribution with high-energy tail flows into a local magnetic bump, an electrostatic potential well is observed to form around the bump. When the plasma flow is guided into a local magnetic dip, a potential hill is observed there. The results attributed to the highly-anisotropic electron distribution are contrary to those for supersonic collisionless plasma flows.

1. INTRODUCTION: Electrostatic potential structures of plasma flows along inhomogenous magnetic fields may have strong effects on particle acceleration and transport in magnetized plasmas. Main works on this topic have been concentrated on electrostatic potential formation in supersonic plasma flows, which is predominated by ion dynamics. Self-consistent potential variations in magnetic dips were analyzed in conjunction with a thermal barrier.[1] Measurements on electrostatic potential in supersonic plasma flows along magnetic dip and bump demonstrated the formation of potential well and hill , respectively.[2] A field-aligned potential difference in a high-speed plasma flow with anisotropic ion velocity distribution has been investigated,[3-5] and a potential increase along converging magnetic field lines is clearly observed.[6]

Here we present potential profiles along a plasma flow with highly-anisotropic electron distribution in regions of local magnetic mirrors. Measured results contrast well with aforementioned results in case of the supersonic plasma flow.

2. EXPERIMENTAL METHODS AND RESULTS: A plasma with density $n \gtrsim$

10^9 cm^{-3} is produced by an argon-gas discharge between a 5-cm-diam grid anode A and an oxcide cathode K at one end of a straight 15.7-cm-diam stainless-steel vacuum chamber which is grounded electrically together with A. The argon gas is fed near K and the gas pressure around the machine center P_{Ar} is in the range 4×10^{-3} - 5×10^{-2} Pa. The plasma diffuses along a strong magnetic field B(=1-4 kG) and is terminated by a floating 5.2-cm-diam endplate at Z=250 cm from A. Local magnetic bumps with mirror ratio $1<R_m(=B_m/B_0)<3.2$ and magnetic dips with $0.1<R_m<1$ can be produced between A and the endplate, axial half-widths of which are below 30 cm. Here B_0 represents a value at the uniform region and B_m at the bump top or dip bottom. The plasma potential ϕ is measured by an axially-movable small emissive probe.

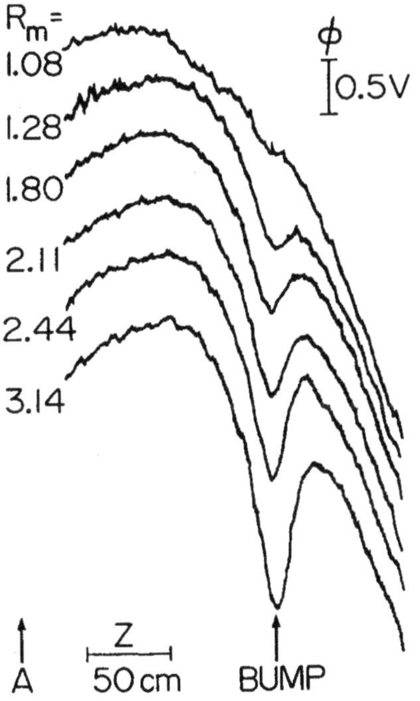

Fig.1. Axial potential profiles for magnetic bumps. B_0=1.24 kG, $P_{Ar}=4.5\times10^{-3}$ Pa. A is anode and the other arrow indicates bump top.

The electron energy distribution function parallel to the magnetic field F_e is obtained from a derivative of the voltage V_p - current I_p characteristics of the Langmuir probe. In the uniform magnetic field (R_m=1.0), F_e is almost Maxwellian with temperature $T_e \gtrsim 2$ eV for a higher $P_{Ar}(>1\times10^{-2}$ Pa). As P_{Ar} is gradually reduced, a high energy tail(>10 eV) appears in F_e, resulting in a two-component Maxwellian type distribution with a sufficiently high tail-density for $P_{Ar} \gtrsim 4\times10^{-3}$ Pa. The ion temperature is estimated to be $T_i \lesssim 0.5$ eV. Main results described below are obtained under this gas-pressure region, where the electron energy distribution f_e is highly anisotropic with respect to the B-field lines.

Figure 1 gives axial potential profiles in case of the magnetic bump. The monotonously decreasing profile along the plasma flow is observed under the almost uniform magnetic field, which is considered to result from an ambipolar diffusion. When R_m is increased, the potential well is formed in the region of the magnetic bump. The potential-well depth $\Delta\phi$ is of the order of $0.1<-\Delta\phi/\bar{T}_e<0.2$ for large $R_m(\approx 3)$, where $\bar{T}_e(\lesssim 12$ eV) is an average temperature of the non-Maxwellian F_e for $R_m=1.0$. Axial variations of F_e are measured for a typical $R_m(=2.79)$ as shown in Fig.2. Almost no change of F_e is recognized in the uniform region between the plasma source and the magnetic bump. The density of high-energy tail is, however, observed to increase considerably along the converging magnetic field around the magnetic bump and in turn decrease as the magnetic field diverges downstream.

For the case of magnetic dip, on the other hand, an electrostatic potential hill appears in the region of magnetic depression as demonstrated in Fig.3. The potential-hill height $\Delta\phi$ increases when R_m decreases, i.e., the magnetic depression becomes large, which is smaller than the maximum value $(-\Delta\phi)$ of the potential-well depth for the

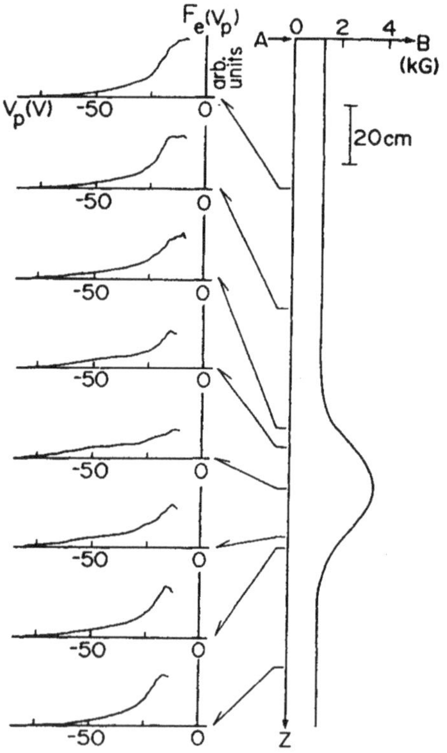

Fig.2. Variations of energy distribution along plasma flow. $R_m=2.79$, $P_{Ar}=4.5\times10^{-3}$ Pa.

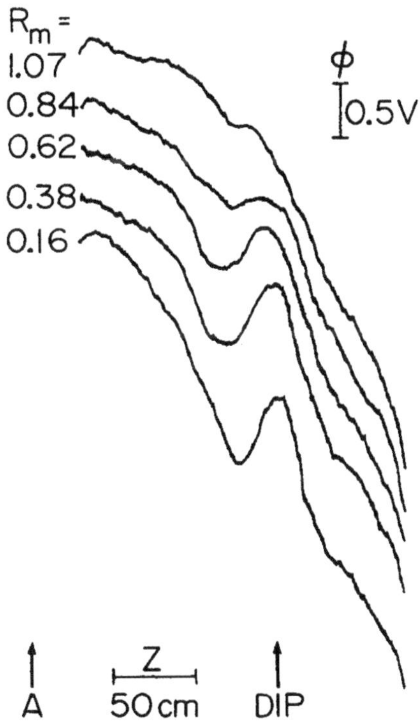

Fig.3. Potential profiles for magnetic dips. B_0=3.71 kG, P_{Ar}=6.7×10^{-3} Pa. Dip bottom is shown by arrow.

magnetic bump in Fig.1. F_e does not almost change except the region of nonuniform magnetic field in a typical magnetic configuration (R_m=0.28), as presented in Fig.4. However, the density of high-energy tail drastically decreases along the plasma flow in the region of magnetic depression.

It is noticed that the potential well for the magnetic bump and the potential hill for the magnetic dip are gradually diminished when P_{Ar} is increased and finally disappear around 5×10^{-2} Pa, where $F_e(f_e)$ is almost Maxwellian(isotropic).

3. DISCUSSIONS: According to measured results mentioned above potential profiles around the magnetic bump and dip are quite different from those obtained previously for the case of supersonic plasma flow. The highly anisotropic f_e and almost isotropic ion distribution f_i seem to be closely related to a mechanism for the local potential formation observed. Let us discuss a self-consistent scenario of electrostatic potential formation in a collisionless plasma flowing along a simply converging magnetic field B(Z) with a minimum B_0 at the entrance and a maximum B_m at the exit. We consider the case where most of the electrons with a smaller average pitch angle with respect to the B-field lines pass through, but most of the ions with a larger average pitch angle are reflected by the magnetic barrier in the absence of

electrostatic potential, $\phi(Z)=0$. Then a spatially decreasing potential profile $\phi(Z)$ is expected to form in order to maintain the charge-neutral condition beyond the magnetic barrier. For the case of simply diverging magnetic field with a maximum B_0 and a minimum B_m, on the other hand, the density spatially decreases due to $\vec{\nabla}_{/\!/}\vec{B}$ acceleration of charged particles in the absence of electrostatic potential. This local-density decrease is enhanced as their pitch angles become large. Thus , a monotonously increasing potential is expected to form to maintain the charge -neutral condition everywhere in case of the plasma flow with anisotropic electrons and almost isotropic ions.

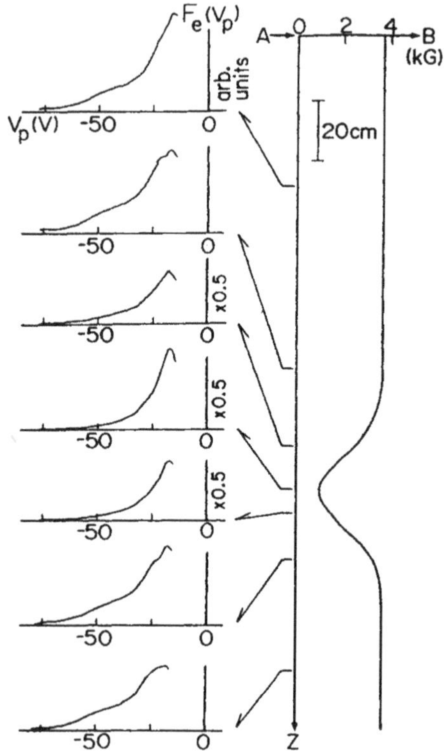

Fig. 4. Variations of energy distribution along plasma flow. $R_m=0.28$, $P_{Ar}=6.7\times10^{-3}$ Pa.

We can calculate the local electron and ion densities $n_{e,i}(Z)$ in both cases of converging and diverging magnetic fields by using Liouvile's theorem and conservation law of particle's energy ε and magnetic moment μ.[7] Self-consistent potential differences $\Delta\phi$ between the exit and entrance of the B-field configurations are obtained by substituting $n_{e,i}(Z)$ into Poisson's equation. For simplicity, we apply a beam approximation with monoenergetic electrons and ions, which is represented by $f_{e,i}=C_{e,i}\delta(\mu-\mu_{e,i0})\delta(\varepsilon-\varepsilon_{e,i0})$. Here $C_{e,i}$ are normalizing constants. The result under the quasi-neutral approximation is as follows:

$$\Delta\phi = \{\varepsilon_{e/\!/0}\varepsilon_{i/\!/0}/(\varepsilon_{e/\!/0}+\varepsilon_{i/\!/0})\}(\tan^2\theta_{i0}-\tan^2\theta_{e0})(1-R_m)$$
$$\simeq \varepsilon_{i/\!/0}(\tan^2\theta_{i0}-\tan^2\theta_{e0})(1-R_m) \quad \text{for } \varepsilon_{e/\!/0}\gg\varepsilon_{i/\!/0}, \qquad (1)$$

where a subscript "0", and $\theta_{e,i0}$ ($=\arctan v_{\perp 0}/v_{/\!/0}$) stand for a value at the entrance, and pitch angles of electrons and ions, respectively. Alfvén and Fälthammer derived the same equation under a more restricted condition.[3] Since $\varepsilon_{e/\!/0}(\sim T_e)\gg \varepsilon_{i/\!/0}(\sim T_i)$ and $\theta_{i0}\gg\theta_{e0}$ in our experiment, Eq.(1) gives a negative value around $-T_i(R_m-1)/e$ for $R_m>1$ (magnetic bump) and a positive value around $T_i(1-R_m)/e$ for $R_m<1$ (magnetic dip). These value predicted are consistent with the measured $\Delta\phi$ in Figs.(1) and (3).

4. CONCLUSIONS AND ACKNOWLEDGEMENTS: Electrostatic potential well and hill are observed to form when a low-pressure DC-discharge plasma with non-Maxwellian electron energy distribution flows into regions of magnetic bump and dip, respectively. The results are quite different from those reported previously for the case of supersonic plasma flow. The highly-anisotropic electron distribution in contrast with the ion distribution is attributed to a mechanism of the potential formation observed, which is qualitatively explained by a self-consistent theory under the quasi-neutral approximation.

We are indebted to H. Ishida for his technical supports.

REFERENCES:

1. J. Kesner et al., Nucl. Fusion 21(1981), 1265.
2. Y. Suzuki et al., J. Phys. Soc. Jpn. 55(1986), 1568.
3. H. Alfvén and C. -G. Fälthammer, Cosmical Electrodynamics(Oxford Univ. Press, London, 1963), 2nd ed., p.162.
4. H. Persson, Phys. Fluids 6(1963), 1756 and 9(1966), 1090; Y. Serizawa and T. Sato, Geophys. Res. Lett. 11(1984), 595 and Phys. Fluids 29(1986), 2753; H. Washimi and I. Katanuma, Geophys. Res. Lett. 13(1986), 897.
5. R. Geller et al., J. Plasma Phys. 12(1974), 467.
6. N. Sato et al., Phys. Rev. Lett. 61(1988), 1615.
7. R. Hatakeyama et al., Plasma Phys. 22(1980), 25.

EXPERIMENT ON STANDING SHOCKS AND LOCALIZED FLUCTUATIONS IN A COLLISIONLESS PLASMA

Tadao HONZAWA

Department of Electric and Electronic Engineering,

Utsunomiya University, Utsunomiya 321, Japan

A standing shock is shown to be formed in a collisionless plasma by letting a streaming plasma interact with perpendicular magnetic fields. Further, results on the shock structure and fluctuations localized in the shock region are presented.

INTRODUCTION

For the last three decades the whole aspects of standing plasma shocks at the boundaries of the magnetospheres of the earth and other planets on the day side have been clarified by many rocket observations[1]. Further, Axford et al.[2] predicted theoretically that a standing plasma shock was possibly formed at the boundary of the solar plasma region (heliosphere). In all these cases, it may be undoubted that magnetic fields perpendicular to plasma flows play an important role in the formation of standing shocks. Whenever a plasma flow interacts with a perpendicular magnetic field, however, an electric field is expected to be built because of the charge separation in the plasma flow.

In this paper, a standing shock is shown to be possibly formed in a collisionless plasma by making a plasma flow interact with a perpendicular magnetic field, though a low-β plasma is here used. Furthermore, observation of low frequency fluctuations locally excited in the front part of the shock is reported. A brief description on a mechanism of the shock formation is also made.

EXPERIMENTAL METHODS

Experiments were carried out by using a conventional double plasma device[3].

Argon gas was introduced into the device up to P $\lesssim$ 3x10^{-4} Torr. A plasma was produced in the driver plasma region by a dc discharge. However, a target plasma was here produced in a quite different manner as described below. A negatively biased mesh grid separated the two plasmas thus produced. The ion component of the driver plasma was injected into the target plasma region through the separation grid. Then, they were neutralized by electrons emitted from heated filaments there. In this way, we could obtain only a low density plasma with a density as low as 10^5-10^6 cm^{-3} on the target plasma side. Ion plasma frequency for such a plasma is estimated to be comparable to or less than 50 kHz. In such plasmas the mean free paths for interparticle collisions were comparable to the size of the device ($\lambda \sim$ 20 cm).

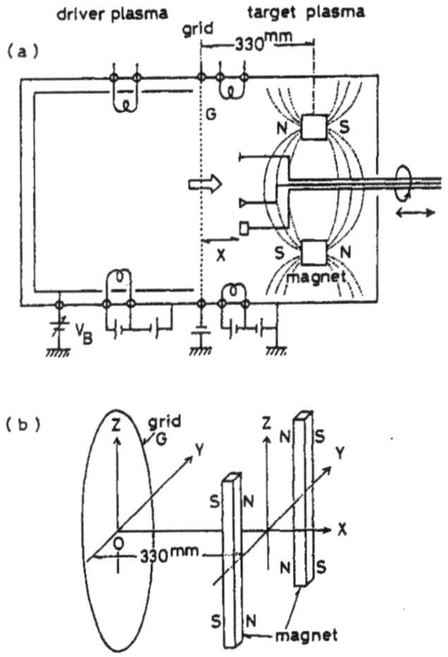

Fig.1.

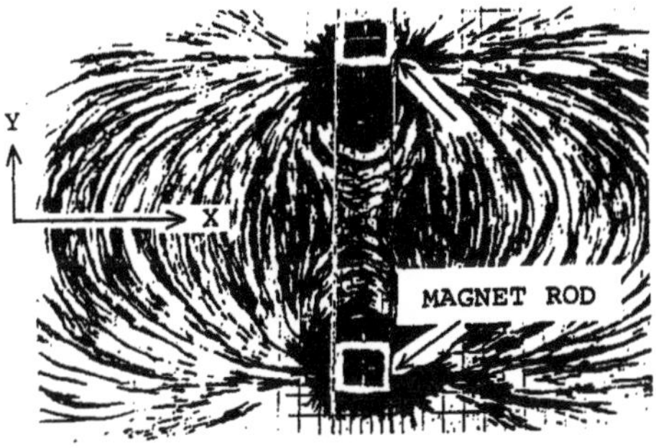

Fig.2.

For shock formation, we set a pair of magnet rods perpendicularly to the axis of the chamber (x-axis) and 33 cm apart from the grid surface in the target plasma chamber, as shown in Fig.1(a). Magnetic fields thus generated were observed to have only the x- and y-components near the x-axis and hence to be confined on the x-y planes. Shapes of magnetic lines of force are shown in Fig.2. Since the plasma particles moved along the x-axis, they could interact with the y-component of magnetic field B_y in front of the magnet rods. The maximum value of B_y was measured to be about 40 Gauss on the x-axis and to increase with increasing y. The fields B_y could have heavy influence on the motion of the electron-component of plasma flow, but they were not strong enough to affect the orbits of the ion-component.

RESULTS AND DISCUSSIONS

To realize a standing shock in the above device, all plasma particles must stream toward the magnetic barrier from the upstream side. By this reason, streaming ions were injected from the driver plasma and were soon after neutralized there by thermal electrons emitted from heated filaments. After that, all the plasma particles could interact with the perpendicular fields B_y. Thus, the orbits of plasma electrons were largely altered and they were locally accumulated in front of the magnetic barrier. In this experiment, using a movable probe positively biased with respect to a plasma, spatial changes of the electron saturation currents along the x-axis were measured, as shown in Fig.3. If the electron temperature T_e is assumed to be uniform throughout the region, these curves show the spatial distributions of the electron density N_e. In fact, analyses of probe characteristic curves showed that T_e was roughly independent of axial positions and had values from 1 to 2 eV. Curves in Fig.3 demonstrate that in the case of V_B = 8 V (ion beam energy $W_b \simeq 8$ eV), for example, N_e is kept almost constant at distances x < 15 cm, but it largely increases around x ≃ 18 cm, that is, a number of electrons are accumulated there. Here, it should be noted that when the magnet rods are removed, N_e is almost uniform

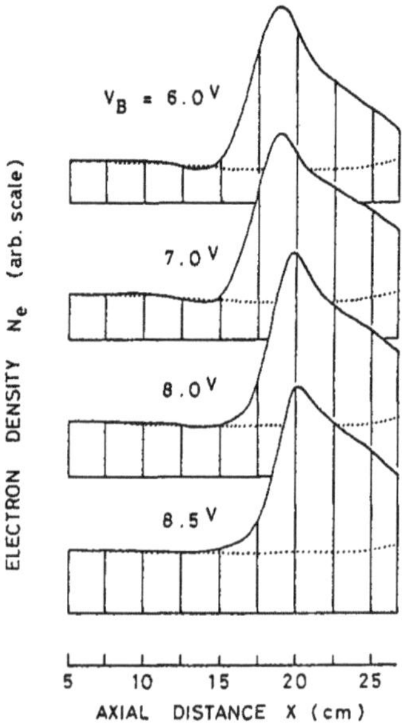

Fig.3.

throughout all distances x, as shown in dotted lines in Fig.3. These results give evidence that the local accumulation of electrons is attributed to the effect of B_y. In addition, the data show that the lengths L for the gradients of N_e-increases are roughly 2 to 3 cm (L ∿ 2-3 cm). After N_e reaches a maximum, it gradually decreases with increasing x. In this experiment, such spatial distributions of N_e were also measured at various distances y.

On the other hand, we could measure the spatial distributions of ion density N_i through ion saturation currents received by a probe. Results obtained are shown in Fig.4. Here, dotted curves correspond to the spatial distributions of N_i in the absence of magnet rods. From these it is found that N_i always decreases with increasing x in both cases with and without a magnetic barrier. The decrease of N_i is found to be due to the ion deceleration by a sheet of ion cloud formed near the grid. On the other hand, the effect of magnetic barrier always gives rise to a minimum in N_i around x ∿ 18 cm and a maximum around x ∿ 23 cm. The occurrence of the ion density minimum and maximum is closely related to the local accumulation of electrons described above.

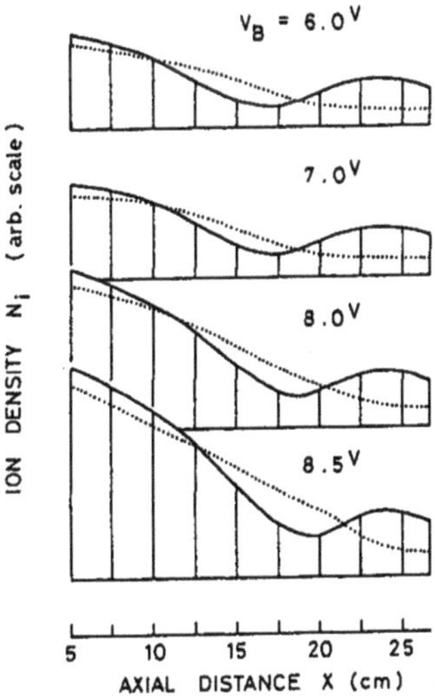

Fig.4.

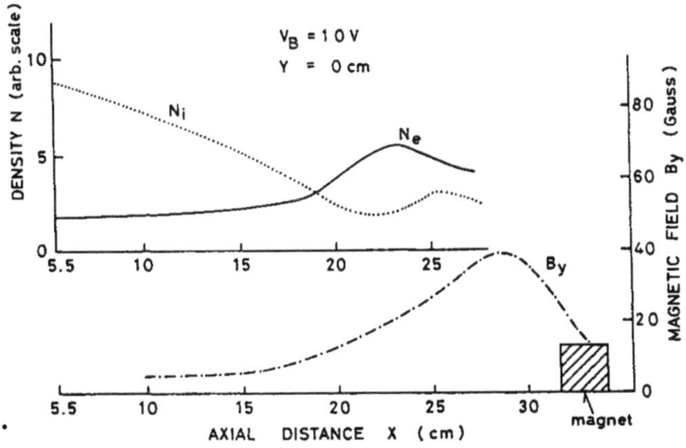

Fig.5.

Axial distributions of N_e and N_i are compared with each other in Fig.5. In this figure, the magnitude of the field B_y is also shown as a function of x. These data reveal that the maxima of N_e and N_i appear at axial positions, where B_y has a magnitude as small as 20-30 Gauss and gradually increases with x. Moreover, it is noted that the N_i-maximum always appears a little behind the position of the N_e-maximum. By the way, the electron Larmor radius r_{Le} is estimated to be about 3 mm around the position of the N_e-maximum, while the ion Larmor radius r_{Li} to be about 1 m. Combination of the axial distributions of N_e and N_i gives some information on the entire picture of a standing shock formed in front of the magnetic barrier. A typical example of two-dimensional patterns of the electron density changes on the x-y plane is shown in Fig.6. The pattern demonstrates that the electron accumulation occurs at positions along a curved line just like a magnetic line of force. The ion distributions were also observed to have a similar form. From these results we can conclude that a bow-shaped standing shock could be realized in a laboratory collisionless plasma under

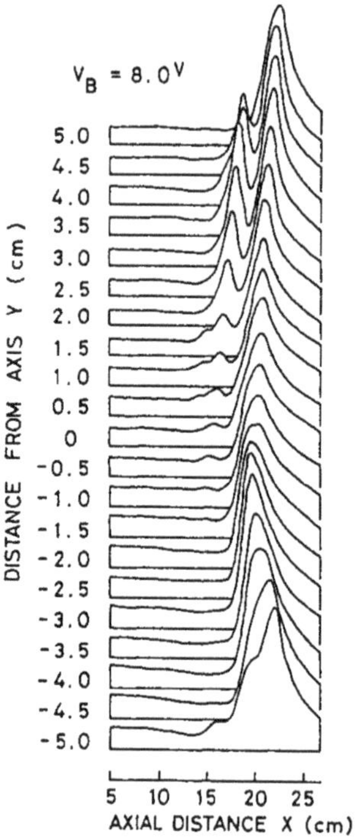

Fig.6.

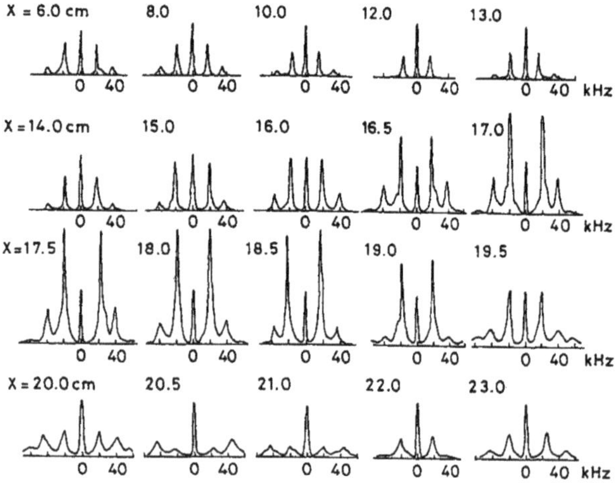

Fig.7.

an appropriate condition.

Next, localized fluctuations were observed to be associated with the shock formation. For studying these frequency spectra of such fluctuations were measured at various positions on the x-y plane. Typical examples of spectra observed at various x on the x-axis are shown in Fig.7. These spectra demonstrate that most intense fluctuations have frequencies as low as f ≃ 20 kHz and are localized in front of the N_e-maximum (see Fig.8). In Fig.8, the N_e-distribution is also given for comparison. The result suggests that the steep gradient of the N_e-increase may be closely related to the excitation of these low frequency fluctuations.

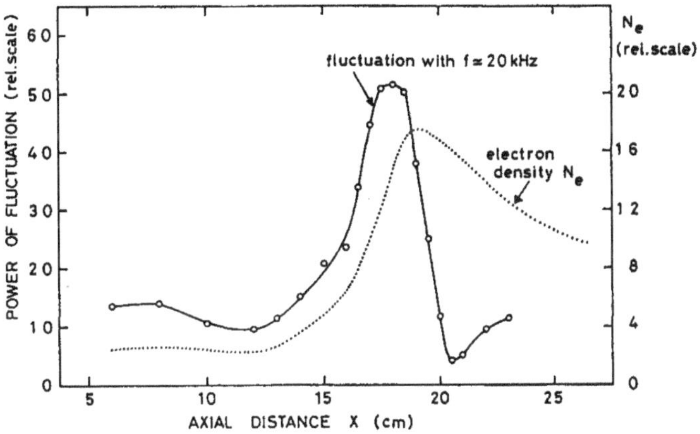

Fig.8.

REFERENCES

1. For example, C.T.Russell, *Collisionless Shocks in the Heliosphere*, edited by B.T.Tsurutani and R.G.Sone (American Geophysical Union, 1985) p109-130.

2. W.I.Axford, A.J.Dessler, and B.Gottlieb, Astrophys. J. 137, 1268 (1963).

3. R.J.Taylor, K.R.MacKenzie, and H.Ikezi, Rev. Sci. Instrum. 43, 1675 (1972).

WEAK DOUBLE LAYERS IN THE TRIPLE PLASMA DEVICE KIWI

T. Klinger and A. Piel,
Institut für Experimentalphysik, Universität Kiel, Germany

1. INTRODUCTION: In some cases only two- and sometimes three-dimensional potential measurements allow a correct interpretation of potential structures [1]. Various experimental investigations of two-dimensional double layer structures were performed in unmagnetized and magnetized plasmas [1, 2]. U and V shaped double layer contours were reported [3, 4] as well as more irregular structures [5]. Frequently, potential dips and multiple double layer structures are the consequence of two dimensional plasma potential profiles.
The present contribution discusses studies of double layer phenomena in the region of fringing magnetic field lines in a triple plasma device. The nonequlilibrium plasma state is achieved by suprathermal electron injection into the magnetized background plasma (the background plasma is produced by ionization). As a consequence we observe complicated double layers with potential jumps even below and above the trapped electron temperature. Especially a double layer formation ($e\Phi_0/k_BT_e \leq 1$) with its high potential region on the grid side cannot be explained in simple terms. There are different possible mechanisms for the formation of such double layer structures: Double layers may be formed by ion injection along converging magnetic field [4]. Alternatively a highly turbulent plasma state with increased effective ionization may be responsible for stationary potential jumps [6].
In the following Langmuir, HF, and emissive probe measurements are presented which were performed to study the conditions for double layer formation in the described situation. For certain discharge parameters the double layer becomes unstable and strong, low frequency, nonlinear potential relaxations propagate along the plasma column. These soliton-like instabilities have been shown to interact effectively with other types of plasma waves *i.e.* ion drift-waves [7] and have some similarities with potential relaxations in Q-machines [8].

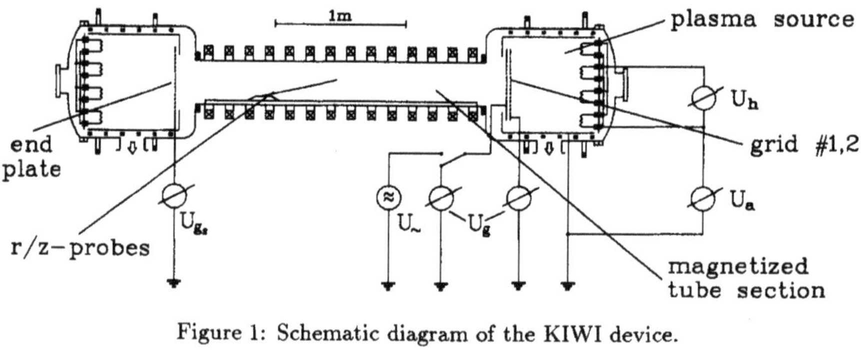

Figure 1: Schematic diagram of the KIWI device.

2. EXPERIMENTAL ARRANGEMENT: The KIWI-experiment is a magnetized triple plasma device (see Fig. 1) which consists of two independent plasma chambers (92 cm dia. ×

90 cm,'magnetic box') and a central tube section (30 cm dia. × 200 cm) immersed in a set of 14 water-cooled magnetic field coils ($B = 0 \ldots 0.15$T). Two compensation coils are installed to adjust the magnetic mapping from the plasma chambers into the tube section. A double plasma mode is achieved by activating only one source chamber (magnetic box) to produce a dense, quiet, and homogenous plasma ($N_e \leq 10^{11}\text{cm}^{-3}, T_e = 2\text{eV}$). The tube section acts as target chamber. The plasma chamber and the tube section are separated by one or two extraction grids with a transparency of about 50%, which can be biased independently for controlled injection of electrons or ions. At the far end of the tube section an additional grid (end grid) is installed as a plasma loss surface. Langmuir-probes are used for density, temperature and floating potential diagnostics. In addition, shielded Langmuir probes were used as hf antennas. Plasma potential profiles were measured using emissive probes in the usual manner [9, 10]. Dynamical measurements are performed by recording time series of probe signal fluctuations, *i.e.* temporal changes of electron and ion saturation current and floating potential.

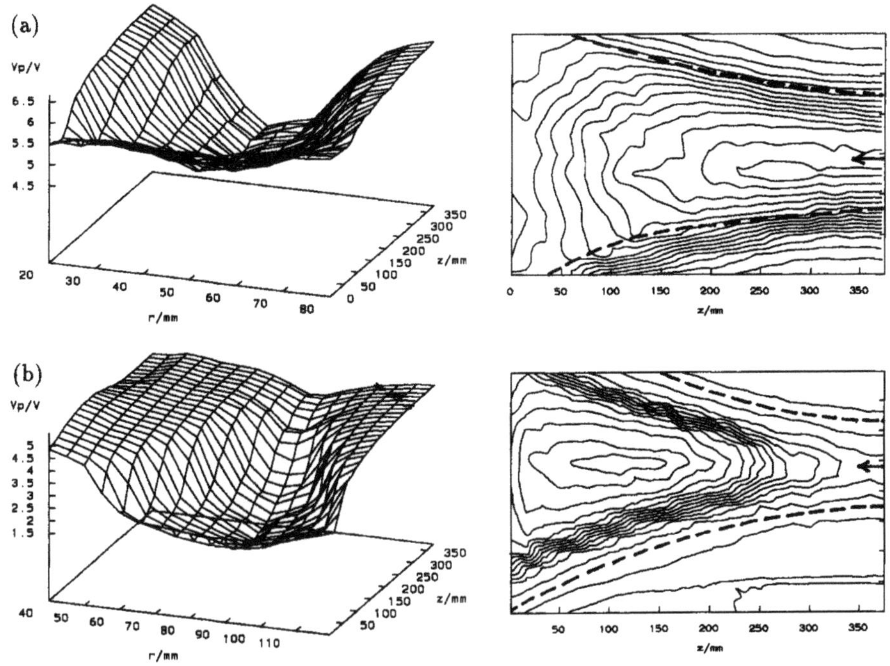

Figure 2: Two different 2d potential structures. The grid is located at $z = 0$mm, the center of the tube section is indicated by an arrow.
(a) $P = 5.0 \cdot 10^{-4}$mBar (b) $P = 6.0 \cdot 10^{-5}$mBar.

<u>3.A Steady State Properties:</u>For positively biased extraction grids $U_g \approx 5$V) thermal electrons as well as a suprathermal electron beam and nonthermalized primary electrons are injected into the tube section. Strongly dependent from the neutral gas pressure and the magnetic field configuration, two basically different potential structures are

observed (Fig. 2). For the subsequent discussion, the fringing magnetic field lines are included in the contour plots of Fig. 2 (dashed lines). There are two basic types of double layers: (A) high potential (B) low potential region on the grid side. In Fig. 3, typical axial potential shapes for different discharge conditions are selected. Even triple layer structures with large potential dips are found. For low neutral gas pressure

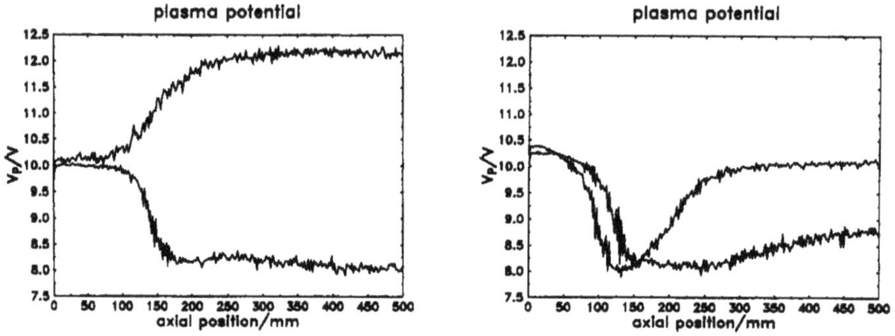

Figure 3: Typical axial potential shapes. The positively biased grid is located at $z = 0$.

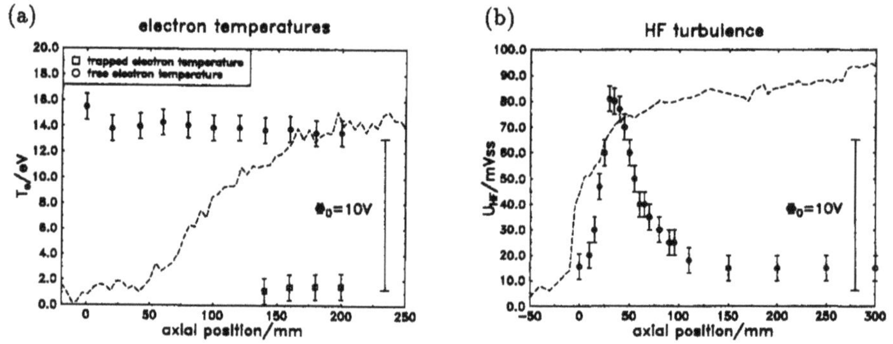

Figure 4: Probe measurement results. (a) Temperatures of free and trapped electrons. (b) HF turbulence amplitude. The corresponding plasma potential step ($e\Phi_0/k_BT_e \approx 5$ is included (dashed line).

$P = 1 \cdot 10^{-5}$mBar...$1 \cdot 10^{-4}$mBar (Fig. 2(a)), both double layer phenomena are coexistent, type (A) in the central region and type (B) near to the plasma edges. In contrast, for gas pressure above $P = 1 \cdot 10^{-4}$mBar only type (B) with V shaped contour occurs apart from the grid sheath of a residual potential drop $\Delta\Phi \approx 0.5$V (Fig. 2(b)). The double layer with low potential region on the grid side is expected in case of suprathermal electron injection into a background plasma and is well known from double plasma experiments [11, 12]. Potential steps up to $e\Phi_0/k_BT_e \approx 10$ were achieved for these double layers. For this situation, reliable Langmuir probe measurements have been possible. Fig. 4(a,b) shows results from diagnostic measurements in the double layer region. Free and trapped electron species are clearly distinguishable (Fig. 4(a)). The free electron temperature approaches values about $T_{e,f} \approx 14$eV, whereas the trapped electrons have low temperature

$T_{e,t} \approx 2$eV and are reflected by the double layer potential step. The ion density is determined to $n_e \approx 5 \cdot 10^9\text{cm}^{-3}$. The high frequency turbulence measurements are performed with hf probes and a broad band hf amplifier. The hf power spectrum is extended over a broad frequency range $f = 100 \ldots 600\text{Mhz} = \omega_{pe}/2\pi$. A strong maximum in the HF amplitude due to two-stream instabilities is located in the high potential region of the double layer near to the transition region. These measurements confirm earlier findings of wave phenomena associated with strong double layers [13]. To conclude, double layer type (B) is identified as strong, beam-type double layer. It is more difficult to characterize double layer type (A). The potential step is always weak $e\Phi_0/k_B T_e \lesssim 1$ and the plasma density is much lower ($n_e < 1 \cdot 10^9\text{cm}^{-3}$) than above. An important point is that no high frequency instabilities are found. However, for certain discharge parameters, there are very large low frequency fluctuations which will be discussed in the following.

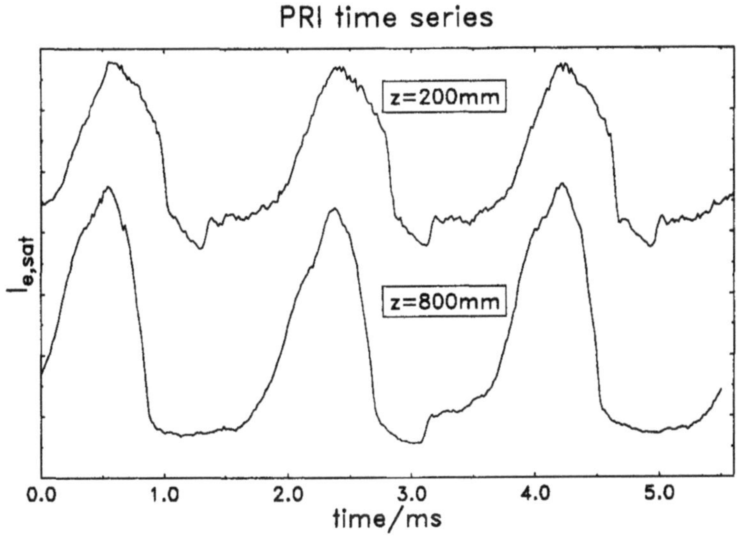

Figure 5: Low frequency oscillations of the plasma density ($\propto I_{e,sat}$). Time series for two different axial positions are displayed. A steepening effect is clearly observable.

3.B Dynamical Aspects: In the context of dynamical aspects, the attention is drawn to the double layers stability properties. Although the general stability of double layers is an unsolved problem, many observations of longitudinal low frequency instabilities have been described [14]. Such instabilities were formerly believed to be of the 'standing ion-acoustic wave' type. Improved time resolved measurements [8], as well as particle simulations [15] revealed the important role of double layers in context of these low frequency waves. They occur as 'moving double layers' periodically travelling along the plasma column with a phase velocity $(2-3)C_s$ and can be characterized as 'potential relaxation instability' (PRI). The low frequency longitudinal waves in the KIWI-device show a lot of similarities with this type of instability. The time evolution of the plasma density is displayed in Fig. 5 for two different axial positions ($z = 200, 800$mm). The time evolution for the potential is very similar. Phase sensitive measurements of the potential evolution allow to estimate wavelength and phase velocities by using two-point correlation methods [16]. We obtain a phase velocity of approximately $(2-3)C_s$ and wavenumbers $k_{||} = O(0.001\text{cm}^{-1})$. These values also agree with previous PRI-measurements [8]. In addition, the slight steepening effect caused by propagation (Fig. 5) is a hint for the solitary character of the PRI.

4. DISCUSSION AND CONCLUSIONS: In the KIWI-experiment three different discharge parameter determine the formation, potential step and polarity of double layers: The grid bias, the neutral gas pressure, and the magnetic field configuration. Complicated two dimensional potential structures were found in a region of converging magnetic fields with electron beam injection (Fig. 2(a,b)). The details of some observations are far from being understood but the basic mechanism for double layer formation may be the following: For low neutral gas pressure, positive grid bias, grounded end grid and appropriate magnetic configuration, the converging magnetic fields reflect injected thermal electrons (outside the loss cone, mirror ratio $(B_2 - B_1)/B_1 \approx 0.5$). The suprathermal beam is injected into the long plasma column and is responsible for weak ionization (mean free path $\lambda_m \approx 2\text{m}$). Ions are accelerated towards the end grid or reflected by the potential wall. This phase space condition is sufficient for the formation of a double layer with the polarity (A) (Fig. 2(a)). The described situation is closely related to known experiments on ion beam reflection in magnetized plasmas [4]. For much higher neutral gas pressure ($> \times 10$) and positively biased grid and end grid, the injected electron beam partially thermalizes and produces trapped thermal electrons. Again, ions are accelerated until they are trapped between the grid and the positive potential step. The conditions for the double layer type (B) are fulfilled.

Unstable weak double layers are observed for very low grid bias $U_g \approx 0\text{V}$ and a magnetic configuration with very weak mapping of nonthermal electrons into the plasma column. Therefore the formation of stable double layers with polarity (A) is suppressed and only nonstationary shock waves occur. They propagate with plasma expansion velocity into the tube section.

In conclusion, the existence of complicated spatial potential structures in inhomogenous magnetic fields have been demonstrated experimentally. The phenomena may be explained in terms of reflection and ionization effects. In our case strongly turbulent plasma states may be excluded as mechanism for double layer formation. However, the explanation of several interesting details (for example the negative potential dip) is still leaking. Appropriate particle-in-cell simulations are in progress.

5. ACKNOWLEDGEMENTS: We thank Dr. T. Pierre for valueable discussions and V. Rohde for his contributions to design and optimisation of the KIWI device. This work has been supported by the Deutsche Forschungsgemeinschaft under contract Pi185/2-1.

References

[1] N. Hershkowitz, *Space Sci. Rev.*, **41**, 351–391, (1985).

[2] M. A. Raadu, *Phys. Rep.*, **178**, 25–97, (1989).

[3] K. D. Baker, N. Singh, L. P. Block, R. Kist, W. Kampa, and H. Thiemann, *J. Plasma Phys.*, **26**, 1–27, (1981).

[4] R. L. Stenzel, M. Ooyama, and Y. Nakamura, *Phys. Fluids*, **24**, 708–718, (1981).

[5] P. Coakley and N. Hershkowitz, *Phys. Fluids*, **22**, 1171–1181, (1979).

[6] M. Guyot and Ch. Hollenstein, *Phys. Fluids*, **26**, 1596–1605, (1983).

[7] T. Klinger and A. Piel, Low-dimensional attractors arising from the interaction of drift waves with low frequency instabilities, In *Proc. of the 1992 ICPP*, Innsbruck, Austria, (1992).

[8] S. Iizuka, P. Michelsen, J. Juul Rasmussen, R. Schrittwieser, R. Hatakeyama, K. Saeki, and N. Sato, *Phys. Rev. Lett.*, **48**, 145, (1982).

[9] J. S. Levine and F. W. Crawford, *J. Plasma Phys.*, **24**, 359–383, (1980).

[10] J. R. Smith, N. Hershkowitz, and P. Coakley, *Rev. Sci. Instrum.*, **50**, 210–218, (1979).

[11] A. N. Sekar and Y. C. Saxena, *Plasma Phys. and Contr. Fusion*, **27**, 673–690, (1985).

[12] B. H. Quon and A. Y. Wong, *Phys. Rev. Lett.*, **37**, 1393–1396, (1976).

[13] S. Torvén and L. Lindberg, *J. Phys. D: Appl. Phys.*, **13**, 2285, (1980).

[14] H. Schamel, *Physics Reports*, **140**, 161–191, (1986).

[15] P. L. Gray, S. Kuhn, T. L. Crystal, and C. K. Birdsall, Particle simulations of the unstable states of a collisionless single-ended plasma device, In R. Schrittwieser and G. Eder, editors, *Second Symposium on Plasma Double Layers and Related Topics*, pages 266–271, (1984).

[16] J. M. Beall, Y. C. Kim, and E. J. Powers, *J. Appl. Phys*, **53**, 3933–3940, (1982).

ELECTRON CYCLOTRON EMISSION FROM DOUBLE LAYERS IN MAGNETIZED PLASMAS

T. Mieno,[a] M. Oertl,[b] R. Hatakeyama and N. Sato

Department of Electronic Engineering, Tohoku University, Sendai 980, Japan

ABSTRCT: Electromagnetic emissions around an electron cyclotron frequency are enhanced when electrostatic one-dimensional double layers are formed in collisionless magnetized plasmas. For the case of a moving double layer, the emission intensity well correlates with the current flowing between a plasma and a positively biased target, which is modulated by a relaxation oscillation. When a stationary double layer is formed by two plasma sources with different potentials, the emission intensity increases with an increase in the potential difference ϕ_D of the double layer until $e\phi_D/T_e \approx 50$, where electron beam component is clearly observed, saturating thereafter.

1. INTRODUCTION: An electrostatic double layer in plasmas[1] induces various kinds of instabilities. Electrostatic instabilities around the ion and electron plasma frequencies,[2] and low frequency instabilities[3] in the vicinity of the double layer have been observed in past experiments. The auroral kilometric radiation (AKR), which is measured by artificial satellites and the ground facilities, is emitted from V-shaped double layers along the field line of the earth.[4] Concerning the radiation mechanism, direct excitation by a non-maxwellian electron distribution function,[5] mode conversion from an electrostatic wave[6] and electron cyclotron maser effect[7] have been discussed. These investigations indicate that an electron beam from negative potential side to positive potential side of the double layer is expected to drive instabilities in the electron cyclotron range of frequencies. Few laboratory experiments to simulate AKR or electromagnetic emission from a double layer, on the other hand, have been carried out.

Here, the electromagnetic emission around the electron cyclotron frequency f_{ce} is investigated when a one-dimensional double layer is formed in collisionless magnetized plasmas. The electron cyclotron emission is observed in two types of double layers. The first case is related to the moving double layer,[8] which is formed between a hot plate and a positively biased end plate in a single-ended Q-machine. The second case is related to the stationary double layer[1,3] which is formed by the double-plasma operation of the Q-machine, where both hot plates produce plasmas and the potential difference is applied between them.

---

[a] Present address: Department of Physics, Shizuoka University, Shizuoka 422, Japan.

[b] Present address: Institute for Zoology, University of Innsbruck, A6020 Innsbruck, Austria.

2. EXPERIMENTAL APPARATUS AND METHODS: The experiment is performed in a Q-machine at Tohoku University,[1,3] as shown in Fig.1. The plasma is produced by contact ionization of potassium atoms at surface of a 3.5cm-diam tantalum hot plate with temperature of about 2300K and is confined by a uniform magnetic field (B~0.2T). Plasma density n is in the range of 10^8-$10^9 cm^{-3}$ and the electron and the ion temperature T_e, T_i are about 0.2eV. The background gas pressure is about $1.3x10^{-4}$ Pa, where collisions between the plasma and neutral particles can be neglected. Plasma parameters including a space potential are measured by movable wire-type probes. Electron cyclotron emissions are received by a horn antenna (the size of the rectangular horn is 60mm x 75mm, theoretical gain is 10.7dB for 6 GHz) and a dipole antenna, which are located near vacuum surface of the stainless-steel vessel (about 8cm from the axis). The electrostatic stationary double layer is produced by the double plasma operation as shown in the figure. Two plasmas are produced by two hot plates located at both ends of the vessel (3m apart). By applying potential difference ϕ_0 between the two hot plates, one of which is electrically grounded, together with the vacuum vessel a double layer is produced in the plasma column. The moving double layer is produced by removing the left-hand hot plate and inserting an axially-movable cold target with 10cm in diameter, where an external bias voltage V_T is applied with respect to the grounded hot plate. Usually, the distance between the hot plate and the target is 1.6m.

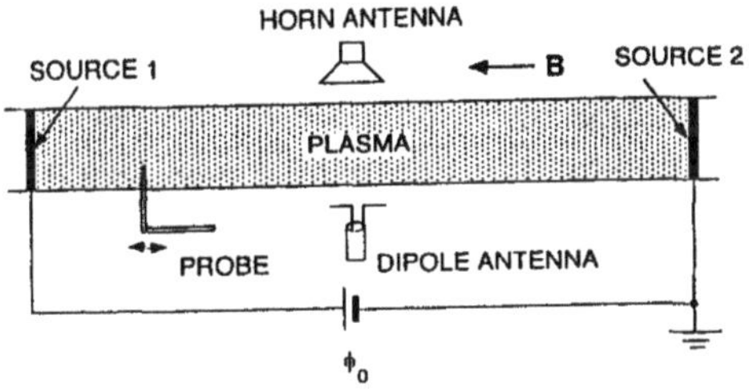

Fig. 1 Schematic of experimental set up for the experiment of stationary double layer. For the experiment of moving double layer, the source 1 is grounded and a movable metal target with 10cm in diameter is inserted instead of the source 2.

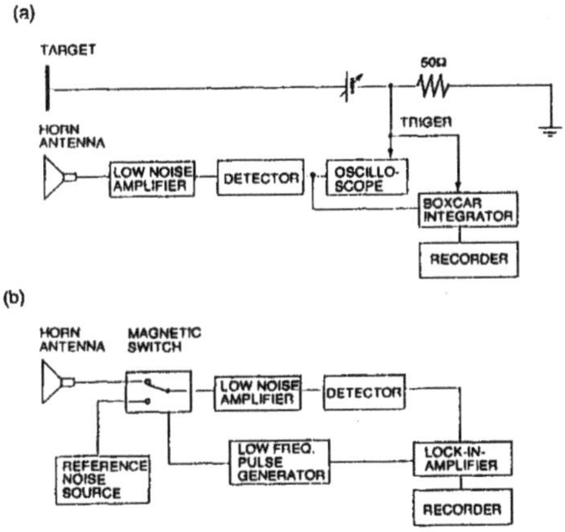

Fig. 2 Schematic diagrams of the detection circuit of the microwave signal (~6GHz) for the moving double layer experiment (a) and for the stationery double layer experiment (b).

Since the detected microwave signal is very weak, it is difficult to measure it by a usual receiver. Sufficient sensitivity can be obtained by means of the radiometer technique first described by Dicke.[9] Schematic detection diagrams for the micro-

wave signal are shown in Fig.2. High S/N ratio and time variation of the microwave power are obtained by the sampling method of a boxcar integrator in case of the moving double layer as shown in Fig.2(a), because the detected signal is synchronously modulated with the relaxation oscillation (~1kHz) of the target current I_T. In case of the stationary double layer, the lock-in method is used with a mechanical switch, which connects the circuits to the horn antenna and a reference noise source alternatively [see Fig.2(b), switch frequency: ~7Hz]. The reference noise source is a terminator for microwave circuit with 50Ω. To amplify the signal and to reduce the noise figure, a low noise preamplifier (gain 20dB, and noise figure 3.6 at 6GHz band) is used.
As a tuner and a detector of the microwave signal, a spectrum analyzer (HP-8555B) is used.

3. EXPERIMENTAL RESULTS:

The moving double layer is formed between a hot plate and a positively biased end plate.[8] The potential jump is created at the hot plate and it moves toward the target around an ion flow speed. When the double layer reaches the target, a strong target current is induced, and again a double layer is formed at the hot plate, which restricts the target current.
Figure 3 shows the time-averaged frequency spectrum of the detected power P_μ for V_T=

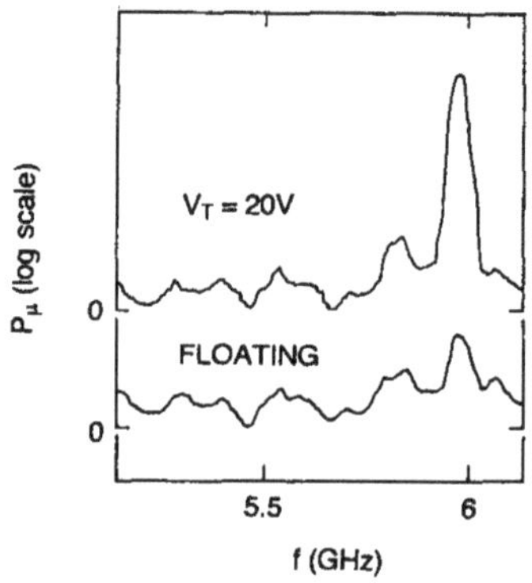

Fig. 3 Typical frequency spectra of the detected microwave signal when the target bias is positive or floating. Applied magnetic field is 0.2T.

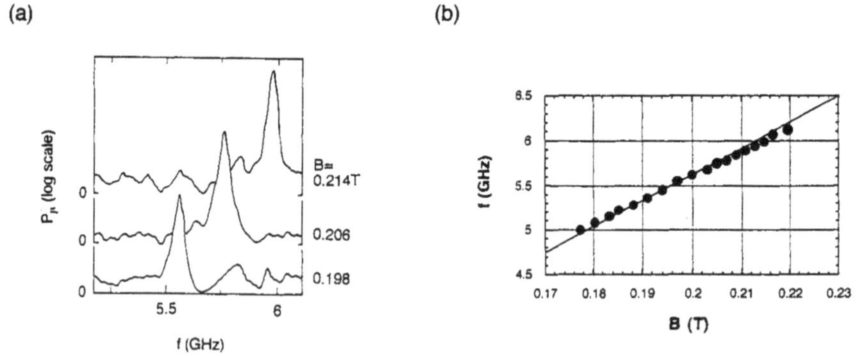

Fig. 4 Frequency spectra for several magnetic fields (a) and frequencies of the detected peak signal versus the applied magnetic field (b). A solid line corresponds to the electron cyclotron frequency.

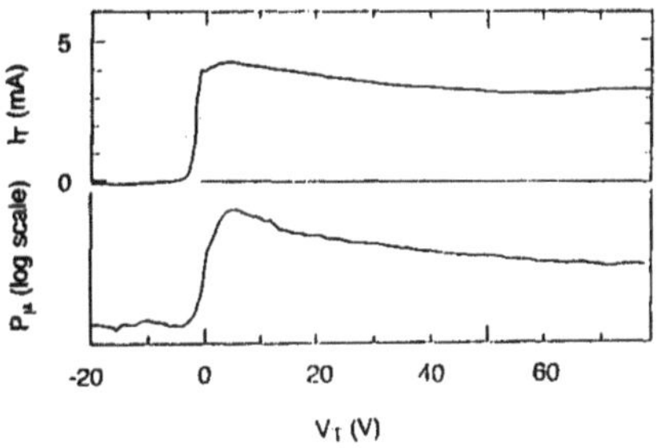

Fig. 5 Time-averaged target current and time-averaged microwave power as a function of the target voltage.

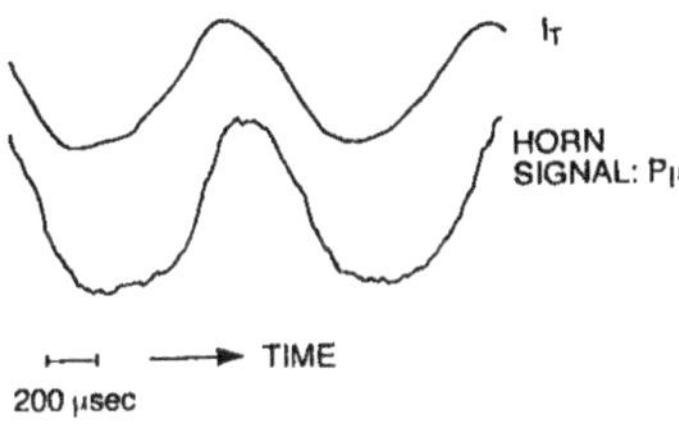

Fig. 6 Phase difference between the target current and the microwave power of the horn signal during the relaxation oscillation.

20V and floating potential. A finite P_μ due to black body radiation around the electron cyclotron frequency f_{ce} is detected in the absence of the moving double layer.[10] When the positive bias is applied to the target, the signal is extremely increased. Frequency spectra and peak frequencies are measured as a function of the applied magnetic field, as given in Fig.4. The peak frequency increases linearly with an increase in the magnetic field and almost equals to the expected electron cyclotron frequency which is indicated by a solid line. The plasma frequency is about 300MHz ($\ll f_{ce}$). Figure 5 presents dependences of the target current I_T and time averaged microwave power P_μ on V_T. P_μ correspondently increases as I_T starts to be driven, and the strong correlation between P_μ on I_T are observed. Figure 6 shows time variations of the emission amplitude and the target current during the relaxation oscillation (f~1kHz) associated with the moving double layer. The horn signal is clearly modulated by the relaxation oscillation and there is a little phase difference between the two modulation signals. Spatial resolution measurements of the microwave emission are tried in order to find the movement of the emission source. However, the reflection of the microwave at the metal surface of the chamber obscures the position of the source. Although a dipole antenna is used to decide the direction of the emission source, the result is not clear.

Let us mention the second case related to the stationary double layer. Figure 7 shows a typical distribution of the space potential along the magnetic field for the case where the potential difference, ϕ_0=10V, is applied between the two hot plates. When such a one-dimentional double layer is formed, an electron-beam component around $e\phi_0$ in the energy distribution parallel to the magnetic field $f_{//}(E)$ is observed at the high-potential side of the double layer. Figure 8 shows frequency spectra of the microwave for various ϕ_0. The emission is greatly enhanced until ϕ_0=10V when the applied potential is gradually increased. The emission intensity saturates for ϕ_0>10V, where f is observed to be broadened in the

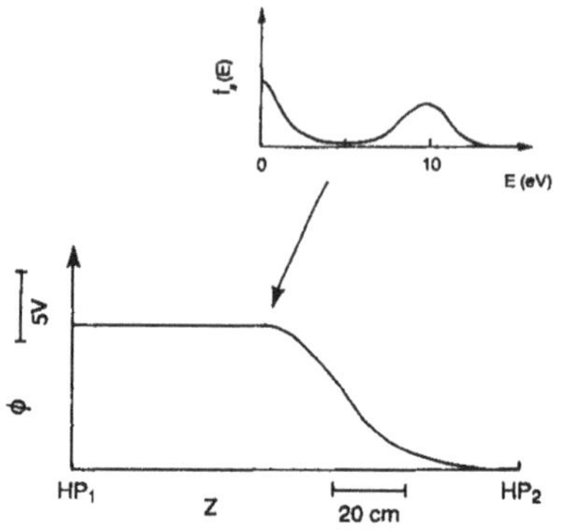

Fig. 7 A typical distribution of the plasma potential along the magnetic field when the stationary double layer is formed (ϕ_0=10V). An electron energy distribution function parallel to the magnetic field on the high-potential side is also shown in the upper figure.

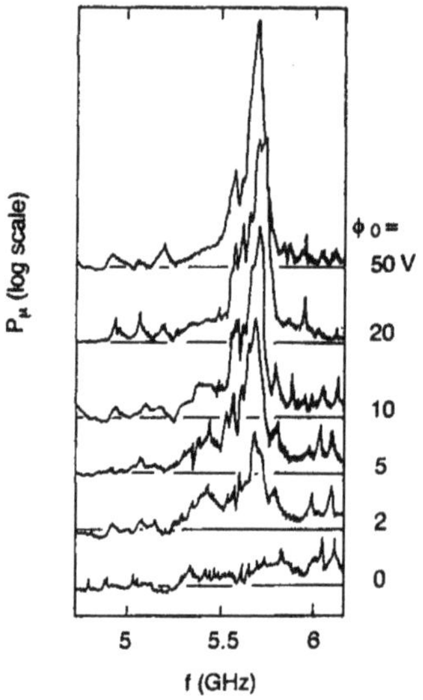

Fig. 8 Microwave frequency spectra detected by the horn antenna for various applied potentials between the hot plates in the stationary double layer experiment.

energy range of eϕ_0. The reason of the P_μ saturation is considered due to a saturation of the beam flux by this relaxation of $f_{//}$.

4. DISCUSSIONS AND CONCLUSION:

The experiment yields clear evidence of the enhanced microwave emission around the electron cyclotron frequency when the moving and stationary double layers are generated in the plasma. The close temporal correlation between the electron current in the plasma and the intensity of the radiation, and the phase difference between them in case of the moving double layer suggest that the microwave is predominantly excited at a local region of the double layer. It is not clarified experimentally, however, where the source of this emission is around the double layer in spite of the effort such as perturbing the magnetic field to change the local cyclotron frequency or shifting the double layer position. A possible way to achieve the spatial resolution may be a construction of absorbent on the metal wall. Anyway, it is considered that the electron currents in the plasma cause the strong microwave radiation. Since the double layer is one dimensional, the bunching effect of the electron beam is neglected. The direct excitation of electromagnetic wave by the non-maxwellian electron distribution[5] or the mode conversion from the electron-beam-excited electrostatic wave to the electromagnetic wave[6] are candidates for the excitation mechanism of this microwave emission.

In conclusion, the electromagnetic wave emission is extremely enhanced when the double layers are formed. The electron current leading to the electron beam generation on the high-potential side certainly gives rise to the emission.

REFERENCES:

1. N. Sato, *Proc. Sympo. Plasma Double Layers, Risø, Denmark, 1982* (published by Risø National Lab., 1982) p.116. N. Sato, R. Hatakeyama, S. Iizuka, T. Mieno, K. Saeki, J. J. Rasmussen and P. Michelsen, Phys. Rev. Lett., **46** (1981) 1330.
2. S. Torven, J. Phys. D, **13** (1980) 2285.
3. N. Sato, R. Hatakeyama, S. Iizuka, T. Mieno, K. Saeki, J. J. Rasmussen, P. Michelsen and R. Schrittwieser, J. Phys. Soc. Jpn., **52** (1983) 875.
4. J. E. Borovsky, J. Geophys. Res., **93** (1988) 5727.
5. F. W. Crawford, Nuclear Fusion, **5** (1965) 73. K. Mitani, H. Kubo and H. Tanaka, J. Phys. Soc. Jpn., **19** (1964) 211. J. L. Green, D. A. Gurnett and R. A. Hoffman, J. Geophys. Res., **84** (1979) 5216.
6. H. Oya and A. Morioka, J. Geophys. Res., **88** (1983) 6189.
7. J. L. Hirshfield and J. M. Wachtel, Phys. Rev. Lett., **12** (1964) 533.
8. S. Iizuka, P. Michelsen, J. J. Rasmussen, R. Schrittwieser, R. Hatakeyama, K. Saeki and N. Sato, Phys. Rev. Lett., **48** (1982) 145. S. Iizuka, P. Michelsen, J. J. Rasmussen, R. Schrittwieser, R. Hatakeyama, K. Saeki and N. Sato, J. Phys. Soc. Jpn., **54** (1985) 2516.
9. R. H. Dicke, Rev. Sci. Instr., **17** (1946) 268.
10. A. F. Kuckes and A. Y. Wong, Phys. Fluids, **8** (1965) 1161.

SOME EQUILIBRIUM AND STABILITY CHARACTERISTICS OF STRONG ION SHEATHS

S. Mukherjee, P.I. John & P.K. Kaw
Institute for Plasma Research,
Bhat, Gandhinagar - 382 424: INDIA.

Abstract

Experiments are carried out in the strong ion sheath ($e\Phi/T_e \gg 1$) in argon plasma. The equilibrium sheath does not conserve ion flux and is electron free. A new way of potential measurement shows deviations from Child's law. Potential oscillations are observed with high pressure. A model based on electron impact ionization and charge exchange is presented for the equilibrium sheath. A local theory for the oscillation is also presented.

Introduction

A sheath is a natural consequence of interaction of a surface with plasma. When a metallic surface in contact with the plasma is biased highly negative ($e\Phi/T_e \gg 1$), ions are attracted and electrons driven away thus forming a strong ion sheath. Thus the ion sheath is non-neutral and strong electric fields are present there. Though a lot of work has been done and considerable literature exists on sheath properties [1-6], however there are certain aspects like validity of Child's law, conservation of ion flux, the role played by ionization and charge exchange, sheath stability subjected to variation in plasma parameters, etc., which require further investigation. This paper attempts to answer some of the above questions. It describes a new method for potential measurement in the sheath. A model based on electron impact ionization of neutrals and charge exchange of ions with neutrals is prescribed which explains the behaviour of electron and ion fluxes in the sheath to the plasma. A local theory of ionization instability explains some features of the potential oscillations.

Experimental setup and Measurement techniques

The experimental device [fig 1] consists of a cylindrical chamber made of stainless steel pumped down to 3. 10^{-5} torr and operated at 1. – 2. 10^{-4} torr by filling with argon.

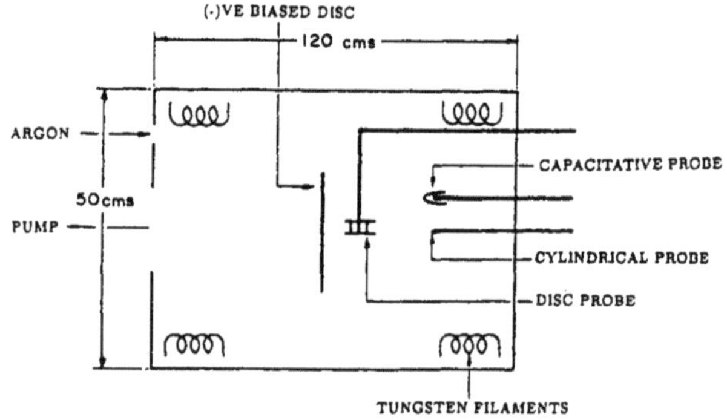

Fig. 1. EXPERIMENTAL SETUP

Plasma is produced by accelerating thermoionic electrons, which undergo collisions with argon atoms. Probe measurements show $T_e \sim 2-4$ eV, and plasma density $7.\ 10^7 - 6.\ 10^8$ per cc. The ions are cold and hence the plasma is not in equilibrium. Plasma potential is 2 volts and floating potential is -20 volts. The sheath is created by biasing a disc to large negative potentials ~ -1 kV. Plasma parameters are varied by changing the pressure and the discharge voltage applied between the filament and the chamber.

Ion flux is measured using a disc probe biased to -70 V, put inside a teflon cylinder so that its collection side faces the plasma and is parallel to the plate. An identical disc probe biased to +40 V and collection side facing the plate is used to measure electron flux. The teflon cylinder is kept long enough to assure the collection of only directed components of flux. The probes are biased such as to ensure that in the plasma they are in the saturation current regions.

We describe a new method of potential measurement in the sheath. We define the floating potential inside the sheath as that potential at which a probe collects zero ion current. As the sheath is electron free, no electron current contribution is there. We used a small cylindrical probe, with the probe tip not seeing the directed ion flux. The probe is biased negative to create an ion sheath around it. Then the plate is biased negative till the probe current becomes zero. It becomes zero because the ions in the probe sheath leave the probe and get collected by the plate. Thus the probe bias must be equal to the local potential created by the biased plate. The probe bias is then initially adjusted for various probe positions so that a plate bias of -1 kV gives zero probe current. Thus the potential profile corresponding to -1 kV is obtained.

To observe potential fluctuations a capacitative probe[7] is used, which is sensitive only to potential fluctuations. The pressure is increased and the discharge voltage changed to create the potential oscilltions on top of the equilibrium profile.

Results & Discussion

Ion flux measurements show that it decreases as the collector is moved from the plasma to the biased plate. This observation [fig 2] is in contradiction to the existing theoretical

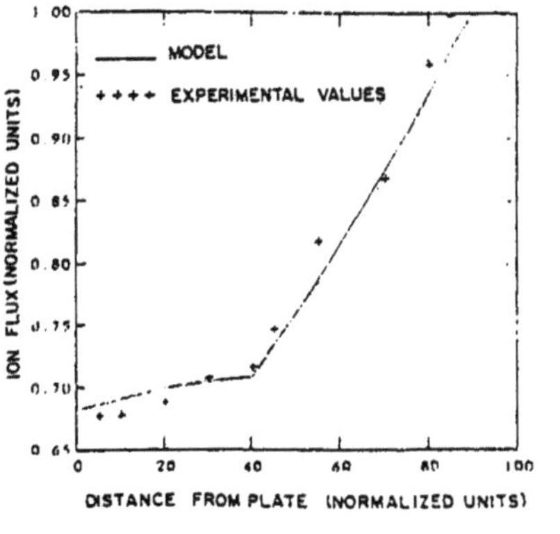

Fig. 2. PROFILE OF ION FLUX. $\Phi_{PLATE} = -1$ kV

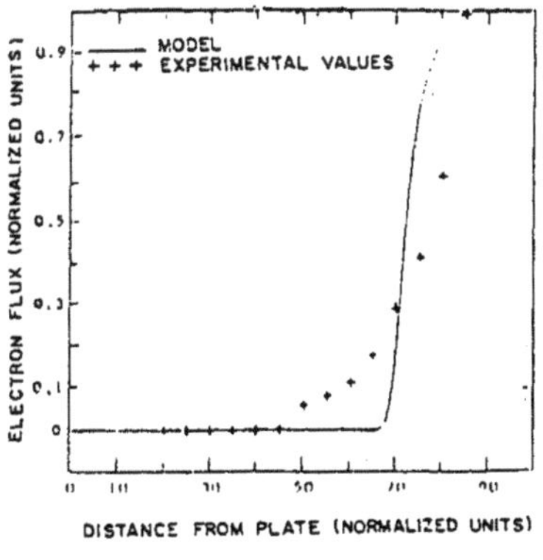

Fig. 3. PROFILE OF ELECTRON FLUX. $\Phi_{PLATE} = -1$ kV

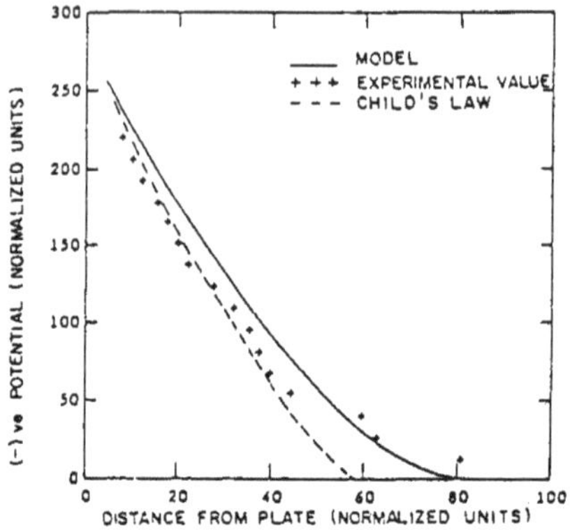

Fig. 4. POTENTIAL PROFILE. $\phi_{PLATE} = -1$ kV

assumption, and our model takes account of it. The sheath is observed to be electron free and electron flux starts growing as one moves from the plasma to the plate [fig 3]. The potential profile, deep inside the sheath matches with Child's law but deviates from it as one moves to the plasma, showing the effect of non-conservation of ion flux on it [fig 4].

The sudden transition from a stable to an oscillatory potential behaviour is observed with increase in pressure and corresponding adjustment of discharge voltage. The features of the oscillations are
(a) for $p > 3.\ 10^{-4}$ torr oscillations observed; (b) oscillations are observed on the plate bias and the discharge voltage but with a low amplitude compared to their dc values; (c)when plate bias becomes equal to the plasma potential, no oscillations are observed; hence sheath is necessary for initiation of this oscillations; (d) with more negative plate potentials lower discharge voltage initiates the oscillations; (e) the oscillations penetrate to greater distances than equilibrium potential profile in the plasma; (f) ion flux also shows the oscillations; (g) the oscilllations are not pure sinusoids; (h) frequency drops from 3 kHz to 1 kHz as pressure increases from $3.\ 10^{-4}$ to 10^{-3} torr; (i) with increased pressure peak of potential fluctuation shifts to the plate and penetration of potential fluctuation is also deeper [fig 5]; (j)ion flux fluctuations and potential fluctuations does not peak at the same spatial location [fig 6].
The above features showing a strong dependance on pressure indicates that a significant role is played by ionization. The local theory predicts the possibility of an ionization instability to be the cause for the oscillations.

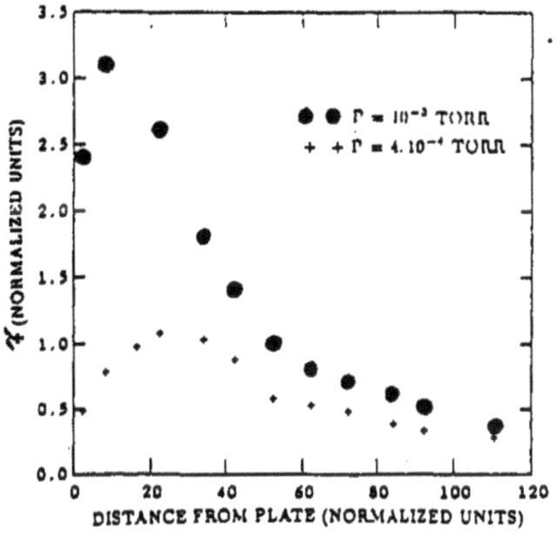

Fig. 5. PROFILE OF POTENTIAL OSCILLATION, Φ_{PLATE} = −500 V

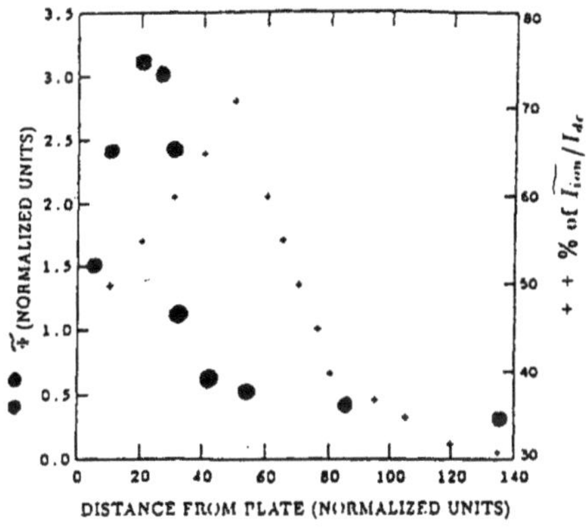

Fig. 6. COMPARISON OF POTENTIAL & ION FLUX FLUCTUATIONS P = 6.10⁻⁴ TORR, Φ_{PLATE} = −600 V

Theoretical Modeling

We use fluid equations for the ions and electrons. The electrons are assumed to satisfy the Boltzman's distribution. Ions are taken to be cold. The ions which start from the sheath-plasma interface (denoted by suffix 1) undergo a charge exchange reaction producing high speed neutrals and ions at rest. These newly produced ions (denoted by suffix 2) further get accelerated by the field in the sheath. Some assumptions are taken here. The newly produced ions undergo a further charge exchange reaction but the ions which are now produced does not contribute to the flux or potential. In both the ion continuity equations a term is kept which is proportional to the electron density and produces ions by electron impact ionization. The cross sections of the various processes are assumed to be constant during the entire range of operation. In the plasma there is a balance between electron impact ionization and charge exchange. At the sheath edge the potential is zero but there is a finite field to accelerate the ions so as to satisfy the Bohm's criterion. The electron density

$$N_e = exp(-\Phi) \tag{1}$$

where $\Phi = -e\phi/kT_1$, $N_e = n_e/n_o$, n_o being the ion density in the undisturbed plasma. For simplicity it is assumed that all the new ions are produced at a location $Y = Y1$ and the entire region of operation is divided in two sections (a)$Y > Y1$ and (b)$Y < Y1$, where $Y = 0$ at the sheath-plasma interface.

For $Y > Y1$, which is the region between the sheath plasma interface and the birth of new ions, the ion continuity equation

$$\frac{d}{dY}(N_1V_1) = \alpha\, exp(-\phi) - \beta\,(N_1V_1) \tag{2}$$

and momentum equation

$$V_1\frac{dV_1}{dY} = \frac{d\Phi}{dY} - \beta\, {V_1}^2 \tag{3}$$

where $Y = x/(kT_e/4\pi n_o e^2)^{\frac{1}{2}}$,
$V = u/(kT_e/m_i)^{\frac{1}{2}}$, $N_1 = n_{i1}/n_o$,

$\alpha = n_n \sigma_{electronimpact} v_{the}/(4\pi n_o e^2/m_i)^{\frac{1}{2}}$,
and $\beta = n_n \sigma_{chargeexchange}(kT_e/4\pi n_0 e^2)^{\frac{1}{2}}$.
The system is finally closed by Poisson's equation,

$$\frac{d^2\Phi}{dY^2} = N_1 - exp(-\Phi) \tag{4}$$

Though there is a net reduction in ion density by charge exchange process.for those ions which originate from the plasma, the new ions which are produced start contributing only from $Y = Y1$ and onwards. A much better way will be to do a kinetic theory or by a much detailed fluid model. In the detailed fluid model the entire sheath region is divided into finite but large number of sections and the complete set of fluid equations are solved in each section. This work is curently under progress.

For $Y < Y1$, the region between the creation of new ions and the plate, one has to take the contribution from both the ion species. The ion continuity and momentum equations of species 1 are

$$\frac{d}{dY}(N_1V_1) = \alpha\, exp(-\phi) - \beta\,(N_1V_1) \tag{5}$$

$$V_1\frac{dV_1}{dY} = \frac{d\Phi}{dY} - \beta\, V_1{}^2 \tag{6}$$

For species 2 the ion continuity and momentum equations are

$$\frac{d}{dY}(N_2V_2) = \alpha\, exp(-\phi) + \beta\,(N_1V_1) - \beta\,(N_2V_2) \tag{7}$$

$$V_2\frac{dV_2}{dY} = \frac{d\Phi}{dY} - \beta\, V_2{}^2 \tag{8}$$

The system is finally closed by Poisson's equation,

$$\frac{d^2\Phi}{dY^2} = N_1 + N_2 - exp(-\Phi) \tag{9}$$

These set of coupled equations are solved numerically using the IMSL routine DGEAR. Profile of Φ vs. Y are compared with the experimental results. From the profile of Φ other quantities like ion & electron flux are obtained by substituting the values α and β which correspond to the experimental parameters. The evaluated ion flux matches to a great extent with our experimental results.

In the limit of $\Phi \gg 1$ one approaches the region near the plate and hence one can neglect the contribution of electrons in electron impact ionization from both the ion continuity equations. Also in the regions near the plate the field is much stronger hence in the loss terms proportional to beta can be neglected in the momentum equations. Thus on adding the ion continuity equations

$$\frac{d}{dY}(NV) = 0 \tag{10}$$

and the momentum equations

$$\frac{d}{dY}(V^2) = 2\frac{d\Phi}{dY} \tag{11}$$

where $NV = N_1V_1 + N_2V_2$ and $V^2 = V_1{}^2 + V_2{}^2$ The system is finally closed by Poisson's equation,

$$\frac{d^2\Phi}{dY^2} = N \tag{12}$$

These set of equations simply tells that both the species of ions that are present behave as a single species and hence are acted on by the same field simultaneousely. From the above equations NV = constant, which is taken to be unity because they start from the plasma with N=1 & V=1. Also $V = (2\Phi)^{\frac{1}{2}}$ because the constant of integration is unity, from Bohm's criterion, which is much less than Φ. Hence
$(\frac{d^2\Phi}{dY^2}) = (2\Phi)^{-\frac{1}{2}}$ whose solution gives the dependance of Child's law.

$$Y = [\Phi_{plate}^{3/4} - \Phi(Y)^{3/4}]2^{5/4}/3 \tag{13}$$

Thus our set of equations reduces to Child's law deep inside the sheath.

From the ion continuity equation we know that the effect of ionization is dominant in the regions of weak potential, because deep inside the sheath electrons are mostly absent. So while doing a linear perturbation analysis we assume that gradients in density and velocity of ions are small and hence can be neglected. We also restrict the entire analysis in the transition region only. Allowing the perturbating variables to have the following dependance $exp[-i(\omega t - kY)]$ we get the following dispersion relation.

$$\Omega^2 P + i\Omega Q + R = 0 \tag{14}$$

where $\Omega = \omega - kV_0$, $P = k^2 + exp(-\Phi_0)$, $Q = 3\beta PV_0 - \alpha exp(-\Phi_0)$ & $R = ik[ikN_0 + \beta N_0 + 2\beta\alpha V_0 exp(-\Phi_0)] - 2P\beta {V_0}^2$
The suffix 'o' denotes the equilibrium values. The value of Q is negative in the region of operation. The roots of the eqn. [14] are complex with the imaginary part being positive indicating an instability. In the absence of electron impact ionization and charge exchange reaction and in the plasma we get the ion acoustic wave dispersion relation $\omega^2 = \frac{k^2}{1+k^2}$. Eqn. [14] indicates that ionization and charge exchange drives the stable ion acoustic plasma waves to an unstable growing mode. The local theory indicates the presence of the oscillations. As one progresses deep inside the sheath this analysis is not valid and one should do a detailed eigen-mode analysis which is currently under progress.

Conclusion

This paper gives the experimental results on non conservation of ion flux and its effect on potential profile in the strong ion sheath. Stability of the sheath is also investigated. A theory based on electron impact ionization for equilibrium and stability is presented.

We are thankful to Dr. R. Singh and Dr. S. Chaturvedi for useful discussions and to Mr. C.K. Adak for assistance in experimental work.

References

1. L. Tonks & I. Langmuir, Phys. Rev., **34**, (1929), p876
2. D. Bohm, in: The Characteristics of Electrical Discharges in Magnetic Fields, eds. A. Gutherie and R.K. Wakerling (McGraw Hill, New York, 1949), p77
3. V. A. Godyak & N. Sternberg, IEEE Transactions On Plasma Science, **18**, (1990), p159
4. J. L. Nightingale, E.R. Ault & A. A. Mondeli, Phys. Fl., **53**, (1982), p886
5. S. Meassick, M.H Cho & N. Herskowitz, IEEE Transactions on Plasma Science, **13**, (1985), p115
6. T.E. Sheridan & J. Goree, IEEE Transactions on Plasma Science, **17**, (1989), p884
7. N. Benjamin, Rev. Sci. Instr., **53**, (1982), p1541

MAGNETIC FIELD EFFECT ON THE ION ENERGIES AT THE CATHODE AND ANODE SIDES OF A DOUBLE LAYER IN A MERCURY DISCHARGE

G. Petraconi, H.S.Maciel, I. Castro and M. Uehara[a]

Depto. de Física do ITA/CTA,Cep:12225, São José dos Campos-Brazil
(a) UNIVAP-CEP:12245, São José dos Campos-Brazil

Introduction

A peculiar feature of free double layers formed in low pressure arc discharge is the strong heating of the walls, especially on the anode side of the double layer. This indicates that the particles in that region have high energies. Accordingly, high-energy ions directed to the walls have been measured by Maciel and Allen in a mercury discharge[1]. The purpose of the present work is to extend the ion energies analysis by including the effect of a magnetic field produced by a single Helmholtz coil. The discharge apparatus is similar to that one used by Maciel and Allen. It is observed that the magnetic field reduces the value of the critical current at which current limitation takes place and the double layer is formed. In this case the double layer, once formed, remains always at rest near the anode end of the coil. A multigrid eletrostatic energy analyser placed flush with the wall was built for measurements of ion energies. A longitudinal motion of the coil allows the double layer to be displaced along the tube because it keeps the same relative position with respect to the coil. By that means we can make measurement on both sides of the layer. The radially directed energies of the ions, the floating potential of the wall and the dispersion of ion energies in the positive column can be inferred from the current-voltage characteristic of the analyser. Various characteristics are presented in the following items to show the effect of the magnetic field (B) on these parameters. It is observed that the ion flux to the wall increases in the anode side of the layer and decreases in the cathode side of the layer as the intensity of B is increased. Aditionally the characteristics show a heating of the ions on the anode side and a cooling of the ion in the cathode side as described by Maciel and Allen but in our case this effect is enhanced by the magnetic field.

Experimental apparatus

The experiments were performed in a tube discharge similar to that of Maciel and Allen.The central region is a pyrex tube 1.0m long and 20 mm internal diameter. Measurements were made with mercury vapour pressure, in the central region, in the range of (2 - 5)x10^{-4} torr and discharge current of 2A.

Experimental Results and Discussion

The schematic diagram of figure 1.a depicts the experimental region. The reference grid GI of the analyser, see figure 1b, is at the floating potential of the wall. A retarding voltage ϕ is applied at the grid GII. This voltage is variable in order to sample the ions and it is representative of the ion energy per unit of charge. At GIII, a voltage is applied which is large enough to reflect all the electrons.

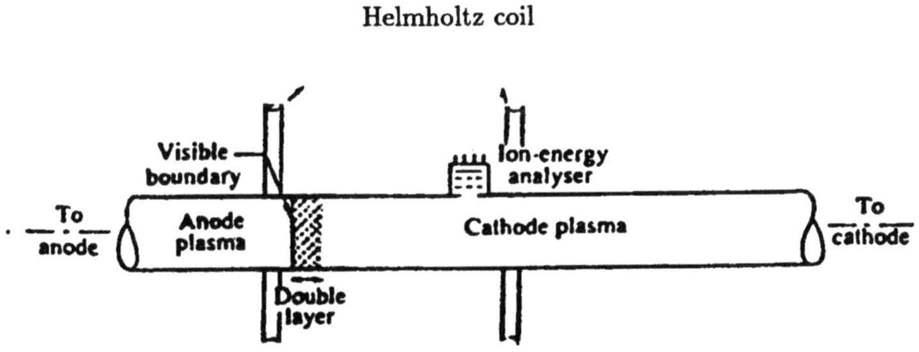

Figure 1a: Schematic diagram of the experimental region. The Helmholtz coil is movable in the axial direction

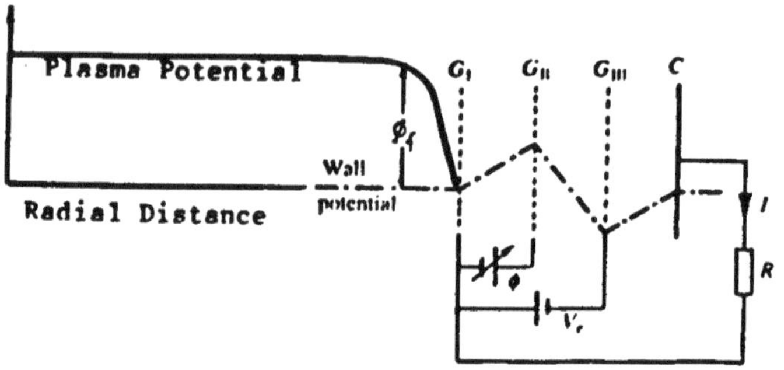

Fig. 1b: Schematic plot of the radial distribution of potential across the positive column and idealized plot of potential distribution across the analyser

Figure 2 shows the experimental characteristics of ion current versus voltage in three cases. Comparisons of curve I (before layer formation) with curve II(at the anode side of the layer) and curve III(at the cathode side of the DL), show a remarkable effect of the layer on the distribution of ion energies in the surrounding plasmas and on the ion flux to the wall. The extraordinarily high energies that the ions can achieve (above 70 eV) in the anode plasma are to explained in terms of transfer of e¬ergy from the electrons accelerated by the DL. The parameter ϕ_f (voltage drop in the wall sheath)- see figure 1b - is negligible (less than 5 V) on the anode and cathode side (after layer formation) for a magnetic field of 50 gauss.

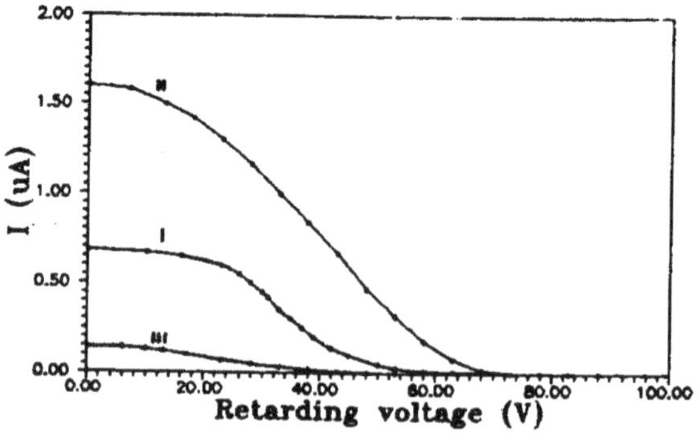

Figure 2 - Ion current versus Retarding voltage
Cathode and anode side-P=2x10^{-4}torr - B=50 gauss

The effect of magnetic field on the characteristics on both plasmas is shown in the figures 3 and 4 for a vapour pressure of $2\text{x}10^{-4}$ torr. In all cases we observe a significant effect of ion heating in the anode plasma and an ion cooling in the cathode plasma produced by the magnetic field. In the curves that were obtained in the presence of the double layer, the analyser is at a distance of 4.0 cm from the DL.

Curve	B (gauss)
1	0
2	13
3	19
4	26
5	38
6	51
7	64
8	114

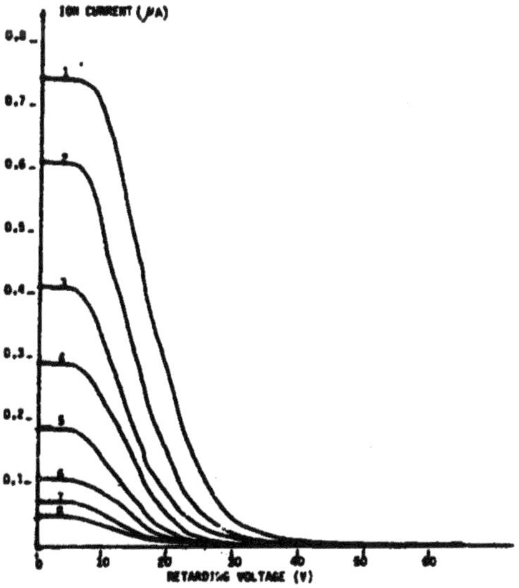

Figure 3 - Ion current (μA) versus retarding voltage (ϕ) - CATHODE SIDE
curves 1 and 2(whithout double layer)
curves 3,4,5,6,7 and 8 (with double layer)

Measurements of current-voltage characteristics in the anode region are shown in the figure 4. The effect of the magnetic field is to encrease the radial flux and the dispersion of ion energies in this region. The double layer forms for a magnetic field as low as 15 gauss with vapour pressure of $2\text{x}10^{-4}$ torr, and discharge current of 2A.

Curve	B(gauss)
1	0
2	13
3	19
4	26
5	38

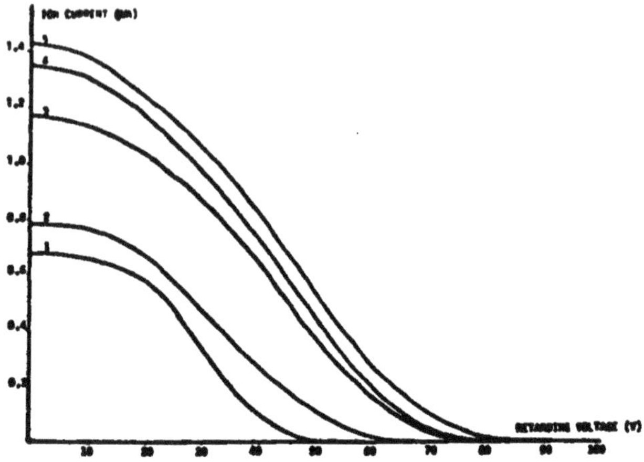

Fig. 4 - Ion current (μA) versus retarding voltage [ϕ(V)] - ANODE SIDE
curves 1 and 2(whithout double layer)
curves 3,4 and 5 (with double layer)

Figure 5 shows the value of the maximum ion current for different positions of the analyser with respect to the layer. At the cathode side this current is much smaller than at the anode side and it doesn't varie apreciably with this distance, but, at the anode side the ion current increase with the distance from the layer.

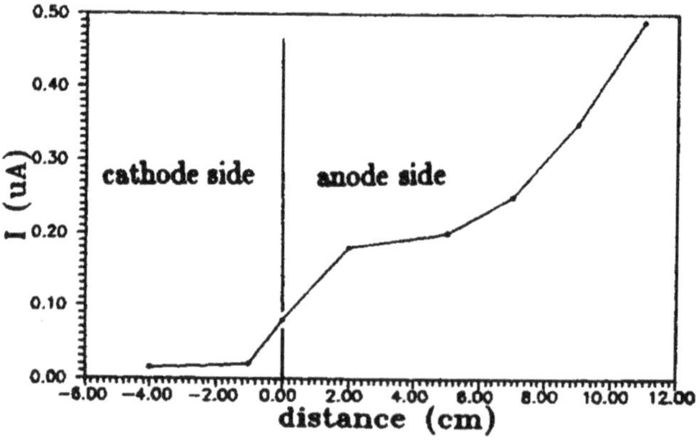

Figure 5 - Total ion current versus distance
B=77 gauss - P = 5.0x10^{-4} torr

Conclusions

We have reported some preliminary results of measurements of current-voltage characteristics of an ion energy analyser which collects ion current directed to the wall of the discharge tube. Double-layer was induced to form by increasing the longitudinal magnetic field produced by a Helmholtz coil placed in the region of analyser. This field affects both the ion current and the ion energies radially directed to the walls. Measurements on both sides of the layer were possible to be taken because the layer followed the longitudinal displacement of the coil. The main qualitative effect of the magnetic field is to decrease the radial flux and the dispersion of ion energies on the cathode side of the layer and to increase the same parameters on the anode side of the layer. This effect on the cathode side, is similar to that observed in a positive column in the absence of the double layer. On the anode side, however, the electrons which are accelerated across the layer are thermalized by the instabilities in that region, the ionization is enhanced and a more dense and hot plasma is produced in the anode side. This effect leads to higher ion flux in the anode plasma than in the cathode plasma. The increasing of the ion current with the magnetic field in the anode plasma is a new and interesting phenomenon. This behaviour deserves careful analysis and it will be object of further work.

REFERENCE

1- Homero Santiago Maciel and J.E. allen, *J. Plasma Physics.* **42** (1989), part 2,p. 321-352.

ELECTROSTATIC INSTABILITIES INDUCED BY A NEGATIVELY BIASED GRID AT THE PLASMA BOUNDARY

G.Popa, R.Schrittwieser*, D.Ruscanu and E.Stamate

Plasma Physics Department,
"Al.I.Cuza" University, Iasi, RO-6600, Romania
*Institute for Ion Physics, University of Innsbruck,
A-6020, Austria

Introduction

Grids are widely used in plasma physics especially then when a separation between electrons and ions is needed as in energy analyzers and ion sources. Grids are also used as exciters or as detectors for different kinds of plasma waves. According to the grid bias it might draws either electrons or ions and corresponding sheath is formed between unperturbed plasma and that grid /1,2/. In such circumstances some electrostatic instabilities might appear.

Thus, a positive biased grid used as a cold plate in a single ended Q-machine excites so called potential relaxation instability (PRI) /3/ which might at its turn to excite ion-acoustic waves travelling in the plasma leaking through the grid /4/. In a similar experimental device but with a negative bias of the grid a new electrostatic instability has been reported which was explained that time as ion space charge instability /5/.

Recently it was shown that a negative r.f. sheath resistance associated with electron inertia in a diode electron rich sheath might sustained a coherent sheath instability near electron plasma frequency /6/. A similar negative r.f. sheath resistance associated to the change of the sheath thickness could explain an other sheath-plasma instability in the range of ion plasma frequency /7/. Moreover, by adding of some external reactive elements of circuit as a capacitor or inductance, the negative r.f. sheath resistance might sustain also electrostatic instability in a grid-plasma system /8/.

Experimental arrangement

The experiments were performed in double-plasma devices from University of Iasi /9/ and University of Innsbruck /10/.

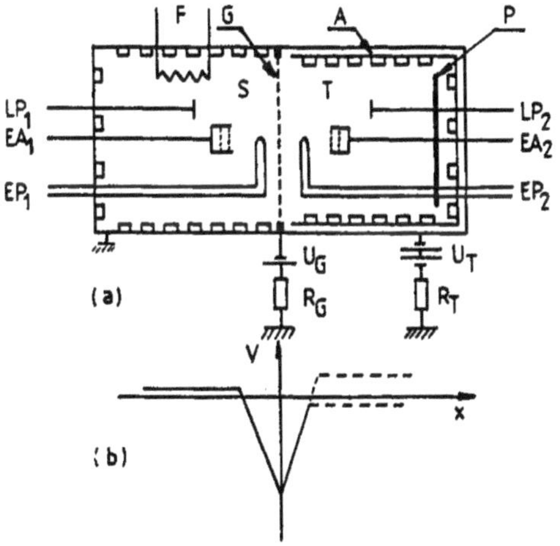

Fig.1.a) Experimental set-up; b) Sketch of the plasma potential along the axis.

Both devices are equipped with multiple magnets for surface plasma confinement as shown in fig.1. Excepting the diameters (35 cm and 45 cm respectively) all other parameters are alike. Argon gas is used at a pressure of about 10^{-4} mbar. The plasma is produced by a dc discharge between filament and the wall only in the source chamber (S). Plasma in the target chamber (T) is produced only as diffuse plasma that leaks through the separation grid (G) (mesh made of 0.09 mm stainless-steel wires spaced 0.1 mm apart). The wall of the target chamber is biased separately (U_T) and could be used as a large collector so that the separation grid and the target chamber could be considered as a large electrostatic analyzer /11/.

The wall of the source chamber is grounded and discharge voltage was kept constant at 60 V. Typical parameters of the source plasma are $n_e = 10^7 - 10^8$ cm^{-3} $T_e = 1.2 - 2.5$ eV and $T_i = T_e/10$. Grid bias (U_G) was also changed in the range from 0 to - 300 V. Plasma parameters were measured in both chambers by plane Langmuir probes (LP 1 and LP 2), small electrostatic analyzers (EA 1 and EA 2) and emissive probes (EP 1 and EP 2) which are axially and radially movable. In addition in the target chamber there is a large disk (P) which cover the whole cross section of the chamber and could be biased separately. The disk could be moved in axial direction up to the grid level so that the volume of the target chamber could be continuously varied.

Experimental results

As well as in the experiments of Barrett and Greaves we have observed an instability of which main features were already presented /12/. Unlike their experiments in our case the target chamber is used as the collector of the large electrostatic analyzer formed by G and T system /11/. In this case the "collector" current I_T depends on the "collector" voltage U_T as the characteristic presented in fig.2a. It shows a large hysteresis which correspond to the transition from stable (ion current AB) to unstable regime (electron current BC) and reverse, from unstable (CD) to stable (EA) regime. This current is about one order of magnitude smaller than the ion grid current (I_G) of which dependence on U_T is presented on the same figure. The I_G presents also hysteresis and the difference between the currents of the two branches B'C' and A'C' corresponds to the ions passing through the grid and are collected by target chamber as I_T when $U_T < 0$. The small difference between the above I_G value of two branches shows that the main part of ions comming from source chamber is collected by separation grid in spite of its large transparency (80 %).

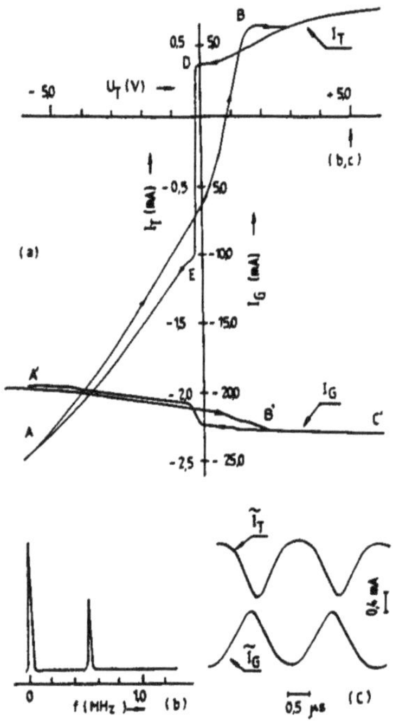

Fig.2

Moreover, the electron current measured by target chamber when $U_T > 0$ shows that not only ions but also electrons pass through the grid in spite of its very large negative bias and large Debye lenght comparative to grid mesh size. Consequently a very low plasma density might be found in the target chamber /11/.

The above hysteresis loops in both the I_T and the I_G characteristics show that a redistribution of the particles in the target chamber should be involved at the transition from stable to unstable regime. Moreover a dynamic structure has to be considered in the latter in connection with the mechanism of the instability which might be followed as fluctuations of both the I_T and the I_G. Typical frequency spectra is presented in fig.2b. Both the phase relation and amplitude of the I_T and the I_G current oscillations are presented in fig.2c.

In order to identify where the space charge is formed and how it behaves during the instability the axial distributions of both the local potential (V_p) and the ion saturation current (I_o) of the EA were registered on both sides of the G for stable (full curves) and unstable regime (dashed curves). Thus plasma potential in the source chamber ($x < 0$) measured by EP1 does not change at the transition between stable and unstable regime (fig.3a) while the potential measured by EP2 shows a net jump towards positive plasma potential in the target chamber ($x > 0$) when the instability is setting in. Moreover, mean value of the plasma potential (dashed curve) shows a peak at the transition to the ion space charge sheath in front of G. Also, the thickness "d" of the ion space charge sheath is almost the same on both sides of G and corresponds fairly well to the theoretical value given by Child-Langmuir relation :

$$j_G = \frac{4\sqrt{2}}{9} \cdot (e/m)^{1/2} \frac{U_G^{3/2}}{d^2}$$

where: j_G is the density of the grid current taken from the I_G (Fig.2a), U_G is the grid bias (-85 V), e and m positive ion charge and mass respectively. This value is shown by arrows on Fig.3a.

More information could be obtained from axial distribution of the I_o measured by two grids electrostatic analyzers EA1 in the S and by EA2 in the T respectively. The first grid bias was -9 V, second grid bias was -15 V and collector bias -90 V. Thus, the presheath in the source chamber is larger for unstable regime when ions are reflected back from the target. Moreover, in this case the ions could not penetrate farther then about 20 mm into the target and a

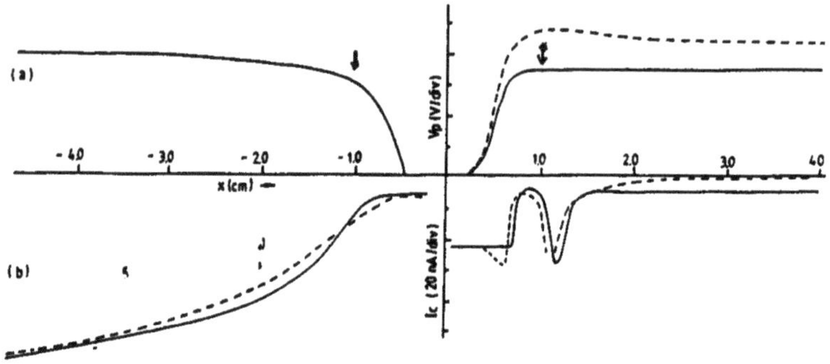

Fig.3

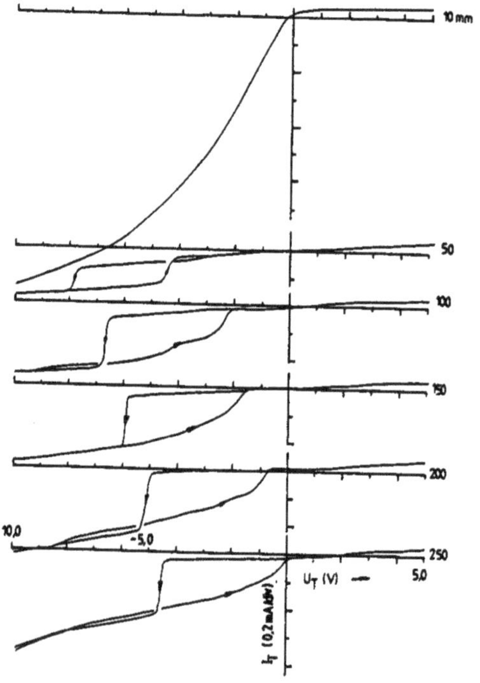

Fig.4. I_T versus U_T grid-plote distance as parameter, U_a = -140 V; n_a = 1.1.10^7 cm^{-3}.

fine structure is formed over that distance. This structure is slightly different for stable regime but an uniform ion density is measured in the rest of the target chamber.

This result suggest a resonant phenomenon between ions bounceing in a rather symmetric potential well around G and the PRI instability. The latter is similar to a plasma diode where both ions and electrons are leaking through the grid so that a virtual cathode is formed at the sheath edge in the T while the anode of the target chamber biased positive acts as an anode of the plasma diode. The potential fluctuates in the whole chamber while the ions travel only in a rather small distance (about 20 mm) from the grid. The frequency of the instability corresponds to the ion transit time in the above ion sheath around the grid. A simple confirmation of the model could be obtained by shifting the P towards the G. In this case both the I_c characteristic and the instability are almost the same until the distance between P and G is smaller than about 35 mm. For smaller distances the instability disappears and I_c characteristic becomes a usual collector characteristic of an electrostatic analyzer (Fig.4).

REFERENCES

1. F.F.Chen - Introduction to Plasma Physics, Metnem, 1974
2. L.Schott - Phys.Fluids, B3 (1991) 236
3. S.Iizuka, P.Michelson, J.J.Rasmussen and R.Schrittwieser, Phys. Rev. Lett. 48 (1982) 145
4. N.Sato, G.Popa, E.Märk, E.Mravlaq and R.Schrittwieser, Phys.Fluids, 19 (1976)
5. G. Popa, N. Sato, E. Märk, R. Schrittwieser and E. Mravlaq, J. Phys. D: Appl. Phys. 9 (1976) 397.
6. R. L. Stenzel, Phys. Fluids B1 (1989) 2273
7. N. Ohno, A. Komori, M. Tanaka and Y. Kawai, Phys. Fluids B3 (1991) 228.
8. D.Ruscanu, G.Popa and V.Anita, Proc.XIV - SS-IS-PIG, Sarajevo (1988) p.465
9. G.Popa, Symp.on Plasma Double Layers, Riso Nat.Lab. (1982) R-472, p.82
10. M.Oertl and R.J.Armstrong, Rep.82/2, Univ. of Tromso, Norway
11. G.Popa and R.Schrittwieser, Int.Conf.on Plasma Phys., Innsbruck (1992) III-1757
12. P.J.Barrett and R.G.Greaves, Phys.Fluids, B1 (1991) 1776

ON THE ANODE DOUBLE LAYER FORMED IN A DC GLOW DISCHARGE

M.Sanduloviciu and B.Ianopol
Department of Plasma Physics, "Al.I.Cuza" University
6600 Iasi, Rumania

INTRODUCTION. The phenomenon taking place in the contact region between a positive metal electrode and a gaseous, conductor present also today a great interest both for the fundamental studies as well as for different applications of the plasma physics (plasma contactors) /1/. In an extensively paper Langmuir /2/ hase shown that in the next vicinity of an anode in a low pressure d.c. gas discharge an electrical double sheath, which delimities two plasmas, one enriched in positive ions and the other in electrons, is formed. In that paper Langmuir has described for the first time in plasma physics a double space charge structure and also proposed a theoretical explanation of this phenomenon. Although very old, the problematic related to the processes occuring in the vicinity of an anode can not be considered as resolved, a fact evidenced by numerous recent published experimental and theoretical papers /1,3/. Starting from the above mentioned situation the physics of the DL today also remains probable, the most important question of the plasma physics /4/. This assertion is also justified by the fact that, very probable, all kinds of the so called selfsustained time dependent phenomena observed in all types of plasma devices are at their origin energy reservoirs localized in DLs. In the present paper we describe experimental results concerning the unstable behaviour of a d.c. glow discharge, produced in a metal tube, related to the double layer formed in front of an anode.

EXPERIMENTAL DEVICES AND RESULTS. The experimental device is schematically presented in Fig.1. The low pressure gas discharge was produced between two metallic electrodes a cathode (Al - disc 5 cm in diameter) and an anode (W - disc 1.5 cm in diameter) placed in a metal tube without galvanic connection with the two electrodes. The working gas was Ar at a pressure arround 10^{-1} Torr.

Information about the processes implied in the generation of the globular anode spots formed after the ignition of the d.c. discharges was obtained considering the tube wall W as a third electrical connection to the gaseous conductor connected between cathode and anode. A first problem is related to the processes which determin the appearance of positive electrical charges at the inner surface of the tube isolated from the cathode and the anode. Contrary to the widely accepted explanation which considers that the tube surface is charged after ambipolar diffusion we start with the experimental proved fact that the most important charging mechanism is related to the presence of a secondary electron emission with a coefficient greater than one /5/. This secondary electron emission is produced by electrons originating at the cathode surface and accelerated in the cathode potential fall. These electrons form a divergent beam of fast electrons which partly collide the inner surface of the discharge tube. Because the emitted secondary electrons are collected by the anode the potential of the W is drawing near the anode potential. When the anode is placed at a sufficient great distance from the cathode so that its surface is not collided by fast electrons in front of it a globular space charge structure bordered by a double layer (DL) is formed. The different forming phase of this structure can be investigated changing the tube potential V_W in respect to the V_A potential. By using an electrical probe which penetrates through the center of the anode one can measure the potential drop developed in front of it for different potential drops between the tube wall and the anode. The result is similar to those published in /5/. Another

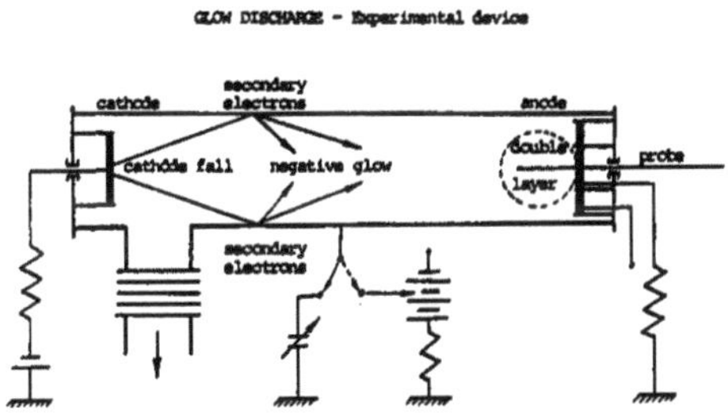

Fig.1

way to obtain information on the processes taking place during the forming process of the DL in front of the anode was by plotting the current-voltage characteristic between the tube wall and the anode, using a supplementary d.c. power supply. In this way it is possible to change continously the current collected by the anode and consequently to determine different states of the anode spot. Such a $(I-V)_W$ characteristic is shown in Fig.2.

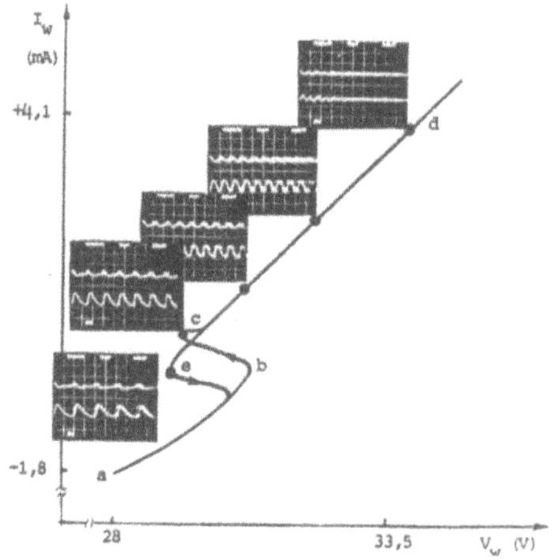

Fig.2

Observations performed on the visible images of the DL formed in the vicinity of the anode and also on the current collected by the anode related to different points of this characteristic have shown that in the ab portion of the $(I-V)_W$ characteristic in front of the anode, an anode spot is present. In the moment when the V_W achieves a certain value, the anode spot spontaneously change its shape and geometrical dimensions turning into a nearly spherical anode glow (fireball). Simultaneusly the current collected by the anode increase from the value corresponding to the point b to the value corresponding to the point c. In the same time the potential difference between the tube wall and the anode decreases. The allure of the characteristic corresponds to a S shaped differential negative resistance. The transition of the anode current between the values specified by the point b and c can be explained taking into account that the fireball dimension is much larger than that of the previosly formed anode spot. Taking into account that the fireball is bordered from the surrounding plasma by a DL and that the positive side of this acts as a virtual anode (whose potential can be greater than the potential of the anode /3/) its area becomes greater after this transition and consequently the d.c. current intensity suddenly increase /6/. The simultaneously observed decreasing of the potential difference between the tube wall and the anode is produced by the fact that a greater electron collecting area determines a decreasing of the internal ohmic resistance, so that the redistribution of the potential drop between the tube wall and the anode, and external resistance (15 kΩ) inseriated with the second power supply is readjusted in the way shown by the $(I-V)_W$ characteristic.

In our experiment the transition to the fireball simultaneously takes place with the periodical modulation of the current collected by the anode. This unstable behaviour of the fireball appears when the potential drop over the DL becomes equal to the ionization potential of the gas. The presence of oscillations proves that this condition is satisfied after this transition. In the $(I-V)_W$ characteristic the appearance of oscillations determines a decrease of the mean d.c. current collected by anode.

For explaining this last fact, one present in fig.3 the dynamical $(I-V)_W$ characteristic. This characteristic show the

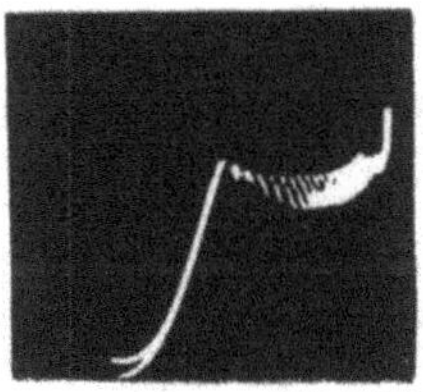
Fig.3

spontaneus increase of the current when the fireball appears and also the transition to the unstable region of the fireball. From this figure it is evident that the maximal values of the current oscillations do not excced the value of the dc current. Consequently the mean value of the current collected by the anode decreases after the appearance of the oscillations. Afterwards the increase of the wall potential also produces an increase of the anode current.

The triggering mechanism of the current modulations can be clarified by measuring the alternative current which flows from the metal tube to the surroundings and from here to the anode. In fig.4 we present comparatively the alternative current intensities I_c flowing from the d.c. power supply, to the cathode, the modulation current intensity I_a collected by the

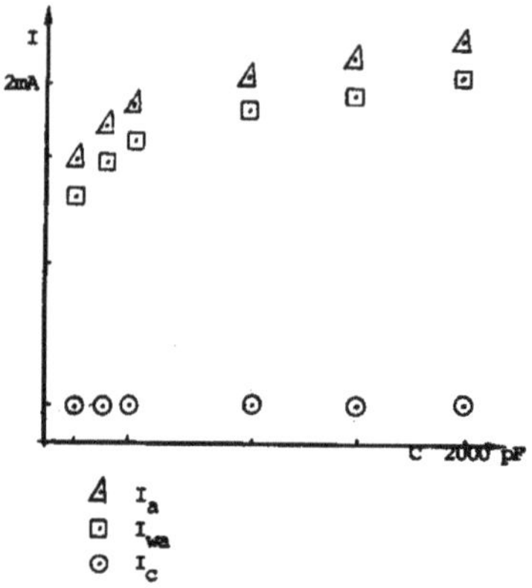

Fig.4

anode, and the alternative current I_{wa} capacitively flowing from the tube wall to the surrounding. This result shows that the alternative current flowing from the tube to the ground and those collected by the anode are approximativelly equal whereas the cathod current I_c is much lower than the both of the above mentioned currents. This fact prove that during the oscillating state the most important current flows capacitively between the tube wall and the anode. Considering these experimental results it is evidently that the anode DL disruption process can be triggered by secondary electrons originating at the inner surface of the tube wall. Briefly speaking the dynamics of the anode DL is the following. During the anode DL forming phase this store electrical energy originating from the external power supply. This energy is accumulated in the electrical field simultaneously developed with the produced space charge separation which is finished with the formation of the anode DL. As long as the potential drop V_{DL} over the DL is lower than the ionization potential of the gase V_i, the anode DL remains in a stable state. In the moment when $V_{DL} = V_i$ the DL disruption processs is started /7/. Taking into account that during the variation of the potential difference between the tube wall and the anode the flux formed from secondary electrons emitted at the inner surface of the tube wall and collected by the anode changes periodically, it is evident that the V_{DL} can be triggered by the variation of the secondary electron flux. This triggering mechanism determine the appearance of oscillations when the periode of the alternative current collected by the anode is equal which the time interval in which a whole evolution of the unstable DL dynamics takes place. As was shown /5/ the oscillations, usually known as anode instabilities, are determined by the reactive elements of an oscillatory circuit propre to the discharge /5/. In agreement with this fact the equivalent circuit of our experimental device is shown in fig.5. In this figure the ohmic and reactive behaviour of different parts of the discharge is specified. The capacitive behaviour of the DL is not considered in this equivalent circuit because its value is very large with respect to the inseriate C_W value.

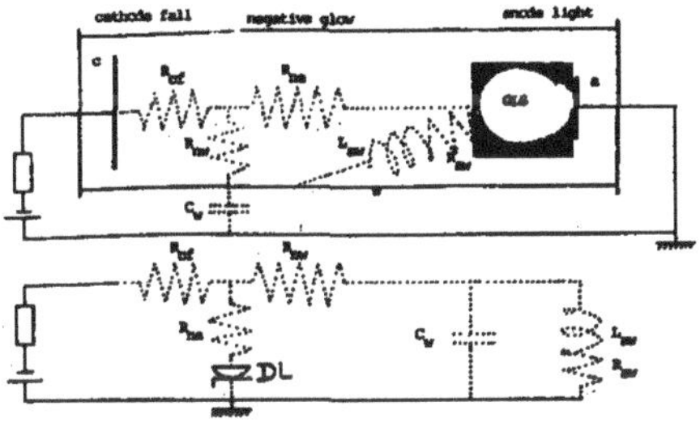

Fig.5

REFERENCES

1. Bin Song, N.D'Angelo and R.Merlino (1991) J. Phys. D. Appl. Phys., **24** 1789 and **25** 938
2. I.Langmuir **33** (1929) Phys. Rev. 955
3. K.G.Emeleus **52** (1982) Int. J.Electr.407
4. H.Alfven (1986) IEEE Trans. on Plasma Sci. (dec.1986)
5. M. Sanduloviciu, E. Lozneanu, **28** (1986) Plasma Phys. Contr. Fus., 585
6. I.E.Allen **27** (1985) Plasma Phys. Contr. Fus., 1343
7. D. Alexandroaei, M. Sanduloviciu, A**122** (1987) Phys. Lett.173

SELF-ORGANIZATION PHENOMENA AT THE ORIGIN OF THE DOUBLE LAYERS

M.Sanduloviciu, E.Lozneanu and V.Melnig
Department of Plasma Physics, "Al.I.Cuza" University
6600 Iasi, Romanie

INTRODUCTION. As is known the notion of double sheath was firstly used in the plasma physics by Langmuir /1/ after investigations performed on the boundary regions between electrodes and a current carrying gaseous conductor. Such a double sheath, now known as double layer (DL), developed, for example, near the anode can be very well distinguished from the other parts of the gaseous conductor since the light phenomena present here attribute it a well delimited visual aspect. In agreement with Langmuir, the anode spot boundary is a double sheath with an inner positive space charge and an outer negative space charge, the potential drop being of the order of the ionizing potential. In spite of the well localized luminous phenomena which prove, visually, the presence of the DL, the model proposed by Langmuir for the explanation of the double sheath appearance does not consider the implication of elementary processes in its generation mechanism. Afterwards much experimental investigations on the generation and dynamics of DL are analyzed in terms of a collisionless plasma model /2,3/. As a consequence, the most theoretical explanations, today widely accepted, start with the assertion that the quantum processes do not play an essential role in the DL phenomenology. One remarks, however, that very important DL studies performed in the last time /4,5/ use plasma devices where the DL presence is related to well localized light phenomena. Considering the above mentioned facts, it is evident that the DL phenomenology is a controversed problem of the plasma physics /6/. It is sure that for solving this problem one must start from an rigorous analysis of all experimental results obtained both in devices where the plasma is collisional (gas discharges) and in devices where usually the plasma is considered collisionless.

An answer to the discrepancy between the presumed two different DL physical scenarios is possible only by a new, definition of the DL. This definition must start with the reasons listed by Allen /4/ but also with the experimentally proved fact that the DL is a primeval form of a "cellular membrane" (evidenced clearly by the globular DL) which assures to the regions delimited by it, the qualities of an autonomous body /7,8/. The essential distinction between the above mentioned two different opinions on the DL physical scenario can be resumed as follows: one of them considers the appearance of a DL as a passive response of a collisionless plasma to the presence of a locally acting electric field produced, for example, by a positive electrode /10/; the other·opinion, imposed by observation on DLs formed in collisional plasmas, considers the DL appearance as a result of a self-organization process finished with the genesis of an autonomous body (the DL), whose viability is assured, in a current carrying gaseous conductor, by the processes taking place inside of it /8,9/. These processes assure, themselves, the condition for its (and consequently of the DL) existence. Essential in the last mentioned model is the assertion that, usually dynamical laws are not sufficient to explain the opposite electrical space charge separation taking place during the DL generation and afterwards the explanation of its unstable behaviour.

EXPERIMENTAL METHODS AND RESULTS RELATED TO THE DOUBLE LAYER PHYSICAL SCENARIO. There are different possibilities to perform investigations on the phenomena related to the DL physical scenario. One of them establishes an indirect connection between the global effects related to the DL generation and dynamics and the allure of the I-V characteristic of the gaseous conductor which contains the DL /5,9/. Another investigation modality uses electrical and spectral measurements performed in the region where the DL is formed /9,11/. The spectral investigations are able to establish a direct relationship between the spatial distribution of the cross section corresponding to different elementary processes (excitations and ionizations) and the resulting space charge distribution which determines the DL potential profile. Such a relation can be observed in a region where electrons are accelerated, a premise always necessary for the DL generation.

Experimental investigations performed on the processes taking place inside the region where the DL potential drop V_{DL} is formed has proved a direct connection between the spatial arrangement of the spectral lines and the variation of the electron concentration /9/. This fact confirms /12,13/ that excitations produced by electrons accelerated in the V_{DL} determine an accumulation process of the electrons which must be considered at the origin of the DL generation /12 /. We remark that the DL forming process comprises different phases during of which its position is also changed. Initially, the necessary electron acceleration is produced in the next vicinity of the positive electrode after applying the positive bias (or in a local constriction of the gaseous conductor). The following evolution phase af the DL generation is related to the fact that the excitation processes of the neutrals are produced with maximal probalities, at certain distances from the positive electrode (when its positive bias is sufficiently high). As a consequence, the electron concentration is locally changed since the part of electrons which have lost their kinetic energy in excitation processes is accumulated. Taking into account their Maxwellian distribution, the electrons with high thermal energy which produce excitations are firstly accumulated and afterwards the electrons with small thermal energy are trapped by the potential barrier already formed. Both processes determine the appearance of a well-localized negative space charge formed by thermalized electrons. The part of electrons with sufficiently high energy, able to penetrate the negative potential barrier, and the electrons which have not produced excitations, gain, by acceleration in the developed electric field, a kinetic energy sufficient to produce positive ions. Taking into consideration the energy dependence of the ionization cross section /5/, the maximum rate of the ion production occurs at a certain distance (nearer the anode) from the well localized negative space charge. Owing to their high mobilities, the electrons producing ionizations and the resulting electrons are quickly collected by the anode so that between the negative space charge and the anode a plasma enriched with positive ions appears. The fact that, inside the DL, the rise of positive potential is accompanied by a concomitant increase of the electron concentration /9/ , may be explained by space distribution of excitation rates corresponding to different energy levels. The final space charge arrangement is also determined by the electrostatic forces that act between the well localized negative space charge and the positive ions. It is important to point out that the DL internal structure is maintained in dynamical equilibrium between the number of electrons accumulated after inelastic collision (excitations), the number of positive ions produced and bounded by electrostatic forces with the negative space charges and different particle loosing processes as diffusion, recombination and other. It is evident that the maintenance of the DL structure is assured by the presence of an electrical current which traverses the gaseous conductor.

When the production rate of positive ions is sufficiently high the positive side of the DL is able to play the role of a virtual anode, placed at a certain distance from the metal anode, a fact which produce a shelling off process from the anode surface so that the DL becomes electrode-free. In such a situation the DL position depends on the anode potential /11/. The electrode-free DL is able to assure itself the condition for its maintenance (a sufficiently high local potential drop so that the variation of the cross section of the quantum processes produces the electrical charge separation which determines, in its turn the above mentioned potential drop). A very important fact which demonstrates that only external imposed electrical fields are not able to explain the DL generation scenario, is the observation that the positive potential of the positive side of the DL can be higher than that of the anode /12,14/. This confirms that only the establishment of a simple electrical contact between two plasmas having different potentials is not sufficient to determine the appearance of an DL /6/. An example of the above discused phenomenon is the formation of a DL at the border of nearly spherical anode spot. Applying a positive bias on the anode, one creates the condition for the acceleration of the thermalized electrons from the plasma. This acceleration is produced in a well localized region in the next vicinity of the anode. Consequently the described space charge separation related to the excitation and ionization rate takes place. Thus, the initially formed anode sheath transforms in a DL. The assertion that internal processes are essential in the DL generation is justified, in this experiment, by measurements which prove that the potential at the positive side of the DL is higher than the anode potential /14/. As a result of this evolution, the DL is shifted at a certain distance from the anode surface forming the "head" of a positive column. The so formed electrode-free DL acts as virtual anode which locally assures the electron acceleration necessary for the DL maintenance. Taking into account the described DL charging mechanism it is evident that a DL is able to act as an accelerator of those particles which have not produced inelastic collisions /16/.

Most evidently, the behaviour of a DL to act as an autonomous body is proved by the phenomenon known as ball of fire. The ball of fire is an electrode-free globular DL which usually appears in low voltage arcs and also in thermionic converters /15/. The ball of fire can be also observed when the potential difference between the electrodes is much less than the ionization potential of the gas. A very important experimental observation is the fact that the ball of fire appears frequently in a pulsating state /8,15/ and that during this state the mean potential in the ball of fire "nucleus" is higher than the anode potential. This proves that essential for the ball of fire appearance and maintenance are the internal causes able to produce a positive ion accumulation in the nucleus. Such a positive nucleus appears after a diffusion conditioned space charge separation takes place in region of the gaseous conductor where a local heating process takes place /8/.

The problem if the same phenomenology is also valid for collisionless plasmas can be resolved only by measurements on the neutral concentrations inside the DL. Our experimental results (pressure measurements performed inside of a globular DL and by investigation, for low pressures, of the recombination spectrum emitted from a DL) prove that the neutral concentration inside a DL is much greater than in the surrounding plasma. This confirms the assertion /9/ that the confinement process related to the presence of electrostatic forces acting between the two electrical layers with opposite sign locally assures, by recombinations, the matter concentration required for the above described DL phenomenology. The visual images of the DL also observed in devices where the plasma is considered collisionless (for example in Tokamak) prove this assertion.

REFERENCES

1. I.Langmuir **33** (1929) Phys. Rev. 955
2. P.Burger **36** (1965) J.Phys.D.Appl.Phys., 1938
3. H.Schamel **3** (1986) Phys. Rep. Rev.Phys. Lett. 161
4. I.E.Allen **27** (1985) Plasma Phys. Contr. Fus., 1343
5. Bin Song, N.D'Angelo and R.Merlino (1991) J. Phys. D. Appl. Phys., **24** 1789 and **25** 938
6. R.Schrittwieser (1992) Symp.on Atomic and Surface Phys. Italy
7. M.Sanduloviciu **29** (1987) Plasma Phys. Contr.Fus.1687
8. M.Sanduloviciu (1991) XX ICPIG, Pisa, 265
9. M.Sanduloviciu and E.Lozneanu (1992) ICPP Innsbruck, 1769
10. N.Sato (1984) ICPP, Lausanne. Proc.Inv. Papers p.555
11. M.Guyot and Ch.Hollenstein **26** (1983) Phys.Fluid. 1596
12. M. Sanduloviciu, E. Lozneanu, **28** (1986) Plasma Phys. Contr. Fus., 585
13. M. Sanduloviciu, (1987) ICPP Kiev, Proc. Invited Papers, 293
14. K.G.Emeleus **52** (1982) Int. J.Electr.407
15. G.Musa **36** (1988) An. St. Univ. Iasi, 49
16. D.A.Briant, R.Bingham and V.de Angelis (1991) Phys.Rev. Lett. (in print)

THE BALL LIGHTNING AS A GLOBULAR DOUBLE LAYER RESULTING AFTER A SELF-ORGANIZATION PROCESS

M.Sanduloviciu
Department of Plasma Physics, "Al.I.Cuza" University
6600 Iasi, Rumania

In the last time many experimental and theoretical studies concerning the physical scenario presumed to explain the appearance of that natural phenomenon, known as ball lightning, were published /1-4/. Since the ball lightning is a very haphazard phenomenon, up today direct experimental investigations were not possible, so that only qualitative reports of eye-witnesses and many photographs have given information about their appearance conditions and properties.

As is known, Kapitza published in 1955 a paper /5/ in which he proposes a model which considers the ball lightning as being formed by short radio waves generated in natural atmospheric processes. More recently Xue-Heng Zheng /3/ proposes a quantitative analysis of the ball lightning considering it as a shell expelled out from an air-filled cavity by the ponderomotive forces appearing when a large amount of radiant energy is stored in plasma. Under such conditions, the plasma is pushed to the periphery to form an electrical conductive shell for stopping energy from spreading away.

Another ball lighting model was recently proposed by B.B.Kadomtsev /2/ which considers it as a gas accumulator supplied with energy by usual lightning and stored as chemical energy.

Most recently, Ohtsuky and Ofuruton have published a paper /4/ which describes the production, in natural pressure conditions, of fireballs using microwave interference. These fireballs exhibit certain properties qualitatively similar to eye-witness observations of ball lightnings such as motion against the wind and the ability to intact pass through a wall.

Other older experiments concerning the artificial production off the ball lightning are those in which the globes are formed in dilute mixtures of gases in air, in some radiofrequency gas discharges and luminous spheres produced from metal wires by electric discharge/1/. None of these is yet adequately understood /1/.

In the present paper I will call the attention that today there is a great number of experimental investigations performed on globular very luminous space charge structures observed under certain conditions in different plasma devices which are able to give essential information for the understanding the ball lightning phenomenology. Resuming also the content of already published papers /6-8/, in the following I briefly present the most important conclusions of such experimental investigations. These experimental results are able to explain the structure, stability, dimensions, luminosity, energy, life time of ball lightning and, additionally, the motion against the wind and its apparent penetration through a dielectric surface

As concerning the explanation of the ball lightning structure it was shown /9/ that globular space charge structures observed in laboratory as fireballs in low voltage arcs or as plasmoides in HF discharges /10/ behave as autonomous bodies resulting after a self-organization process. Such body appears in regions where a current concentration produces a strong heating process so that a diffusion-driven electrical space charge separation takes place /9/. Owing to their high mobilities, the electrons leave this heated region so that a positive nucleus which behaves as virtual anode makes possible the evolution of the self-organisation process described in /9,11/. The result of this self-organization process is the appearance of a spherical double layer which delimitates the positive nucleus against the surroundings. The stability of the fireball was explained /9/ considering the electrostatic forces which act between the two adjacent layers having opposite electric signs (acting similar to the surface tension in liquids) so that the spherical shape corresponds to a stable equilibrium characterized by minimal value of the potential energy. In our opinion it is very probable that in the region where a usual linear lighting strikes the Earth surface there are similar conditions as those in gas discharges. The strong heating process taking place in this region produces a plasma and, as a consequence of the above mentioned different diffusion rates of electron and positive ions, the appearance of a positive nucleus.(see the model proposed by Kadomtsev /2/). After its appearance, this nucleus initiates all the processes described in /9,11/, so that in the final state, a globular space charge structure, delimitated by a double layer, is formed.

As concerning the energy supplying mechanism and the related lifetime of the ball lightning, it was shown that a globular charge structure delimitated by a double layer is able to absorb, at resonance, electromagnetic energy /10/. Such a spherical space charge structure behaves as a serial oscillatory circuit /10/, whose eigen frequency depends on the dimensions and of other parameters of the plasma inside of it and also on the gas nature. We point out that a positive nucleus bordered by a spherical double layer is able to change its dimensions (and internal structure) so that always a self-acordable mechanism assures the maximal energy absorption from external electromagnetic sources.

Supposing that the ball lightnings have similar properties, it is evident that their lifetime depends on the presence of external source of electromagnetic energy. Such source are usual linear lightning which always generates damped electromagnetic oscillations. Consequently the ball lightning produced in the above mentioned way becomes able to maintain its existence as long as the external electromagnetic energy sources are present. A temporal succession of more linear lightning can explain the large life time of ball lightnings observed in certain cases.

The above described physical scenario, presumed to act also during the ball lighting generation, can also explain its motion against the wind and also the penetration through a dielectric surface. Thus,the experiments performed on the double layer dynamics prove that the ball lightning displacement always takes place in the direction where the potential drop across the double layer becomes equal with the ionization potential of the working gas. Evidently, for a spherical double layer, the potential drop across it does not reaches simultaneously the same value at the entire spherical surface so that the direction of the ball lightning motion depends only on the regions of the spherical double layer where the potential drop reachese such value. Because the energy supplying mechanism is assured by absorption at resonance of electromagnetic energy which also locally assures the heating process necessary for the maintenance of the electrical space charge separation, is it evident that such a system is able to reform at the other side of a dielectric surface in the moment when this region heated by high frequency reaches such a surface.

The well localized light phenomena proper to the ball lightning can be explained by considering the great excitation rate existent at the spherical double layer border related to the self-organization mechanism taking place during the generation and maintenance of the spherical double layer.

Concluding, I consider that the recent obtained knowledge on the physical scenario which explains the appearance of fireballs in plasma devices is able to give essential information on similar phenomena appeared under natural conditions.In my opinion the ball lightning phenomenology initially comprises the appearance of a positive plasma (the nucleus of the ball lightning) which acts as a virtual anode. Such a nucleus appears in that "points" of the Earth surface where certain natural causes, as for example the strike of a linear lightning (or radiowaves antinodes /5/), produce a local strong heating process so that a diffusion driven electrical charge separation takes place. In the following phase of the ball lightning evolution, phenomena similar to those which determine the appearance of fireballs in laboratory, takes place. As is known, this phenomenon involves a self-organization process ended with the appearance of an electrical double layer whose stable equilibrium corresponds to a spherical shape. After its formation the life of the ball lightning is assured by electromagnetic energy absorption produced at resonance, from external natural sources as damped electromagnetic waves which accompanies every linear lightning.

REFERENCES

1. S. Singer **350** (1991) Nature 108
2. B.B. Kadomtsev **13** (1990) Comm. Plasma Phys. Contr. Fus. 277
3. X.H. Zheng **148** (1990) Phys. Lett. 463
4. Y.H. Ohtsuky and H. Ofuruton **350** (1991) 139
5. P.I. Kapitza **101** (1955) Dokl. Akad. Nauk 245
6. M. Sanduloviciu (1988) Int. Symp. on Phys. **Ionized Gases,** Sarajevo 477
7. M. Sanduloviciu (1991) XXth ICPIG Pisa 267
8. M. Sanduloviciu (1991) Int. Workshop on Plasma Phys. Pichl Austria, Book of Abstracts 36 .
9. M. Sanduloviciu (1991) XXth ICPIG Pisa 265
10. M. Sanduloviciu **29** (1987) Plasma phys. Contr. Fus. 1687
11. M. Sanduloviciu and E. Lozneanu (1992) ICPP Innsbruck 1769.

A MECHANISM OF PLASMA POTENTIAL DEFORMATION DUE TO LOCALIZED LARGE-AMPLITUDE ELECTRON PLASMA WAVES

Naoyuki Sato,[a] Rikizo Hatakeyama, and Noriyoshi Sato, Department of Electronic Engineering, Tohoku University, Sendai 980, Japan

ABSTRACT: A mechanism of plasma potential deformation due to nonlinear electron acceleration and ponderomotive force is demonstrated by formulating a plasma-sheath equation. The calculated result is consistent with experimental potential distributions in the presence of localized large-amplitude electron plasma waves.

1. INTRODUCTION: Behaviors of a plasma in high-frequency electrostatic and electromagnetic fields provide many topics in basic plasma, space plasma and fusion plasma physics. Electrostatic potential formations as a result of interactions between radio-frequency (rf) fields and plasmas are relevant to particle acceleration[1], plasma heating and plasma confinements[2,3]. It is basically important to clarify mechanisms of plasma potential deformation in connection with plasma-wave phenomena. In case of large-amplitude wave propagation, however, a few kinds of nonlinear effects simultaneously act on a plasma at least. It is usually difficult to prove what kind of nonlinearities plays a causal role in observed phenomena such as plasma potential deformation.

We have observed a clear potential deformation along the magnetized collisionless plasma flow in the presence of localized large-amplitude electron plasma waves generated externally. The mechanism of potential deformation is considered to result from the electron acceleration due to the nonlinear electron wave and/or the ponderomotive force on electrons. Here we propose a model based on the experimental results and calculate potential distributions which is strongly deformed by nonlinear waves.

2. A MECHANISM AND FORMULATION: Let us briefly describe experimental results in order to argue a mechanism of plasma potential deformation. A typical interferometer wave pattern of extremely large-amplitude electron plasma waves is presented together with plasma potential ϕ, electron temperature T_e and density n profiles in Fig. 1, where an external rf potential V_{RF}=8 V is applied to a ring exciter in a single-ended Q machine. In spite of almost no change of ϕ in the region of z<0, there appears a drastic decrease in ϕ, with a peak generation near the exciter ($z \simeq 0$) in the region of z>0, resulting in a large-scaled potential drop, $\Delta\phi=\phi(\equiv\phi_0, z<0)-$

$\phi(z>0) \simeq 20T_{eo}/e$. Here $\phi_o \simeq -0.5$ V, the potassium plasma density $n_o \simeq 5\times10^7$ cm^{-3} and the electron temperature $T_{eo} \simeq 0.2$ eV in the absence of the rf field. We can also find an increase in T_e, $\Delta T_e = T_e(z>0) - T_e(z<0) \simeq 30T_{eo}$ and a decrease in n, $\Delta n = n(z<0) - n(z>0) \simeq 0.2n_o$. The characteristic scale length of $\Delta\phi$ is several ten times the Debye length λ_{eo}, being nearly equal to the theoretical width of classical double layer.

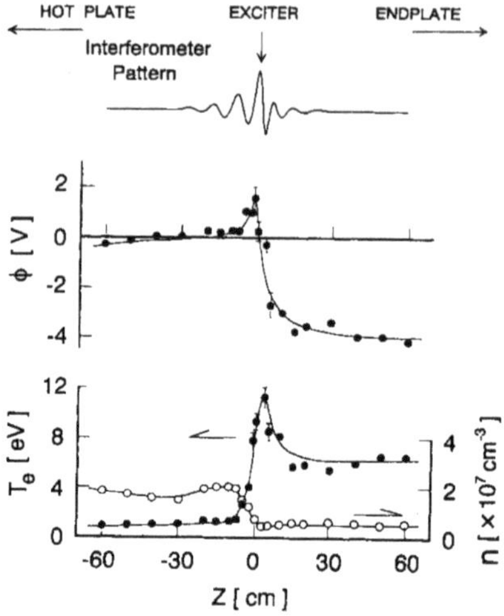

Fig.1. Spatial profiles of plasma potential, electron temperature and plasma density.

Since the waves are strongly damped spatially, the rf field is localized near the exciter, providing a ponderomotive force repelling electrons. Thus a potential increase is expected to appear in the region of localized rf field. A part of the ions are reflected by this local increase in ϕ, resulting in the n decrease in the downstream region. On the other hand, the enhanced wave damping gives rise to a number of detrapped electrons[4], generating a bump on tail in the electron energy distribution. With an increase in V_{RF}, this bump is deformed into a generation of appreciable group of electrons with high energy. The result is clearly observed in the region of z>0. The electron energy increase implies a decrease of the electron density in the region of z>0 because the electron flux is conserved. The potential drop should be generated in the region of z>0 in a way to accelerate the ions passed through the local potential increase at $z\simeq 0$, providing a necessary decrease of the ion density for the charge neutrality.

To formulate the scenario mentioned above we take account of axial electron and ion energy distribution functions f_e and f_i under the collisionless condition, which are greatly modified by the nonlinear waves. In front of the hot plate, electrons start with a half-Maxwellian axial energy distribution. On the other hand, ions start with a truncated Maxwellian distribution accelerated by the sheath potential drop ϕ_u in front of the hot plate. Low energy ions are reflected by the local potential ϕ_p at z=0 and low energy electrons by the net value of ϕ_p minus ponderomotive potential Φ_p at z=0. For simplicity, Φ_p at z=0 are estimated from the relation[5]

$$n_p/n_o=\exp\{e(\phi_p-\phi_o)/T_{eo}-e\Phi_p/T_{eo}\} \quad (1)$$

in the region of $z \lesssim 0$ under the assumption that the rf fields are localized at z=0. Here n_p is the plasma density at z=0. According to the observation of electron acceleration, electrons passed through z=0 are assumed to flow with higher electron temperature T_e into the region of z>0, keeping the flux constant throughout the column. For $z \geq 0$, we suppose a monotonously decreasing potential along the plasma flow, being followed by a sheath potential drop in front of the endplate (z=L). Most of electrons are reflected by this potential decrease while all ions are accelerated and are absorbed by the endplate, where the net electric current is zero. Then, we can determine $f_{e,i}$ as a function of ϕ at a given axial position z by taking account of particle acceleration, deceleration, and reflection due to the electrostatic and ponderomotive potential profiles supposed above. Determining charged particle's fluxes by integrating $f_{e,i}$ with respect to energy, and using flux constant and currentless conditions on the border of exciter position (z=0), we can obtain the potential ϕ_c at the endplate position as follow,

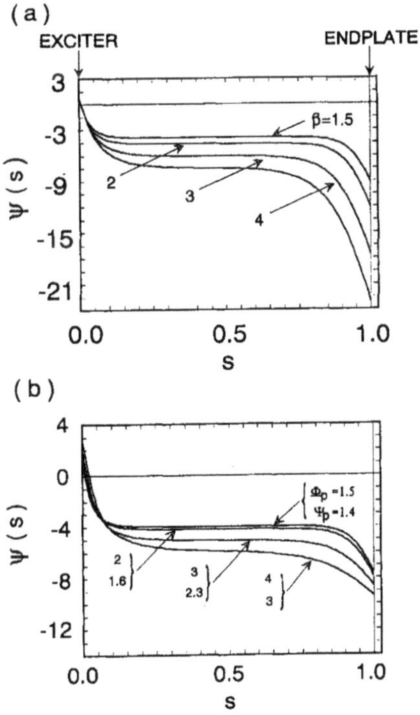

Fig.2. Effects of (a) electron temperature ratio β and (b) normalized ponderomotive potential Φ_p and local potential Ψ_p on potential distribution $\Psi(s)$ at (a) Φ_p=1.1, Ψ_p=0.9, (b) β=1.1. α=1.2, γ=150.

$$\phi_c=T_e/e\{\ln\sqrt{\rho}/\alpha-(1-T_{eo}/T_e)(\phi_u-\Phi_p)e/T_{eo}+(T_{eo}/T_e-2)e\phi_p/T_{eo}\}, \quad (2)$$

where α is the density ratio of electron to ion at the hot plate n_{se}/n_{si} (>1 under the electron rich condition) and ρ ($\equiv m_e/m_i$) is the mass ratio. This equation shows that ϕ_c decreases as T_e and/or ϕ_p increase. Substituting the local densities $n_{e,i}(\phi)$ which is obtained by integrating $f_{e,i}$ into one-dimensional Poisson's equation $d^2\phi(z)/dz^2=4\pi e[n_e(\phi)-n_i(\phi)]$ yields

$$d^2\Psi(s)/ds^2=1/2\gamma^2[\sqrt{\rho/\beta}\exp\{(\Psi-\Psi_c)/\beta-\Psi_p\}$$
$$\times\{1-2\delta\,\mathrm{erf}\sqrt{(\Psi-\Psi_p+\underline{\Phi}_p-\Psi_u)/\beta}+\mathrm{erf}\sqrt{(\Psi-\Psi_c)/\beta}\}$$
$$-\exp(-\Psi)(1-\mathrm{erf}\sqrt{\Psi_p-\Psi})], \quad (3)$$

where $\delta=1$ $(\Psi_p\geq\Psi>\Psi_p-\underline{\Phi}_p+\Psi_u)$ or 0 $(\Psi_p-\underline{\Phi}_p+\Psi_u\geq\Psi>\Psi_c)$, $\Psi\equiv e\phi/T_{eo}$, $\Psi_{p,u}\equiv e\phi_{p,u}/T_{eo}$, $\underline{\Phi}_p\equiv e\Phi_p/T_{eo}$, $\beta\equiv T_e/T_{eo}$, $s=z/L$ and $\gamma=L/\lambda_{eo}$. We numerically solve Eq.(3) as a boundary-value problem in the region between s=0 and 1 as in Ref. 6.

3. RESULTS AND DISCUSSIONS: Calculations are performed by solving a system of first order ordinary differential equation with variable-step and -order method. Figure 2 gives dependencies of $\Psi(s)$ on β, $\underline{\Phi}_p$ and Ψ_p, where the values of $\underline{\Phi}_p$ and Ψ_p are determined so as to be convenient for the comparison with experimental results as described later. In the potential profile predicted there appears a region of zero potential gradient midway between s=0 and 1 when γ exceeds several ten times $L/\lambda_e(e\phi_p/T_e)$. Then we can consider the midway region and regions with steep potential gradient to be plasma and sheath, respectively. With increasing β, Ψ decreases anywhere except for $s\simeq 0$, while the plasma region and sheath widths become narrow and large, respectively. For a fixed β the whole structure of $\Psi(s)$ does not change drastically, but the potential gradients near s=0 and 1 become small when $\underline{\Phi}_p$

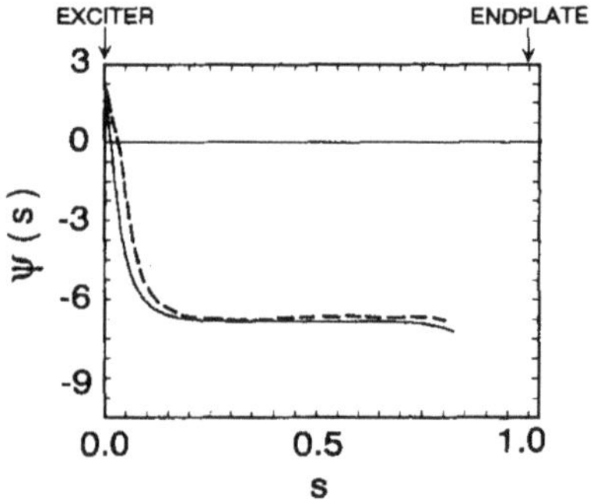

Fig.3. A solid and broken lines denote a calculated and experimental distributions, respectively, for parameters given in table 1. The former is fitted to the latter expect for the endplate sheath region.

	Calculation	Experiment
Data line	————	- - - - - -
T_{eo} [eV]	0.4	0.2~0.5
n_o [cm^{-3}]	4.3×10^6	$4\sim10\times10^6$
L [cm]	90	90
$\underline{\Phi}_p$	3.93	3~5 (wavy underline)
Ψ_p	2.3	1.95
Ψ_u	-1.13	-1.3
α	1.1	1~3
β	2.6	2~4
ρ	1.39×10^{-5}	1.39×10^{-5}

(wavy underline) Values are obtained by using Eq.(1).

Table 1. Comparison of parameters for calculated and experimental potential distributions.

and Ψ_p are increased. The direct comparison between a calculated and a typical experimental potential distribution is presented in Fig. 3 in order to fix an appropriate range among the variety of parameters. Here it is noted that we restrict a fitting section to the region between exciter and plasma. The parameters applied to this calculation and the experiment are summarized in Table 1, the values of which are comparable to each other. Dependencies of $\Delta\Psi$ [$=\Psi_u - \Psi(s=0.5)$], which is obtained from such data as shown in Fig. 2, on (a) β, (b) Φ_p and (c) Ψ_p are given in Fig. 4. The $\Delta\Psi$ linearly increases with increasing β. The increasing rate is found to be enhanced as both Φ_p and Ψ_p become large. In case of (b), however, the $\Delta\Psi$ linearly decreases with increasing Φ_p for the fixed Ψ_p. According to Figure 4(c) $\Delta\Psi$ becomes large with an increase in Ψ_p for the fixed Φ_p. Needless to say, these $\Delta\Psi$ closely relate to endplate potential Ψ_c[6]. Namely the parameter dependence of $\Delta\Psi$ in Fig. 4 is deduced from Eq.(2) for fixed Ψ_u and β, where the relation between Φ_p and Ψ_p is given in Eq.(1). Physically speaking, it is considered that Φ_p and Ψ_p increases correspond to enhancements of ion and electron rich conditions in the downstream plasma, resulting in an increase and a decrease of electrostatic potential, respectively. In experiments it is impossible to control Φ_p and Ψ_p independently though Ψ_u ($\simeq\Psi_o$) is almost unchanged even for a wide change of experimental parameters.

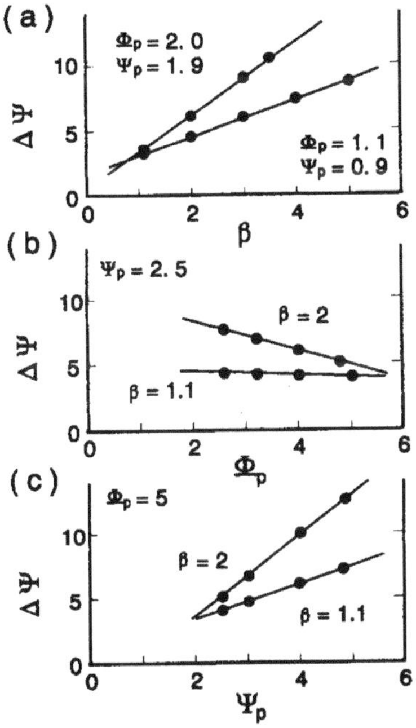

Fig.4. Dependencies of calculated plasma potential difference $\Delta\Psi$ on β, Φ_p and Ψ_p with (a) Φ_p and Ψ_p, (b) β and (c) β as parameters. $\alpha=1.2$, $\gamma=200$.

In Fig. 5(a) β is plotted versus the rf potential applied to the exciter, which is the value at z=50 cm in a temperature profile as shown in Fig. 1. Since we suppose that Ψ_p is formed by only ponderomotive force on electrons, Ψ_p is derived from experimental data when electron plasma waves can not propagate because of Landau damping and the wave heating does not occur (nonadiabatic condition). Scaling between Φ_p and Ψ_p with increasing eV_{RF}/T_{eo} is given by substituting experimental results of n_p and ϕ_p into Eq.(1). The Φ_p is almost proportional to Ψ_p and is larger than Ψ_p. Thus, the calculated $\Delta\Psi$ is compared with the experimental $\Delta\Psi$ in Fig. 5(b), where error bars in the theory are derived from deviations in the estimate of Φ_p. The result is reasonable in meanings that $\Delta\Psi$ increases with an increase in eV_{RF}/T_{eo}, and an increase in Φ_p and/or β for fixed eV_{RF}/T_{eo}, as already mentioned in Fig. 4.

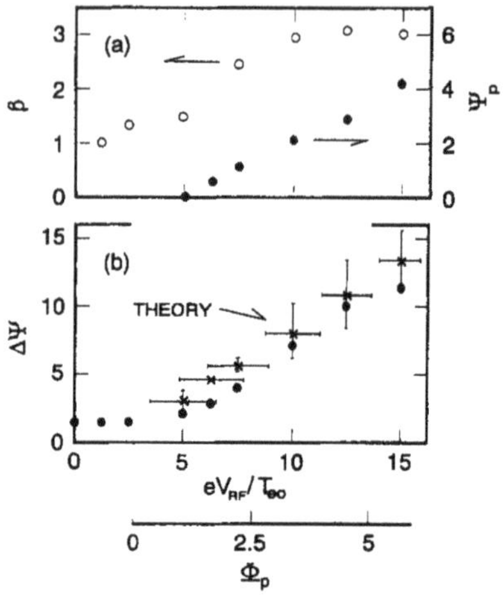

Fig.5. (a) β, Ψ_p, and (b) $\Delta\Psi$ as a function of normalized rf potential eV_{RF}/T_{eo}. Crosses in (b) denote calculated values.

4. SUMMARY: We have calculated the potential distribution modified by the nonlinear effects of localized large-amplitude electron plasma waves in a collisionless plasma flow. A plasma-sheath equation is formalized by taking account of spatial variations of charged particles' energy distribution functions, which result from nonlinear electron acceleration and/or ponderomotive force on electrons. The predicted result is consistent with our experimental results, which gives a steep potential drop just behind the wave exciter, being followed by a flat-potential region (plasma).

ACKNOWLEDGEMENT: We thank T.Watanabe for his interest in our work. One of us (Naoyuki Sato) expresses his thanks to H.Mase for his continuous support and helpful discussions.

REFERENCES: [a]Permanent Address: Department of Electrical and Electronic Engineering, Ibaraki University, Hitachi 316, Japan.

1. V.N.Aldokhin *et al.*, Sov.Phys.Tech.Phys. 30, 880(1985).
2. D.P.Grubb *et al.*, Phys.Rev.Lett. 53, 783(1984).
3. H.Fujita *et al.*, J.Phys.Soc.Jpn. 57, 504(1987).
4. T.P.Starke and J.H.Malmberg, Phys.Rev.Lett. 37, 505(1976); N.Sato *et al.*, Phys.Rev.Lett. 37, 1684(1976).
5. B.Bezzerides *et al.*, Phys.Fluids 21, 2179(1978).
6. L.A.Schwager and C.K.Birdsall, Phys.Fluids B2, 1057(1990).

INTERMITTENT STRUCTURES IN ASSOCIATION WITH RAYLEIGH TAYLOR TURBULENCE IN TOROIDAL MAGNETISED PLASMA

Y.C. Saxena, & A.K. Singh, Institute for Plasma Research, Bhat, Gandhinagar, 382 424, INDIA.

ABSTRACT: In an experiment carried out in an inhomogeneous plasma in a purely toroidal magnetic field, we have investigated the characterstics of the low frequency ($\leq$ 20 kHz) turbulence. The turbulent fluctuations have been identified as caused by R-T instability and R-T driven drift instability. The conditionally averaged potentials reveal the formation of intermittent structures in presence of the turbulence in the plasma.

INTRODUCTION: Magnetically confined, homogeneous as well as inhomogeneous, low β-plasmas are subjected to two types of low frequecy ($\omega \ll \Omega_i$), electrostatic flute ($\mathbf{k}.\mathbf{B} = 0$) instabilities viz. the velocity shear drive Kelvin-Helmholtz (K-H) instability and the curvature induced Rayleigh-Taylor (R-T) instability. In nonlinear regimes these instabilities are known to give rise to vortex like structures in plasma flow [1]. The vortex motion results from the process of self organisation from the initially random motion, accompanied by strong spatial and temporal intermittence. Flute or convective cell type of perturbations have been investigated in great details theoretically and numerically and can be studied by an analysis restricted to two dimensions in plane perpendicular to the magnetic field. Presence of large vortices in turbulence has been demonstrated in experimental investigations of the nonlinearly saturated stage of K-H instability [2]. In this paper we report a preliminary study of intermittent coherent structures in presence of plasma turbulence in a low β–plasma embeded in a purely toroidal magnetic field. The general behavior of plasma turbulence in this system has been reported elsewhere [3].

EXPERIMENTAL SYSTEM: The experiment was carried out in the toroidal device BETA [3]. The plasma column in the device was about 18.0 cms in diameter, terminated by a poloidal annular ring acting as a limiter. Plasma was produced by bombardment of neutral gas by energetic electrons, emitted from a hot cathode. The plasma production and confinement was achieved by a toroidal magnetic field, which can be varied in the range of 200 Gauss to 1 kGauss. The vacuum vessel was evacuated to a base pressure of 10^{-6} torr and filled with Argon at a pressure of 10^{-4} Torr. Plasma density was $n_e \sim 10^{11}cm^{-3}$, at the center of plasma column. Plasma temperature was found to be $T_e \sim 4$ eV. The radial profiles of plasma density and floating potentials are shown in figure 1. The density and floating potential both peak at the center of the device near the cathode locations, and fall as we go radially away from the center. Thus the plasma is characterized by strong density and potential gradients in the radial direction with a scale length ~ 6.0 cm. The curvature of the magnetic field provides an effective gravity with $g \approx C_s^2/R$. The entire plasma cross-section can be devided into three regions (marked as regions I,II and III on figure 1) from the point of view of the density and potential profiles. The effective g and the density gradient are antiparallel to each other in the radially outer loactions making this region unstable to R-T instability. The velocity shear modifies this effect [4].

RESULTS AND DISCUSSION: Detailed studies of the density and potential fluctuations for magnetic fields from 200-1000 Gauss indicate that the fluctuations are caused by the R-T and R-T induced drift instability. The fluctuations at low toroidal magnetic fields($\leq$200 gauss) exhibit coherent multimode spectrum. The spectrum becomes turbulent at higher magnetic fileds. The presence of velocity shear due to nonuniform radial electric field modifies the R–T instability in the outer region [3].

In order to investigate for the possible vortex structures the potential and density fluctuations have been measured on a spatial grid of 15×24 grid points, in a plane perpendicular to the magnetic field i.e. in the r–z plane. The measurements are made at 1 kG magnetic field where the plasma is in a fully developed turbulent state. The turbulence is characterized by its power spectrum and probability distributions. The spatial and temporal correlations also provide information about the turbulence.

The probability distribution function (PDF) for the potential and density fluctuations were found to be highly non Gaussian in the system every where except in the shadow of the limiter. The shape of these PDFś also suggest the bursty nature of certain events at amplitude close to 60 %. The correlation times are short $\sim 50 - 200\mu s$. The non Gaussian PDF together

with broadband spectrum and small correlation times, suggest the presence of intermittent coherent structures in the turbulence [5].
In order to look for these structures the cross-correlation measurements over a grid of 15×24 probe location was obtained in a plane perpendicular to the magnetic field. The contours of equal cross-correlations $< \phi(t)\phi(t+\tau)) >$ can be taken as a measure of the strctures of streamlines of the flow These contours should reveal the presence of coherent structures having lifetime within the cross correlation times. These contours reveal the formation of large closed eddies in plasma, in reg II and III. The size of the eddies is almost as large as the length over which plasma density has sharp gradients $\sim$ 5.0 cms. The region I does not have any close contours. The reference probe lies outside the close structure. However change in location of reference probe may not have any significant effect on the shape of these contours. The cross-correlations are insensitive to the sign of the quantity under considerations and it preserves only the phase information. Therefore no information about the sign of these eddies can be inferred from these studies. The conditional averaging [6] was, therefore, used in order to gain some more insight about the dynamics of the organised structures in the turbulence.
The conditionally averaged potentials were obtained, after applying different conditions on the amplitude at reference probe, over the entire plane perpendicular to the magnetic field. Time resolved contours of these potentials were plotted on this 2–dimensional plane. Since $V_D = \frac{\mathbf{E}\times\mathbf{B}}{B^2}$ and $\mathbf{E} = -\nabla\phi$, these contours represent the actual streamlines of flow of the plasma particles. The plasma can flow along these lines.
Figure 2 shows the contours of conditional averaged potentials for $\phi_1 = -1.5\phi_{RMS}$. The ϕ_{RMS} observed on the reference probe was $\sim$ 2.2 volts, which corresponds to a fluctuation level of $\sim$ 40%. The plot at time t=0 shows the presence of closed structures in the central region of plasma. These structures have negative values of potential and extend to the entire region of sharp density gradient i.e. from r=-4.0 cms to r= +4.0 cms. Besides these, structures with positive potentials were also seen in the top left corner.
At a later instance of time these positive contours move downward and the negative contours start disappearing from the central region. Few of the positive valued lines extend upto the radially outboard side and get opened

up. The plasma particles can flow along these open lines to the outer region. The point to be remembered here is that the gravity points in the radial outward direction. The situation, therefore is similar to the formation of bubbles and density lobes when a heavier liquid falls through a lighter one. In this situation similar closed eddies are formed which rises in the fluid. The structures become elongated in the vertical direction (perpendicular to the density gradient). The radial extent of these elongated structures are very small $\sim 1cm \sim 1/5th$ of the scale length of the density variation. Thus there is a strong shear in the velocity due to sharp gradients in the potential across these structures in the radial direction. These structures then break into smaller structures which move vertically down in the frame. This observation appears to be similar to what has been described by Cowley et al [7]. In their calculations they have considered the effect of temperature gradient on the vortex motion, resulting from turbulence due to ion temperature gradient. It has been shown that a large single eddy elongated in the direction perpendicular to the temperature gradient is unstable and breaks into smaller eddies via excitation of secondary instabilities with higher frequencies and smaller wavelengths. In the present experiments presence of ion temperature gradients are not known. The system is characterized by a strong density and potential gradient. The effect of these gradients on the vortex dynamics have not been studied theoretically or experimentally. Similar results have also been obtained by Kaw et al. [8]. In this case they showed that an initial elliptical vortex formed as a result of some primary instability, was unstable and generated secondary instabilities, which broke the larger eddy into several smaller eddies. The study reported here can not give any information about the secondary instabilities. The smaller structures are also observed to coalesce with the bigger one with same sign and form a bigger structure. This process takes typically 75–100μs. This type of coalescence is observed for both the positive as well as negative structures.

The plots at further delays show, that the entire central region gets occupied by the positive valued structures and the negative structures are pushed away from this region. The negative structures, however, start reappearing in the top left corner as time progresses. This appearance and disappearance of these bigger structures from the center takes typically 100μs. The $\mathbf{E} \times \mathbf{B}$ time scales for the magnetic field of 1 kGauss and electric field 1.0 Volt/cm (the observed dc electric field), would be in this range for distances of the order of circumference of the minor cross section. The structures also reveal that initial closed structures open up and get extended upto the radial outer locations, opening a channel for the flow of plasma particles in the outward direction. This suggests that the transport must be getting enhanced due to the turbulence exhibited by the instabilities present in the system. This open channel however is found to exist only for time of the order of the correlation time.

CONCLUSIONS: Study of two dimensional low frequency turbulence in a plasma paced in a purely toroidal magnetic field reveals presence of intermittent potential structures with life times of the order of correlation times and sizes of the order of density gradient scale sizes. Negative and positive potential eddies appear alternately on time scale of the order of correlation time. Breaking of structures elongated in direction perpendicular to the density gradient and coalescence of smaller structures into bigger ones is also observed.

REFERENCES:

1. P.K. Shukla, M.Y. Yu, H.U. Rahman & K.H. Spastchek (1984) Phys. Rep. **104**,227 and references therein.
2. T. Huld, A.H. Nielsen, H.L Pécseli, J.J. Rasmussen (1991) Phys. Fluids, **B3**, 1609, and references therein.
3. G. Prasad, D. Bora and Y.C. Saxena (1992) Geophys. Res. Letts. **19**, 241 & 245.
4. P.N. Guzdar, P. Satyanarayan, J.D. Hubba and S.L Ossakow (1982) Geophys. Res. Letts. **9**, 547.
5. R.H. Kraichnan (1990) Phys. rev. Lett. **68(5)**, 575.; A. Pumikr et al. (1991) Phys. Rev. Lett **66(23)**, 284.; R. Benzi et al. (1991) Phys. Rev. Lett **67(17)**, 2299.
6. R. Blackwelder (1977) Phys. Fluids **20**, 232.; R.J. Adrian (1979) Phys. Fluids **22**, 2065.; H.L. Pécseli and J. Trulsen (1989) Phys. Fluids **B1**, 1616.
7. S.C Cowley, R.H. Kulsrud and R. Sudan (1991) Phys. Fluids **B3**, 2767.
8. P.K. Kaw, R. Singh, A. Das, and N. Chakraborty (1991) Proc. TCM on Research Using Small Tokamaks, (Washington, IAEA) 308.

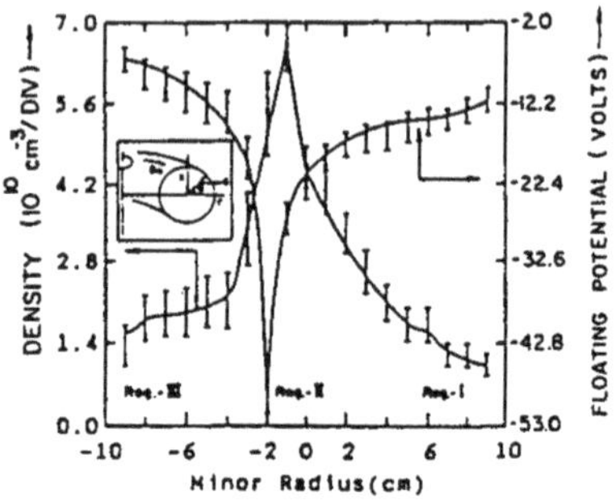

Figure 1: The radial profiles of plasma density and potential

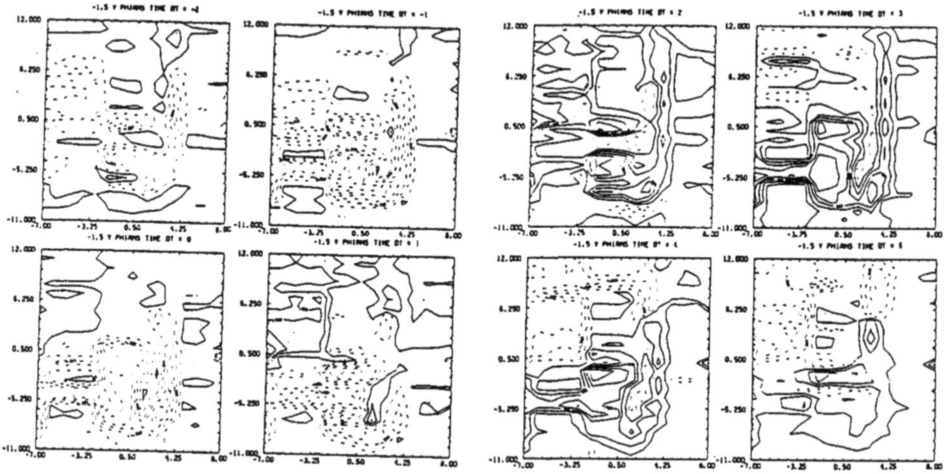

Figure 2: Contours of conditionally averaged potentials with a condition $\phi = -1.5 volts \phi_{rms}$. See text for details.

ON THE INTERACTION OF IONIZATION WAVE AND DOUBLE LAYER IN A POSITIVE PLASMA COLUMN

I.SIRGHI, M.VASILACHE and G.POPA
Plasma Physics Department, "Al.I.Cuza" University
RO-6600 Iasi, Romania

Introduction

A large number of paper are dealing with the properties of ionization waves that are either self or externally excited in positive plasma column (PC) /1-4/. However, most of them are neglecting the effects of axial ends of the PC.

Thus, it is known that ionization wave groups could be frequently excited from a place situated at the cathode end of the PC relatively independent of the position of the exciter /5/. Also, when the anode end of the PC is reached by the front of the ionization waves a secondary phenomenon takes place consisting of almost synchronous oscillations of the potential of whole plasma column. This was called echo phenomenon which provides the background of the feedback mechanism for self-excited ionization instabilities /6/. Recently, influence of the anode end of the PC on the propagation of the ionization wave groups was studied /7/.

Since very frequently at the plasma boundaries space charge structures appear consisting of two adjacent space charge of opposite sings known as a double layer (DL) it is possible that some of the aspects discussed above to be related with these structures. Thus, the place from which ionization wave groups are developing could be the DL that exists at the cathode end of the PC. Also, the echo phenomenon could be produced by DL that might exist in certain condition at the anode end of the PC.

These assumptions are sustained by our experimental results on the interaction between DL and ionization waves, which would be described in this paper.For this purpose an "artificial" stationary DL was produced using a transverse magnetic field acting locally on a stationary PC and its interaction with ionization wave was studied.

Experimental setup and results

Experimental setup is shown in fig.1. A glow discharge is

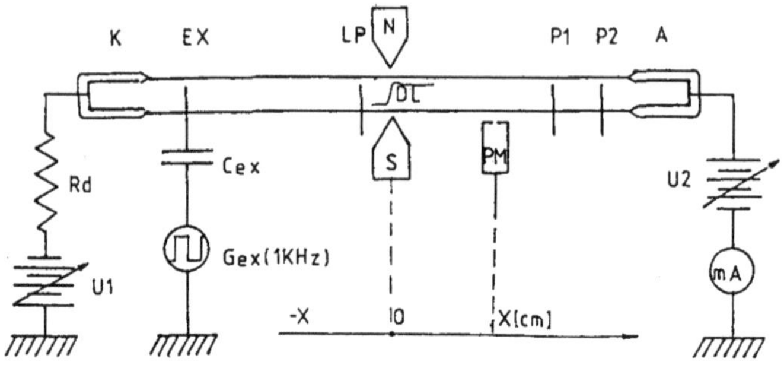

Fig.1. Experimental set-up, U_1, U_2 power suplies, K-cathode, A-anode, PM-photomultiplier, LP-Langmuir probes, P_1, P_2-cylindrical probes.

mentained in a discharge tube (inner diameter 2.7 cm, lenght 90 cm). The PC is kept free from self-excited ionization waves by adjusting the pressure of helium gas and the d.c. current. An ionization wave group is excited at each potential jump applied to a cylindrical Langmuir probe (EX, installed 5 cm away from cathode) from a rectangular waveform generator. The fluctuations of optical radiation associated with the ionization waves are followed by a photomultiplier (PH) which can also move along the tube axis.

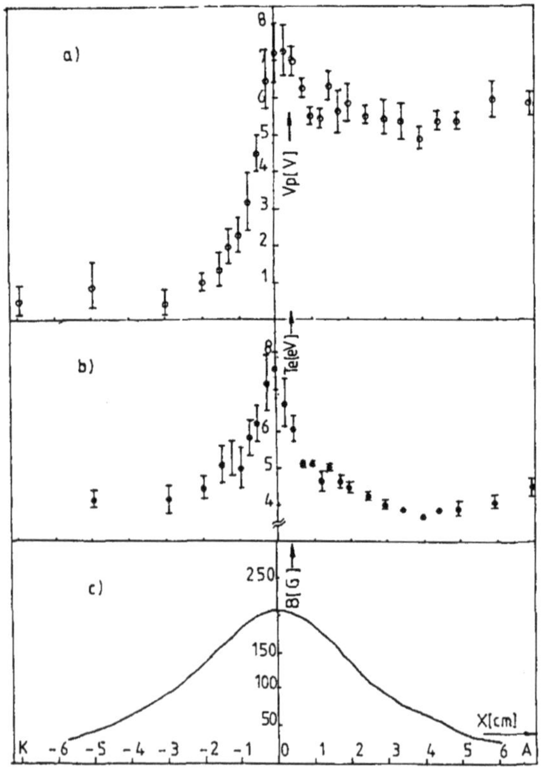

Fig.2.a,b,c

A transverse magnetic field with an adjustable strength having a distribution like that shown in fig.2c is acting on a local region which could also be moved along the PC.A cylindrical Langmuir probe (LP) provides information about both the plasma potential and the electron temperature at different places in the region where magnetic field is acting. The structure of the plasma potential which appears in the region of the magnetic field is shown in fig. 2a. It corresponds to a DL structure. The distribution of the electron temperature in this region is also shown in fig.2b.

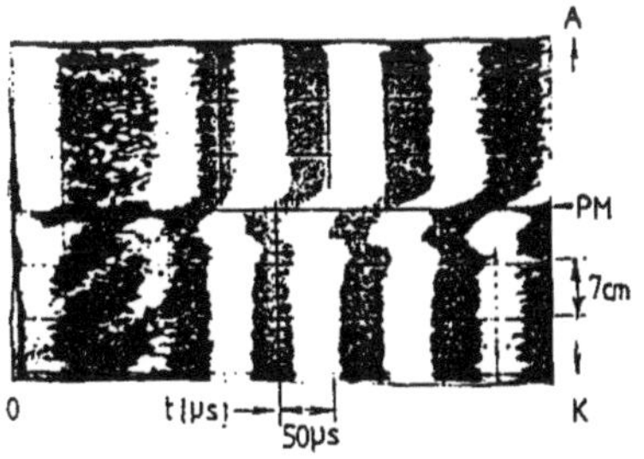

Fig.3.Space-time display for optical signal received by PM fixed at 50 cm away from cathode and the DL is moving

Using Stirand's technique /8/ the phase shift produced when the ionization wave is passing through the DL was analyzed (fig. 3). In this case the PM is fixed and the DL is moving by moving of the magnetic field. In general the ionization wave that has passed through the DL undergone a phase shift which increases with increasing of the potential drop on DL.

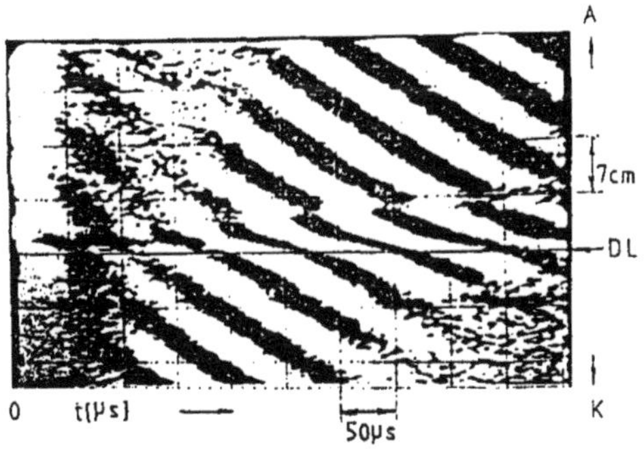

Fig.4. Space-time display for optical signal received by PM which s moving along the PC. The DL as fixed 50 cm away from cathode

Evolution of an ionization wave group in presence of the steady DL can be followed both on space-time diagram shown in fig.4 and on box-car diagrams presented in fig.5. In both diagrams fluctuations of the optical radiation were processed. If in the former case only the above phase shift of the ionization wave comming from cathode side of the PC and passing through the DL, is shown a clear wave group streaming from the anode side of the DL can be observed in the latter.

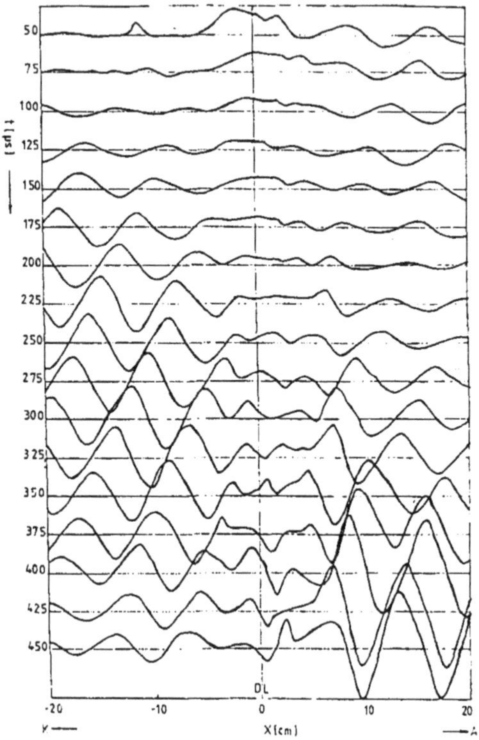

Fig.5. The box-car diagrams for optical signals received by PM for different instantf time. The DL is fixed at 50 cm away from cathode.(p = 0.76 torr, I = 75 mA)

This wave is excited by externally signal which disturbs the DL (first two traces of fig.5), and propagates towards the anode. Later on when the ionization wave group, comming from the cathode region, is passing through the DL, the both enhancement and phase shift can be measured. There is a distance (about 2 cm) along DL where light fluctuations are minimum. In this region the ionization group comming from cathode end of the PC is attenuated very rapidly on the cathode side of the DL and it appears again on the anode side of the DL (Fig.6).

More informations are presented in fig.7 where a space-time diagram is shown using the electrical signal piked up by a fixed LP when the DL is moving. Thus when the DL is between the K and LP, the LP is receiving sequently the signals corresponding to the first wave group, the echo produced by the first wave group and the second wave group excited by the echo /10/. When the DL is between LP and the A an interference can be observed between the signals corresponding to the first wave group and an other echo which is steaming from the cathode side of the DL. The phase

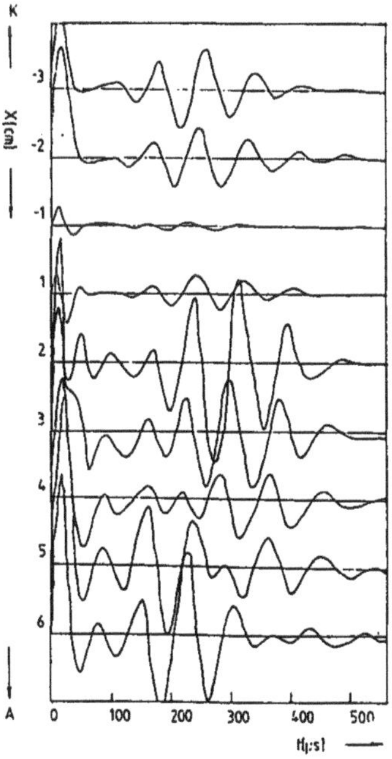

Fig.6. The oscillograme of optical signal received by PM at different places on the DL (p = 0,76 torr, I = 75 mA)

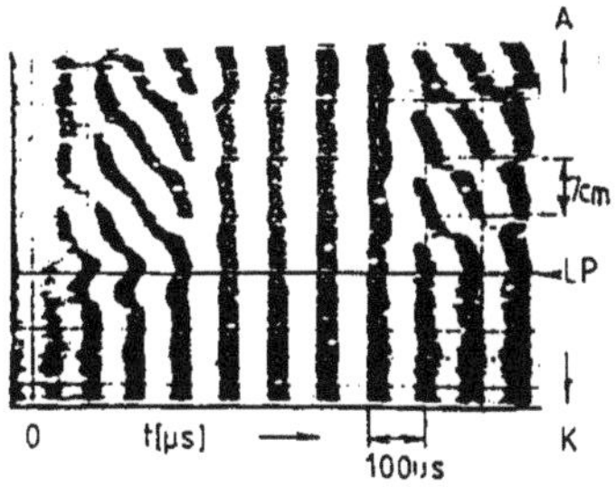

Fig.7. Space-time display for electrical signal piked up by Langmuir probe fixed at 47 cm away from cathode when DL is moving.(p= 0,76 torr,I= 75 mA)

of this echo signal corresponds to the wave group phase when this reaches the cathode side of the DL. Also, we can observe that the echo signal produced by the wave group that reaches the anode end of the CP is not affected by the DL.

Conclusions

From the experiments presented above result that the DL can play a key role in given an explanation of different characteristics of ionization waves self or externally excited in the PC.

Thus, the stationary DL is the place from where a ionization wave group can develops when it is perturbed. Sametimes, the DL acts as an amplifier for ionization wave that are passing through it. In a first phase, the DL absorbs the energy from incoming wave group that reaches the cathode side of the DL and an echo is produced. In a second phase the DL releases a greater amount of energy to a second wave group which develops on the anode side of the DL.U

References

/1/. N. L. Oleson and A. W. Cooper, Adv. Electron. Phys., 24, 255 (1968).

/2/. L. Pekarek and V. Krejci, Czech. J. Phys. B12, 296 (1962).

/3/. S. W. Rayment, J. Phys. D Appl. Phys. Vol. 7 (1974).

/4/. I. Grabec and I. Poberaj, Z.Naturforsh, 25a, 287 (1970):

/5/. G. Popa, N. Dumitrascu and M. Sand..loviciu, Buletinul inst. Politehnic Iasi, (1988)

/6/. G. Popa and N. Dumitrascu, An. St. Univ. Iasi, 34, 1988)

/7/. K. Ohe and T. Kimura, J. Appl. Phys. 70 (12), 7282, (1991).

/8/. O. Stirand, V. Krejci, L. Laska, Rev. Sci. Instrum.,37 1841, (1966).

/9/ G.Popa, K.Ohe and N.Dumitrascu, J.Phys.D: Appl.Phys. (1989)

CURRENT-FREE DOUBLE LAYER IN AN EXPANDING PLASMA WITH TWO ELECTRON POPULATIONS

R.L. Stenzel and G. Hairapetian, Department of Physics, University of California, Los Angeles, CA 90024-1547, USA.

1. INTRODUCTION: The free expansion of a plasma into vacuum has received much attention in laser-plasma physics, plasma releases in space and the filling of wakes. One interesting aspect is the acceleration of ions to high energies due to a moving ambipolar electric field at the expansion front. In the present work[1] a new ion acceleration process has been observed involving the formation of a stationary potential double layer in a plasma with two electron populations, a cold Maxwellian and a shell of energetic electrons. The double layer potential is determined by the effective temperature of the energetic electrons. At the double layer cold electrons are reflected, energetic electrons are decelerated, and ions are accelerated. The double layer is located where the densities of cold and energetic electrons, each obeying the Boltzmann relation $n_{(z)} = n_0 \exp(e\phi_{(z)}/kT)$, become equal. The double layer is current-free since the expansion into vacuum is ambipolar.

2. FURTHER INFORMATION: As shown in Fig.1 the experiments are performed in a large vacuum chamber (1m diam. × 2m length, base pressure 3×10^{-7} Torr) in which neutral gas (Ar)

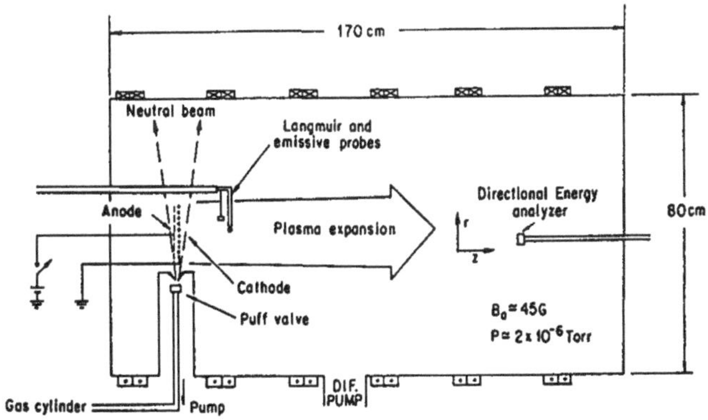

Fig.1 Schematic of the experimental setup.

is pulsed radially into a gridded anode-cathode gap. The pulsed discharge plasma ($n_e \approx 10^{10}$ cm^{-3}, $kT_e \approx 4$eV $\approx 10kT_i$) expands axially along a uniform dc magnetic field ($B_0 \approx 45$G) into high vacuum. The plasma potential is diagnosed with emissive probes, the ion energy and distributions are obtained from a directional energy analyzer[2], and the electron density is found from Langmuir probes and a microwave resonator probe[3].

The plasma expansion is mainly driven by a small percentage ($\lesssim 5\%$) of primary electrons whose energy corresponds to the discharge voltage ($V_{dis} \approx 80$V). As shown in Fig.2 a potential drop of $\Delta\phi \approx V_{dis}$ exists at the expanding plasma boundary. It decelerates the electrons and accelerates the ions so as to result in an ambipolar plasma expansion. As the expansion proceeds a distinct potential jump evolves and settles down to a fixed position. Thus, a stationary double layer has developed.

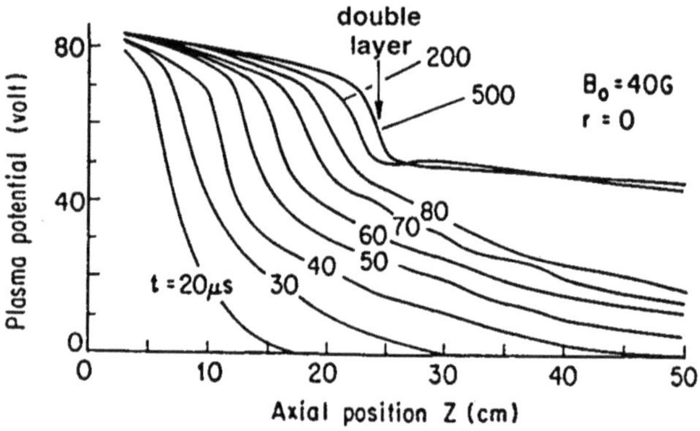

Fig.2 Axial potential profiles at different times after start of plasma expansion.

The characteristics of the double layer are demonstrated in Fig.3. The potential drop $\Delta\phi \approx 25$V corresponds approximately to the axial electron energy $\frac{1}{2}m < v_z^2 > \approx 27$eV of the isotropic distribution of mono-energetic primary electrons [$\frac{1}{2}m(v_x^2 + v_y^2 + v_z^2) \approx 80$eV]. This potential drop is large enough to reflect most of the low energy electrons. Ions gain energy while primary electrons loose kinetic energy which they had acquired from the energy source driving the discharge. The momentum gained by the ions balances that lost by the electrons so as to

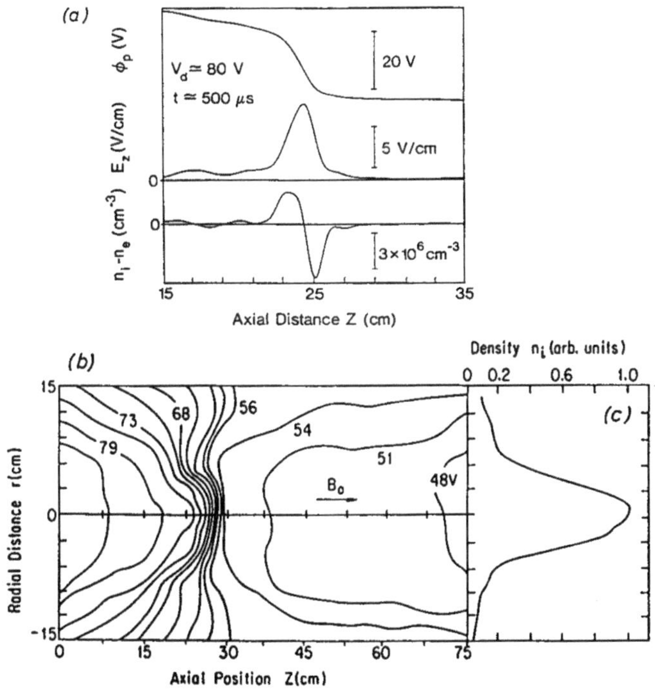

Fig.3 Characteristics of the double layer.
(a) Potential, electric field and space charge density on axis.
(b) Two dimensional equipotential contours $\phi(r,z)$ = const.
(c) Radial density profile at z = 60cm.

establish a stationary double layer. The space charge imbalance is small, $(n_i - n_e)/n_e \sim 10^{-3}$. The two-dimensional potential profile indicates a double layer with $\underline{E} \parallel \underline{B}$ in the central plasma column of approximately 10cm diameter at 0.5 n_{max}. The double layer appears to focus the initially diverging ion flow. The magnetic field is too weak to directly affect the ion motion.

The double layer potential and location depend on the density ratio of energetic to Maxwellian electrons, n_{tail}/n_{Maxw}. Figure 4 shows the plasma potential and density vs. axial position at different ratios n_{tail}/n_{Maxw} in the source which is varied by changing the neutral

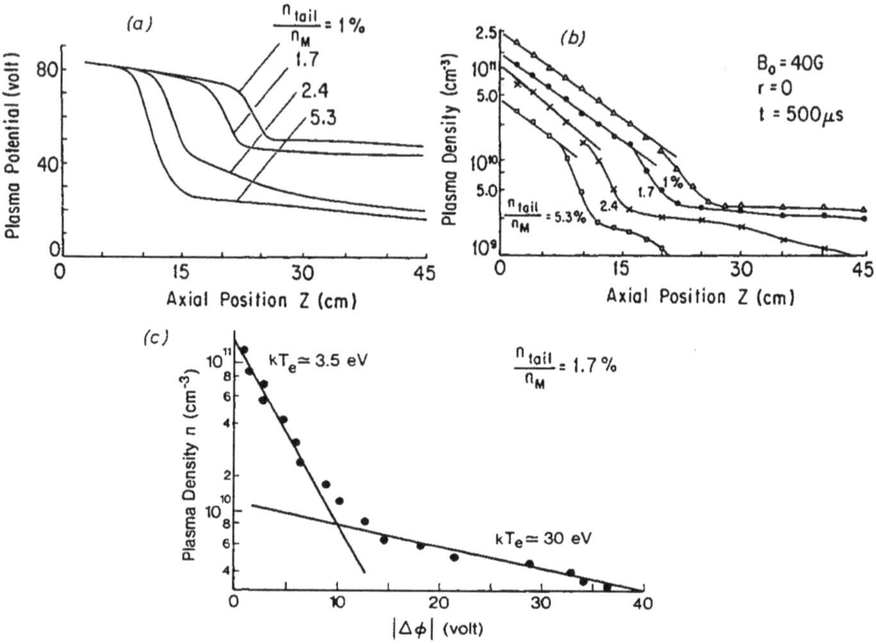

Fig.4 (a) Shift in the double layer position due to changing the density ratio of tail to Maxwellian electrons.
(b) Logarithmic plot of density vs. position at different n_{tail}/n_{Maxw}.
(c) Density vs. potential, ℓn n vs. Δϕ, obtained from (a) and (b) upstream of the double layer, showing the two electron populations following the Boltzmann relation. The double layer starts to form where $n_{cold} \approx n_{hot}$.

pressure ($\propto n_{Maxw}$) at a fixed discharge current ($\propto n_{tail}$). The double layer weakens and shifts away from the source as n_{tail}/n_{maxw} is decreased. Plotting ℓn n vs. Δϕ for n_{tail}/n_{Maxw} = 1.7%, Fig. 4(c) shows that the Boltzmann relation is satisfied for the two electron populations, and that the double layer onset occurs where the rapidly dropping cold electron density becomes comparable to the smaller but slowly decreasing tail electron density (Δϕ ≈ 10V, z ≈ 18cm). These observations are consistent with earlier theoretical predictions[4].

The double layer remains unaffected when the expanding plasma stream reaches the end of the chamber wall (Δt ≤ 200 μs). In order to avoid the build-up of neutral gas the plasma source is pulsed (t ≤ 1ms). However, a steady-state source in free space can be expected to produce a steady-state double layer, hence a continuous beam of ions. This process is different from the ion acceleration in the expansion front which produces a transient ion beam.

The expanding cold ions drift relative to the trapped Maxwellian electrons which can result in sound instabilities. Figure 5 summarizes some of the properties of the observed

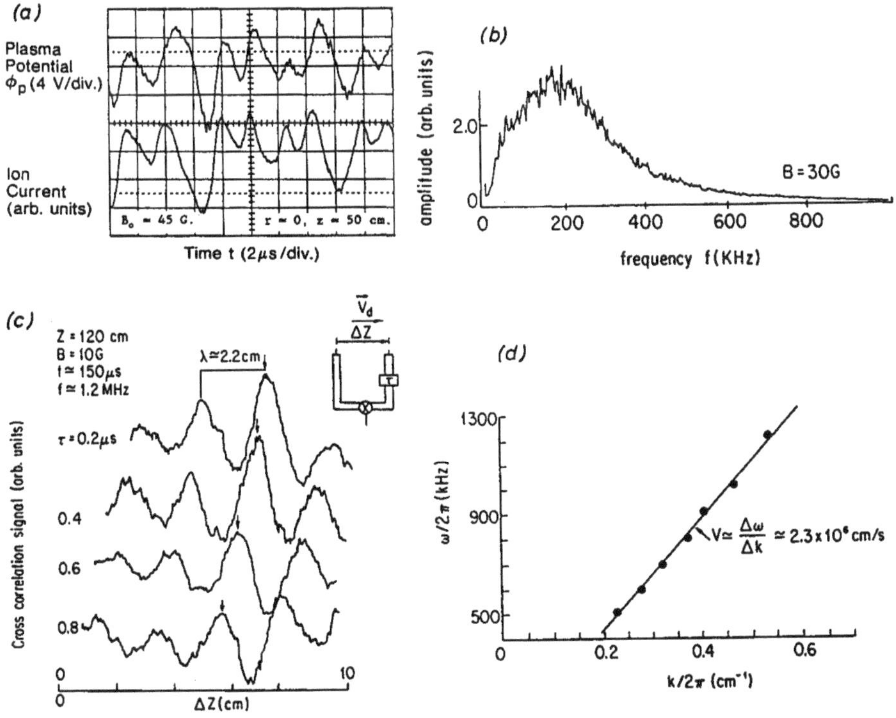

Fig.5 Characteristics of low frequency fluctuations in an expanding plasma.

(a) Strong correlation between density and potential fluctuations vs. time indicates a Boltzmann relations.

(b) Typical frequency spectrum exhibits maximum near the lower hybrid frequency $(f_{ce}\, f_{ci})^{1/2}$.

(c) Narrowband cross-correlation vs. position at different delay times τ indicates wave propagation along the drifting ions.

(d) Dispersion ω vs. k shows a constant phase velocity which exceeds the ion streaming velocity.

fluctuations. Low frequency fluctuations in density and potential are correlated, as expected from Boltzmann's law. They grow along the beam ($\delta n/n$ = 5 ... 50% for z = 0 ... 80cm), exhibit a spectral peak near the lower hybrid frequency, $(f_{ce}\, f_{ci})^{1/2} \propto B_0$, and propagate in the direction of, but faster than the drifting ions ($v_i \simeq 1.6 \times 10^6$ cm/s, $\omega/k \simeq 2.7 \times 10^6$ cm/s). The fluctuations exhibit little dispersion ($\omega = kv$, where $v \simeq$ const.), they maximize on axis ($\nabla p = 0$), and can have short correlation lengths ($\ell_{corr} \approx \lambda_\parallel$). Aside from the basic questions about the instability mechanism the fluctuations grow to such large amplitudes that they break up the initially laminar plasma flow into a bursty flow of plasmoids. For example, in Fig.5a the axial fluctuating electric field, $E_{zmax} \simeq (\partial\phi/\partial t)/v_z \simeq (12V/\mu s)/(2.7\ cm/\mu s) \simeq 4V/cm$ is of the same order as the double layer electric field ($E_z \simeq 8V/cm$, Fig.3a). Strong, nonlinear interactions between co-moving fields and ions can be expected. This regime of strong turbulence requires further investigation.

In summary, a current-free double layer has been observed in an expanding plasma with two electron populations. The double layer separates cold and hot electrons and provides for an energy transfer from electrons to ions. A continuous ion beam is generated.

This work was supported in part by NSF grants ATM 90-12709, PHY91-02132, and NASA grant NAGW-1570.

REFERENCES:

1. G. Hairapetian and R.L. Stenzel, *Phys. Rev. Lett.* **61** (1988), 1607; *Phys. Rev. Lett.* **65** (1990), 175; *Phys. Fluids* **B3** (1991), 899.
2. R.L. Stenzel, W. Gekelman, N. Wild, J.M. Urrutia and D. Whelan, *Rev. Sci. Instrum.* **54** (1983), 1302.
3. R.L. Stenzel, *Rev. Sci. Instrum.* **47** (1976), 603.
4. B. Bezzerides, D.W. Forslund and E.L. Lindman, *Phys. Fluids* **21** (1978), 2179.

OBSERVATIONS OF BUNEMAN MODES AS LEADING PHENOMENA OF A STRONG DOUBLE LAYER

Y. Takeda, K. Yamagiwa[a] and H. Inuzuka[b]
Department of Physics, College of Science and Technology
Nihon University, Kanda Surugadai 1-8, Tokyo 101, Japan
[a] Faculty of Science, Shizuoka University
Oya 836, Shizuoka 422, Japan
[b] Faculty of Engineering, Shizuoka University
Johoku 3-5-1, Hamamatsu 432, Japan

A rapidly growing wave burst around the real frequency of the Buneman instability, $f_B = 2^{-4/3} (m_e/M)^{1/3} f_{pe} \simeq 350$ MHz, was detected at the onset of a high-current plasma discharge. The measurement is performed by means of a capacitively coupled probe. It is coincident accurately with abrupt rise of a strong E-field pulse which has the negative polarity and amplitude, typically 10 kV/cm, where m_e and M denote the electron- and ion mass and f_{pe} is the electron plasma frequency.

1. Introduction

The Buneman instability is of great interests, since in its nonlinear stage the formation of virtual cathodes and electric double layers is suggested by analytic theories and computer simulations.[1,2] Further it has recently attracted renewed interests in connection with the acceleration of particles during the impulsive phase of solar flares, and in other energetic phenomena of cosmic objects.[3] We have observed for the first time the spatial evolution of a potential hump generated near the cathode into a strong Buneman double layer (or an asymmetric electron hole) with the reverse polarity[4] in the initial phase of a high-current straight plasma discharge. The behavior of the electric fields is in accordance with the context of a nonlinear stage of the Buneman instability.[5] But decisive experimental verification of the above scenario in its nonlinear stage has not been done yet.

Hence we have tried to detect ultra-high-frequency fluctuations given by the Buneman instability, as a precursory phenomenon of nonlinear potential structures such as a solitary pulse and double layer that are generated by a plasma current along the magnetic field, by means of a capacitively-coupled probe.[6]

The experimental arrangements, procedures, and main diagnostic tools of the present experiment are described in Sec. 2. The highlights of the present observation of UHF fluctuations given by the nonlinear Buneman mode are shown in Sec. 3. Discussions on the observations and the conclusion are given in Sec. 4.

2. Experimental arrangements and the method of measurement of UHF fluctuations given by Buneman modes

The experimental apparatus and the axial arrangement of diagnostic tools are shown in Fig. 1.

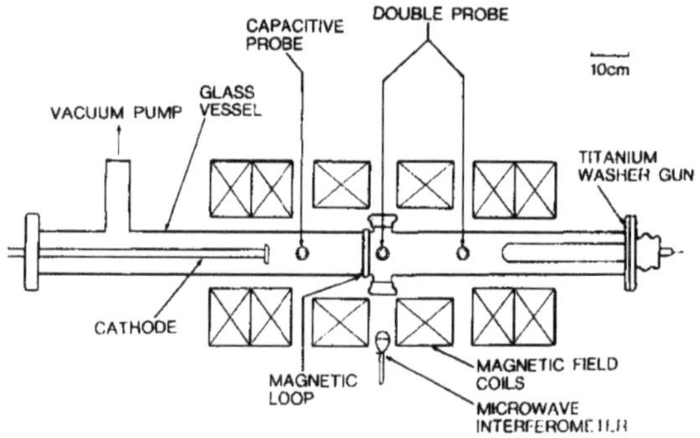

Fig. 1 Experimental apparatus and the axial arrangement of diagnostic tools

We have drawn a high discharge current along the magnetic field with a preexisting deuterium plasma produced by a titanium washer gun. The configuration of the magnetic field is a magnetic mirror with mirror ratio 1.2 and the field intensity at the mirror point is typically 1.5 kG. The discharge is ignited by applying a high voltage V=13-18 kV from a capacitor with C=2.2 μF between the cathode (aluminum disk 50 mm in diameter) and the cylindrical muzzle of the plasma gun (43 mm in diameter and 34 cm long) after a suitable delay time, typically 20 μs from firing the gun. Additional details of the experimental apparatus and the plasma diagnostics that include the parallel electric field measurement via an optically-isolated transmission system are described in the previous paper.[5]

The electron density is a key plasma parameter which determines the drift velocity of electrons for excitation of the current driven instability and sensitively affects its behavior in its nonlinear stage. Thus the electron density was monitored by means of a microwave interferometer, which is arranged at the center of the apparatus and operated at the frequency 69 GH_z.

The capacitive probe is fabricated from an open-ended semirigid 50 Ω coaxial cable that is electrically insulated by a thin quartz tube, 4 mm in o.d. (Fig. 2). It is inserted into a discharge vessel (pyrex glass) at the measuring port arranged 10 cm in front of the cathode. Fluctuations (>400 MHz) picked up by the capacitive probe are converted to the lower intermediate frequency range (0-200 MHz) by heterodyne detection using a fixed frequency local oscillator (f_l = 500 MH_z) and a balanced mixer. The IF signal of the mixer is then amplified by a wide-band amplifier and finally recorded on a high-speed digitizing oscilloscope (HP54510A). The schematic drawing of the capacitive probe and the block diagram of the receiving system is shown in Fig. 2.

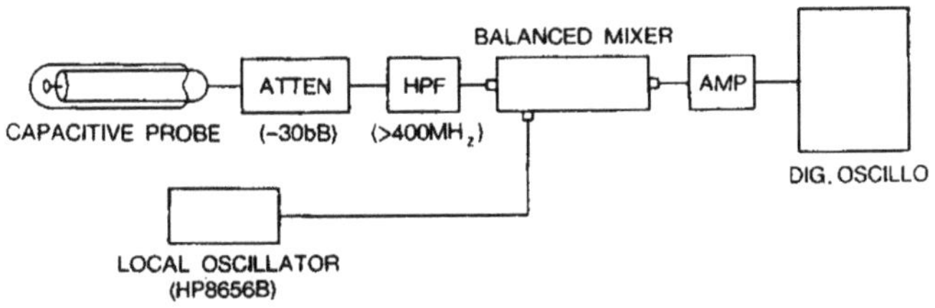

Fig. 2 The block diagram of a receiving system of UHF fluctuations picked up by the capacitive probe.

The accurate measurements of the delay time due to signal transmission and processing in the measuring systems of UHF fluctuations and the parallel E-field are exceedingly important in identifying the former as the Buneman mode, which leads to the formation of an electron hole and a Buneman (or electron-type[4]) double layer. The overall delay time in the measuring system of the capacitive probe is carefully measured by using another UHF signal generator (HP8657B) with pulse modulation and finally determined to be 130 nsec. Therefore a digitized IF signal of UHF fluctuations picked up by the capacitive probe is traced taking into account the (differential) delay time of the transmission, 40 nsec with respect to that of the simultaneously obtained parallel E-field.

3. Simultaneous observations of the electric field and UHF fluctuations at the onset of a straight plasma discharge

Fig. 3 shows a typical time profile of the discharge current together with temporal variation of the electron density at the midplane, and data sets of the parallel electric field and of the IF converted UHF fluctuations picked up by the capacitive probe, which were simultaneously obtained in the same discharge shot. The mean electron density is determined to be 2.4 x 10^{12} /cm^3 at the onset of the discharge (Fig. 3 (b)). Thus the real frequency of the Buneman instability, $f_B = 2^{-4/3} (m_e/M)^{1/3} f_{pe}$, is 368 MH_z for the preexisting deuterium plasma, where m_e and M denote the electron- and ion mass and f_{pe} is the electron plasma frequency. The highlights of the present observation are that the IF signal(Fig.4 (b)) of the capacitive probe shows a rapidly growing wave burst at t = 21.82 μs with the frequency (500-160) MH_z (see Fig. 5) corresponding to the above real frequency of the Buneman instability and the negative E-field pulse with amplitude nearly 10 kV/cm breaks out coincident accurately with rapid growth of the UHF wave burst.

The comparison of the early temporal behavior of the discharge current with data of the parallel E-field and of IF converted UHF fluctuations shows that these phenomena occur at the onset of the discharge and accordingly the total plasma current, I is as small as 0.3 kA. The question whether the measured current density, which is estimated from the total current at the onset of the discharge, can trigger the Buneman mode or not is discussed in section 4.

Fig. 5 shows a temporal variation of the peak frequency of the IF converted UHF wave bursts calculated in terms of the Wigner distribution[8] for non-stationary data. Each point of Fig. 5 was obtained from the frequency- and time analysis of 32 data points. The peak frequency of the instantaneous IF spectrum extends to 150-200 MH_z initially, and then slightly shifted to the lower frequency after t > 21.85 μs.

4. Discussions and conclusion

In the first place we should discuss the question whether the very low measured current at the onset of the discharge can trigger the Buneman mode which gives rise to the observed UHF wave bursts coincident accurately with abrupt rise of the electric field pulse with the negative polarity, or not. Hence the current density

$j = I/ 2\pi r_p \delta r$ is estimated to be nearly 66 A/cm^2 on the assumption that the plasma current is carried by a thin skin current with depth δr comparable to the collisionless skin depth c/ω_{pe}=0.3 cm. Mantei et al.[7] have shown that the critical current density to excite the Buneman instability is lowered by a factor of 2 taking into account the finite electron to ion temperature ratio of an initial plasma e. g., $j_{crit} = 0.5\ n_e e v_{et}$ for $T_e/T_i = 2$, where v_{et} is the electron thermal velocity. Thus the critical current density for excitation of the Buneman instability is estimated to be j_{crit} = 60 A/cm^2 and it is slightly less than the above skin current density, 66 A/cm^2, where the electron temperature of the initial plasma is taken to be 40 eV. This estimation is realistic and is thought to be not so far from the current density attained in the initial phase of the discharge.

We present another evidence which proves that the observed UHF wave bursts are given by the Buneman modes. For a discharge shot with a lower electron density, n_e= 0.7 x 10^{12} /cm^3 at the onset of the straight plasma discharge, the UHF burst as shown in Fig. 3 was not observed due to cut off by the high-pass filter (Fig. 2), while the negative E-field pulse with amplitude,10 kV/cm was clearly detected.

In conclusion, we have observed the UHF wave bursts given by the Buneman mode as leading phenomena of a strong double layer with the reverse polarity. The detection of these UHF wave bursts or spikes supports the excitation of Buneman modes due to the high skin current, with thickness comparable to the collisionless skin depth c/ω_{pe}, at the onset of a high voltage straight plasma discharge.

REFERENCES:

1. N. G. Belova, A. A. Galeev, R. Z. Sagdeev, and Yu. S. Sigov, Pis'ma Zh. Eksp. Teor. Fiz. **31**(1980), 551 [Sov. Phys. JETP Lett. **31**(1980), 518].
2. N. Singh, and R. W. Schunk, J. Geophys. Res. **88**(1983), 10081.
3. V. S. Berezinskii, S. V. Bulanov, V. A. Dogiel, V. L. Ginzburg(editor) and V. S. Ptuskin, *Astrophysics of Cosmic Rays*, (North-Holland 1990) p. 482.
4. V. I. Maslov, in *Nonlinear World* (Proc. IV International Workshop on Nonlinear and Turbulent Processes in Physics, Kiev, USSR, Oct. 1989)Vol. 2, p. 898.
5. Y. Takeda and K. Yamagiwa, Phys. Fluids B **3**(1991), 288.
6. J. A. Schmidt, Rev. Sci. Instrum. **39**(1968), 1297.
7. T. D. Mantei, F. Doveil, and D. Gresillon, Plasma Phys**18**(1976), 705.
8. J. S. Bendat and A. G. Piersol, *Random Data, Analysis and Measurement Procedures*, 2nd ed., (John-Wiley&Son1986) p. 457.

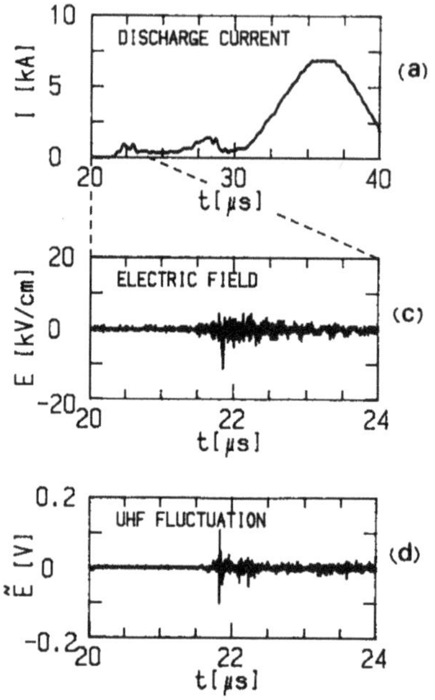

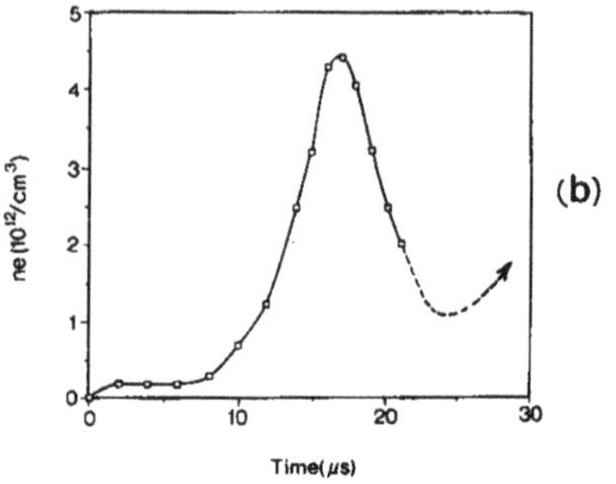

Fig. 3 Typical data sets of a high-current straight plasma discharge. (a) the discharge current, (b) the electron density, (n_e=2.4 x 10^{12}/cm^3 at t= 21.0 μsec), (c) the electric field (parallel to B) measured with a double probe at the center, and (d) the IF signal of UHF fluctuations picked up by the capacitive probe in front of the cathode.

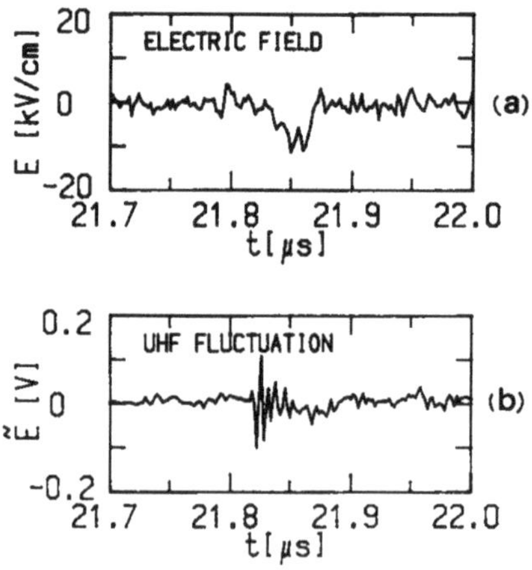

Fig. 4 Expanded time profiles of Fig. 3 (c) and (d) around the UHF wave burst. (a) and (b) are the electric field and the IF signal during 0.3 μsec, respectively.

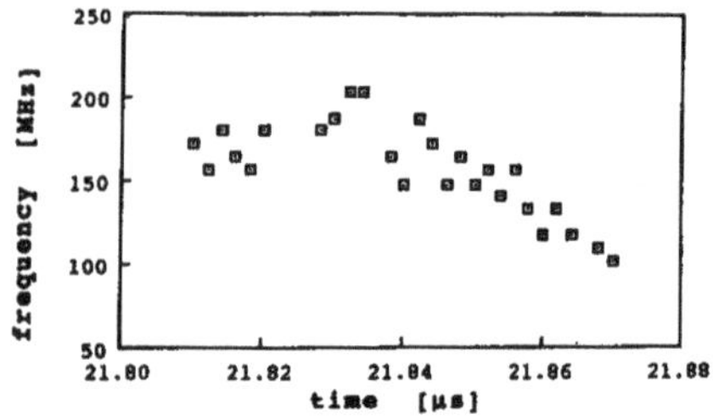

Fig. 5 Temporal variation of the peak frequency of the IF converted UHF wave burst (Fig. 4 (b)) analyzed in terms of the Wigner distribution during t = 21.80 - 21.88 μsec.

ON THE CONNECTION BETWEEN THE NEGATIVE RESISTANCE BEHAVIOR AND THE DYNAMICS OF A PLASMA DOUBLE LAYER

S.J.Talasman and M.Sanduloviciu

Department of Plasma Physics, "Al.I.Cuza" University
6600 Iasi, Romania

INTRODUCTION. The phenomenon of space charge separation, which is implied in the appearance of a double layer, is based on a series of processes that are still not enough investigated and understood. The simple presence of an electric field due to a positive biased electrode immersed in a collisionless plasma column /1/ or the voltage difference between two adjacent collisionless plasma /2/ are not sufficient to explain the multitude of phenomena that accompany the appearance and then, the unstable behaviour of a double layer (DL).

The appearance, the disappearance and the unstable behaviour of a DL are highly reflected in the current-voltage characteristics and the space distribution of the light emitted from plasma. To explain the patterns of such curves is necessarily to take into account not only the plasma particles dynamics but also the quantum processes, that means space charge separation produced by inelastic collisions in a region where (after electrons acceleration) the excitation and ionization rates become space dependent /4/. This might be considered as an useless thing in the case of collisionless plasma, but even in this case the light phenomenon observed into the DL region prove that the collisions can not be neglected. The present paper refers to the phenomena that take place during the appearance and then during the unstable state of a three-dimension double layer (3DDL) obtained in a nearly collisionless plasma.

EXPERIMENTAL RESULTS. The experimental device principally consists (Fig.1) of a 20 cm length and 6 cm diameter glass tube (GT) having at one end a positively biased plate (A) and at the other end a radiofrequency plasma source (PS) /3/. The diameter of A is equal to the inner diameter of the GT. The plasma obtained by the help of a radiofrequency glow discharge diffuses through two grids G_1 and G_2 and forms in GT a plasma column. The voltage supply U_1 is used for biasing the electrode A and U_2 for biasing the plasma from the source with respect to the grid G_1. Plasma was investigated by the help of an axial movable double probe (DP) penetrating through the middle of A, by the help of a photomultiplier (PM) and of a capacitive probe (MF).

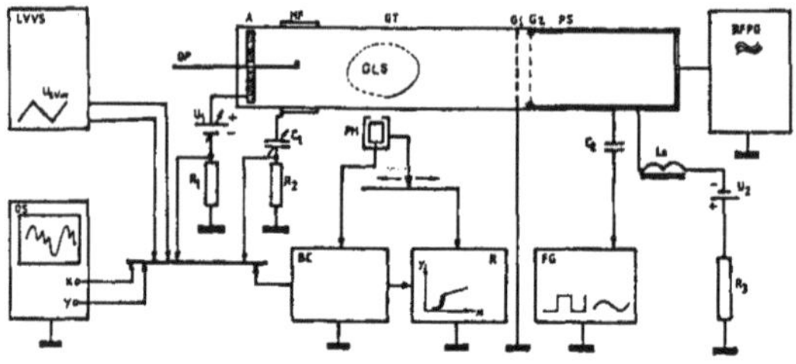

Experimental device

RFPG : radiofrequency power generator; PS : plasma source; G1,G2 : grids; GT : glass tube; MF : metal foil; DP : double probe; A : anide; LVVS : linear variable voltage supply; BC : bix-car; FG : function generator; PM : photo-multdplier; GLS : globular light structure.

Fig.1

For voltages U_1 between 50 V and 100 V and U_2 between 10 V and 30 V one obtained at a certain distance from A, a nearly globular light structure (GLS) (Fig.2,a). With the increasing of pressure one can obtain even two such structures (Fig.2,b).

Fig.2

a) b)

In the unstable state of the GLS, oscillations of the current flowing through plasma column appear. These oscillations are nearly harmonically in the case of a single GLS. In the case of two GLSs, depending on U_1, the current oscillations exhibit a period doubling phenomenon and a route to chaos (Fig.3). The time-resolved axial distribution of the light emitted from the GLS evidences a pulsation of it in the unstable state (Fig.4). This is also evidenced plotting the time-resolved axial distribution of the axial electric field, using the DP (Fig.5) (consisting of two tantalum wires of 0.1 mm in

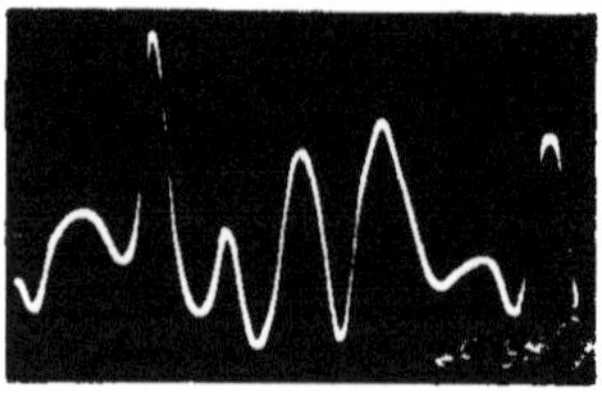

Fig.3

diameter, having lengths of 3 mm and a space separation of 2.5 mm). The DP was connected to a high input impedance differential amplifier and the output signal was supposed to be

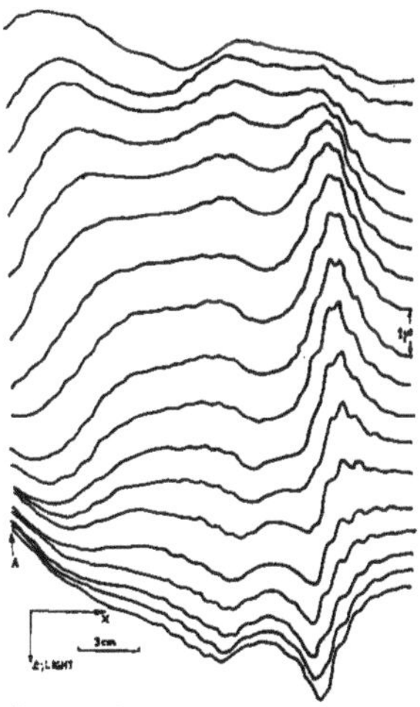

Time resolved axial distribution of the light emission

Fig.4

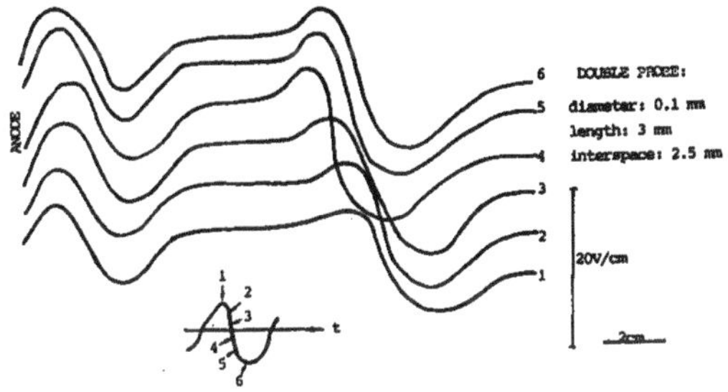

Time resolved electric field axial distribution

Fig.5

proportional to the electric field. The current-voltage characteristics were plotted for different U_1 voltages (Fig.6). The linear variable voltage excursion was ± 15 V. As can be seen from these curves there are three regions of "negative" resistance. The first is related to the transition to the unstable state of the first GLS and the other two regions are related to the appearance and the disappearance of the second

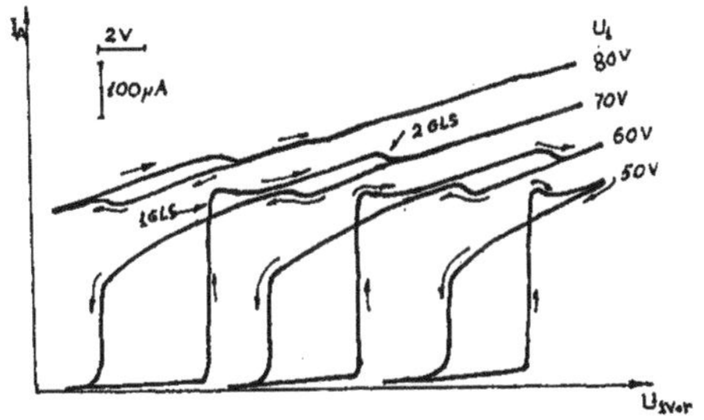

Current-voltage characteristics

Fig.6

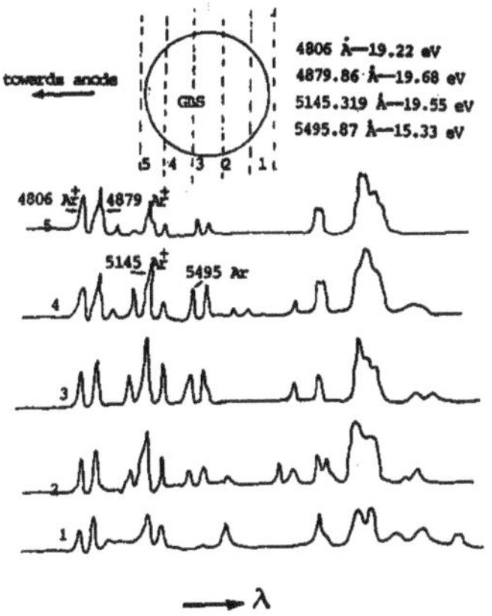

Qualitative light spectrum distribution within GLS.

Fig.7

GLS. The time averaged spectral investigation of the light emitted from different regions of the GLS, reveals that different quantum processes with different weights, depending on the region, are present (Fig.7). One worked in Ar in a pressure range of (10^{-2} - 10^{-3})Torr. The neutral mean free path in this case is about 10 cm. The electron concentration was of order 10^{9} cm^{-3} and the Debye length was about 0.1 mm.

THE ANALYSIS OF THE RESULTS. As can be seen from Fig.5, the globular light structure is surrounded by a double layer having its positive side inwards and its negative part outwards. In the unstable state of the GLS this 3DDL periodically expands disrupts and reforms. This variation is accompanied by a time-space variation of the integral light emitted from the GLS border. The suddenly increase of the current takes place in the moment when the GLS becomes unstable because the periodical DL disruptions involved in a stage of the unstable process are accompanied by an electrons and positive ions releasing process /5/. Taking into account the fact that the maximum value of the oscillation is limited at the maximum potential value from the I-V characteristic, it is clear that the mean d.c. current intensity decrease in the oscillatory state. This determines the first observed negative branch of the I-V characteristic. With the increase of the anode voltage the amplitude of oscillations also increases and in the same time the 3DDL together with GLS become better structured. The second "negative" branch of the I-V characteristic can be explained considering the transition to a new oscillation mode (Figs.3, 4). The fact that in a certain anode voltage interval (lower than that for which a second GLS appears) the oscillation pattern remains sinusoidal, prove the ability of the GLS to preserve its internal structure. The same conclusion could be stated considering the oscillations pattern observed when two GLSs are presented with a tendency to maintain their internal ordered structure. Because the appearance of the second GLS implies a change in the oscillation pattern, the main value of the current decreases, a fact evidenced by the second "negative" slope of the I-V characteristic. The oscillatory state corresponding to the last pattern reveals a coupling between two selforganized oscillatory systems. This coupling is performed through the current. If the anode voltage decreases one observe that the last oscillatory state is maintained for even smaller anode voltage than the appearance one (a hysteresis loop). This can be explained considering the tendency of the second GLS to maintain its internal space charge structure corresponding to the selforganized state /6/.

CONCLUSIONS

1) In a plasma device such like that presented in this paper, where the plasma is nearly collisionless, it is possible to obtain a selforganized structure. The border of this structure is a nearly spherical double layer whose dynamics assures the "viability" of the GLS. This "viability" is possible only in a pulsatory state during of which a periodical exchange of matter and energy with the surrounding plasma take place.

2) The driving processes taking place during the double layer generation are mainly of quantum nature like excitations, and ionizations. Other processes as desexcitations (to which are due the discrete spectral emissions) and the recombinations are also important in the double layer dynamics because in this way neutral particles in the fundamental state are produced which in turn make possible the existence of previous mentioned driving processes.

3) The stability of such a GLS is assured by electrostatic forces appeared after internal space charge separation sustained by quantum processes.

4) The allure of the I-V characteristic is related to the double layer generation and dynamics, giving an informations about the processes involved in the DL phenomenology.

REFERENCES

1. N.Sato (1984) ICPP, Lausanne, Proc.Inv.Papers, p.555
2. R.Schrittwieser (1992) Symp.on Atomic and Surface Phys., Italy

3. S.Talasman, M.Sanduloviciu (1986) Proc.of Int.Conf.on Plasma Science and Technology, Beijing, China
4. M.Sanduloviciu (1992) Romanian Journ. of Physics (in print)
5. M.Sanduloviciu, E.Lozneanu (1992) ICPP, Innsbruck Austria, p.1771
6. M.Sanduloviciu, D.Alexandroaei, S.Talasman, B.Ianopol (1992) ICPP, Innsbruck Austria, p.1956

CAVITON INDUCED DOUBLE LAYER-LIKE POTENTIAL STRUCTURE

T. Tanikawa
School of High-Technology for Human Welfare, Tokai University
317 Nishino, Numazu, Shizuoka 410-03, Japan
and
A.Y. Wong
Department of Physics, University of California, Los Angeles
Los Angeles, California 90024-1547, U.S.A.

A new mechanism of generating double layer-like potential structure at the location of a density cavity created by intense, localized electron plasma oscillation--caviton-- is experimentally investigated in a high-power microwave-plasma system.

1. INTRODUCTION

Cavitons are localized electron plasma waves (EPWs) trapped inside density cavities which are generated by the ponderomotive forces associated with the localized waves themselves.[1] Cavitons manifest various interesting phenomena, such as the existence of trapped caviton eigenmodes[1] and the emission of various electromagnetic radiation as well as electrostatic waves (linear EPWs and ion acoustic waves).[2,3]

Here, we wish to report another interesting phenomenon associated with cavitons, namely the formation of double layer-like potential structure following the formation of cavitons. Our observation is made in an inhomogeneous plasma which is irradiated by a high-power microwave beam. Resonance absorption is utilized to generate a caviton at the critical layer, where the local plasma frequency matches the frequency of the incident microwave. A negative potential well is initially formed at the location of the caviton. It appears that this potential dip initiates the formation of a double layer-like potential structure by repelling the background electrons.

Our observaation may help us understand some important problems in space and laboratory plasmas, such as the formation mechanism of double layers in highly density modulated plasmas and the control of the plasma space potential using rf fields.

2. EXPERIMETAL SETUP

Our experiment was performed in the Large Caviton Device (LCD) which consists of a large (~1.4 m diameter, ~2.3 m length) vacuum vessel (see Fig. 1). A plasma was produced by a pulsed, high-power hf trasmitter ($f_{hf} \simeq 400$ kHz, $P_{max} \simeq 15$ kW) connected to a discharge loop antenna (~80 cm diameter) located near one end of the chamber. With the maximum transmitter power of nearly 15 kW, a plasma density of $n \sim 3 \times 10^{11}$ cm^{-3} can be obtained for argon gas with the neutral pressure of $p_n \simeq 3 \times 10^{-4}$ Torr. The experiment was performed in a quiescent ($|\delta n/n| < 0.3\%$) afterglow plasma. The plasma was mainly produced at one end of the chamber near the discharge loop antenna and as it flowed to the other end, a density gradient was naturally created along the axis of the chamber. The typical experimental parameters for an Ar plasma are: the critical density $n_c \simeq 1 \times 10^{11}$ cm^{-3}, $T_e \simeq 6T_i \simeq 1.5$ eV, the density gradient scale length $L \equiv n(\partial n/\partial z)^{-1} = 50\text{-}100$ cm $= (1.7\text{-}3.5) \times 10^4\lambda_D$, and $\nu_{en}/\omega_{pe} \simeq 4 \times 10^{-4}$. The parameters for the microwave pulse to drive a caviton are: $f_o = 2.815$ GHz, the peak power $P_o \le 500$ kW, the pulse duration $\Delta t_{rf} = 0.15\text{-} 1.2\ \mu s = (4.2 \times 10^2 - 3.4 \times 10^3)f_o^{-1}$, and the normalized field intensity $|E_o|^2/(8\pi n_c T_e) = 0.05\text{-}0.5$.

Plane (1 mm diameter), one-sided Langmuir probes were employed to measure various plasma parameters, such as the plasma density, the electron temperature and the plasma space potential. In most cases, the measurements were carried out just after the incident microwave pulse was terminated in order to avoid nonlinear probe effects which could be caused by strong rf fields. Electron and ion distribution functions were measured by using retarding-grid energy analyzers.

3. EXPERIMENTAL RESULTS AND DISCUSSION

An example of a fully developed double layer-like potential structure associated with a caviton is shown in the lower frame of Fig. 2 (closed circles). For this case, the original critical layer is $z \simeq 45$ cm; therefore, the density cavity near $z = 45$ cm is the caviton due to the resonantly excited EPW (see the upper frame in Fig. 2).

The sequence of the development of the potential structure is as follows: Upon the injection of the microwave pulse, an intense EPW is resonantly excited at the critical layer. This inherently localized EPW can strongly accelerate electrons which belong to the tail part of the electron distribution function due to the transit time acceleration mechanism.[4] The plasma space potential of the entire plasma then quickly rises 10 to 20 V above the initial potential of a few volts above the ground potential, depending on the incident microwave power, since fast electrons escape rapidly out of the plasma region. Meanwhile, a caviton starts to form due to the ponderomotive force associated with the localized EPW. After the ions follow the electron movement, and the quasi-neutrality condition is attained ($t \geq 1/f_{pi}$), a negative potential well appears at the location of the caviton [$n_e \propto \exp(e\phi/T_e)$]. It appears that this potential dip repels background electrons, causing the expansion of the low potential region. Although it is very difficult to measure the plasma space potential profile inside the microwave pulse, the temporal development of the potential structure can be followed by varying the duration of the microwave pulse and by measuring the space potential just after the microwave pulse is terminated so as to avoid the adverse probe effect. The initial development of the negative potential well is shown in the lower frame of Fig. 2 (open circles).

We note here that we have some evidence of the generation of an ion phase-space hole at the location of the caviton (or the location of the initial negative potential dip) which propagates towards the direction of the side with the higher density at close to the ion acoustic speed. The situation might be similar to the one discussed by Hasegawa and Sato.[5]

In Fig. 3, the potential difference between the high and low potential regions, $\Delta\phi_p$, of the caviton induced double layer-like potential structure is plotted as a function of the duration of the incident microwave pulse, Δt_{rf}. Although $\Delta\phi_p$ saturates at $\Delta t_{rf} \simeq 0.45\ \mu s$, $\Delta t_{rf} \simeq 0.7\ \mu s$ is required for the formation of a fully developed potential structure.

Our measurements show that the electron and ion distribution functions are basically consistent with the observed potential structure. For example, the electron distribution function in the underdense region (or the higher potential region) shows the anisotropic feature: only when the probe faces the lower potential side, it collects fast electrons (see Fig. 4). The cold part of the distribution also shows an apparent heating effect because the initial rise of the entire plasma space potential provides better electron confinement.

The incident microwave power dependences of the slope of the cold part, T_{eC}, and the hot part, T_{eH}, of the electron distribution function are shown in Fig. 5. Figure 6 shows the incident microwave power dependence of $\Delta\phi_p$. Comparing Fig. 6 with Fig. 5, it can be seen that the power dependence of $\Delta\phi_p$ scales with that of T_{eH}.

4. CONCLUSION

A double layer-like potential structure or a multiple double layer[6] is observed to develop following the formation of a caviton. A negative potential dip which forms $\geq 1/f_{pi}$ after the injection of the microwave beam appears to initiate the formation of a double layer-like potential structure by repelling background electrons. The observed particle distribution functions are consistent with the observed potential structure.

REFERENCES

1. T. Tanikawa, A.Y. Wong, and D.L. Eggleston, Phys. Fluids 27, 1416 (1984).
2. T. Tanikawa, L. Scurry, and A.Y. Wong, Bull. Am. Phys. Soc. 31, 1579 (1986).
3. P.Y. Cheung, A.Y. Wong, T. Tanikawa, J. Santoru, D.F. DuBois, H.A. Rose, and D. Russell, Phys. Rev. Lett. 62, 2676 (1989).
4. A.Y. Wong, P. Leung, and D. Eggleston, Phys. Rev. Lett. 39, 1407 (1977).
5. A. Hasegawa and T. Sato, Phys. Fluids 25, 632 (1982).
6. P. Coakley and N. Hershkowitz, Phys. Fluids 22, 1171 (1979).

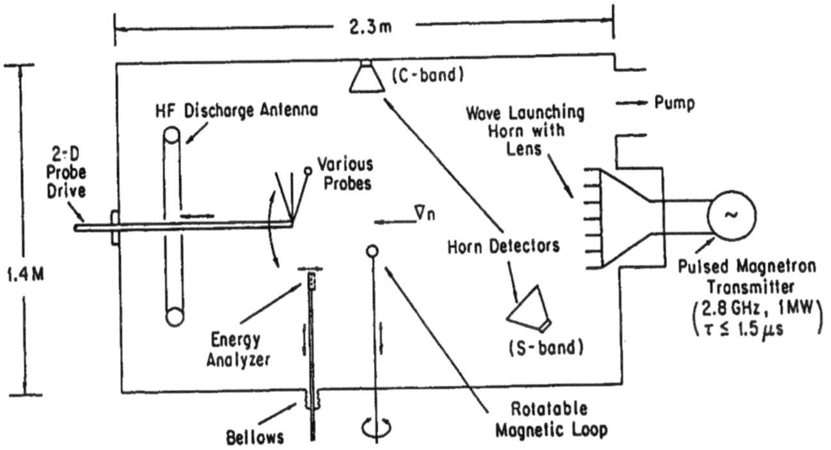

FIG. 1. Schematic diagram of the experimental apparatus.

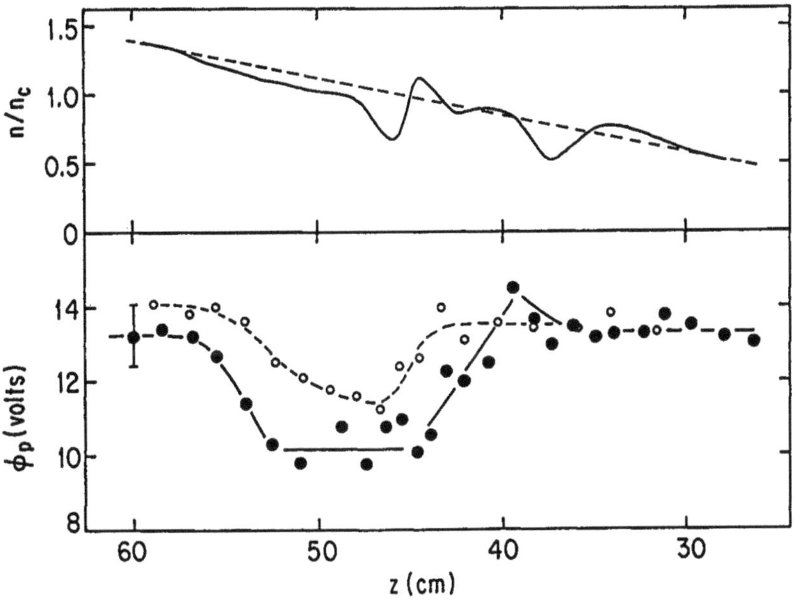

FIG. 2. Profiles of the plasma space potential developed around the cavitons. •: $\Delta t_{rf} \simeq 0.75\ \mu s$; o: $\Delta t_{rf} \simeq 0.15\ \mu s$. The peak power of the incident microwave pulse was ~160 kW. The potential measurements were performed at ~0.4 μs after the termination of the microwave pulse. The full I-V characteristic of a cold Langmuir probe was taken for each data point. The density profile for the $\Delta t_{rf} \simeq 0.75\ \mu s$ case is shown in the upper frame. A negative potential well appears at the location of the caviton. The region of negative potential expands as time goes by.

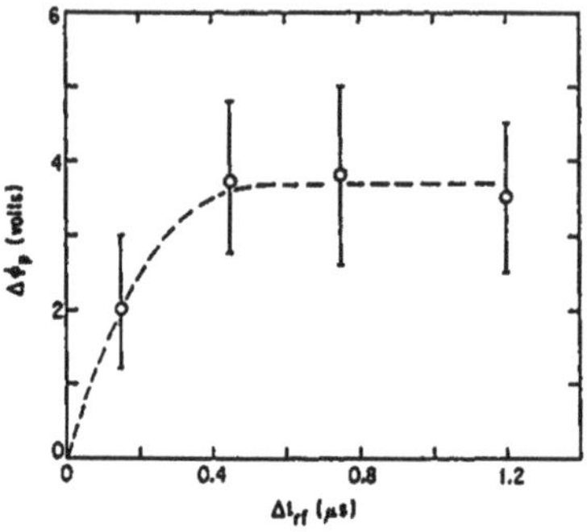

FIG. 3. $\Delta\phi_p$ vs. duration of the incident microwave pulse Δt_{rf}. The peak power of the incident microwave pulse was ~150 kW.

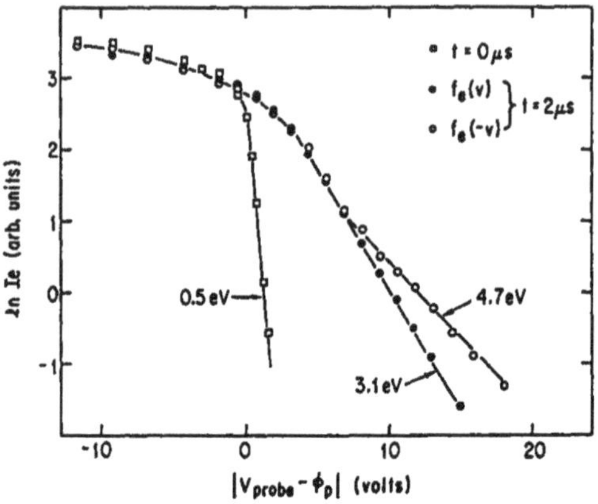

FIG. 4. Electron distribution function measured in the underdense region (the higher potential side) of the caviton.
o: for electrons moving toward the lower density side.
•: for electrons moving toward the higher density side.
The hot electrons coming from the caviton region are clearly seen.
□: distribution function just before the turn-on of the microwave pulse. The distribution is symmetric in both directions.

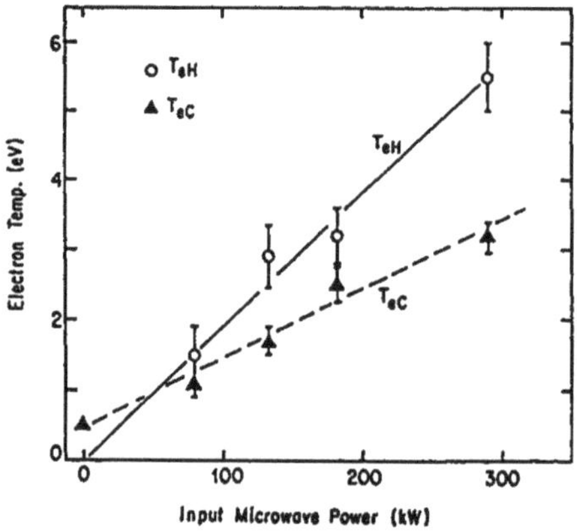

FIG. 5. Hot electron temperature T_{eH} vs. input microwave power P_{IN}. The temperature of cold electrons is also shown as a function of P_{IN}. The duration of the microwave pulse, Δt_{rf}, was ~1.2 μs. The measurements were performed at ~0.8 μs after the termination of the incident microwave pulse. The probe was located below the caviton and collected electrons moving toward the lower density side.

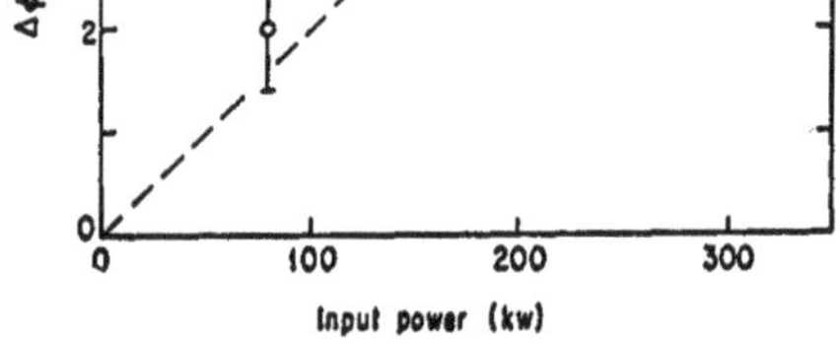

FIG. 6. $\Delta\phi_p$ vs. input microwave power P_{IN}. The duration of the microwave pulse, Δt_{rf}, was ~1.2 μs. The measurements were performed at ~0.8 μs after the termination of the incident microwave pulse.

CONTRIBUTED PAPERS

THEORY

KINETIC ION EFFECTS ON NUMERICALLY COMPUTED LARGE–AMPLITUDE RAREFACTIVE ION–ACOUSTIC SOLITONS

S. Baboolal[a,b], R. Bharuthram[b,c] and M. A. Hellberg[b]

Departments of Computer Science[a] and Physics[c], University of Durban–Westville and Plasma Physics Research Institute[b], Department of Physics, University of Natal, Durban, South Africa.

In the past we studied arbitrary–amplitude rarefactive double layers in a plasma comprising two Boltzmann electron species and multi–kinetic ion species where the latter followed the Schamel distributions for the trapped and free populations. Here we make a similar study of rarefactive solitons. The effects of ion temperature and a second ion species are examined. Also a quantitative comparison is made with results from arbitrary–amplitude fluid theory for the same plasma composition.

In positive ion plasmas we generally find good correspondence between the two approaches. We suggest that departure from fluid theory may be attributed to ion trapping and/or Landau damping.

However, in the presence of negative ions we find that the potential amplitudes are some three–fold bigger than the corresponding structure computed from fluid theory for solitons moving at the same velocity. This effect persists over wide variation of the ion trapping parameter and moreover, it is shown that an effect such as inverse Landau damping for example, cannot account for such large wave enhancement. Rather, our computations show that it arises precisely because of the form of the negative–ion distribution function suggested in earlier works: in particular that, at the point where the potential $\phi = -\phi_m$ ($\phi_m > 0$) the function is a shifted Maxwellian (corresponding to free particles). When we relax this requirement, i.e. consider it to be a function of the energy $E = \frac{1}{2}u^2 - \phi$ instead of the shifted energy $E' = \frac{1}{2}u^2 - (\phi + \phi_m)$ we find excellent agreement with fluid theory, except where ion trapping is significant.

1. Theory: The basic model equations employed are the the double–Boltzmann equations for the cold (normalized density: n_c with equilibrium value n_{c0}, temperature: T_c) and hot (normalized density: n_h with equilibrium value n_{h0}, temperature: T_h) electrons, where $T_{eff} = T_c T_h/(n_{c0}T_h + n_{h0}T_c)$ is the effective electron temperature, and the one–dimensional Vlasov equation for each ion species. In the wave frame moving with Mach number M ($s = x - Mt \Rightarrow \dot{s} = v - M \equiv u$, with $\dot{x} = v$) the latter is for the $k-th$ ion species

$$u\frac{\partial f_k}{\partial s} - \sigma_k \frac{1}{Q_k}\phi'(s)\frac{\partial f_k}{\partial u} = 0. \tag{1}$$

Here $f_k = f_k(s,u)$ is the time–independent distribution function of the $k\!-\!th$ species with x the spatial coordinate normalized to the Debye length $\lambda_{de} = (T_{eff}/4\pi n_0 e^2)^{1/2}$, where n_0 is the total electron density at equilibrium w.r.t. to which are normalized the electron and ion densities. The time variable t has been normalized to the inverse ion plasma frequency $\omega_{pj}^{-1} = (m_{ij}/4\pi n_0 e^2)^{1/2}$ of some $j\!-\!th$ positive ion species, the velocity v is normalized to the cold–ion limit of the ($j\!-\!th$ ion species) sound speed $C_{s0} = (T_{eff}/m_{ij})^{1/2}$ and ϕ is the electrostatic potential ($= e\tilde{\phi}/T_{eff}$ with $\tilde{\phi}$ the unnormalized potential). $\phi'(s)$ denotes the total derivative w.r.t. s of the time–independent potential $\phi(s)$, σ_k is the signed charge multiplicity and Q_k is the ratio of the mass (m_{ik}) of the $k\!-\!th$ ion species to the normalizing positive–ion component mass (m_{ij}). These normalizations and notations correspond to previous studies[1].

For singly charged ions in a negative potential perturbation ϕ where $-\phi_m \le \phi \le 0$, for some $\phi_m > 0$, we employ the following distribution functions for the $k\!-\!th$ positive (2) or negative (3) ion species with drift velocity u_{k0}

$$f_k(s,u) = \left[\frac{Q_k\theta_k}{2\pi}\right]^{\frac{1}{2}} \begin{cases} exp\left[-\frac{Q_k\theta_k}{2}\left\{s_u\left(u^2 + \frac{2\phi}{Q_k}\right)^{\frac{1}{2}} - u_{k0}\right\}^2\right] & , u^2 > -\frac{2\phi}{Q_k} \\ exp\left[-\frac{Q_k\theta_k u_{k0}^2}{2}\right] exp\left[-\frac{\beta_k Q_k\theta_k}{2}\left(u^2 + \frac{2\phi}{Q_k}\right)\right], & u^2 \le -\frac{2\phi}{Q_k} \end{cases} \tag{2}$$

$$f_k(s,u) = A_k\left[\frac{Q_k\theta_k}{2\pi}\right]^{\frac{1}{2}} \begin{cases} exp\left[-\frac{Q_k\theta_k}{2}\left\{s_u\left(u^2 - \frac{2\phi}{Q_k}\right)^{\frac{1}{2}} - u_{k0}\right\}^2\right] & , u^2 > \frac{2(\phi+\phi_m)}{Q_k} \\ F\, exp\left[-\frac{Q_k\theta_k u_{k0}^2}{2}\right] exp\left[-\frac{\beta_k Q_k\theta_k}{2}\left\{u^2 - \frac{2\phi}{Q_k}\right\}\right], & u^2 \le \frac{2(\phi+\phi_m)}{Q_k} \end{cases} \tag{3}$$

$f_k(s,u) = R.H.S.\ of\ (3)\ above$ with $u^2 - 2\phi/Q_k$ replaced by $u^2 - 2(\phi+\phi_m)/Q_k$. (3′)

Here $s_u = sign\ (u) = \pm 1$, $\theta_k = T_{eff}/T_{ik} = 1/T_k$ and $Q_k = m_{ik}/m_{ij} = 1/m_k$ where T_{ik} is the unnormalized ion temperature with T_k its normalized value. Further, β_k is the trapping parameter for the $k\!-\!th$ species and is defined as the ratio of the free ion temperature to the trapped ion temperature. A_k is a normalization constant computed as before[1]. Also, we remark that it is easily verified that the functions (2)–(3) which are functions of the energy, $E_k = \frac{1}{2} Q_k u^2 \pm \phi$, where the + (–) sign applies to positive (negative) ions, satisfy the stationary Vlasov equation. Moreover, the forms chosen for E_k in (2) – (3) give drifting Maxwellians for the free ($u^2 > \ldots$) particle distributions and Maxwellians for the trapped particles ($u^2 \leq \ldots$) at ∞ where $\phi = 0$. For negative ions this contrasts with previous forms[3,4] (some albeit given for positive ions in a positive potential), in particular (3′), which are expressed in terms of the shifted energy $E' = \frac{1}{2} Q_k u^2 - (\phi + \phi_m)$ and thus reduce to such distributions at the wave minimum. The additional factor F in (3) is chosen so that continuity is maintained at the free–trapped particle separatrix $u^2 = 2(\phi + \phi_m)/Q_k$ in velocity space. It evaluates to

$$F = exp\ [(\beta_k \theta_k Q_k/2)\{2\phi_m/Q_k\} - (\theta_k Q_k/2)\{2\phi_m/Q_k - 2s_u(2\phi_m/Q_k)^{\frac{1}{2}} u_{k0}\}].$$

We adopt the convention that the wave is propagating with Mach number M through a non–drifting plasma, as we had in our fluid description[1]. Thus we take $u_{k0} = M$ for all ion species k. Further with the boundary conditions

$$\phi, \frac{\partial \phi}{\partial x} \to 0;\ n_c(n_h) \to n_{c0}(n_{h0});\ N_k \to N_{k0};\ k=1,2,\ldots\ at\ x = \pm\infty, \tag{4}$$

and the charge neutrality condition

$$1 = n_{c0} + n_{h0} = \sum_k \sigma_k N_{k0}\ at\ x = \pm\infty, \tag{5}$$

for the normalized electron and ion densities (N_k, $k=1,2,\ldots$), the Poisson equation (in the stationary frame) reduces[1] to the Sagdeev or pseudo–potential (V) form,

$$\frac{1}{2}\left\{\frac{d\phi}{ds}\right\}^2 + V(\phi,\phi_m,M) = 0, \text{ with } V(\phi,\phi_m,M) = -\int_0^\phi N(\theta)\, d\theta \tag{6}$$

where $N(\phi)$ is the density function given by

$$N(\phi) = n_c + n_h - \sum_k \sigma_k N_k(\phi) \tag{7}$$

and the ion densities $N_k(\phi)$ are determined numerically[1] through velocity–space integrations of the distribution functions $f_k(s,u)$, $k=1,2,\ldots$. Now soliton solutions are obtained by first solving the pseudo–particle reflection condition,

$$V(\phi,\phi_m,M) = 0 \tag{8}$$

for ϕ by means of iterations satisfying the constraint $\phi = \phi_0 = -\phi_m$ with various chosen but fixed M. Then, with the pair (ϕ_0, M) known the soliton profile may be computed as before[1,2].

2. Solitons in positive ion plasmas: We have computed typical Sagdeev and real potential profiles of rarefactive solitons in a positive single–ion–species plasma with $n_{c0} = 1/10$, $T_h/T_c = 30$ and ion temperature $T_i = 0.06$. We find here, that for a wide range of the trapping parameter, viz., $0.01 \leq \beta \leq 1.00$ there is negligible change in the respective profiles and in their departures from fluid theory with adiabatic ions. This is in contrast to the case for double layers where, we recall[1] for instance, that there is a significant drop in the potential amplitude for the $\beta = 0.50$ case, and moreover, even for $\beta = 0.20$ the double layer structure becomes narrower than the $\beta = 0.01$ structure. Thus, trapped positive ions appear to have negligible effect on rarefactive soliton structures, suggesting that these are essentially fluid phenomena.

In Fig 1a we show the effect of varying the ion temperature T_i on a typical ($M = 1.2$) soliton structure in the same plasma. The characteristic damping[2] with increasing ion temperature is evident. In fluid theory[2] we had explained this as the result of decreasing wave dispersion. Thus, the kinetic results here are essentially similar to the fluid theory results, but departs from these in as much as there is sharper damping for $T_i > 0.06$, which we attribute to Landau damping: e.g. at $T_i = 0.06$ the fluid (positive) phase velocity is $C_{sf} = 1.086$ whilst the complex kinetic velocity is $C_{sk} = (1.101, -0.002)$ and, at $T_i = 0.1$ their values are 1.140 & $(1.179, - 0.018)$ respectively, computed from previously derived dispersion relations[1] for the long wavelength limit.

In Fig 1b we show the effect of a lighter ion density N_{10} on the amplitude of a $M = 1.2$ soliton in the same plasma with small trapping paramters $\beta_1 = \beta_2 = \beta = 0.01$. Whilst the characteristic damping seen previously[2] is evident the fluid results tend to damp faster until a cut–off[2] point. It seems that Landau damping [e.g. at $N_{10} = 0.02$, $C_{sf} = 1.146$, $C_{sk} = (1.138, - 0.033)$] although noticeable is overshadowed by what we believe is greater dispersion in the kinetic case, requiring stronger nonlinearity for balance.

3. Solitons in a negative–ion plasma: We consider an (Ar^+, F^-) negative–ion plasma with $n_{c0} = 5/6$, $T_h/T_c = 5$, the −/+ ion mass ratio = 0.4756, $N_{10}(F^-) = 1.6$, & $T_1 = T_2 = T_i = 0.035$ typical of recent experiments[2]. The trapping parameters are again set

small to $\beta_1 = \beta_2 = \beta = 0.01$. Figs. 2a & 2b show typical ($M = 3.0$: here higher actual sound speeds force larger M) soliton Sagdeev and real potential profiles using the present (3) and previous[3,4] form (3′) of the distribution function. These are shown together with the fluid theory calculations. We see that form (3′) exaggerates potential sizes considerably: they give some three times the fluid theory amplitudes. Figs. 2c & 2d giving the variation of potential amplitude ϕ_m with ion temperature $T_i = T_1 = T_2$ and with negative–ion density N_{10} show that this is by no means an isolated case.

Also, we find here that inverse Landau damping is insignificant in every case: for example for $T_1 = T_2 = 0.02$, $C_{sk} = (2.462, 0.000)$ in Fig 2c, and, for $N_{10} = 1.6$, $C_{sk} = (2.478, 0.000)$ in Fig 2d. Thus inverse Landau damping cannot account for wave enhancement due to (3′). We also did the computations employing a shifted Maxwellian in place of (3), i.e. neglected trapped particles, and again found good agreement with fluid theory and (3). For example, for $N_{10} = 1.6$ in Fig 2d we find $\phi_0\ (Maxwellian) = -0.807$, $\phi_0\ (K) = -0.808$ and $\phi_0\ (F) = -0.810$. Finally, we observe that the variation of ϕ_m with ion temperature and with negative ion concentration follows the familiar patterns discussed before[2].

4. Conclusion: The use of the usual kinetic ion distribution functions in positive ion plasmas gives results in agreement with fluid theory. Only minor departures occur due to ion trapping even over a wide range in the trapping parameter. The characteristic damping of amplitudes with ion temperature and light ion contamination largely follow fluid theory predictions[2]. However, in negative ion plasmas the use of the usual negative ion distribution function results in large departure in amplitudes from fluid theory. But, a modification of it suggested here produces results in good agreement with fluid theory. Again the characteristic damping with ion temperature and negative ion concentration follows fluid theory predictions[2], with (3) and (3′) agreeing only qualitatively.

REFERENCES

1. S. Baboolal, R. Bharuthram and M.A. Hellberg, *Phys. Fluids B* **2**, 2259 (1990).
2. ———— *J. Plasma Phys.* **41**, 341 (1989) & *J. Plasma Phys.* **44**, 1 (1990).
3. H. Schamel, *Plasma Phys.* **14**, 905 (1972); *Z. Naturforsch* **38a**, 1170 (1983); *Physica Scripta* **20**, 306 (1979); *Physica Scripta* **T2/1**, 238 (1982).
4. M.K. Hudson, W. Lotko, I. Roth and E. Witt, *J. Geophys. Res.* **88A**, 916 (1983); S. Sutradhar and S. Bujarbarua, *J. Phys. Soc. Jpn.* **56**, 139 (1987).

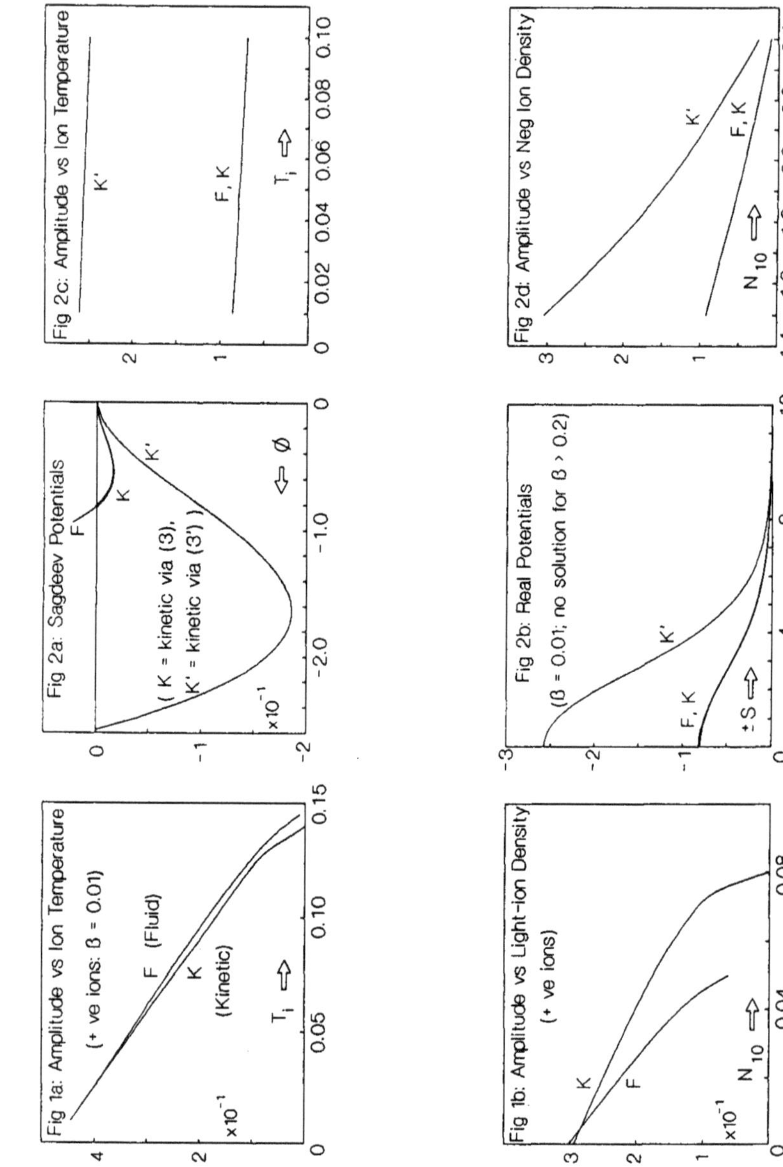

Fig 1a: Amplitude vs Ion Temperature
(+ ve ions: ß = 0.01)
F (Fluid)
K (Kinetic)
T_i ⇨
×10⁻¹
Fig 1b: Amplitude vs Light-ion Density
(+ ve ions)
K
F
N_{10} ⇨
×10⁻¹
Fig 2a: Sagdeev Potentials
F
K
K'
(K = kinetic via (3),
K' = kinetic via (3'))
⇦ Ø
×10⁻¹
Fig 2b: Real Potentials
(ß = 0.01; no solution for ß › 0.2)
K'
F, K
± S ⇨
Fig 2c: Amplitude vs Ion Temperature
K'
F, K
T_i ⇨
Fig 2d: Amplitude vs Neg Ion Density
K'
F, K
N_{10} ⇨

MULTICOMPONENT DOUBLE LAYERS

P.Carlqvist, Department of Plasma Physics, Alfvén Laboratory, The Royal Institute of Technology, S-100 44 Stockholm, Sweden

ABSTRACT: A model of a multicomponent double layer is considered. For the sake of simplicity it is assumed that the layer is steady, strong, and non-relativistic. The multicomponent double layer is composed of several different kinds of positively and negatively charged particles. On the basis of the model a current condition is derived for the layer which describes the proportions of the current densities of the different charge components that are possible. The internal structure of the double layer is also investigated. It is demonstrated that the density distribution of each of the components is of the same shape as the density distributions of the components of a corresponding two-component layer. Similarly it is shown that the potential distribution within the layer is of the same shape as that of the two-component layer. The potential drop across the layer is found to be a function of the thickness of the layer and of the different current densities. It is pointed out that the multicomponent double layer may give rise to a selective acceleration of particles so that the abundance of elements among the accelerated particles need not necessarily be the same as the abundance of elements in the ambient plasma. This property may be of some significance especially in connection with double layers occurring in cosmic plasmas.

1. INTRODUCTION: In most theoretical work on double layers it is generally assumed that the layer is built up by electrons and one species of positive ions. This type of double layer model is often well applicable to double layers that are studied in laboratory experiments comprising only one kind of gas like Hg or H. However, if we want to study double layers in cosmic plasmas the above mentioned type of model may be questionable and in some cases even inappropriate. The reason for this is that cosmic plasmas are generally composed of a whole set of chemical elements and so the double layers occurring in these plasmas will be built up by more than two kinds of charged particles. Hence, there is a need of a theory describing the physics of such *multicomponent double layers*. In the present paper we shall study the properties of a simple type of multicomponent double layer.

2. THE MODEL: We consider a model of a one-dimensional, strong, and steady double layer. The layer is limited by the anode border A and the cathode border C and has the thickness d. Outside the borders there is plasma. An x-axis is introduced so that it is perpendicular to the borders with $x = 0$ at A and $x = d$ at C. Several species of both positive and negative particles are accelerated through the layer by the electric field $E = -dV/dx$. The positive and negative particles are supposed to enter the layer at A and C respectively with zero velocity. The charge distribution in the double layer, caused by the accelerated particles, is consistent with a potential drop V_{DL} across the layer. We define the potential to be $V = 0$ at C so that the potential at A is $V = V_{DL}$. It is assumed that the potential drop is so small that no particles

attain relativistic velocities in the layer. Furthermore it is supposed that there are no collisional or other losses in the layer.

3. THE CURRENT CONDITION: From the conservation of energy we find the velocities of the positive and negative particles in the layer

$$v_{p\mu} = \left(\frac{2q_{p\mu}V_1}{m_{p\mu}}\right)^{1/2} \quad \text{and} \quad v_{n\nu} = -\left(-\frac{2q_{n\nu}V}{m_{n\nu}}\right)^{1/2} \quad . \tag{1 a,b}$$

Here q and m denote charge and mass while $p\mu$ and $n\nu$ are subscripts representing the different species of positive and negative particles respectively. It is to be noticed that $q_{n\nu} < 0$. We have also introduced the quantity $V_1 = V_{DL} - V$.

As a result of their motion across the double layer the different kinds of positive and negative particles give rise to the current density components

$$i_{p\mu} = q_{p\mu} n_{p\mu} v_{p\mu} \quad \text{and} \quad i_{n\nu} = q_{n\nu} n_{n\nu} v_{n\nu} \tag{2 a,b}$$

where n denotes the particle density. Since we are considering a steady double layer both $i_{p\mu}$ and $i_{n\nu}$ must be constant.

The potential distribution across the layer is consistantly described by the density distributions of all the species of charged particles accelerated through the layer and by Poisson´s equation. Hence we obtain

$$\frac{d^2V}{dx^2} = -\frac{1}{\varepsilon_0}\left[\sum_{\mu} q_{p\mu}n_{p\mu} + \sum_{\nu} q_{n\nu}n_{n\nu}\right] \quad . \tag{3}$$

Using Equations (1) and (2) we find

$$\frac{d^2V}{dx^2} = -\frac{1}{\sqrt{2}\,\varepsilon_0}\left[\sum_{\mu} \frac{m_{p\mu}^{1/2} i_{p\mu}}{q_{p\mu}^{1/2} V_1^{1/2}} - \sum_{\nu} \frac{m_{n\nu}^{1/2} i_{n\nu}}{(-q_{n\nu})^{1/2} V^{1/2}}\right] \quad . \tag{4}$$

Multiplying Equation (4) by $dV/dx = -dV_1/dx$ and integrating we get

$$\left(\frac{dV}{dx}\right)^2 = \frac{2\sqrt{2}}{\varepsilon_0}\left[\sum_\mu \frac{m_{p\mu}^{1/2} i_{p\mu} V_1^{1/2}}{q_{p\mu}^{1/2}} + \sum_\nu \frac{m_{n\nu}^{1/2} i_{n\nu} V^{1/2}}{(-q_{n\nu})^{1/2}}\right] - C_1 \tag{5}$$

where C_1 is a constant of integration. Since the double layer is neutral as a whole the electric field at the boundaries A and C must be $E = -dV/dx = 0$. Furthermore the potentials at A are, $V = V_{DL}$ and $V_1 = 0$, and at C, $V = 0$ and $V_1 = V_{DL}$. Using these values and Equation (5) we can express C_1 in two ways, one for each of the boundaries, yielding the equation

$$C_1 = \frac{2\sqrt{2}}{\varepsilon_0}\sum_\mu \frac{m_{p\mu}^{1/2} i_{p\mu} V_{DL}^{1/2}}{q_{p\mu}^{1/2}} = \frac{2\sqrt{2}}{\varepsilon_0}\sum_\nu \frac{m_{n\nu}^{1/2} i_{n\nu} V_{DL}^{1/2}}{(-q_{n\nu})^{1/2}} \quad . \tag{6}$$

From the second and third members of this equation we obtain

$$\sum_\mu \frac{m_{p\mu}^{1/2} i_{p\mu}}{q_{p\mu}^{1/2}} = \sum_\nu \frac{m_{n\nu}^{1/2} i_{n\nu}}{(-q_{n\nu})^{1/2}} \quad . \tag{7}$$

In the special case of a double layer containing only two kinds of particles, viz. electrons and singly ionized positive ions, Equation (7) reduces to

$$\frac{i_{p1}}{i_{n1}} = \left(\frac{m_{n1}}{m_{p1}}\right)^{1/2} \tag{8}$$

where $i_{p1} = i_i$ and $i_{n1} = i_e$ are the current densities and $m_{p1} = m_i$ and $m_{n1} = m_e$ are the masses of the ions and electrons respectively. Equation (8) is identical with the well-known Langmuir condition[1] which prescribes the ratio of the ion current density and the electron current density in a two-component layer. From this special case we see that we can interpret Equation (7) as a more general current condition valid for multicomponent double layers. The equation describes how the different current densities must be related to one another when the masses and charges of the different species of particles present in the double layer are given.

As an example of the applicability of Equation (7) we show in Figure 1 how the current densities depend on one another in the second simplest type of double layer consisting of three components only. The components chosen are electrons, O_2^-, and NO^+ which constitute the same kinds of charged particles as those which are present in the lower ionosphere at a hight of about 70 km.

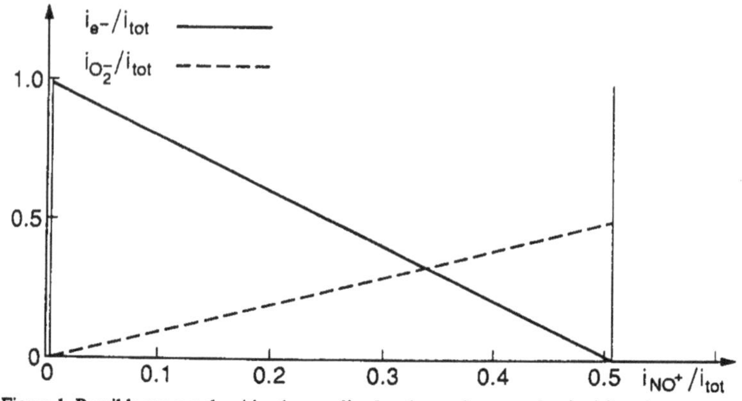

Figure 1. Possible current densities (normalized to the total current density) in a three-component double layer composed of e^-, O_2^-, and NO^+ as given by the current condition. The currents carried by e^- and O_2^- are shown as functions of the current carried by NO^+ which is limited to the interval $4.26 \cdot 10^{-3} - 5.08 \cdot 10^{-1}$.

4. THE POTENTIAL DROP: In order to calculate the potential drop across the multicomponent double layer, V_{DL}, we combine Equations (5) and (6) and find

$$\left(\frac{dV}{dx}\right)^2 = \frac{2\sqrt{2}}{\varepsilon_0}\left[V_1^{1/2}\sum_{\mu}\frac{m_{p\mu}^{1/2}\, i_{p\mu}}{q_{p\mu}^{1/2}} + \left(V^{1/2} - V_{DL}^{1/2}\right)\sum_{\nu}\frac{m_{n\nu}^{1/2}\, i_{n\nu}}{(-q_{n\nu})^{1/2}}\right] \quad . \tag{9}$$

For a two-component layer with $\mu = 1$ and $\nu = 1$ this equation reduces to

$$\left(\frac{dV}{dx}\right)^2 = \frac{2\sqrt{2}}{\varepsilon_0}\left[V_1^{1/2}\frac{m_{p1}^{1/2}\, i_{p1}}{q_{p1}^{1/2}} + \left(V^{1/2} - V_{DL}^{1/2}\right)\frac{m_{n1}^{1/2}\, i_{n1}}{(-q_{n1})^{1/2}}\right] \quad . \tag{10}$$

With $m_{pl} = m_i$, $m_{nl} = m_e$, and $q_{pl} = -q_{nl} = e$ we find that this is the same expression as the one Langmuir obtained for a two-component double layer consisting of ions and electrons.[1] For the two-component layer Langmuir derived the electron current density

$$i_e = \frac{4\sqrt{2}\varepsilon_0}{9} C_2 \left(\frac{e}{m_e}\right)^{1/2} \frac{V_{DL}^{3/2}}{d^2} \tag{11}$$

where C_2 is a numerical constant which he estimated to 1.860. Later Raadu[2] has refined the calculations and obtained the slightly higher value, $C_2 = 1.865$.

Comparing Equations (9) and (10) we note that the form of Equation (9) is the same as the form of Equation (10). Hence we can obtain a solution of Equation (9) by substituting

$$\frac{m_{n1}^{1/2} i_{n1}}{(-q_{n1})^{1/2}} = \frac{m_e^{1/2} i_e}{e^{1/2}} \rightarrow \sum_{\nu} \frac{m_{n\nu}^{1/2} i_{n\nu}}{(-q_{n\nu})^{1/2}} \tag{12}$$

in Equation (11). The potential drop across the multicomponent double layer can then be expressed as

$$V_{DL} = \left[\frac{9d^2}{4\sqrt{2}\,\varepsilon_0 C_2} \sum_{\nu} \frac{m_{n\nu}^{1/2} i_{n\nu}}{(-q_{n\nu})^{1/2}}\right]^{2/3} \quad . \tag{13}$$

In the case of a double layer containing several kinds of positive particles but only one kind of negative particle we may infer from Equation (13) that the potential drop V_{DL} is completely defined by the current density of the negative particles only. The distribution of the current densities of the positive particles does not influence the potential drop provided the current condition is satisfied. For symmetry reasons these statements also hold if the words positive and negative are inter-changed.

Using the general current condition (7) we may also write Equation (13) in the more symmetric form

$$V_{DL} = \left\{\frac{9d^2}{8\sqrt{2}\,\varepsilon_0 C_2}\left[\sum_{\mu} \frac{m_{p\mu}^{1/2} i_{p\mu}}{q_{p\mu}^{1/2}} + \sum_{\nu} \frac{m_{n\nu}^{1/2} i_{n\nu}}{(-q_{n\nu})^{1/2}}\right]\right\}^{2/3} \quad . \tag{14}$$

5. THE INTERNAL STRUCTURE: The potential distribution inside the double layer can be studied by combining Equation (5) and the current condition (7) yielding

$$\left(\frac{dV}{dx}\right)^2 = \frac{2\sqrt{2}}{\varepsilon_0} \sum_{\mu} \frac{m_{p\mu}^{1/2} i_{p\mu}}{q_{p\mu}^{1/2}} \left[(V_{DL} - V)^{1/2} + V^{1/2} - V_{DL}^{1/2}\right] \quad . \tag{15}$$

From this equation at least two properties of the layer may be established. In the first place the equation shows that the distribution of the electric field is symmetric with respect to the centre of the layer. Secondly it is clear that the potential distribution of the multicomponent layer must always be of one and the same shape, independently of how many different components there are that build up the layer. This also implies that the potential distribution must be of the same shape as that found by Langmuir for the two-component layer.[1]

From this latter property and Equations (1) and (2) it follows that the density distributions of all of the positive components are of the same shape. The density distributions of the negative components are also of the same shape but reversed as compared to the distributions of the positive ones.

6. CONCLUSIONS: We have studied a steady and non-relativistic model of a multicomponent double layer being composed of several species of positive and negative particles. On the basis of this model we have derived a current condition which prescribes the possible proportions of the current densities of the different particle species in the layer. We have also obtained an expression which shows how the potential drop across the layer depends on the current densities, charges, and masses of the various particle species. Furthermore we have studied the potential distribution and the distributions of the particle densities in the layer. It has been found that each of these distributions is of the same shape as the corresponding distributions in Langmuir´s two-component layer.

Cosmic plasmas are known to be composed of many different chemical elements. If a double layer occurs in such a plasma we may expect it to be of the multicomponent type discussed above. This type of double layer may accelerate considerable fluxes of different particle species. It is, however, not evident that the abundance of elements among the accelerated particles will be the same as the abundance of elements in the ambient plasma. In case the double layer appears in a partly ionized plasma it is obvious that the accelerated particles will be composed of only those elements in the plasma which are ionized. We are here dealing with a selection that depends on the probability of ionization of the elements. But also in the case of a fully ionized plasma we may expect a *selective acceleration* of charged particles. The reason for this is that the flows of the different elements towards the double layer do not necessarily need to be proportional to the abundances of the corresponding elements in the plasma. The collisional cross-sections of the different elements as well as the current condition of the layer will influence the result. Hence the multicomponent double layer may be considered a mechanism for selective acceleration of charged particles in plasmas.

ACKNOWLEDGEMENTS: I wish to thank Dr. M. Raadu for stimulating discussions and valuable comments.

REFERENCES :

1. I. Langmuir, *Phys. Rev.* **33** (1929), 954.
2. M. A. Raadu, *Proc. Symp. Plasma Double Layers*, P. Michelsen and J. J. Rasmussen (Eds.), Risø, Denmark (1982), 65.

GIGAVOLT-CHARGED SOLITONS IN A STRONGLY MAGNETIZED PLASMA

Tzihong Chiueh

Department of Physics and Institute of Astronomy, National Central University
Chung-Li, Taiwan, Republic of China

Abstract:

New type of magnetosonic solitons, propagating in a relativistic-Alfven-speed plasma, are found to exist. In the soliton rest frame, the soliton can sustain a charge-separation electrical potential on the order of one gigavolt, and an even higher electrical potential when the plasma has a finite temperature. Unlike the conventional magnetosonic solitons, where the magnetic force and the electron dynamics play the sole roles, this type of solitons are mainly determined by the ion dynamics and the electrical force. The length scale in question is larger than the conventional one by the square-root of the ion-to-electron mass ratio. This electric-field-dominant soliton can also exist in the nonrelativistic regime. It is found that the major part of the kinetic energy of the flow, in the soliton frame, can be converted into the electrical potential energy within the soliton.

(I) Introduction:

In a wide range of astrophysical settings, high-speed phenomena seem quite common. They tend to appear with two non-mutually-exclusive characteristics: high Alfven speed and high bulk speed. In some situations, the speed can be close to the speed of light, a regime perhaps uncommon to the solar-system plasmas but occurring in the neutron-star magnetosphere[1] with a relativistic Alfven speed, in the active galactic nuclei[2] with a relativistic bulk speed, and so on. In an environment where the Alfven speed is close to the speed of light, the present work suggests that an interesting object — giga-volt charged soliton — may arise and may transport, and when necessary dissipate, energy effectively. Closer to home, although the Alfven speed is nowhere near relativistic, we can still extend this new-type of highly-charged soliton to the non-relativistic regime[5]. The only requirement is that the change of the flow-aligned momentum of the ions is much greater than that of the electrons. The energy convertion for this new soliton can be so efficient that the major part of the soliton kintic energy become a charge-separation electrical potential energy within the soliton.

When the plasma has a sufficiently high (relativistic/non-relativistic) temperature, i.e. a sufficient thermal spread of velocity in the particle phase space, some of the thermal energy can be converted additionally to the electrical potential energy. This theoretical finding points to the possibility that a type of "transverse (to the magnetic field)" double layer, related to that found by Morawetz[6] three decades ago, but in the relativistic regime, may exist. In the Morawetz's case, a thermal spread of particle velocity is essential in allowing for the existence of a population of reflected ions, which ultimately break the field symmetry between the upstream and downstream regions to form a magnetized double layer. The possibility of this "transverse" double layer is now under investigation.

(II) Theory Outline[3,4]:

In the frame where the field is stationary, there are ion flows and electron flows (as well as flows of other species) separately for the object of interest. As this object is spatially localized, the flow speeds of individual species are expected to be the same far upstream and far downstream. Because the length scale in question is of order of the ion scale (for example, the ion skin depth, c/ω_{pi}, for a non-relativistic soliton), ions and electrons respond to the fields differently, thereby yielding charge separation. Electrons are basically tightly tied to the magnetic field lines, which move more slowly than the particle ions when the force is exactly perpendicular to the field lines. The leading ions pull along the lagged electrons mainly by the electrostatic force. Theoretically, the electron positions generally overshoot and a large-amplitude wave can then be set up. In the case of interest, the lagged electrons are pulled just enough to produce no overshoot, and the field will then become localized. This is how a soliton forms, in distinction with the large-amplitude wave. We begin the theoretical analyses with the conventional two-fluid equations of ions and electrons (with appropriate modifications when Relativity is considered), and the Maxwell's equations.

$$N_\alpha U_\alpha = F_\infty,$$

$$\frac{d}{dx}[(P_\alpha + e_\alpha)U_\alpha U_\alpha] = -c^2\frac{d}{dx}P_\alpha + c^2 q_\alpha N_\alpha\left[\gamma_\alpha E_x + \frac{V_\alpha}{c}B_z\right],$$

$$\frac{d}{dx}[(P_\alpha + e_\alpha)V_\alpha U_\alpha] = c^2 q_\alpha N_\alpha\left[\gamma_\alpha E_y - \frac{U_\alpha}{c}B_z\right],$$

$$E_y = E_\infty,$$

$$\frac{dE_x}{dx} = 4\pi\Sigma q_\alpha \gamma_\alpha N_\alpha,$$

and

$$\frac{dB_z}{dx} = -\frac{4\pi}{c}\Sigma q_\alpha N_\alpha V_\alpha,$$

where the subscript α denotes the species index, and N, P and e are the number density, the pressure and the internal energy; U and V are the x-component and y-component of the flow speed; F_∞ is the particle number flux; E_∞ is a constant, and E and B are the electric field and the magnetic field. Σ is the summation over the species. The motion in third spatial dimension is null, and hence can be ignored.

The set of equations is known to have three integrals associated with the conservation laws of the time component and the two spatial components of the energy-momentum tensor: T^{01}, T^{11} and T^{21}. It turns out a fourth integral does exist when the conservation law of entropy is taken into account. Let us denote it as S, and

$$S \equiv \frac{1}{2c^3}\Sigma q_\alpha\left[(P_\alpha + e_\alpha)^2 U_\alpha^2\left[1 + \frac{U_\alpha^2}{c^2}\right]\right] + \frac{T^{01}}{c^2}F_\infty q^2\phi \tag{1}$$

where ϕ is the electrical potential, defined to be vanished far upstream. In the non-relativistic limit, this new integral becomes $S = S' + S_0$, where S_0 is a constant, much larger than S', and is associated with the rest mass energy of the flow, and the dynamically important part

$$S' = \frac{F_\infty}{c}\left[\Sigma q_\alpha m_\alpha U_\alpha(2P_\alpha + \frac{1}{2}m_\alpha U_\alpha F_\infty) + (m_i + m_e)F_\infty q^2\phi\right]. \quad (2)$$

Apparently, if the flow momentum changes substantially and is mainly carried by the ions, then only the ion terms contribute to the summations in eqs. (1) and (2). The physical meaning of this new integral is more illuminating in the non-relativistic and zero temperature limit. It states that the sum of the x-component of the ion kinitic energy and the charge-separation electrical potential energy is a conserved quantity. This is an elemental result of Mechanics when the motion is one-dimensional and no magnetic field is present. It is amazing to observe that this result can remain true in situations when the magnetic field is present. Futhermore, as eqs.(1) and (2) apply to the cases with a magnetic field of any strength, this new integral, in the limit ignoring the electron momentum, can be versatile; it allows us to accurately determine the electrical potential in a wide range of physical conditions.

On the other hand, when the change of the electron x-momentum is comparable to the change of the ion x-momentum, the electrical potential will be substantially reduced. This usually occurs when the object of interest has a much smaller length scale, of the electron scale, where although the electrons can respond quickly to the acceleration by the highly localized enhanced magnetic field, the ions can not. The change in the ion speed is small and the change in the ion momentum can only be comparable to that of the electrons in spite of the ion's massiveness. One example is the conventional magnetosonic soliton[7], for which the length scale is of the electron skin depth, c/ω_{pe}. In this example, the charge-separation electric potential is substantially reduced and the associated electric force is negligibly small and can be rightfully ignored.

We now have six relations: $E_y = E_\infty$, the ion continuity equation, $N_i U_i = F_\infty$, and the four aforementioned integrals, but we have also eight unknowns: $E_y, \phi, B_z, N_i, U_i, V_i, P_i$ and e_i, in the regime where the electron inertia can be ignored. Two additional relations are needed to solve the problem. They are the equations of state for relating P_i and e_i to N_i. Relativistically, the pressure P_i and internal energy e_i are defined as

$$P_i \equiv \frac{m_i c}{2}\int dv_x dv_y \frac{v_x^2 + v_y^2}{(c^2 + v_x^2 + v_y^2)^{1/2}} f_i(x, \mathbf{v}),$$

$$e_i \equiv m_i c \int dv_x dv_y (c^2 + v_x^2 + v_y^2)^{1/2} f_i(x, \mathbf{v})$$

with N_i defined as

$$N_i \equiv \int dv_x dv_y f_i(x, \mathbf{v})$$

where $f(x, \mathbf{v})$ is the distribution function in the rest frame of the flow and $\mathbf{v}$ is the ion 4-velocity in that reference frame. Since we have used the conservation of entropy to derive eq.(1), the self-consistency demands that the distribution function $g(x, \mathbf{v})$ truely satisfy the stationary Vlasov equation. (Note that g and f are not the same object, as g is the distribution function measured in a particular global reference frame, whereas f is the distribution function measured in every local reference frame. In other words, the bulk velocity has been filtered out at every position for f, and f and g can be expressed in terms of each other if the bulk velocity is known at every position.) In general, it requires that

$$g(x, \mathbf{v}) = K(q\phi(x) + m_i(1+\mathbf{v}^2)^{1/2})$$

where K is an arbitary function. To facilitate the analysis, we adopt the water-bag model for f:

$$f(x, \mathbf{v}) = \frac{N_\infty}{\pi\beta_t^2\gamma_t^2c^2}, \quad when \quad |\mathbf{v}| < c\beta_t\gamma_t\sqrt{N_i(x)/N_\infty}$$

$$f(x, \mathbf{v}) = 0, \quad otherwise.$$

Here β_t is the ratio of the background ion thermal speed to the speed of light and γ_t is the Lorentz factor of the thermal speed. With this special choice, f can automatically satisfy the stationary Vlasov equation if the boundary of f, proportional to $N_i(x)$, also satisfies the stationary Maxwell's equations and momentum equations.

P_i and e_i are now found to relate to N_i in a nontrivial manner[4] (non-polytropic form). They are, however, reduced to the usual two-dimensional polytropic equations of state in the non-relativistic limit ($P_i \propto N_i^2$) and the ultra-relativistic limits ($P_i = e_i/2 \propto N_i^{3/2}$) .

(III) Main Results:

The closed set of equations can be combined to yield a single equation, which has the form

$$\left(\frac{d\phi}{dx}\right)^2 + V(\phi) = const.$$

where $V(\phi)$ is the Sagdeev potential. $V(\phi)$ becomes singular at some maximum value ϕ_m, below which $V(\phi)$ is regular. Depending on the parameters, such as the magnetosonic Mach number, temperature, soliton speed, etc., the soliton peak amplitude can approach ϕ_m from below. When the peak amplitude equals ϕ_m, the soliton becomes singular, must break up and can no longer exist. The breakup amplitude can be precisely determined from eq.(1).

When $\sqrt{\gamma}\gamma_t\beta_t << 1$, a cold soliton limit, and when $\gamma >> 1$, it follows $\phi_m = 1$ gigavolt. Here, γ is Lorentz factor of the soliton speed. On the other hand, in the hot soliton limit, $\sqrt{\gamma}\gamma_t\beta_t >> 1$, and when $\gamma >> 1$, we have $\phi_m = \sqrt{\gamma}\gamma_t\beta_t/12$ gigavolt. A very hot soliton can contain a much larger electrical potential than 1 gigavolt; however, ϕ_m can also be less than this value when the temperature is not high enough.

In the non-relativistic limit, ϕ_m can also be obtained with a straightforward algebra. As the general expression of ϕ_m for a finite-temperature plasma is rather complicated, we give its value in the zero-temperature limit: $\phi_m = 2m_iU_A^2/q$, where U_A is the Alfven speed far upstream. The critical magnetosonic Mach number in this limit is 2. Like the relativistic case, when a small but finite temperature is taken into account, it is found that the value of ϕ_m is reduced as a result of the decreasing magnetosonic Mach number.

(IV) Few areas of application:

We have discovered an interesting collective object that must propagate exactly perpendicular to the magnetic field to exist. How, in practice, can we, or the Nature, create such an odd object that demands so much for it to exist? Perpendicularity of propagation to a uniform magnetic field may seem a very stringent requirement. It is only so when the propagation speed is far below the speed of light. When the propagation speed is close to the speed of light, even when the propagation direction is not perpendicular to the magnetic field, there always exists a reference frame, where the parallel component of the magnetic field can be Lorentz-transformed away and the propagation becomes exactly perpendicular. The condition under which this can occur is that the speed of propagation v_p satisfies $v_p sec\theta > c$, where θ is the angle between the original direction of the magnetic field and that of the propagation. This particular reference frame is also called the Hoffman-Teller frame. With this consideration of practicality, a relativistic soliton is perhaps more realistic than a nonrelativistic one.

Relativistic solitons exist in a magnetized plasma which has a relativistic Alfven speed. In the Nature, the neutron-star magnetosphere is one of the few places that provide such an environment. The field strength is about 10^{12} gauss near the surface and decreases gradually outwards. If such solitons are excited and propagate outwards across the magnetosphere, it is natural to expect that the magnetosonic Mach number should increase outwards as the magnetosonic speed decreases. The solitons become steepened and may eventually break up in the outer magnetosphere when the singularity occurs. In the context of the Gamma-Ray-Burst, we propose that solitons are the key agent that transprots energy from the neutron star to the outer magnetosphere[7]. This scenario may account for the few-tens-MeV gamma ray observed in the bursts.

Another application of the relativistic soliton is concerning the coherent particle acceleration by a single soliton. The acceleration is across the magnetic field. As no particle can be trapped by the soliton forever, unlike the Surfatron[8], there exists a upper limit to which the highest energy a particle can acquire. Nevertheless, as we have more control on the shape, the speed, etc. of the soliton than the wave of the Surfatron, this advantage of the soliton accelerator may compensate for its disadvantage of the finite energy gain, and may lead to a more realistic acceleration device.

Acknowledgements:

This work is supported by the grants of the National Science Council, NSC 80-E-sp-008-02 and NSC 81-0208-M-008-513.

REFERENCES:

1. Michel, F.C., **Theory of Neutron Star Magnetosphere**, U. Chicago Press (1991).
2. Blandford,R.D., Netzer,H. and Woltjer,L., **Active Galactic Nuclei**, Springer Verlag

(1990).
3. Chiueh, T., Phys.Rev.Lett. **63**, 116 (1989).
4. Chiueh, T., Phys Rev. A **44**, 6944 (1991).
5. Adlam, J. and Allen, J., Philos. Mag. **3**, 448 (1958).
6. Morawetz, C.S., Phys.Fluids **4**, 988 (1961).
7. Chiueh, T., Chinese Jl. Phys., **30** (1992).
8. Katsouleas,T. and Dawson, J., Phys.Rev.Lett. **51**, 392 (1983).

A SIMPLE MODEL OF GENERATION OF DOUBLE LAYERS IN BOUNDED COLLISIONAL PLASMAS

Carmen Iuliana Ciubotariu-Jassy*
Physics Department, Polytechnic Institute
Strada Sfîntu Lazăr no. 21
F4-15, 6600 Iaşi, Romania

Abstract. The electric field distribution is investigated in a "collisional extended Pierce diode model" during irradiation by strong laser light.

1. INTRODUCTION AND MODEL

We consider a steady state single-emitter plane plasma diode in equilibrium with an external radiation which produces pairs of charged particles in the gas between the two parallel electrodes (continuous optical discharge"COD",Fig. 1)[1,2]. It corresponds to a situation where the system is collisional for the electrons, i.e., the electron mean free path is small with respect to the distance L between the electrodes. The electrons leave the emitter with constant density n_0 and velocity v_0, and are absorbed when hitting either electrode. Ions are assumed to be infinitely massive ($\mu_i \ll \mu_e$), forming a background that neutralizes the charge density of the electrons at the electrode from which the electrons are injected. The electrodes are held at the potential difference U which corresponds to a stationary current ("collisional extended Pierce diode model"). The ioniza-

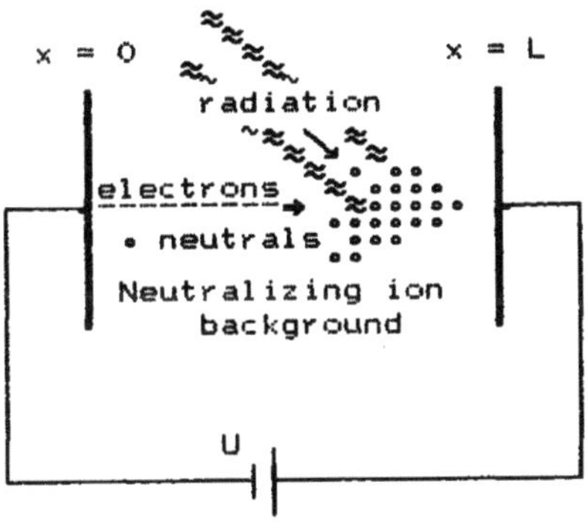

Fig. 1. Collisional plasma diode.

*Present address: Laboratoire de Physique des Gaz et des Plasmas, Universite Paris Sud, Bat. 212, 91405 Orsay Cedex, France.

$$-\int_{0_+}^{L_-} E(x)dx = U, \qquad -\int_{0_+}^{x} E(x)dx = \varphi(x), \tag{7}$$

with $\frac{dN}{dt}(x)$ the stationary ionization rate, x the position from the left electrode,n_o both the "reference ion background" density and electron emission density, and $x = 0_+$ is a position infinitesimally close to the right of the emitter surface $(x = 0)$[4]. In order to close the system of equations we need the momentum transfer equation,

$$m_e v_e \frac{dv_e}{dx} = -\frac{1}{n_e}\frac{dp_e}{dx} - eE - m_e D_e, \tag{8}$$

or the corresponding moment equation obtained from electron Boltzmann equation. A phenomenological collision term $(-m_e D_e \equiv -m_e v_e/\tau)$ was included in the latter equation to simulate the effect of damping. Instead of the informations induced by the equation (6) we prefer, for simplicity, the equation (3) completed with the parameters dN/dt and μ_e given by simulations and experiments. It is worth to note that non and "semi" self-consistent[5] or self-consistent[6] microscopic simulations of the electron transport by Monte-Carlo or particle in cell technique[7] in DC or RF discharges have been performed[8].

In the following we present the equations governing the electric field distributions for various spatial profiles of the ionization rate dN/dt. We consider that these profiles can be obtained practically from the shape of the laser beam or can be deduced from the microscopic simulations[9].

3. THE HOMOGENEOUS IONIZATION RATE CASE

We first apply the equations (1)-(7) to the case of uniform ionization rate,

$$\frac{dN}{dt} = R = \text{consatnt}. \tag{9}$$

Inserting (9) into the equation (2), we obtain

$$j_e(x) = eRx + j_o, \quad j_o \equiv j_e(0) = n_o e v_o, \tag{10}$$

and from Eqs. (3) we find

$$n_e(x) = \frac{j_o}{e\mu_e E} + \frac{R}{\mu_e E}\, x. \qquad (11)$$

In order to compare our results with the equations obtained in the case of collisionless plasma diode[10], we introduce the a-dimensional quantities:

$$\alpha^2 = \frac{q\rho_o L^2}{\varepsilon_o m v_o^2} \equiv \frac{\omega_{po}^2 L^2}{v_o^2}, \qquad \rho' = \frac{\rho}{\rho_o}, \qquad v' = \frac{v}{v_o},$$

$$x' = \frac{x}{L}, \qquad E' = \frac{qL}{mv_o^2} E, \qquad U' = \frac{q}{mv_o^2} U, \qquad (12)$$

$$R' = \frac{1}{n_o \omega_{po}} R, \qquad \mu_e' = \frac{mv_o}{qL} \mu_e,$$

where $\omega_{po} = (n_o e^2/\varepsilon_o m)^{1/2} \equiv (q\rho_o/\varepsilon_o m)^{1/2}$ is the electron plasma frequency of the beam, m=electron mass, $q = -e$, $\rho_o = -en_o$, $\rho = -en_e$. We note that the charge of the electron has been included into the electric field, so that the electric field E' is of the opposite sign as the physical electric field.

Now, from the equation (11) (dropping the primes), we obtain

$$\rho = \frac{1}{\mu_e E} + \alpha \frac{R}{\mu_e E}\, x, \qquad (13)$$

and the equation (4) $[\frac{dE}{dx} = \alpha^2(\rho - 1)]$ becomes

$$E\frac{dE}{dx} = \alpha^2 \frac{1}{\mu_e} + \alpha^3 \frac{R}{\mu_e} x - \alpha^2 E. \qquad (14)$$

Since $\alpha^3 \frac{R}{\mu_e} \neq 0$, substituting into equation (14)

$$x = u - \frac{1}{\alpha R}, \qquad \frac{dE}{dx} = \frac{dE}{du}, \qquad (15)$$

we get the homogeneous differential equation

$$EdE = (\alpha^3 \frac{R}{\mu_e} x - \alpha^2 E)du, \qquad (16)$$

and by means of the substitution $E = u\gamma$, where γ is the new unknown function. it is transformed to an equation with variable separable. The solution is

$$x=-\frac{1}{\alpha R}+C\left[\left|-\mathcal{E}^2-\alpha^2\mathcal{E}+\alpha^3\frac{R}{\mu_e}\right|\left|\frac{-2\mathcal{E}-\alpha^2-[\alpha^4+(4\alpha^3R/\mu_e)]^{1/2}}{-2\mathcal{E}-\alpha^2+[\alpha^4+(4\alpha^3R/\mu_e)]^{1/2}}\right|^{\frac{\alpha^2}{(\alpha^4+4\alpha^3\frac{R}{\mu_e})^{1/2}}}\right]^{-\frac{1}{2}}, \tag{17}$$

where C is determined by the condition

$$\mathcal{E}(0) \equiv \mathcal{E}_o = \alpha RE_o. \tag{18}$$

E_o represents the electric field at the injection plane, and may be either positive or negative. As an example, Fig. 2 shows the spatial dependence of the "electric field" $\mathcal{E}$ for a particular choice of parameters

α, R, μ_e, C.

It is seen that the four real roots

$\mathcal{E}_1$, $\mathcal{E}_2$, $\mathcal{E}_3$, $\mathcal{E}_4$,

for a given x may lead to the formation of double layers. A slow variation of parameters explains the LF instability of DL, as it was observed in some laboratory and numerical experiments (see [11]). The enlarged version of this work considers the cases of stationary ionization wave and a Gaussian profile for R.

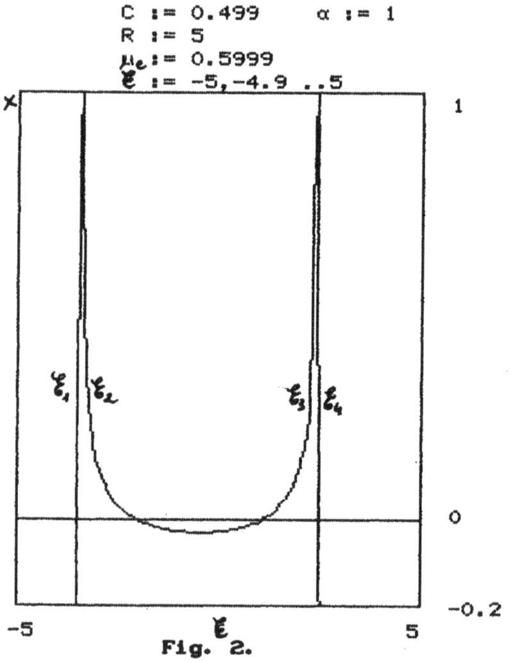

Fig. 2.

ACKNOWLEDGEMENTS

The author wishes to thank Professors A. M. Pointu, M. Fitaire, M.Sanduloviciu and G. Popa for encouragement. Special thanks are due to Dr. Michales Chryssos for forcing me to clarify,in my own mind,as well as on paper, the concepts which have been presented here.

REFERENCES

1. J. Uhlenbush, Invited paper, XVI International Conference on Phenomena in Ionized Gases, Dusseldorf 29th August - 2nd September 1983, Eds. W. Botticher, H. Wenk, E. Schulz-Gulde, p. 119.
2. Yu. P. Raiser, JETP Letters **11** (1970), 120.
3. C. Carloff, E. Krametz, J. A. Schafer, J. Uhlenbush, Contributed paper,XVI International Conference on Phenomena in Ionized Gases, Dusseldorf 29th August - 2nd September 1983, Eds. W.Botticher, H. Wenk, E. Schulz-Gulde, Volume 2, p. 330.
4. H. Horhager, S. Kuhn, Phys. Fluids **B2** (1990), 2741; p. 2743.
5. J. P. Boeuf, E. Marode, J. Phys. D: Appl. Phys. **15** (1982), 2169.
6. K. Kitamori, T. Tagashira, in Proc. of the 9th ESCAMPIG, Lisbon, Portugal, Ed. C. M. Ferreira, European Physical Society, p. 277.
7. R. W. Boswell, I. J. Morey, J. Appl. Phys. **52** (1988), 21.
8. J. P. Boeuf, Ph. Belenguer, in "Non Equilibrium Processes in Partially Ionized Gases" edited by Mario Capitelli and J. Norman Bardsley, NATO ASI SERIES: SERIES B: PHYSICS, Volume 220, p. 1.
9. Ph. Belenguer, J. P. Boeuf, Phys. Rev. **A41**(1990),4447; p.4456.
10.W. S. Lawson, Phys. Fluids **B1** (1989), 1483.

DOUBLE LAYER WITHIN THE BOUNDARY SHEATH OF PLASMAS WITH NEGATIVE IONS

R. Deutsch and E. Räuchle, Institut für Plasmaforschung, Universität Stuttgart
Pfaffenwaldring 31, D - 7000 Stuttgart - 80, Germany

1. Introduction.

Double layers are observed in the cathode fall region of plasmas with negative ions by Gottscho et al. [1]. They determined experimentally the electric field distribution in the plasma boundary sheath in the neighbourhood of the cathode. In the experimental conditions the collisions between ions and neutrals are dominant in the sheath. The mean free path is of the order of the thickness of the cathode fall region, hence the collisionless sheath theory is suited to describe the field inside the sheath. This theory is extended here to plasmas with negative ions and used to describe the structure of the sheath.

In the bulk plasma and in the presheath the negative ions gain momentum from the electric field and by collisions with other particles. If the negative ion density is less than a critical value, the mean momentum of a negative ion gained by Coulomb collisions with positive ions is larger than the momentum obtained from the electric field and losed by friction. In this case the negative ions are captured by the positive ion flow moving together through the presheath into the collisionless sheath.

During the penetration into the sheath the negative ions are then reflected by the electric field. Because of this reflection a double layer inside the sheath is formed.

The theoretical results are in good agreement with the experiment.

2. Negative ion capture by the positive ion flow in the bulk plasma and in the presheath

A quasineutral unmagnetized plasma in contact with a plane electrode (fig.1) is considered. It is assumed that inside the sheath the particle density is low enough that collisions can be neglected. The processes in the bulk plasma and in the presheath, which determine the ion density and the particle flow from the presheath into the sheath, strongly depend on the collisions; therefore in the presheath and in the bulk plasma just these processes are considered. The changes in the mean momentum of the negative ions are dominated by the elastic collisions with positive ions and neutrals characterized by the frequencies ν_{-+} and ν_n. (The influence of the friction forces due to charge exchange processes can be also included in ν_n).

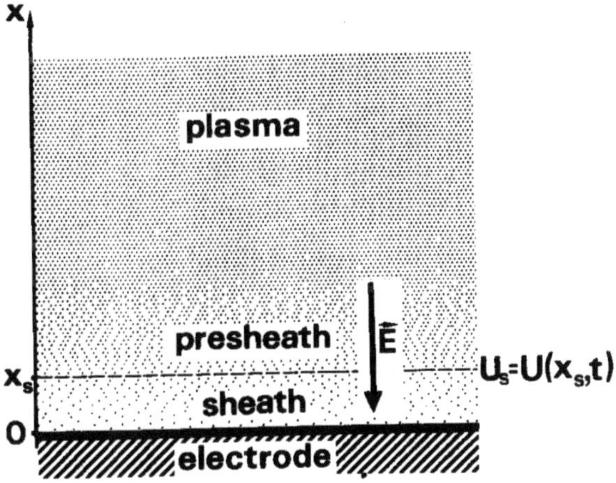

Fig. 1 Geometry and notations.

If

$$n_- < \left(1 - \frac{\nu_n m_+}{\nu_{-+} m_-}\right) n_+. \tag{1}$$

with the mean densities $n_{-,+}$ in the presheath and the bulk plasma and the masses $m_{-,+}$ of the negative and positive ions, the momentum transfer from the positive ions to the negative ions by Coulomb collisions is larger than the momentum change due to the electrostatic force and the retarding force of the collisions of the negative ions with neutrals. In this case the negative ions move in the same direction as the positive ions to the negatively charged electrode. Eq. (1) is the condition for the capture of the negative ions by the positive ion flow which is accelerated in the electric field of the plasma.

3. The model of the collisionless sheath.

At the interface between the sheath and presheath (at $x = x_S$ in fig.1) the positive ions are moving from the presheath into the sheath with an average velocity $v_{S+} < 0$ and the negative ions with $v_{S-} < 0$. At $x = x_S$ the velocity distributions are approximated by shifted Maxwellian distributions plotted in fig. 2 for negative ions for $T_e \gg T_i$. The velocity spread around the mean velocity is characterized by the ion temperature T_i.

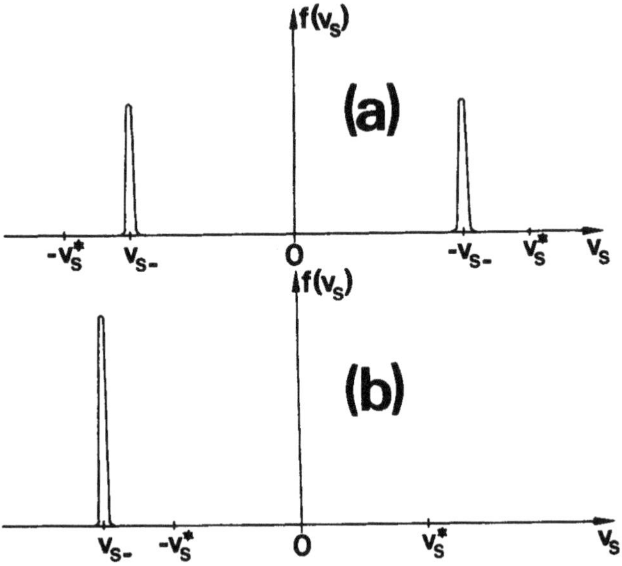

Fig. 2 Velcity distribution of the negative ions at $x = x_S$ with ion reflection (a) and without ion reflection (b) in the sheath.

The ion densities calculated from the shifted Maxwellian distribution functions and the electron density calculated from the Boltzmann distribution function are introduced into the Poisson equation. For the potential U(x) results:

$$\frac{d^2U}{dx^2} = \frac{n_S q}{\varepsilon_o}\left\{ \frac{1}{1+\alpha} \exp\left[\frac{q\left(U(x) - U_S\right)}{kT_e}\right] + \right.$$

$$+ \frac{\alpha}{1+\alpha}\sqrt{\frac{m_-}{2\pi kT_i}} \int_{z^*}^{\infty} \frac{(z - v_{S-}) \exp\left[-\frac{m_- z^2}{2kT_i}\right] dz}{\sqrt{(z - v_{S-})^2 + \frac{2q}{m_-}\left(U(x) - U_S\right)}} -$$

$$\left. - \sqrt{\frac{m_+}{2\pi kT_i}} \int_{v_{S+}}^{\infty} \frac{(z - v_{S+}) \exp\left[-\frac{m_+ z^2}{2kT_i}\right] dz}{\sqrt{(z - v_{S+})^2 - \frac{2q}{m_+}\left(U(x) - U_S\right)}} \right\} . \qquad (2)$$

with $z^{*} = v_{S-} + \sqrt{\frac{2q}{m_-}\left(U_S - U(x)\right)}$, the density ratio $\alpha = n_{S-}/n_{Se}$, the elementary charge q, the index e for electrons and the index S for the quantities at $x = x_S$.

4. Generalized Bohm criterion.

Integrating eq. (2) with the boundary conditions $U(x_S) = U_S$ and $\left.\frac{dU}{dx}\right|_{x_S} \approx 0$ a real solution results if

$$\frac{1+\alpha}{\beta_+ - 3/\Theta} < 1 - \frac{\alpha}{\beta_- - 3/\Theta} . \quad (3)$$

with $\beta_\pm = \frac{m_\pm v_{S\pm}^2}{kT_e}$ and $\Theta = T_e/T_i$. The inequality (3) is a generalization of the Bohm criterion for plasmas with negative ions to the case, when the negative ions are captured by the positive ion flow and $\beta_- \gg T_i/T_e$. The parameter values, which satisfy eq. (3), lie outside of the hatched region of the β_+- β_--plane of fig. 3.

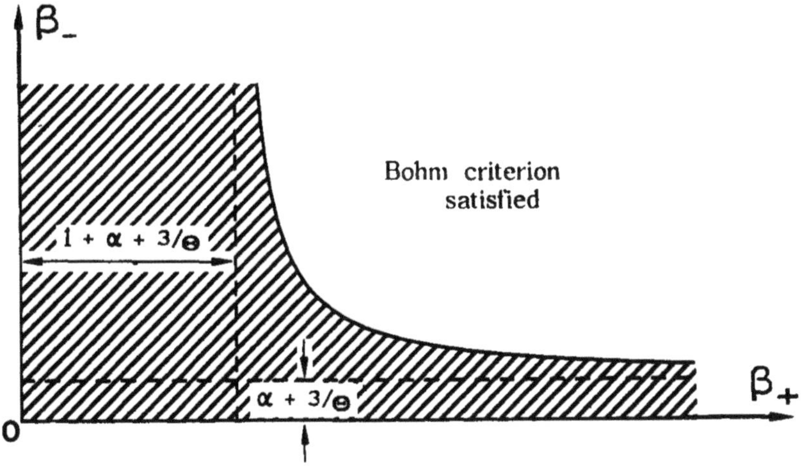

Fig. 3 Bohm limit in the β_+- β_- - plane

The positive ions need to have at $x = x_S$ for a stable sheath a larger velocity than in plasmas without a captured negative ion flow. For $\alpha \rightarrow 0$ eq. (3) coincides with the usual Bohm criterion of plasmas with positive ions only[2-3].

5. Reflection of the negative ions.

Because of the small mass of the electrons the plasma potential is usually positive with respect to the electrodes. In the collisionless sheath the positive ions are accelerated to the electrodes. The negative ions are slowed down and if

$$U(0) - U_S < -\frac{m_-}{2q} v_{S-}^2$$

they are reflected totally (fig. 4) by the field.

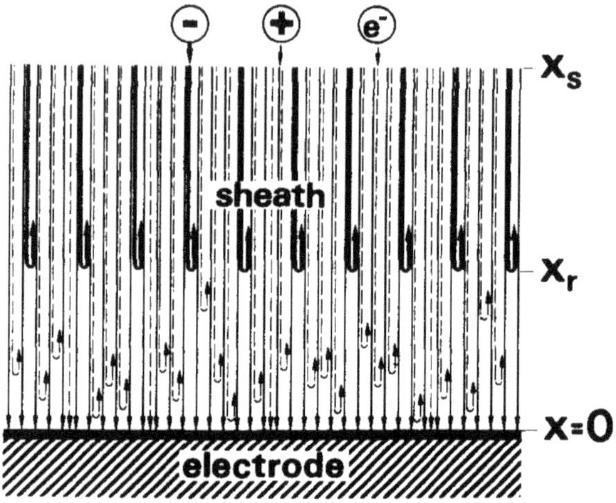

Fig. 4 Reflection of the negative ions in the sheath.

Integrating eq. (2) the electric potential, electric field and charge density as functions of the distance from the electrode are determined. Typical results are given in figs. 5(a) and 5(b). Here the normalized potential is

$$\varphi = \frac{m_+ v_{S+}^2 - 2q(U - U_S)}{kT_e} .$$

The electric field is normalized by $E^* = -kT_e/2q\lambda_D$ with the Debye length λ_D. Results are shown in Fig. 5(a) for α=0.1. E changes monotonically with the normalized distance $\xi = x/\lambda_D$ and the charge density in the sheath is everywhere positive. In Fig. 5(b) the results are

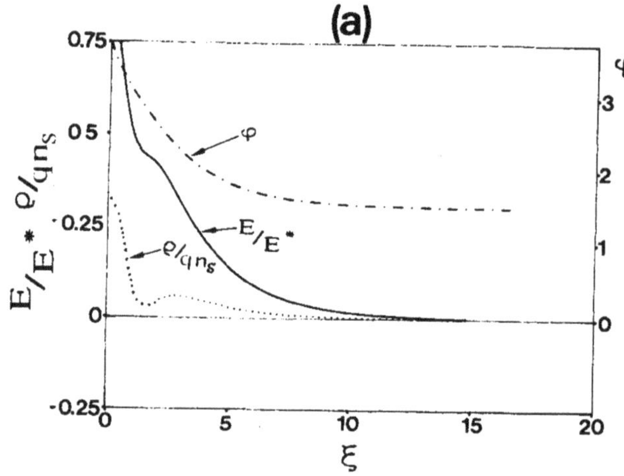

Fig. 5(a) Sheath structure for $\Theta = 100$, $\alpha = 0.1$ and $\beta_- = \beta_+ = 1.5$

shown for a larger negative ion density (α = 0.145). Here a double layer exists inside the sheath with a maximum and a minimum of the electric field. This is a result of reflection of the negative ions.

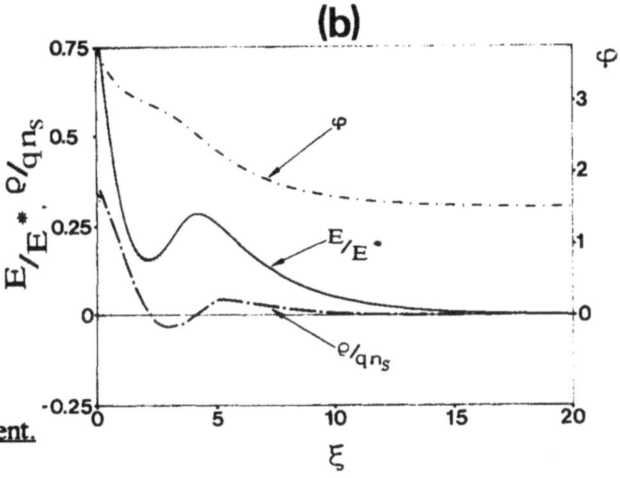

Fig. 5(b) Sheath structure for $\Theta = 100$, $\alpha = 0.145$ and $\beta_- = \beta_+ = 1.5$

6. Comparison with experiment.

The spatial distribution of the electric field in the cathode fall was measured by Gottscho et al.[1] for RF glow discharges at 50 kHz in BCl_3. Double layers are observed. For a comparison of the experimental results with theory, Fig.3b of ref. [1] with $\omega t = \frac{11}{10}\pi$ is reproduced in fig. 6 by the squares . The theoretical curve obtained by integration of eq. (2) is given by the solid line.

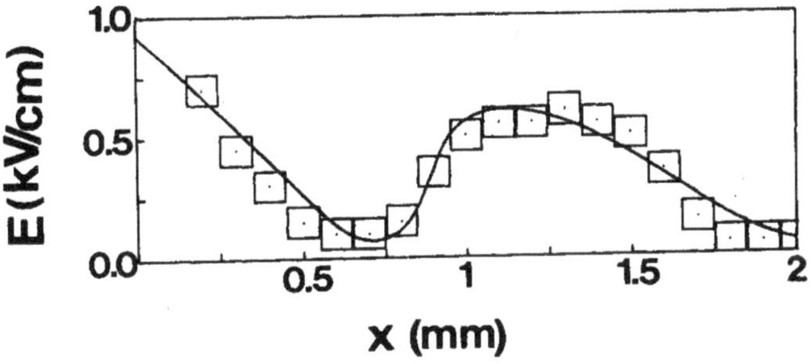

Fig. 6 Experimental points of ref. [1] (squares) and the theoretical curve for the electric field (solid line) with α = 0.727, $\beta_+ = 45$, $\beta_- = 42.75$, $\Theta = 100$, $T_e = 2.1$ eV and $n_e = 1.16 \cdot 10^{10}$ cm^{-3}.

References.

1 R.A.Gottscho, Phys. Rev. **A 36**, 2233 (1987).

2 N.St.J.Braithwaite and J.E.Allen, J. Phys. D: Appl. Phys. **21**, 1733 (1988)

3 K.- U. Riemann, J. Phys. D: Appl. Phys. **24**, 493 (1991)

NONLINEAR STRUCTURES IN A COLLISIONLESS PLASMA DETERMINED BY SELF–CONSISTENT ELECTRON TRAPPING

A.Ya. Ender,[a] S.Kuhn,[b] and V.I. Kuznetsov[a]

[a] *A.F. Ioffe Physicotechnical Institute, Russian Academy of Sciences, 194021 St. Petersburg, Russia.*

[b] *Institute for Theoretical Physics, University of Innsbruck, A–6020 Innsbruck, Austria*

Abstract. It is well known that in a single–emitter collisionless plasma diode ("KDSI") large–amplitude, periodic nonlinear oscillations may occur which alternately exhibit a "slow" (ion) and a "fast" (electron) stage. The fast stage basically consists in the growth of Pierce–type instability into its nonlinear regime and may result in various nonlinear structures crucial to the subsequent slow stage. The present paper deals in detail with the situation in which Pierce instability leads to electron trapping and, subsequently, to an oscillatory final state.

1. Introduction. In a bounded collisionless plasma where both ions and electrons enter from one of the boundaries (the emitter), an aperiodic Pierce–type instability arises at certain moments of the nonlinear oscillation process ("fast stage").[1-3] As a result, various nonlinear structures are formed in the interelectrode plasma and slowly evolve during the subsequent ion stage ("slow stage") of the process. Examples of such structures are: moving virtual cathodes, which strongly limit the electron current and determine the oscillation amplitude; moving double layers, inside which the potential drops strongly and charge separation occurs over a few Debye lengths; strongly accelerated cold ion beams, due to a "fast" transfer of energy from an electron beam to the ions via the electric field, etc. Thus, the Pierce instability is the main source for the formation of these plasma structures in a collisionless bounded plasma, and the type of each structure is determined by the late nonlinear stage of this instability development. Therefore, studying the late stage of Pierce–instability evolution is undoubtedly of self–independent interest.

In the course of the fast stage, a deep positive–potential well for electrons may form. The electrons lose energy in their interaction with the strongly time–dependent electric field which is present during the well's formation, and if the rate at which the potential distribution changes is sufficiently high, part of the electrons approaching the collector are reflected by the repulsive potential "wall" existing there, and are eventually trapped in the well. The trapped electrons can have a substantial effect on the final stage of this process and, hence, also on the character of the subsequent ion stage. The present paper is concerned with a detailed study of the self–consistent electron trapping in a potential well resulting from Pierce–instability development, and of the nonlinear structures arising as a result thereof.

2. The Pierce diode as a model for the KDSI. In the Knudsen Diode with Surface Ionization (KDSI) (also known as "single–emitter collisionless plasma diode"), charged particles are injected from the electrode surfaces. When the fast stage of the oscillation process begins, a positive potential jump accelerating electrons exists near the emitter so that, in fact, an electron beam enters the plasma. For this reason, the fast stage of nonlinear KDSI dynamics can be studied in terms of the Pierce–diode model in which a directed monoenergetic electron beam propagates between plane–parallel electrodes through a uniform positive ion background in a vanishing electric field[4] ("ground state"). The various other "plasma states", which may exist for an ion distribution corresponding to the ground state (which is the fully self–consistent, exactly time–independent state), have been analyzed and compared by us both for the KDSI and the Pierce diode by means of the "(η,ϵ) diagram" technique.[5] The state with an electron potential well was found to always exist for both diodes. The growth rates are practically equal, and the potential perturbations associated with the instability have equal shapes for equivalent gaps in both diodes.

3. Estimating the threshold for electron trapping. In calculations of time–dependent electron processes in the KDSI,[6] and later also in analogous simulations in the Pierce diode,[7,8] two different types of Pierce–type instability development were discovered. The first type of fast transition, which coincides with the one found by means of the (η,ϵ) diagram, ends up at a time–independent state in which the potential distribution is characterized by a deep well for electrons but no trapped electrons are present. It was found that for $d \lesssim 1.5\pi\lambda_D$ (where $\lambda_D \equiv [mv_0^2/(4\pi n_0 e^2)]^{\frac{1}{2}}$ is a formal Debye length, with v_0 the velocity and n_0 the density of the unperturbed cold electron beam) no electron trapping occurs as the Pierce instability develops from a very small initial perturbation. For a wider gap, on the other hand, multiply reflected electrons were discovered, such as in Ref. 8, where a Pierce diode with $d = 2.5\pi\lambda_D$ was studied as well. Therefore one may expect electron trapping in potential wells for gap widths exceeding some threshold value which must lie somewhere between $1.5\pi\lambda_D$ and $2.5\pi\lambda_D$.

To estimate this threshold for electron trapping, we have carried out some preliminary calculations, simplified by discarding reflected electrons, for several diode gap widths d/l between $1.5\pi\lambda_D$ and $2\pi\lambda_D$. The shape of the potential well was found to remain practically unchanged and close to a quadratic parabola during the final stage of such a transition. Let us to define the "equilibration rate" of the potential well by

$$\tilde{\omega}_i(t) \equiv \frac{1}{\Phi^1(\infty)-\Phi^1(t)} \frac{d\Phi^1(t)}{dt}, \tag{1}$$

where $\Phi^1(t)$ is the maximum potential at time t. It turns out that, independently of the gap width and, hence, of the linear growth rate, the dimensionless equilibration rate $\tilde{\omega}_i(t)/\omega_p$ always lies between 0.1 and 0.2, which is close to the maximum linear growth rate. In studying the threshold for electron trapping we assume that the potential distribution evolves according to

$$\begin{aligned} \Phi(z,t) &\simeq \Phi_{max}\left[1-\exp(-<\tilde{\omega}_i>t)\right]\left[1-\left[\frac{2z}{d}-1\right]^2\right], \\ \Phi_{max} &= \frac{mv_0^2}{e}\left[\frac{1}{8}\left[\frac{d}{\lambda_D}-\pi+2\right]^2-\frac{1}{2}\right], \end{aligned} \tag{2}$$

with $<\tilde{\omega}_i>$ an average equilibration rate of the potential well, lying in the interval $(0.1 \div 0.2)\omega_p$.

The change in total energy of an electron during its motion from its point of emission $(0,t^0)$ to the point (z,t) is connected with the change in field by the relation[6]

$$G(z,t;v^0) = -e\int_{t^0}^{t} d\tau \frac{\partial}{\partial\tau}\Phi(x,\tau)\Big|_{x=x(\tau;t^0,v^0)} . \tag{3}$$

One may assume that the potential distribution does not change substantially during the electron's transit from the emitter to its point of reflection, so that, as a good approximation, the integration in (3) can be performed along the trajectory corresponding to the field which existed at the time the electron left the emitter. The minimum gap width for which electron trapping sets in is determined by the condition that the first reflection occurs at the collector. Using the law of energy conservation and the condition that at a reflection point the particle's kinetic energy vanishes, we find that the threshold for electron trapping satisfies the following condition:

$$\frac{(2+\kappa)e^{-\kappa}-(2-\kappa)}{2\kappa^2}\left[\left[\frac{d}{\lambda_D}-\pi+2\right]^2-4\right]-\frac{1}{2} \geq 0, \tag{4}$$

where $\kappa \equiv \tilde{\omega}_i d/v_0$. By solving Eq. (4) we found that in the Pierce diode electron trapping may occur only if $d/\lambda_D > (1.6 \div 1.7)\pi$.

4. Time evolution of electron trapping. Our study of the electron–trapping process is based on the numerical method developed in Ref. 6, where it was adopted to calculating the electron stage of the KDSI oscillations. This method permits one to calculate with high accuracy the charged–particle distribution functions in a time–dependent electric field. We have carried out calculations for several values of the gap width. To take into account a thermal spread in electron velocities, the electron emission distribution function is taken to be of the following form:

$$f_e^{0+}(v) = \frac{n_0}{2\Delta v_0}\,\Theta(v-v_0+\Delta v_0)\,\Theta(v_0+\Delta v_0-v)\,, \tag{5}$$

with $\Delta v_0 = 0.02v_0$. The threshold gap width for electron trapping turned out to lie between $1.8\pi\lambda_D$ and $1.9\pi\lambda_D$, thus being in fair agreement with the semi–analytical estimate obtained above.

We have studied the electron–trapping process in detail for $d = 1.97\pi\lambda_D$. The time step is $0.3\omega_p^{-1}$, the spatial meshwidth is $d/35$, and the initial Φ–amplitude equals 1% of Φ_{max}. During the initial stage, the process develops according to linear theory, and the width of the forming potential well turns out to be less than the gap width, cf. curve 1 in Fig. 1. Later, the well becomes deeper and wider. Approximately at the time when the potential minimum Φ^2 achieves it most negative value, reflected electrons appear, cf. point 4 in Fig. 2. Since the value of the negative potential barrier is less than 60% of the electron initial energy, it is obvious that it is the interaction of the electrons with the *time–dependent* electric field which is essential to the process of reflection. The reflected electrons form spatially bounded group, cf. Fig. 3. Moving toward the emitter these electrons loose energy, and when they reach the emitter, a small part of them leaves the gap. The remaining electrons, however, are reflected again and begin to move to the collector. This is how electrons get trapped.

When the reflected electrons move from the center to the emitter, the growth rate of the well's depth increases sharply, cf. Fig. 2. Newly emitted electrons now move into the fast–growing potential well, and because of their interaction with the rapidly changing electric field they may loose a significant amount of energy so as to be reflected by the potential wall near the collector. In fact, a new group of reflected electrons is formed. Some part of them forms the second cloud of trapped electrons, in the same way as described for the first group.

As the process continues, both groups of trapped electrons move in the potential well. They are alternately reflected by the retarding potentials near the electrodes, thus giving rise to potential oscillations at twice the frequency of the electron oscillation. For this reason, the whole electron process does not end up at a time–independent state, but rather leads to an oscillating potential structure whose time–averaged well depth is less than the time–independent one which would occur without trapped electrons, cf. solid and dashed curves in Fig. 2.

5. Final–stage oscillations. This oscillatory behavior of the potential can be explained as follows. As follows from inspection of Fig. 3, the spatial trapped–electron–density distribution may be approximated by

$$n_e^t(z,t) = \frac{N}{z_2(t)-z_1(t)} \Theta[z-z_1(t)]\, \Theta[z_2(t)-z] \,. \tag{6}$$

We now decompose the potential distribution and the electron density into two parts as follows: $\Phi(z,t) = \Phi^0(z) + \tilde{\Phi}(z,t)$ and $n_e(z,t) = n_e^0(z) + n_e^t(z,t)$, where the superscript ^{0} refers to the final stage of the process with reflected electrons discarded. We then find

$$\tilde{\Phi}(z,t) = -4\pi e N d \begin{cases} \left[1 - \frac{z_1+z_2}{2d}\right]\frac{z}{d} & (z < z_1) \\ \left[1 - \frac{z_1+z_2}{2d}\right]\frac{z}{d} - \frac{(z-z_1)^2}{2d(z_2-z_1)} & (z_1 < z < z_2) \\ \frac{z_1+z_2}{2d}\left[1 - \frac{z}{d}\right] & (z > z_2) \end{cases} . \tag{7}$$

Figure 4 shows a typical example of $\tilde{\Phi}(z,t)$. This function is always negative, which means that the trapped electrons tend to decrease the depth of the potential well. When the trapped–electron cloud is closest to the emitter or to the collector, its spatial width Δ reaches its minimum value and the amplitude of the potential perturbation becomes

$$\frac{|\tilde{\Phi}_{min}|}{4\pi e N d} = \frac{\Delta}{2d}\left[1 - \frac{\Delta^2}{4d^2}\right] \ll 1 \,. \tag{8}$$

When the trapped electrons are in the center, their spatial width is greatest and the amplitude assumes its maximum:

$$\frac{|\tilde{\Phi}_{min}|}{4\pi e N d} = \frac{1}{4}\left[1 - \frac{\Delta}{2d}\right] \simeq 1 \,. \tag{9}$$

Thus, the influence of the trapped electrons on the potential distribution is strongest when they are in the center of the diode, and smallest when they are adjacent to one of the electrodes.

Acknowledgements. This work was supported by the Academy of Sciences of the USSR, the Austrian Academy of Sciences, and the Austrian Science Foundation (FWF) under contracts P7005 and P8405.

References

[1]V.I. Kuznetsov and A.Ya. Ender, Zh. Tekh. Fiz. **53**, 2329 (1983) [Sov. Phys. Tech. Phys. **28**, 1431 (1983)].
[2]R.Ya. Kucherov, Z.A. Oganezov, L.S. Timoshenko, and V.K. Tskhakaya, Fiz. Plazmy **15**, 1321 (1989) [Sov. J. Plasma Phys. **15**(11), 766 (1989)].
[3]S. Iizuka, K. Saeki, N. Sato, and Y. Hatta, J. Phys. Soc. Japan **54**, 2516 (1985).
[4]J.R. Pierce, J. Appl. Phys. **15**, 721 (1944).
[5]V.I. Kuznetsov and A.Ya. Ender, Zh. Tekh. Fiz. **47**, 2237 (1977) [Sov. Phys. Tech. Phys. **22**, 1295 (1977)].
[6]V.I. Kuznetsov and A.Ya. Ender, Zh. Tekh. Fiz. **51**, 2250 (1981) [Sov. Phys. Tech. Phys. **26**, 1326 (1981)].
[7]T.M. Burinskaya and A.S. Volokitin, Fiz. Plazmy **9**, 453 (1983) [Sov. J. Plasma Phys. **9**, 261 (1983)]
[8]T.L. Crystal and S. Kuhn, Phys. Fluids **28**, 2116 (1985).

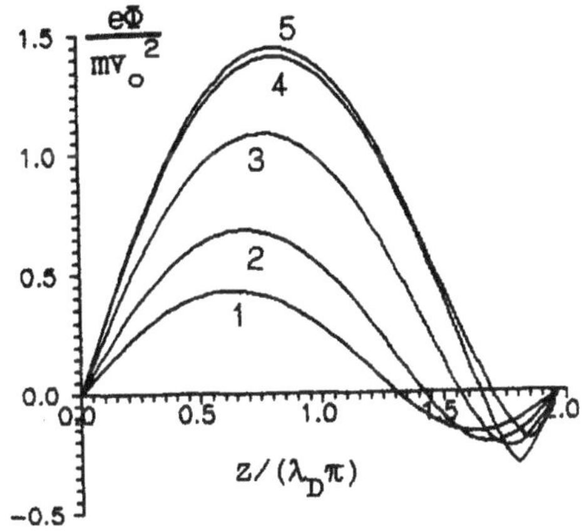

FIG. 1. Potential distributions corresponding to times $t\omega_p = 52.5$ (curve 1), 60.0 (2), 67.5 (3), 72.3 (4), and 74.7 (5) in a Pierce diode with $d/\lambda_D = 1.97\pi$.

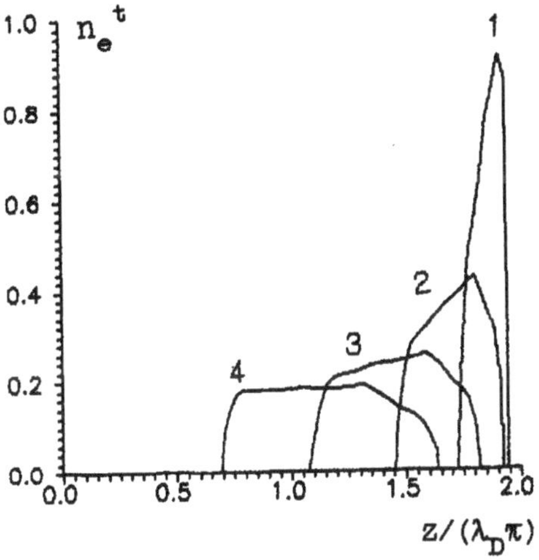

FIG. 3. Spatial trapped–electron–density profiles corresponding to times $t\omega_p = 78.9$ (curve 1), 79.5 (2), 80.1 (3), and 80.4 (4) in a Pierce diode with $d/\lambda_D = 1.97\pi$.

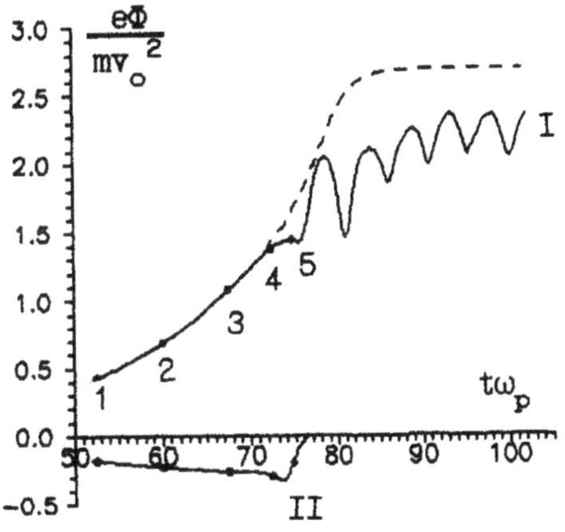

FIG. 2. Time history of the maximum potential Φ^1 (curve I) and the minimum potential (curve II) in a Pierce diode with $d/\lambda_D = 1.97\pi$. Points 1 to 5 correspond to the same times as in Fig. 1.

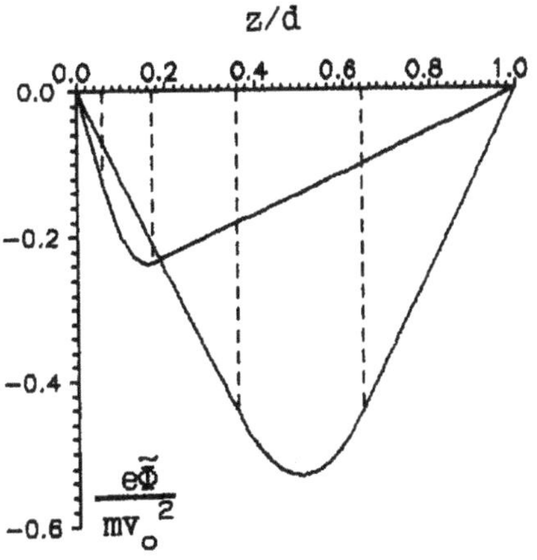

FIG. 4. Typical examples of potential perturbations caused by trapped electrons.

NONLINEAR SURFACE WAVES IN A CYLINDRICAL PLASMA WAVEGUIDE

M. Georgieva, A. Shivarova and I. Urdev, Sofia University, Faculty of Physics, BG-1126 Sofia, Bulgaria

1. INTRODUCTION: As known, the surface waves (SWs) are proper modes of guided plasma systems.[1] Since the SW field amplitudes vary in direction perpendicular to that of the wave propagation, the study of the SW behaviour in the nonlinear case should lead to two-dimensional nonlinear problem.[2] Consequently, the action of the wave on its own behaviour through the plasma nonlinearity should combine the nonlinear effects of self-modulation and self-focussing. Important information for both the self-modulation and the self-focussing of the waves consists in their nonlinear dispersion relation (NLDR).

Here we present a derivation of the NLDR of SW propagation. The considered system consisting of a plasma column surrounded by a dielectric/vacuum (i.e. an open plasma waveguide) is of interest with a view to the experiments in a gas-discharge plasma.[3,4] The mechanisms of striction nonlinearity due to the action of the ponderomotive force and the ionization nonlinearity are involved under the condition of weak nonlinearity.

2. NONLINEAR WAVE EQUATION: We investigate a SW (TM azimuthally symmetric mode $(\partial/\partial\varphi = 0)$: $\vec{E} = (E_r(r), 0, E_z(r))$, $\vec{H} = (0, H_\varphi(r), 0))$ propagating in z-direction along a cylindrical plasma column $(r < R)$ bounded by a dielectric $(r > R)$. The wave field variation is of the type $\propto\exp(-i\omega t + ikz)$ where k and ω are the wavenumber and the frequency, respectively.

In the plasma, the set of the Maxwell's equations reduces to the following equations for the electric $(\vec{E})$ and magnetic $(\vec{H})$ fields of the wave:

$$\begin{aligned} \Delta\vec{E} + (\omega^2/c^2)\varepsilon_0\vec{E} &= -(1/\varepsilon_0)\vec{\nabla}\,\vec{\nabla}\cdot(\delta\varepsilon\vec{E}) - (\omega^2/c^2)\delta\varepsilon\vec{E} \\ \Delta\vec{H} + (\omega^2/c^2)\varepsilon_0\vec{H} &= -(\omega^2/c^2)\delta\varepsilon\vec{H} + i(\omega/c)(\vec{\nabla}\delta\varepsilon)\times\vec{E}. \end{aligned} \qquad (1)$$

The plasma permittivity presented as $\varepsilon = \varepsilon_0 + \delta\varepsilon$ includes a nonlinear contribution $\delta\varepsilon$ to its equilibrium value ε_0. The righ-hand sides of eqs.(1) are nonlinear terms in which the linear field solutions are introduced, assuming weak nonlinearity. The same holds for the nonlinear part of the plasma permittivity which, taken in the form

$$\delta\varepsilon = \mp \alpha |\vec{E}(r)|^2/E_n^2 \tag{2}$$

with $\alpha = 1 - \varepsilon_0$ accounts for the two mechanisms:[2,5] (i) striction nonlinearity (sign "+") and, (ii) ionization nonlinearity at recombination (sign "-") and diffusion (sign "+") controlled regimes. E_n is the normalizing field amplitude with a particular value for each type of nonlinearity.

Having in mind the form of the linear solutions[1] for the SW field, eqs.(1) reduce to:

$$\begin{aligned} &\frac{d^2E_z}{dr^2} + \frac{1}{r}\frac{dE_z}{dr} - \left[k^2 - \frac{\omega^2}{c^2}\varepsilon_0\right]E_z = P_z(r) \\ &\frac{d^2H_\varphi}{dr^2} + \frac{1}{r}\frac{dH_\varphi}{dr} - \left[k^2 - \frac{\omega^2}{c^2}\varepsilon_0 + \frac{1}{r^2}\right]H_\varphi = Q_\varphi(r) \end{aligned} \tag{3}$$

where the nonlinear terms are:

$$\begin{aligned} &P_z(r) = -\left(k_{\perp 0}^2/\varepsilon_0\right)A\left[\beta I_1(x)\frac{d\delta\varepsilon}{dx} - \left(1 - \beta\right)I_0(x)\delta\varepsilon\right] \\ &Q_\varphi(r) = -i\left(k_{\perp 0}\omega/c\right)A\left[I_0(x)\frac{d\delta\varepsilon}{dx} + \left(1 - \beta\right)I_1(x)\delta\varepsilon\right]. \end{aligned} \tag{4}$$

Here A is the amplitude of the linear E_z-field component at the axis ($r = 0$) and $k_{\perp 0} = \left[k_0^2 - \left(\omega^2/c^2\right)\varepsilon_0\right]^{1/2}$ describes the linear field decay in radial direction; $\beta = k_0^2/k_{\perp 0}^2$, $x = k_{\perp 0}r$ and $\delta\varepsilon = \gamma F(x)$ with $\gamma = \mp\alpha\left(|A|^2/E_n^2\right)$ and $F(x) = I_0^2(x) + \beta I_1^2(x)$. The left-hand sides of eqs.(3) involves the total wave field amplitude $E_z = E_{z0} + \delta E_z$, $H_\varphi = H_{\varphi 0} + \delta H_\varphi$ and wavenumber $k = k_0 + \delta k$, i.e. linear parts and nonlinear contributions to them. The corresponding equation for the fields in the region, occupied by the dielectric can be obtained from eqs.(3) by replacing $\varepsilon_0 \to \varepsilon_d$, $P_z(r) \to 0$, $Q_\varphi(r) \to 0$, $k_{\perp 0} \to k_{\perp d0}$, $x \to y = k_{\perp d0}r$.

3. SOLUTIONS FOR THE WAVE FIELDS: The solution of eqs.(3) for the E_z-electric field component and for the magnetic field component H_φ in the plasma ($r < R$) are:

$$E_z(r < R) = A[I_0(x) + W(x) + T(x)] \quad (5)$$

$$H_\varphi(r < R) = -i(\omega\varepsilon_0/k_{\perp 0}c)A[I_1(x) + V(x) + S(x)] \quad (6)$$

where

$$W(x) = (k_0\delta k/k_{\perp 0}^2)\, xI_1(x),$$

$$V(x) = (k_0\delta k/k_{\perp 0}^2)\left[xI_0(x) - 2I_1(x)\right],$$

$$T(x) = (1/\varepsilon_0)\left[I_0(x)\int x\delta\varepsilon G(x)dx - (1/\gamma)K_0(x)\int x(\delta\varepsilon)^2 dx\right],$$

$$S(x) = (1/\varepsilon_0)\left[I_1(x)\int x\delta\varepsilon G(x)dx + (1/\gamma)K_1(x)\int x(\delta\varepsilon)^2 dx\right]$$

with $G(x) = I_0(x)K_0(x) - \beta I_1(x)K_1(x)$.

The relevant results for the field components in the dielectric $(r > R)$ are:

$$E_z(r > R) = B[K_0(y) + Z(y)] \quad (7)$$

$$H_\varphi(r > R) = i(\omega\varepsilon_d/ck_{\perp d0})B[K_1(y) + Y(y)] \quad (8)$$

with

$$Z(y) = -(k_0\delta k/k_{\perp d0}^2)\, yK_1(y),$$

$$Y(y) = -(k_0\delta k/k_{\perp d0}^2)\,[yK_0(y) + 2K_1(y)].$$

The constant B is related to A by the linear boundary conditions.

4. NONLINEAR DISPERSION RELATION: The NLDR is derived, introducing expressions (5)-(8) into the boundary conditions

$$\{E_z\}\big|_{r=R} = 0, \qquad \{H_\varphi\}\big|_{r=R} = 0, \quad (9)$$

applied at the plasma-dielectric interface $r = R$. Written in the form

$$D = D_0 + D_1 = 0, \quad (10)$$

the NLDR includes

$$D_0 = \frac{k_{\perp 0}}{\varepsilon_0}\frac{I_0(\tilde{x})}{I_1(\tilde{x})} + \frac{k_{\perp d0}}{\varepsilon_d}\frac{K_0(\tilde{y})}{K_1(\tilde{y})}, \quad (11)$$

i.e. the well-known[1] linear dispersion law of the SWs ($D_0 = 0$) and a nonlinear contribution to it given by the term

$$D_1 = \frac{k_{\perp 0}}{\varepsilon_0}\frac{I_0(\tilde{x})}{I_1(\tilde{x})}\left[\frac{W(\tilde{x}) + T(\tilde{x})}{I_0(\tilde{x})} - \frac{V(\tilde{x}) + S(\tilde{x})}{I_1(\tilde{x})}\right]$$

$$+ \frac{k_{\perp d0}}{\varepsilon_d}\frac{K_0(\tilde{y})}{K_1(\tilde{y})}\left[\frac{Z(\tilde{y})}{K_0(\tilde{y})} - \frac{Y(\tilde{y})}{K_1(\tilde{y})}\right] \tag{12}$$

where $\tilde{x} = k_{\perp 0}R$ and $\tilde{y} = k_{\perp d0}R$. The latter results in:

$$\delta kR = \frac{k_{\perp 0}}{k_0}\frac{\varepsilon_d}{I_1(\tilde{x})}\left[T(\tilde{x}) - \frac{I_0(\tilde{x})}{I_1(\tilde{x})}S(\tilde{x})\right]\left\{\varepsilon_d\left[\frac{I_0^2(\tilde{x})}{I_1^2(\tilde{x})} - 1 - 2\frac{I_0(\tilde{x})}{\tilde{x}I_1(\tilde{x})}\right]\right.$$

$$\left. + |\varepsilon_0|\left[\frac{K_0^2(\tilde{y})}{K_1^2(\tilde{y})} - 1 + 2\frac{K_0(\tilde{y})}{\tilde{y}K_1(\tilde{y})}\right]\right\}^{-1} \tag{13}$$

with

$$T(\tilde{x}) = \mp\left(\alpha|A|^2/\varepsilon_0E_n^2\right)\left[\mathcal{P}(\tilde{x})I_0(\tilde{x}) - \mathcal{M}(\tilde{x})K_0(\tilde{x})\right],$$

$$S(\tilde{x}) = \mp\left(\alpha|A|^2/\varepsilon_0E_n^2\right)\left[\mathcal{P}(\tilde{x})I_1(\tilde{x}) + \mathcal{M}(\tilde{x})K_1(\tilde{x})\right]$$

and

$$\mathcal{P}(\tilde{x}) = \int^{\tilde{x}} xF(x)G(x)dx; \qquad \mathcal{M}(\tilde{x}) = \int^{\tilde{x}} xF^2(x)dx.$$

Thus, the nonlinear contribution δk described by relation (13) is:

$$\delta kR = \mp\left(\alpha|A|^2/E_n^2\right)\ DK \tag{14}$$

where

$$DK = \frac{k_{\perp 0}}{k_0}\frac{\varepsilon_d}{|\varepsilon_0|}\frac{\tilde{x}\mathcal{M}(\tilde{x})}{I_1^2(\tilde{x})}\left\{\varepsilon_d\left[1 + \frac{\varepsilon_d}{|\varepsilon_0|}\kappa\right]\frac{I_0^2(\tilde{x})}{I_1^2(\tilde{x})} - \left(\varepsilon_d + |\varepsilon_0|\right)\right.$$

$$\left. - 2\frac{\varepsilon_d}{\tilde{x}}\left(1 - \kappa\right)\frac{I_0(\tilde{x})}{I_1(\tilde{x})}\right\}^{-1} \tag{15}$$

with $\kappa = k_{\perp 0}^2/k_{\perp d0}^2$, gives the effect of the wave dispersion behaviour.

In a quasi-static approximation ($k_{\perp 0} = k_{\perp d0} \equiv k_0$, $\kappa = 1$) and for a plasma column bounded by vacuum ($\varepsilon_d = 1$), relation

(15) results in :

$$DK = \frac{\tilde{x}M(\tilde{x})}{I_1^2(\tilde{x})}\left[\left(1 + |\varepsilon_0|\right)\frac{I_0^2(\tilde{x})}{I_1^2(\tilde{x})} - |\varepsilon_0|\left(1 + |\varepsilon_0|\right)\right]^{-1}. \qquad (16)$$

The futher simplifications for thin cylinder approximation lead to the following result for the nonlinear wavenumber shift:

$$\delta kR = \mp \frac{k_0^3R^3|\ln k_0R|}{2\left(|\ln k_0R| - 2\right)}\frac{|A|^2}{E_n^2}. \qquad (17)$$

5. CONCLUSION: The nonlinear dispersion relation which governs the propagation of the azimuthally-symmetric surface mode at high-power excitation along an open cylindrical plasma waveguide is obtained. The nonlinear mechanisms of striction nonlinearity (due to the action of the ponderomotive force) and of ionization nonlinearity (caused by the additional heating and, thus, ionization in the wave field (and balanced either by volume recombination or ambipolar diffusion)) are considered in parallel. A general procedure based on introducing the effects of these nonlinear mechanisms into the plasma permittivity is applied. The nonlinear change of the wavenumber which depends on the wave amplitudes, is determined by both the peculiarity of the wave dispersion behaviour and of the normalizing field amplitude. It is established that the expression describing the wave dispersion behaviour has the same form, irrespective of the type of nonlinearity. The latter is specified by both the value of the normalizing field amplitudes E_n (different for striction and ionization nonlinearities) and the different signs in the expression for nonlinear contribution to wavenumber δk. The upper sign refers to the ionization nonlinearity at recombination controlled regime, whereas the lower one should be taken into account when striction nonlinearity or ionization nonlinearity at diffusion controlled regime predominate. The determination of the nonlinear wavenumber in explicite form in the whole frequency range of the waves requires additional numerical calculations.

The obtained results can be considered as a basis for studying of self-modulation (soliton formations)[3] and self-focussing (formation of nonlinear waveguide channels) in an open plasma waveguide. Their application for explaining the deviations[4] of the wave dispersion from the linear dispersion law observed recently in surface wave sustained plasma[6] (microwave discharges created by propagating SW at high-power excitation) would be also of interest.

ACKNOWLEDGEMENTS: This work was supported by the National Foundation for Scientific Research in Bulgaria (project n∘F27). One of us (A. Sh.) wishes to acknowledge the Alexander von Humboltdt Fondation for an award of a Fellowship and Prof. H. Schlüter for the kind hospitality.

REFERENCES:

1. A. Shivarova and I. Zhelyazkov, *in:* Electromagnetic Surface Modes, ed. A.D. Boardman, Wiley, Chichester, (1982), 465.
2. A. Shivarova, *in:* Spatial Dispersion in Solids and Plasmas, ed. P. Halevi, Elsevier, Amsterdam (1992), ch. 10, in press.
3. D. Grozev, A. Shivarova and S. Tanev, *J. Plasma Phys.* 45 (1991), 297.
4. G. Golubyatnikov, A. Kostrov, A. Shivarova, E. Tatarova and D. Zamfirov, Abstract NATO Advanced Research Workshop on Microwave Discharges: Fundamentals and Applications (Vimeiro, 1992), 57.
5. M. Georgieva and A. Shivarova, NATO Advanced Research Workshop on Microwave Discharges: Fundamentals and Applications (Vimeiro, 1992), in press.
6. M. Moisan and Z. Zakrzewski, *in:* Radiative Processes in Discharge Plasma, NATO ASI series (ed. J.M. Proud and L.H. Luessen), Plenum Publ., N.Y. (1987), 381.

'BOHM' SHEATH CRITERIA FOR DUSTY PLASMAS

M.A. Hellberg*, K. Thomas*, R.L. Mace*, R. Bharuthram*†, P.K. Shukla‡

* Plasma Physics Research Institute, Department of Physics, University of Natal, 4001 Durban, South Africa

† Department of Physics, University of Durban-Westville, Durban, South Africa

‡ Institut für Theoretische Physik, Ruhr-Universität Bochum, Bochum, Germany

1 Introduction

Recently considerable interest has developed in the study of dusty plasmas, in which, in addition to ions and electrons, a tenuous component of massive dust-particles (usually with a large negative charge) occurs. These dusty plasmas are found in cometary regions and planetary magnetospheres, and in addition are observed in low temperature industrial devices, where the dust is generated by sputtering off the walls.

It is of interest to investigate the electrostatic sheath that may form at the edge of such a dusty plasma. In this paper we consider the sheath stability, modelling our study on the classic Bohm sheath criterion, which requires that the ion velocity exceeds the ion-acoustic speed at the sheath-edge. We investigate three models of dust and ion behaviour, and thus density distribution functions, which represent three different time scales.

Using the linearized approach of Bohm we obtain similar criteria in each case. These may be interpreted in terms of conditions on the electrostatic self forces, thermal energies and kinetic energies of the components at the sheath-edge. Preliminary attempts at numerical modelling of the nonlinear sheath problem have thus far not been successful in generating a Bohm-like criterion.

2 The Bohm Sheath Criterion

To set the scene, we remind ourselves briefly of the Bohm model[1]. We consider electrons of temperature T_e (measured in energy units) forming a Boltzmann distribution,

$$n_e = n_{eo} \exp(e\phi/T_e), \tag{1}$$

the subscript o applying to variable values at the sheath-edge. The cold (singly-charged) positive ions of mass m_i satisfy steady-state equations of continuity and of energy conservation, from which one can find an ion density distribution.

Substituting into Poisson's equation, one then introduces typical ion-acoustic normalizations, i.e. densities with respect to the density at the sheath-edge, n_o, speeds u with respect to the ion-acoustic speed, $C_s = (T_e/m_i)^{\frac{1}{2}}$, distances with respect to the Debye length, $\lambda_D = (T_e/4\pi n_o e^2)^{\frac{1}{2}}$ and potentials with respect to $(-T_e/e)$ (thus $\chi = -e\phi/T_e$). The Mach number of the infalling ions at the sheath-edge, μ, is given by $\mu = u_o/C_s$.

The result is the nonlinear Poisson equation in normalized variables,

$$\chi'' = -e^{-\chi} + \left(1 + \frac{2\chi}{\mu^2}\right)^{\frac{-1}{2}}. \tag{2}$$

This equation can be integrated once to yield

$$\frac{1}{2}(\chi'^2 - \chi_o'^2) = (e^{-\chi} - e^{-\chi_o}) + \mu^2 \left\{ \left(1 + \frac{2\chi}{\mu^2}\right)^{\frac{1}{2}} - \left(1 + \frac{2\chi_o}{\mu^2}\right)^{\frac{1}{2}} \right\}. \quad (3)$$

Following Bohm, one assumes as boundary conditions at the sheath-edge that both potential and electric field vanish (i.e. $\chi_o = \chi_o' = 0$). In that case equation (3) reduces to

$$\frac{1}{2}(\chi'^2) = (e^{-\chi} - 1) + \mu^2 \left\{ \left(1 + \frac{2\chi}{\mu^2}\right)^{\frac{1}{2}} - 1 \right\}. \quad (4)$$

If we assume that the normalized potential $\chi \ll 1$, the right hand side may be expanded, and as the left hand side is non-negative, one obtains at quadratic order in χ a condition which needs to be satisfied, viz.

$$\mu^2 > 1. \quad (5)$$

This criterion, that the speed of the incoming ions should exceed the ion-acoustic speed, can also be expressed in the form that, at the sheath-edge, the kinetic energy of the infalling cold ions should exceed the thermal energy of the electrons for a stable monotonic sheath to exist.

3 Linear Studies

The dusty plasma is assumed to be infinite, collisionless and unmagnetised, and consists of inertialess electrons, singly charged ions and massive dust particles (density n_d) carrying a large negative charge (typically $M_d \geq 10^4$ and $| Z_d | \geq 10^2$).

In all cases we shall seek a stationary monotonic potential solution to represent the sheath. Quasineutrality is assumed at the sheath-edge.

We shall consider 3 cases, which effectively involve different assumed values of m_i and M_d/m_i and thus represent different timescales in the sheath-formation:

3.1 Case I:

We consider first the shortest timescale, which is nevertheless long enough that the electrons can respond 'instantaneously', while the massive dust particles do not have time to move. The electrons have a Boltzmann distribution; the ions are assumed cold, satisfying equations of continuity and of energy conservation; the uniformly-distributed dust particles are infinitely massive and are stationary. We use the standard Bohm normalizations, specifically normalizing densities with respect to n_{io}, which is larger than n_{eo}. Defining the fractional electron density $\alpha = n_{eo}/n_{io}$ and the fractional 'effective' dust density, derived from the charge density, $\delta = | Z_d | n_{do}/n_{io}$, it follows from quasineutrality at the sheath-edge that $\alpha + \delta = 1$. Poisson's equation then takes the form

$$\chi'' = -\alpha e^{-\chi} - (1 - \alpha) + \left(1 + \frac{2\chi}{\mu^2}\right)^{\frac{-1}{2}}, \quad (6)$$

yielding on integration,

$$\begin{aligned} \frac{1}{2}(\chi'^2 - \chi_o'^2) &= \alpha(e^{-\chi} - e^{-\chi_o}) \\ &\quad -(1 - \alpha)(\chi - \chi_o) + \mu^2 \left\{ \left(1 + \frac{2\chi}{\mu^2}\right)^{\frac{1}{2}} - \left(1 + \frac{2\chi_o}{\mu^2}\right)^{\frac{1}{2}} \right\}. \quad (7) \end{aligned}$$

Hence, for the Bohm boundary conditions, we get

$$\frac{1}{2}(\chi'^2) = \alpha(e^{-\chi} - 1) - (1-\alpha)\chi + \mu^2\left\{\left(1+\frac{2\chi}{\mu^2}\right)^{\frac{1}{2}} - 1\right\}. \tag{8}$$

Assuming that the potential is small, we expand the right hand side and obtain

$$\frac{1}{2}(\chi')^2 = \alpha(1 - \chi + \frac{1}{2}\chi^2 ... - 1) - (1-\alpha)\chi + \mu^2\{(1 + \frac{\chi}{\mu^2} - \frac{1}{2}\frac{\chi^2}{\mu^4} ... - 1)\}, \tag{9}$$

The condition that it be non-negative leads one to the requirement

$$\mu^2 > \frac{1}{\alpha}, \tag{10}$$

where we recall that $\alpha = n_{eo}/n_{io}$.

Thus it is clear that the effect of the inclusion of a stationary uniform background of negatively charged dust, which reduces the electron density to a fraction α of that of the ions, is to increase the minimum speed for the infalling ions by a factor $1/(\alpha)^{\frac{1}{2}}$.

The criterion is similar to the Bohm criterion, and by a suitable choice of characteristic speed, it can be brought even closer to that form.

We have here normalized with respect to a standard ion-acoustic speed, $C_s = (T_e/m_i)^{\frac{1}{2}}$. However, if we choose a slightly different (hybrid) speed for normalization, viz. $C_{sei} = \lambda_e\omega_{pi} = \left(\frac{T_e n_{io}}{m_i n_{eo}}\right)^{\frac{1}{2}}$, the factor α drops out of the inequality, which thus reduces to

$$\mu_{ei}^2 > 1. \tag{11}$$

In other words, by choosing our normalization speed as the hybrid sound speed above, one can cast the criterion in the Bohm form.

3.2 Case II:

In this intermediate case the timescales are such that the motion of the cold ions and the cold dust play similar roles and we introduce the dust mass M_d and speed U_d. We shall again use the standard ion-acoustic normalization used initially for Case I.

The dust density now has a form similar to that of the ions, and Poisson's equation is

$$\chi'' = -\alpha e^{-\chi} - (1-\alpha)\left(1 - \frac{2\epsilon\chi}{\mu^2}\right)^{\frac{-1}{2}} + \left(1+\frac{2\chi}{\mu^2}\right)^{\frac{-1}{2}}, \tag{12}$$

where we have introduced the parameter

$$\epsilon = \frac{|Z_d| m_i u_o^2}{M_d U_{do}^2}, \tag{13}$$

which represents the ratio of kinetic energy to potential energy of an ion compared to that of a dust particle, at the sheath-edge.

On integrating and applying the Bohm boundary conditions, this yields

$$\frac{1}{2}(\chi'^2) = \alpha(e^{-\chi} - 1) - (1-\alpha)\left(\frac{\mu^2}{\epsilon}\right)\left\{\left(1 - \frac{2\epsilon\chi}{\mu^2}\right)^{\frac{1}{2}} - 1\right\} + \mu^2\left\{\left(1+\frac{2\chi}{\mu^2}\right)^{\frac{1}{2}} - 1\right\}. \tag{14}$$

Expanding the right hand side, one finds

$$\begin{aligned}\frac{1}{2}(\chi')^2 &= \alpha(1-\chi+\frac{1}{2}\chi^2...-1)\\ &-(1-\alpha)\left(\frac{\mu^2}{\epsilon}\right)\left\{(1-\frac{\epsilon\chi}{\mu^2}-\frac{1}{2}\frac{\epsilon^2\chi^2}{\mu^4}...-1)\right\}\\ &+\mu^2\{(1+\frac{\chi}{\mu^2}-\frac{1}{2}\frac{\chi^2}{\mu^4}...-1)\}.\end{aligned} \tag{15}$$

At quadratic order this yields a condition

$$\mu^2 > \left(\frac{1}{\alpha}\right)\{1+(1-\alpha)\epsilon\}. \tag{16}$$

When $\alpha = 1$ (no dust) this naturally reduces to the standard Bohm criterion, and for $\epsilon \to 0$, i.e. infinite dust kinetic energy or mass, it reduces to Case I.

One way of trying to interpret this result is to assume that one should get a relation of the form $\mu_{eff} > 1$, where $\mu_{eff} = u_o/(T_e/M_{eff})^{\frac{1}{2}}$, and M_{eff} is the resulting effective mass characteristic of the problem, in the sense that it plays the relevant role in yielding a characteristic sound-like speed.

Assuming this form, we can then substitute in equation (16) to obtain an expression for M_{eff}. In fact it turns out that

$$M_{eff} = \frac{m_i M_d n_{eo} u_o^2}{n_{io} M_d U_{do}^2 + Z_d^2 n_{do} m_i u_o^2}. \tag{17}$$

This is typical of an effective mass weighted relative to the kinetic energies of the two drifting species, ions and dust.

3.3 Case III:

On this longest timescale we assume that the electrons and the ions have independently equilibrated and each forms a Boltzmann distribution, and consider the detailed dynamics of the dust particles only. We then introduce an effective temperature involving the two Boltzmann species, viz. $T_{eff} = n_{io}T_iT_e/(n_{eo}T_i + n_{io}T_e)$. We will normalize densities with respect to n_{io}, potential with respect $(-T_{eff}/e)$, mass with respect to M_d, and hence introduce a dust Mach number $\mathcal{M} = U_{do}/(T_{eff}/M_d)^{\frac{1}{2}}$. Then, defining $\beta_e = T_{eff}/T_e$ and $\beta_i = T_{eff}/T_i$, and writing $Z = | Z_d |$, we obtain Poisson's equation

$$\chi'' = -\alpha e^{-\beta_e\chi} - (1-\alpha)\left(1-\frac{2Z\chi}{\mathcal{M}^2}\right)^{\frac{-1}{2}} + e^{\beta_i\chi}. \tag{18}$$

Integrating and applying the usual Bohm boundary conditions, this yields

$$\frac{1}{2}(\chi'^2) = \frac{\alpha}{\beta_e}(e^{-\beta_e\chi}-1) + \frac{1}{\beta_i}(e^{\beta_i\chi}-1) - (1-\alpha)\left(\frac{\mathcal{M}^2}{Z}\right)\left\{\left(1-\frac{2Z\chi}{\mathcal{M}^2}\right)^{\frac{1}{2}}-1\right\}. \tag{19}$$

The linearized form of this equation yields

$$\begin{aligned}\frac{1}{2}(\chi')^2 &= \frac{\alpha}{\beta_e}(1-\beta_e\chi+\frac{1}{2}\beta_e^2\chi^2...-1) + \frac{1}{\beta_i}(1+\beta_i\chi+\frac{1}{2}\beta_i^2\chi^2...-1)\\ &-(1-\alpha)\left(\frac{\mathcal{M}^2}{Z}\right)\left\{\left(1-\frac{Z\chi}{\mathcal{M}^2}-\frac{1}{2}\frac{Z^2\chi^2}{\mathcal{M}^4}...-1\right)\right\}\end{aligned} \tag{20}$$

This may be reduced at quadratic order to

$$\mathcal{M}^2 > \frac{Z(1-\alpha)}{\beta_i + \alpha\beta_e}, \tag{21}$$

which can be further simplified to yield

$$\mathcal{M}^2 > \frac{Z_d^2 n_{do}}{n_{io}}. \tag{22}$$

This form is reminiscent of the criterion found for Case I before we chose the alternative density normalization. Hence, we now go back to our original equations and normalize densities with respect to the 'effective' dust density (Zn_{do}) obtained from the dust charge density, rather than n_{io}. This affects the temperature normalization, in that we then have $T_{eff,Z} = Z_d^2 n_{do} T_i T_e/(n_{eo}T_i + n_{io}T_e)$ and we normalize potentials by $(-T_{eff,Z}/Ze)$. In that case, the factor Z drops out and, after a fair amount of manipulation, one can show that in this normalization the criterion becomes

$$\mathcal{M}_Z^2 > n'_{io}\beta_i + n'_{eo}\beta_e, \tag{23}$$

where $n'_{jo} = n_{jo}/Zn_{do}$. This expression can then be further reduced, to obtain

$$\mathcal{M}_Z^2 > 1, \tag{24}$$

i.e. in this normalization, we recover the Bohm form of criterion. We note that in this normalization the effective Debye length used is

$$\lambda_{Dd} = \left(\frac{T_{eff}}{4\pi Z_d^2 e^2 n_{do}}\right)^{\frac{1}{2}} \tag{25}$$

and the characteristic velocity is the true dust-acoustic speed $C_{sd} = \lambda_{Dd}\omega_{pd}$.

4 Discussion

At first sight the three criteria we have obtained, viz. equations (10), (16) and (21), look very different from one another. Nevertheless, we found that in cases I and III a simple change in normalization led us to a Bohm-like criterion. This leads us to seek a common thread in the three expressions. In fact if we multiply through each of the three expressions by a factor e^2 and rearrange the terms, we can present the criteria in the following form:

Case I: Boltzmann electrons, drifting ions, stationary dust:

$$\frac{n_{eo}e^2}{T_e} > \frac{n_{io}e^2}{m_i u_o^2} \tag{26}$$

Case II: Boltzmann electrons, drifting ions and dust:

$$\frac{n_{eo}e^2}{T_e} > \frac{n_{io}e^2}{m_i u_o^2} + \frac{n_{do}Z_d^2 e^2}{M_d U_{do}^2} \tag{27}$$

Case III: Boltzmann electrons and ions, drifting dust:

$$\frac{n_{eo}e^2}{T_e} + \frac{n_{io}e^2}{T_i} > \frac{n_{do}Z_d^2 e^2}{M_d U_{do}^2} \tag{28}$$

Thus in each case the criterion for the existence of a stable sheath potential at the edge of a dusty plasma may be written in the following form:
At the sheath-edge the sum (over all Boltzmann species) of the ratios of the electrostatic self force density of the species to the thermal energy density of the species must exceed the sum (over all drifting species) of the ratios of the electrostatic self force density of the species to the kinetic energy density of the species.

Alternatively, one may cast this 'Bohm' criterion in the form

$$\Sigma_{Boltz}\frac{1}{\lambda_{DB}^2} > \Sigma_{drift}\frac{\omega_{pd}^2}{U_{do}^2}, \tag{29}$$

where Σ_{Boltz} represents a sum over all Boltzmann species, Σ_{drift} a sum over all drifting species, λ_{DB} the Debye length for each of the Boltzmann species, ω_{pd} the plasma frequency for each of the drifting species and U_{do} the drift speed of each of the drifting species.

Clearly, each of these formulations reduces to the usual Bohm criterion in the case of a single (electron) Boltzmann species and a single (ion) drifting species.

5 Numerical Calculations

To augment the analytical studies discussed above, we wished to study the nonlinear effects in this problem. It is difficult to simulate the Bohm formulation exactly, as the sheath-edge boundary condition ($\chi_o = \chi_o' = 0$), if substituted into the Poisson equation implies that $\chi_o'' = 0$ also, and the solution to Poisson's equation is then zero everywhere.

As a result we considered two alternative numerical models to describe the sheath formation:
(a) A two-point boundary value problem, in which the potential is fixed at each end ($\chi_o = 0$ and χ_{wall}). This is essentially a diode model. It did not produce interpretable results showing a cut-off at $\mu = 1$.
(b) A model in which the electric field at the sheath-edge was taken as zero ($\chi_o' = 0$), and the potential at the opposite end fixed (χ_w). For Case I this involves integrating equation (8) outwards from the wall. Although finite-width sheaths were found over a wide range of Mach numbers, small amplitude ($\chi_w = 0.01$) calculations produced cut-offs in the sense that for $\mu^2\alpha < 1$ the integrations broke down at $\xi = 20$ to 50 λ_D at values of χ greater than 0.1 χ_w, while above the Bohm cut-off, the integrals yielded very small ($< 10^{-2}\chi_w$) remanent χ at 100 λ_D. For $\chi_w = 1$, the results are less clear, although they point in a similar direction. Further investigations are being carried out.

6 Reference

1. F.F. Chen, *Introduction to Plasma Physics and Controlled Fusion*, Plenum Press, New York (1984).

ON THE GENERAL CONDITIONS FOR DOUBLE LAYER FORMATIONS IN FINITE PLASMAS

N. Jelić, M. Čerček, T.Gyergyek and M.Stanojević
J.Stefan Institute, Jamova 39, 61111 Ljubljana, Slovenia

1. INTRODUCTION

Conditions for existence of stationary double layers basically comprise the requirement for the double layer quasineutrality at the boundaries up to quadratic term, as well as the requirement for overall quasineutrality of double layer [1]. An application of these requirements leads to the prominent set of expressions known as Langmuir and Bohm-Block conditions [2]. A rather general application option, given by Andrews and Allen [1] with a kinetic model resulted in theirs "coupling relations"+ a momentum balance equation. In the case of a strong double layer coupling relations were uncoupled, leading to the Harison-Thompson [3] generalization of Bohm conditions, while momentum balance takes form of a generalized Langmuir condition. On the other hand, parameters of adjacent plasmas are determined by their own boundary conditions including double layer as an integral part of the system. Therefore, quantities entering conditions for a double layer existence like density, temperature, plasma potential and the double layer position, are not, a priori, independent parameters, but rather are variables which should be calculated. This is an extended formulation of the problem of a strong monotonic double l. existence conditions that were chosen in our investigations. Particular attention was paid to the

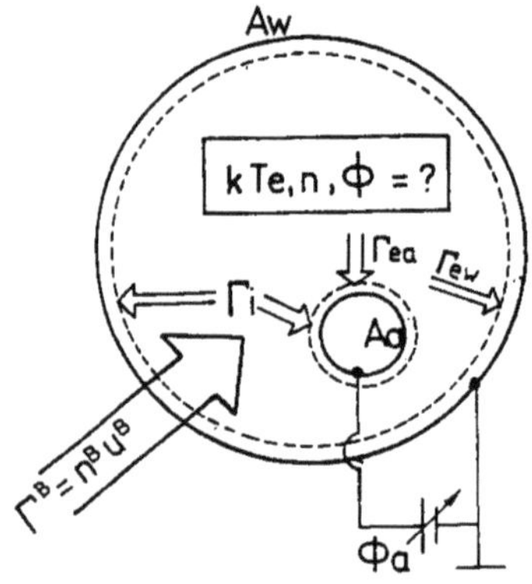

Figure 1. Schematic sketch of a plasma system maintained by an energetic beam of electrons in the presence of absorbing wall and an electrode

anode type [4,5,6,7,8,9,10,11,12] double layers where the basic constraint to the plasma - double layer system is imposed by a specific mechanism for high potential plasma production.

2. BASIC CONSIDERATIONS

We considered a plasma system given schematically in Figure 1. Plasma is maintained in a stationary state in contact with boundaries of area A_w and A_a which represent the wall and electrode (anode), respectively. It was assumed that, in the case when a double layer is present, both low and high potential plasmas could be described by the help of this model. Our task was calculating of plasma potential Φ, plasma density n and electron temperature T_e, for the case when the ion thermal velocities are negligible. Basic equations of our system refere to the particle: $\oint_A \mathbf{\Gamma}_i d\mathbf{A} = n_a \int_V dV \int_{\mathbf{v}} [n_e^B f_e^B(\mathbf{v}) + n_{e2}^{Th} f_e^{Th}(\mathbf{v})] \sigma_i(\mathbf{v}) v d^3\mathbf{v}$, charge: $e \oint_A (\mathbf{\Gamma}_i - \mathbf{\Gamma}_e) d\mathbf{A} = 0$, and energy: $\oint_A (\mathbf{Q}_e + \mathbf{Q}_i) d\mathbf{A} = \frac{1}{2} n_a \int_V dV \int_{\mathbf{v}} m_e v^2 [n_e^B f_e^B(\mathbf{v}) + n_e^{Th} f_e^{Th}(\mathbf{v})] \sigma_i(\mathbf{v}) \mathbf{v} d^3\mathbf{v}$ conservation. Here both beam (superscript "B") and thermal population (superscript "Th") velocity distributions f are normalized to one and n_a denotes the neutral gas density. Main electron population is Maxwellian, with the electron flux density $\Gamma_e = n\sqrt{kT_e/2\pi m_e} \exp(e\Delta\Phi/kT_e)$, where $\Delta\Phi$ is the electrode or wall potential with respect to plasma. The ion velocity at the plasma sheath boundary could be identified with the Bohm velocity $u_i = \alpha\sqrt{kT_e/m_i}$ in all directions relevant for ion losses, where $\alpha(n^B/n^{Th})$ represents a complicated function of the order of unity. Consequently, the ion and electron energy flux densities $\mathbf{Q}_{e,i} = \frac{1}{2} m_{e,i} \int_{\mathbf{v}} f_e(\mathbf{v}) v^2 \mathbf{v} d^3\mathbf{v}$ are $Q_e = (2kT_{e2} + e|\Delta\phi|)\Gamma_e$ and $Q_i = \alpha^2 kT_e \Gamma_i$, respectively. Ionization cross-section is given by $\sigma_I = \frac{1}{2}\vartheta m_e(v^2 - v_I^2)$, where v_I denotes electron velocity corresponding to the first ionization energy level Φ_I, and $\vartheta \sim 0.8 \times 10^{-21}\, m^2/eV$. The change of electron beam energy and plasma density in the region of quasineutrality was ignored. In our approximation expressions for particle and energy conservation could be trivially integrated resulting in a closed set of equations for the calculation of the plasma potential $\Phi = |\Delta\Phi_a| + \Phi_a$ with respect to the wall, plasma density and electron temperature:

$$\frac{e\Phi}{kT_e} = \begin{cases} \ln\mu + \ln\left[a\exp\left(\frac{e\Phi_a}{kT_e}\right) + (1-a)\right], & a \geq a_{cr} \\ \ln\mu + \ln\frac{1-a}{1-\mu a}, & a \leq a_{cr} \end{cases} \tag{1}$$

$$n_e = \frac{1}{1-\mathcal{F}} \frac{\Phi^B - \Phi_I}{\Phi_r} \frac{1}{\alpha} \sqrt{\frac{2e\Phi^B}{kT_e}}\, n^B \tag{2}$$

$$kT_e = \frac{e(\Phi_B - \Phi_I)(1-\mathcal{F})}{(2 + \frac{e|\Delta\Phi_a|}{kT_e})\frac{\Gamma_{ea}}{\Gamma_i} a + (2 + \frac{e|\Phi|}{kT_e})\frac{\Gamma_{ew}}{\Gamma_i}(1-a) + \alpha^2} \tag{3}$$

where $e\Phi^B$ is the beam energy, $a = A_a/A$ is the ratio of the electrode to total boundary, and $\mu = \frac{1}{\alpha}\sqrt{m_i/2\pi m_e}$. Electron fluxes to the wall and electrode are denoted by subscripts w and a respectively. Plasma and anode potentials are equal for:

$$a_{cr} = \frac{1}{\mu} \frac{1 - \mu\exp\left(-\frac{e\Phi_a}{kT_e}\right)}{1 - \exp\left(-\frac{e\Phi_a}{kT_e}\right)} \tag{4}$$

Volume ionization due to thermal motion of electrons is described by the function:

$$\mathcal{F} = \frac{2kT_e + e\Phi_I}{e\Phi_r} \frac{1}{\alpha} \sqrt{\frac{8}{\pi}} \exp\left(-\frac{e\Phi_I}{kT_e}\right) \tag{5}$$

In above formulas the similarity parameter is introduced:

$$\frac{1}{e\Phi_r} \equiv n_a \vartheta \frac{V}{A} \sqrt{\frac{m_i}{m_e}} \tag{6}$$

We calculated a reference value $\Phi_r = 5\,V$ for argon with the density $n_a = 5 \times 10^{20}\,m^{-3}$ (reference pressure $\sim 3 \times 10^{-4}\,Torr$) satisfying $V/A = 1\,cm$.

Eqs. (1) - (3) make a closed set providing beam electron density is known. Independent variable of the system, for a chosen beam density and effective potential Φ_r, is the anode potential Φ_a. This variable could be substituted by another one, i.e. anode current, which is, sometimes, more appropriate for obtaining single valued solution. However, if the beam density is dependent variable of the problem, our system should be extended for one equation more, and appropriately restructured. A general physical constraint of our system is given by the total energy density balance $\epsilon_o(d\Phi/dz)^2 - 2\int(m_i v^2 f_i + m_e v^2 f_e)dv = const.$ This expression should be evaluated at appropriate boundaries of a sheath to obtain an expression for electric field calculation.

3. APPLICATIONS OF THE MODEL

Plasma at the low potential side of a double layer could be modeled by the help of our equations, providing there is a plasma at high potential instead of the anode. There is a geometrical limitation for the total double layer area which may not exceed the critical value given by expression (4). Nevertheless, varying such a small double layer area will strongly influence potential of the low-potential plasma, providing the double layer and wall area ratio is not negligible with respect to the square root of electron and ion mass ratio. This should have consequences on the ionizing beam energy, i.e. plasma production rate, as well as on the plasma losses. However, if the double layer area is negligible our system degenerates to a floating wall - case. For a wide range of effective potentials Φ_r, covering usual laboratory pressures and gases in spherical and cylindrical geometry, plasma parameters depend only on the effective pressure.

On the other hand plasma at high potential side of a double layer could be described by our model providing the wall (Figure 1) is substituted by the low potential plasma with the double layer which accelerates primary electrons to energies $e\Phi^B = e\Phi_{DL}$ determined by the double layer potential Φ_{DL}. After taking into account Bohm and Block critical velocities we rearrange energy density balance in the form of extended Langmuir condition:

$$\frac{\Gamma_{e1}}{\Gamma_{i2}} = \frac{1 - \beta_2\sqrt{kT_{e2}/e\phi_{DL}}}{1 - \beta_1\sqrt{kT_{e1}/e\phi_{DL}}}\sqrt{\frac{m_i}{m_e}} \equiv \Theta\sqrt{\frac{m_i}{m_e}} \tag{7}$$

where Θ defines Langmuir factor, and where $\beta_1(\alpha_1)$ and $\beta_2(\alpha_2)$ are complicated functions of Block and Bohm factors, respectively, both with values of the order of $\sqrt{2}$. Subscripts "1" and "2" are used to distinguish particles which originate from low- and high-potential side of a double layer, respectively. Using expression (7) our Eqs. (1)-(3) could be written in the form:

$$\Phi_{DL} = \Phi_I + \frac{\alpha_2}{\Theta}(1 - \mathcal{F})\Phi_r \tag{8}$$

$$\frac{A_{DL}}{A_a} = \frac{1}{\alpha}\sqrt{\frac{m_i}{2\pi m_e}}\exp\left[\frac{e(\Phi_a - \Phi_{DL})}{kT_e}\right] \tag{9}$$

$$kT_e = \frac{1}{2+\alpha^2}\left[\Phi_a - \Phi_I - \frac{\mathcal{F}(1-\mathcal{F})}{\Theta}\Phi_r\right] \tag{10}$$

where it was supposed that the double layer area could be controlled only if the anode potential remains below the plasma potential. Our next crucial task is to determine Φ_r versus Φ_a range

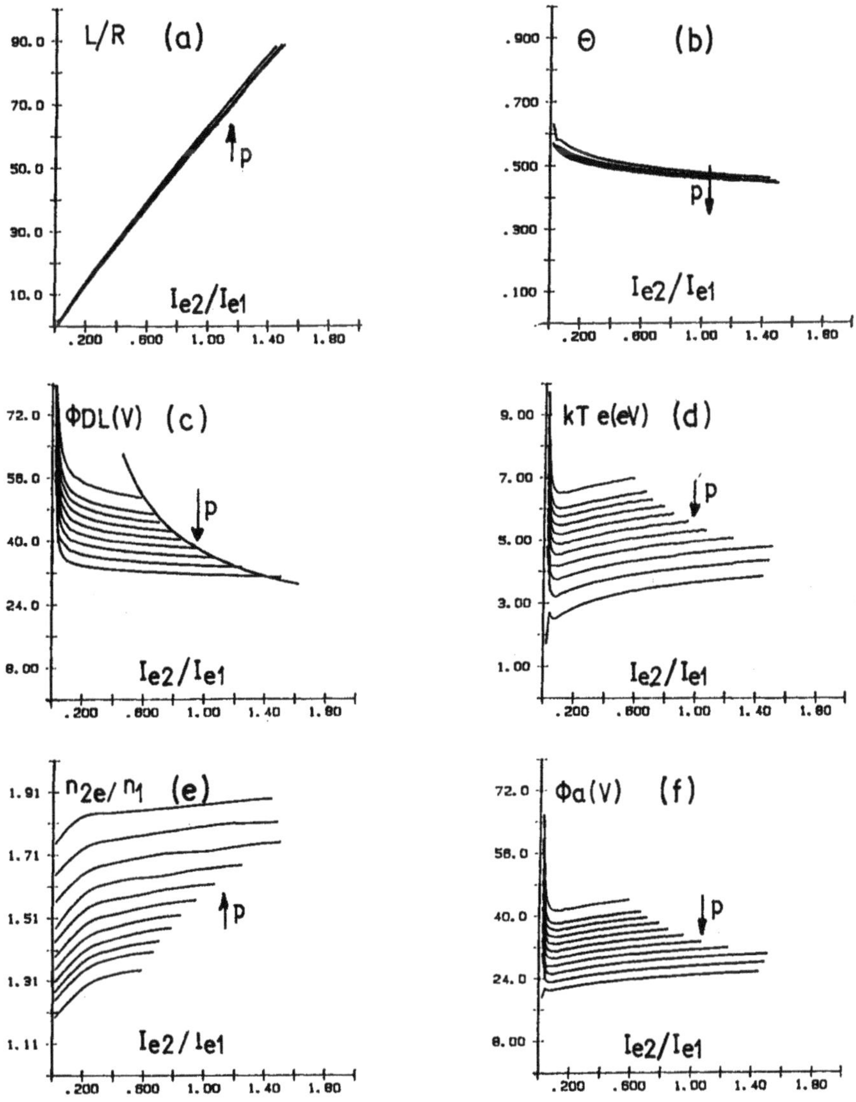

Figure 3. Quantitative comparison of the theoretically predicted electron temperature (a), plasma density (b), double layer potential (c) and anode potential (d) with the experimental data for the case $I_{e2}/I_{e1} = 0.5$.

in which system (7)-(10) has a solution. Under the condition that the electron temperature T_{e1} could be neglected, Langmuir factor becomes simpler and could be written as a function of Φ_r and Φ_a:

$$\left[\alpha(1-\Theta)^2+\gamma\mathcal{F}\right]\frac{1-\mathcal{F}}{\Theta}\Phi_r=\gamma\Phi_a-\left[(1-\Theta)^2+\gamma\right]\Phi_I \tag{11}$$

which definition range should be investigated. Here $\gamma\equiv\beta^2/(2+\alpha^2)$. However, due to contribution involved by the term $\mathcal{F}$ this system is very unpleasantly resisting to a correct handling by analytical methods. Neglecting thermal ionization $\mathcal{F}=0$ and assuming $\alpha=1$ and $\beta^2=2$ makes this problem somewhat easier. In this case a solution of the Eq. (11) for the Langmuir factor should be above the critical value:

$$\Theta_{cr}=1+\frac{1}{3}\frac{\Phi_a}{\Phi_r}+\sqrt{\left(1+\frac{1}{3}\frac{\Phi_a}{\Phi_r}\right)^2-1} \tag{12}$$

It can be seen that lower pressures (higher Φ_r) require higher anode potential to obtain the same value for Θ_{cr}. Therefore, according to Eq. (10) the temperature should be also higher and, consequently, the definition of Langmuir factor give higher double layer potential. All these dependencies are in qualitative agreement with our experimental result given below. Quantitative results obtained with $\mathcal{F}=0$ are also satisfactory providing Φ_a is not too high. Otherwise, the temperature "feedback" should be included.

In Figure 2 solution of our system (4) - (7) is presented for argon plasma in cylindrical geometry, for the case when the radial electron current is ignored. Instead of the anode potential here independent variable chosen is the anode plasma current normalized to the electron saturation current of the low potential plasma. Dependence of the anode plasma length (a), Langmuir factor (b), plasma potential (c), electron temperature (d), plasma density (e) and anode potential (f) is shown for a range of pressures 5×10^{-4} to 10^{-5} *Torr*. It can be seen that the solution is limited by some maximum current which is lower for the value of lower pressures. In addition, it can be shown that there is a maximum power input low $P_{max}\sim const\times I_{e1}$, providing the thermal ionization is taken into account.

4. COMPARISON WITH EXPERIMENTAL RESULTS

In our experiments on double layers [13] plasma parameters on both sides of an anode type double layer were measured under different conditions for the low potential plasma production, as well as for different pressures, anode radii and anode potentials, i.e. for different anode - external circuit currents. Experimental device and setup is described elsewhere [13]. It was obtained that the normalized anode current-voltage characteristics are invariant respectively to the low potential plasma density, as predicted by the theory. Consequently, the most of measurements are performed holding electron saturation current to the double layer constant. In this way it was obtained that the anode plasma lenght does not depend on pressure and follows linear dependence on the anode current, as qualitatively predicted by our model. However, the plasma lenght was about five times shorther than predicted. Other parameters depend weakly on the anode current, but significantly depend on gas pressure and anode radius as predicted by the model. Quantitative comparison of the theoretically predicted electron temperature, plasma density, double layer potential and anode potential with the experimental ones, obtained in a weakly magnetized (70 G) argon plasma column is presented in Figure 3a,b,c,d, respectively, under the condition when low potential plasma parameters are held constant and $I_{e2}/I_{e1}=0.5$.

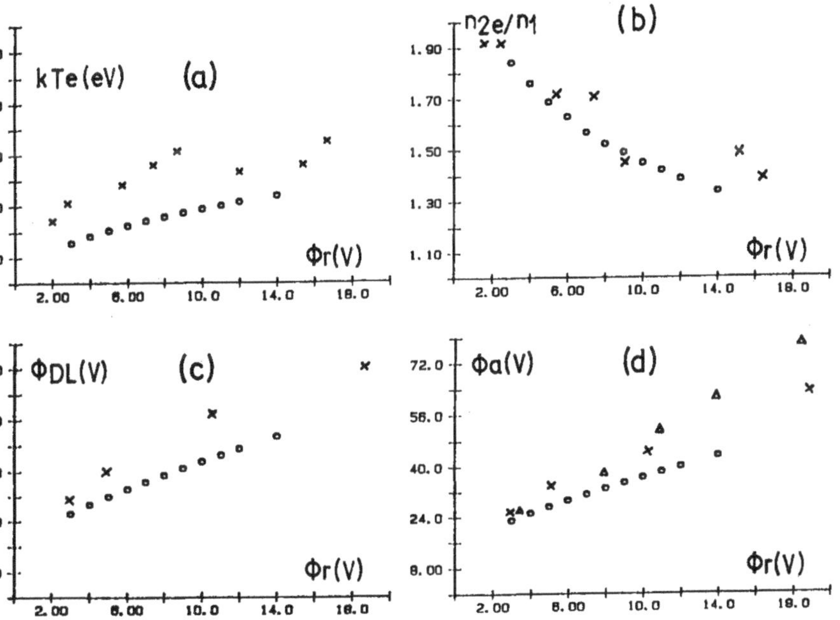

Figure 2. Dependence of the anode plasma length (a), Langmuir factor (b), plasma potential (c), electron temperature (d), plasma density (e) and anode potential (f) on the normalized anode current, for different effective pressures in cylindrical geometry.

A satisfactory agreement is obvious. To obtain better results one should take into account the magnetic field presence which further complicates model by new boundary conditions. Further experimental and theoretical investigations are in progress.

5. REFERENCES

[**1**] J.G.Andrews and J.E.Allen, *Proc.Roy.Soc.Lond.* A. 320, 459 (1971), [**2**] M.A.Raadu and J.J.Rasmunssen, *Astrophys.Space Sci.* 144, 43 (1988), [**3**] E.R.Harison and W.B.Thompson, *Proc. Phys. Soc. Lond.* 72, 2145 (1959), [**4**] S.Torven and D.Andersson, *J.Phys. D*, 12, 717 (1979), [**5**] S.Torven and L.Lindberg, *J.Phys. D*, 13, 2285 (1980), [**6**] D.Andersson, *J.Phys. D*, 14, 1403, (1981), [**7**] D.Andersson and J.Sørensen *J.Phys. D*, 16, (1983), [**8**] H.Fujita, K.Matsuo and S.Jagura, *Phys.Fluids*, 27, 2948, (1984), [**9**] S.L.Cartier and R.L.Merlino, *Phys.Fluids*, 30, 2549, (1987), [**10**] R.Plamondon, J.Teichmann and S.Torven, *J.Phys. D*, 21, 286 (1988), [**11**] H.Amemiya and Y.Nakamura, *Plasma Phys.* 26, [**12**] B.Song, N.D'Angelo and R.L.Merlino, *J.Phys. D*, 24, 1789 (1991) [**13**] N.Jelić, M.Čerček and T.Gyergyek: *Contributed Papers of the 1989 Intern. Conf. on Plasma Phys.*, New Delhi, Ed. by A.Sen and P.K.Kaw, 689, (1989)

THE EFFECTS OF ION INERTIA ON DUST-ACOUSTIC DOUBLE LAYERS

R.L. MACE AND M.A. HELLBERG
Plasma Physics Research Institute, Department of Physics,
University of Natal, Durban, South Africa

1 INTRODUCTION

Space and astrophysical plasmas often comprise a tenuous, highly negatively-charged dust component in addition to one or more heavy ion components and an electron component. In fact, even in our solar system there is a significant dust content (see the review by Goertz 1989) making studies of dusty plasmas imperative if we are to better understand the plasma and associated wave dynamics in the 'near' Earth region.

Just as the nonlinear ion-acoustic wave in ordinary plasmas can give rise to ion-acoustic double layer (DL) potential structures, so the nonlinear dust-acoustic wave (Rao, Shukla and Yu 1990) can give rise to dust-acoustic DLs (Bharuthram and Shukla 1992). However, in the latter it is the effective pressure of two relatively hot ion components which provides the restoring force, rather than the effective pressure provided by two-temperature hot Boltzmann distributed electrons. Because of the large mass of the ions, however, the usual assumption of inertialess Boltzmann distributions (Bharuthram and Shukla 1992) for these components becomes questionable. We note that Rice *et al.* (1992) showed, in a study of rarefactive ion-acoustic solitons, that even electron inertia can play a role in the wave dynamics.

In this paper we investigate the effects of finite ion mass (hence inertia) on dust-acoustic DLs. In particular we address the question of whether the neglect of the ion inertia is generally valid. The latter approximation is often justified by arguing that the dust particles are, on average, very much more massive than the ions and electrons, and hence on the slow dust timescales the ions and electrons respond 'instantaneously'.

While on the face of it this argument seems logical and sound, a deeper analysis of the underlying equations reveals some subtle, often overlooked, points. It will be shown that the large disparity in the dust and ion masses gives rise to some unexpected near-cancellations in the density equations. These serve to restrict the parameter regimes in which dust-acoustic DLs can occur by forcing the (usually cool) ion density to become complex-valued. However, in the regimes in which the inertial model admits double layers, the two theories are found to be qualitatively, and for small enough ion/dust mass ratios quantitatively, similar.

2 BASIC EQUATIONS

The dusty plasma consists of a tenuous dust fluid, a hot ion fluid, a cool ion fluid and a hot electron fluid which, for quasineutrality, satisfy

$$Z_c n_{0c} + Z_h n_{0h} + Z_d n_{0d} = n_{0e}. \tag{1}$$

(We shall assume $Z_c = Z_h = 1$ hereinafter.) The dust component introduces new time and length scales upon which we base our normalizations (Mace and Hellberg 1992): times with respect to the inverse dust plasma frequency; spatial distances by the dust Debye length $\lambda_{Dd} \equiv (T_{\rm eff}/4\pi n_{0d} Z_d^2 e^2)^{1/2}$, where the definition of the effective temperature derives from (cf.

Bharuthram and Shukla (1992)) $n_{0e}/T_e + n_{0c}Z_c^2/T_c + n_{0h}Z_h^2/T_h \equiv n_{0d}Z_d^2/T_{eff}$; velocities by the dust sound speed $v_{sd} \equiv (T_{eff}/m_d)^{1/2}$; number densities by the 'effective' dust density $|Z_d|n_{0d}$; pressures by $|Z_d|n_{0d}T_{eff}$; electrostatic potential by $T_{eff}/|Z_d|e$; electric charge by the magnitude of the dust charge $|Z_d|e$ and masses by the average dust particle mass m_d.

We assume that each plasma component obeys the (normalized) hydrodynamical equations of continuity, momentum and energy. The latter is written in the form

$$\left\{\frac{\partial}{\partial t} + u_j\frac{\partial}{\partial x}\right\} p_j + \gamma_j p_j \frac{\partial u_j}{\partial x} = 0, \tag{2}$$

where $\gamma_j = 1, 3$ determines whether the fluid responds, respectively, isothermally or adiabatically to the changes in the wave potential ϕ (which naturally places limits on the thermal speed of that component). This system of hydrodynamical equations is closed by the (normalized) Poisson equation.

Writing the equations in terms of the moving coordinate $s = x - Mt$ (where M is the Mach number) and imposing the one-sided boundary conditions

$$\phi(s) \to 0; \quad \frac{d\phi}{ds} \to 0; \quad u_j(s) \to 0; \quad n_j(s) \to n_{0j}; \quad p_j \to p_{0j} \qquad \text{as } s \to \infty, \tag{3}$$

enable one to integrate the hydrodynamical equations, from which we arrive at (Baboolal *et al.* 1988; Mace and Hellberg 1992)

$$\frac{1}{2}m_j M^2 \left\{\left(\frac{n_{0j}}{n_j}\right)^2 - 1\right\} + T_j \ln\left(\frac{n_j}{n_{0j}}\right) = -q_j\phi, \tag{4}$$

for the ion and electron densities, and

$$\left(\frac{n_d}{n_{0d}}\right)^2 = \frac{1}{6T_d}\{(M^2 + 3T_d + 2\phi) \pm [M^4 - 2(3T_d - 2\phi)M^2 + (3T_d + 2\phi)^2]^{1/2}\}, \tag{5}$$

for the dust density where, as discussed elsewhere (Mace *et al.* 1991; Mace 1992), the solution corresponding to the negative surd in the above equation is the one that is continuous with $T_d = p_{0d}/n_{0d}$ and corresponds to the correct density branch $n_d = n_{0d}/(1 + 2\phi/M^2)^{1/2}$ at $T_d = 0$.

It is also worth noting that (4) reduces to the Boltzmann distribution for $m_j = 0$. For finite m_j we solve (4) numerically to determine the density n_j as a function of ϕ.

These density functions are used to construct the Sagdeev (or pseudo-) potential which is then used to determine the correct DL potential profile (Baboolal *et al.* 1988; Mace and Hellberg 1992). Numerical results based on this method are presented in the following section.

3 Numerical Results

3.1 *Dust-ion plasma*

Here we assume that the overwhelming majority of the electron component has been collected by the 'large' dust particles. Such a plasma model is applicable to the *F*-ring of Saturn (Bharuthram and Shukla 1992; Northrop 1989). Throughout this subsection, then, we will set $n_{0e} = 0$. Other fixed parameters are: $Z_d = 100$ and $q_c = q_h = 0.01$ (charge normalized w.r.t the magnitude of the dust charge $|Z_d|e$).

Figure 1 illustrates the effects of finite ion mass on the DL Mach number and potential. Although the Mach number M_{dl} is not significantly affected by the ion/dust mass ratio $m_h = m_c$, the DL potential exhibits a stronger dependence on this parameter—especially for mass ratios ranging between $10^{-7} < m_c, m_h < 10^{-4}$. Note that, within the framework of our model, for the parameter values depicted in figure 1 there is an upper threshold ion mass, i.e. ion masses larger than $\sim 10^{-4}$ cannot give rise to DLs because ϕ_{dl} rapidly approaches zero there. We have been unable to obtain finite-amplitude DLs for m_c, m_h larger than $\sim 10^{-4}$. Note further, that DL Mach numbers slightly less than unity are permitted because of our normalizations: we have normalized speeds with respect to the dust-acoustic speed based on a calculation in which the ions were treated as inertialess. Were we to have normalized with respect to the dust-acoustic speed calculated numerically from the 'dispersion relation' obtained via linearization of the original fluid equations we would have found a Mach number that was always > 1 (Mace and Hellberg 1992).

Figure 2 shows the DL Mach number and potential as functions of the cool/hot ion temperature ratio in the inertial and inertialess models. Even for the relatively small ion masses depicted here ($m_c = m_h = 10^{-6}$) we observe a profound decrease in the allowable range of temperature ratios ($\sim 0.08 < T_c/T_h < 0.101$) when compared to the inertialess theory ($0.004 < T_c/T_h < 0.101$). We should stress here that this is the case even though there is very little difference between the curves predicted by either theory in the range where the inertial theory permits DL solutions (see our earlier remarks). We remark here, parenthetically, that on a run with $m_c = m_h = 10^{-7}$ (an order of magnitude decrease in the ion masses) the range of allowable temperature ratios was only slightly increased to $0.07 < T_c/T_h < 0.101$ (see the remark regarding the range of masses in the previous paragraph). At this juncture it is worth pointing out that the reason for the cut-off at the lower end of temperature scale is due to the cool ion density becoming complex-valued, but we defer discussion of this point till later (§4). It is also interesting to note that for these particular parameter values *only* DLs having $|Z_d|e\phi/T_{eff} \leq 1$ (where T_{eff} is an effective temperature of the ions—see Mace and Hellberg (1992)) are allowed by the inertial theory.

3.2 Dust-ion-electron plasma

Here we generalize the results of the previous subsection to incorporate finite electron densities, i.e. there remains a finite electron population even though the dust has become highly negatively charged. The other fixed parameters are the same as those used in the previous subsection, but in addition we have $q_e = -0.01$.

Figure 3 illustrates the effect of finite electron densities on dust-acoustic DLs. As is evident increasing this parameter yields a decrease in both the DL Mach number and potential for all the values of T_e/T_h considered. However, this effect is reduced for large electron temperatures. Therefore the strongest DL profiles could be expected in dust-ion plasmas in which most of the electrons have been adsorbed onto the dust particles. These conclusions disagree with those of Bharuthram and Shukla (1992): they find that for large T_e/T_h both the Mach number and DL potential exhibit regimes where there is an increase with n_{0e} for the value of $T_c/T_h = 0.05$ that they considered. It should be pointed out that we have tried to duplicate figures 5 and 6 of Bharuthram and Shukla (1992) in an inertialess calculation with $T_c/T_h = 0.05$ (albeit outside the existence domain of the inertial theory), but have been unsuccessful. We always find that both the Mach number and DL potential decrease with n_{0e}.

4 Discussion

In this paper we have investigated the validity of the assumption of inertialess ions, by considering double layers in a dusty plasma, and shown that even though the ion and average dust masses may be significantly different, the approximation that the ions are inertialess can yield misleading results, viz. the inertialess assumption drastically exaggerates the parameter space in which double layers can exist. In apparent opposition to this statement, however, for parameters in which inertial theory predicts double layers there is relatively good agreement between the two theories.

Reconciliation of these two 'opposing' statements lies in the following: Firstly, recall that the only difference between the inertial and inertialess theories lies in the definition of the densities of the isothermal plasma components.

Consider the function

$$f(x) = a\left\{\frac{1}{x^2} - 1\right\} + b\ln(x) + c, \tag{6}$$

($a \geq 0$, $b > 0$) whose similarity to equation (4), defining the isothermal particle densities, is obvious when one identifies x with n_j/n_{0j}; a with $m_j M^2/2$; b with T_j and c with $q_j\phi$; and whose zeroes give the required value of $x = n_j/n_{0j}$. For $a = 0$ (corresponding to inertialess theory) $f(x)$ is a monotonic function of x whose range extends from $-\infty$ to $+\infty$ for all values of c (i.e. $q_j\phi$), with $f(x) \to -\infty$ as $x \to 0^+$. Consequently it is always possible to find a root of the equation $f(x) = 0$, namely $x = \exp(-c/b)$, which serves to define the densities of the isothermal components in the inertialess theory.

More generally, (in the inertial theory) a is small, but nonzero. For our purposes it will prove useful to write $f(x)$ in integral form:

$$f(x) = -2a\int_1^x \frac{dt}{t^3} + b\int_1^x \frac{dt}{t} + c. \tag{7}$$

For $x > 1$ the second integral is clearly larger in magnitude than the first and dominates for large x and hence $f(x) \to +\infty$ as $x \to \infty$; however, for $0 < x < 1$ the first integral is (much) larger in magnitude than the second and consequently $f(x)$ is dominated by this term for such x. Indeed $f(x) \to +\infty$ as $x \to 0^+$ for all finite positive a, which is profoundly different from the behaviour of $f(x)$ when $a = 0$ corresponding to the inertialess case. In that case we noted $f(x) \to -\infty$ as $x \to 0^+$. Clearly the behaviour of this function is strongly influenced by the magnitude of a and moreover, for $a > 0$ it *must* possess a global minimum.

It is the existence of this minimum which places an upper limit on the maximum allowable electrostatic potential. For real roots one requires that this minimum at $x_m = \sqrt{2a/b}$ (in terms of our original parameters $x_m = M/(T_j/m_j)^{1/2}$), viz: $f_0 = f(x_m) = b[1 + \ln(2a/b)]/2 + c - a$, lie below the x-axis, i.e. $f_0 < 0$. However, as discussed by Mace and Hellberg (1992), for $a > 0$ this minimum can be raised above the x-axis by increasing $c > 0$ (i.e. $q_j\phi > 0$), rendering complex roots. Clearly this places an upper bound on the allowable values of c. Note that the magnitude of a can really be very small (recalling that $b = T_j < 1$ also) and still raise f_0 so that it is sufficiently near the x-axis to limit severely the maximum value of c because, as $x < 1$ decreases, the magnitude of the first term in (7) increases much faster than the second term decreases, yielding near cancellations for $0 < x < 1$. Note further that although variation in a brings about a substantial change in the value of f_0 through the logarithmic term its effects on the root nearest $x = 1$ (the one that defines the ion densities in the inertial theory) are far less pronounced, which explains the qualitative similarities between the two theories when applied in the region of validity of the inertial theory: both predict similar densities.

5 SUMMARY AND CONCLUSIONS

In summary, we find that when ion inertia is taken into account the allowable parameter regimes for the existence of double layers are reduced—in particular the range of double layer amplitudes is restricted to less than $\sim T_{\rm eff}/|Z_d|e$, i.e. ion inertia prevents the formation of large amplitude DLs in the present model (cf. Bharuthram and Shukla 1992). The primary reason for this restriction is the extremely limited range of cool-to-hot ion temperature ratios allowed in the inertial model ($0.08 \leq T_c/T_h < 0.101$ for ion masses $m_c = m_h = 10^{-6}m_d$; $0.07 < T_c/T_h < 0.101$ for ion masses $m_c = m_h = 10^{-7}$).

The main conclusion of our investigation can be summarized as follows: inertialess theory based on Boltzmann distributed ions is a good approximation when considering dust-acoustic

DLs as long as one bears in mind, based on inertial considerations such as was done here, the *extremely* restricted parameter regimes in which it is applicable.

It is important to point out that our conclusions are valid only for weak (fluid), stationary DLs in a dusty plasma. (Our results have important implications for solitons and other nonlinear structures in fluid dusty plasmas also.) The existence of DLs of larger amplitude, or of a non-stationary nature, cannot be ruled out by an investigation such as this. Indeed large amplitude dusty DLs seem a plausible phenomenon and would probably be amenable to study through a kinetic theoretical treatment. This would hopefully extend the somewhat restricted existence parameter regimes implied in the present model.

References

Baboolal, S., Bharuthram, R. and Hellberg, M.A. *J. Plasma Phys.* **40** (1988), 163.

Bharuthram, R. and Shukla, P.K. *Planet. Space Sci.* **40** (1992), 465.

Goertz, C.K. *Rev. Geophys.* **27** (1989), 271.

Mace, R.L. *PhD thesis: Linear and Nonlinear Electron-Acoustic Waves in Plasmas with Two Electron Components*, University of Natal, Durban, South Africa (1992).

Mace, R.L., Baboolal, S., Bharuthram, R. and Hellberg, M.A. *J. Plasma Phys.* **45** (1991), 323.

Mace, R.L. and Hellberg, M.A. Submitted to *Planet. Space Sci.* (1992)

Northrop, T.G. in *A Variety of Plasmas: Proceedings of the 1989 International Conference on Plasma Physics* (1989), p 91, Indian Academy of Sciences, Bangalore.

Rao, N.N., Shukla, P.K. and Yu, M.Y. *Planet. Space Sci.* **38** (1990), 543.

Rice, W.K.M., Hellberg, M.A., Mace, R.L. and Baboolal, S. in *Proceedings of the International Conference on Plasma Physics* (1992), Innsbruck, Austria.

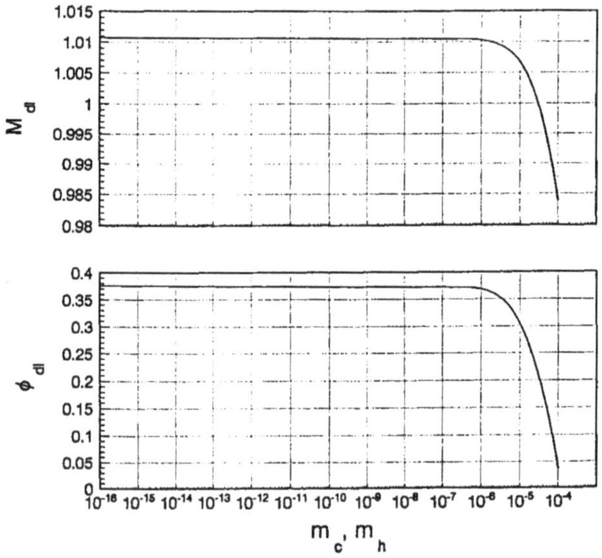

FIGURE 1. DL MACH NUMBER $M_{\rm dl}$ AND ELECTROSTATIC POTENTIAL $\phi_{\rm dl}$ AS FUNCTIONS OF THE ION MASS $m_c = m_h$. Other parameters are: $n_{0c}/n_{0h} = 0.11$; $T_d/T_c = 0$ and $T_c/T_h = 0.09$.

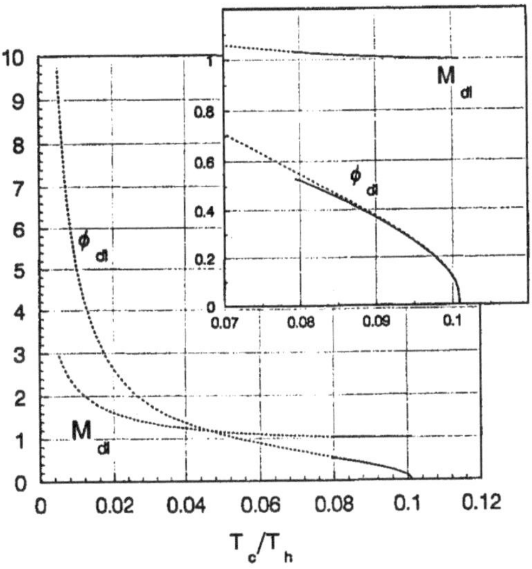

FIGURE 2. DL MACH NUMBERS AND POTENTIALS AS FUNCTIONS OF THE RATIO OF COOL-TO-HOT ION TEMPERATURE. (—) denotes inertial calculation and (−−) denotes the inertialess calculation. Plasma parameters are the same as those used in Figure 1; and $m_c = m_h = 10^{-6}$. The inset shows a zoomed view of the curves in the restricted parameter regime in which the inertial calculations confirm the existence of DLs.

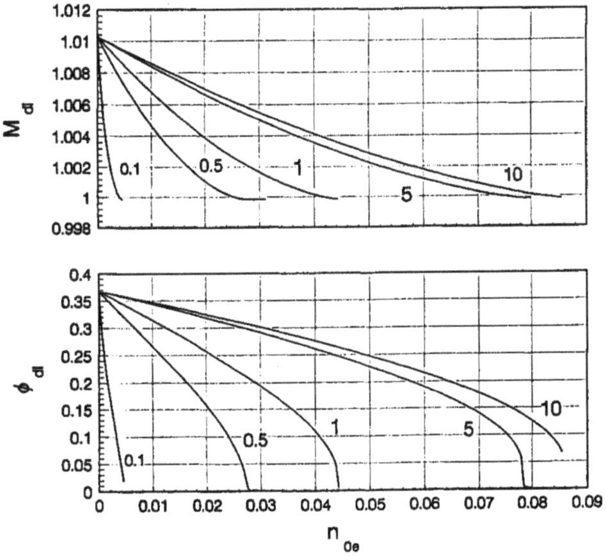

FIGURE 3. DL MACH NUMBER AND POTENTIAL AS FUNCTIONS OF ELECTRON DENSITY. The electron-to-hot ion temperature ratio T_e/T_h is the parameter labelling the curves. The electron mass $m_e = 5.4466 \cdot 10^{-10}$ and other parameters are as in figure 1.

POTENTIAL STRUCTURE OF CATHODE SHEATH IN LOW PRESSURE DIODE DISCHARGES

H.Mase, N.Y.Sato, T.Tanabe and T.Ikehata, Department of Electrical and Electronic Engineering, Ibaraki University, Hitachi 316, Japan

The potential structure of cathode sheath in low pressure discharges, including with cold-cathode and hot-cathode, are discussed by using the improved collisionless sheath theory. The potential profile and the sheath thickness are obtained from Poisson's equation in bipolar space-charge conduction. The magnitude of potential drop is derived from the equations of particle balance and energy balance in the plasma volume. The formation of double layer (DL) is also discussed.

1. INTRODUCTION: In the diode discharges, a boundary layer with large potential fall occurs near the cathode with a few exception /1,2,3/ in hot-cathode discharges. The boundary layer, cathode sheath (CS), plays an important role to sustain the discharge and the potential drop in it is nearly equal to the voltage across the electrodes. As well known, the conventional theories of CS in the cold-cathode discharges are based on the collisional sheath in which the electron multiplication M and self-sustaining condition $\gamma(M-1)=1$ is essential. These theories, however, still involve some principal problems to be solved, when we will derive the self-consistent solution /4/. Moreover, they can not apply to CS in the hot-cathode discharges, because the linear breakdown condition is used and then CS is isolated from the plasma. On the other hand, CS in the hot-cathode discharges have been explained by using the collisionless sheath model. However, description of potential drop is still incomplete.

In this paper, we discuss the cathode sheath formation and the potential structure in both cold- and hot-cathode discharges by improved collisionless sheath theory which includes the plasma balance equations of particles and energies.

2. POTENTIAL STRUCTURE IN BIPOLAR SPACE-CHARGE SHEATH: We consider one-dimensional collisionless sheath formed by the electron current J_e from the cathode and the ion current J_i from the plasma, assuming that their initial velocities are neglectively small (see Fig.1). Poisson's equation in the bipolar space-charge conduction /5/ is given by following equation,

the bipolar space-charge conduction /5/ is given by following equation,

$$d^2\eta/d\xi^2 = -(1/2\sqrt{2})(1/\sqrt{(\eta_0-\eta)}-a_n/\sqrt{\eta})$$
$$\eta=e\phi/kT_e,\quad \xi=x/\lambda_D,\quad a_n=(J_e/J_i)\sqrt{(m/M)} \qquad (1)$$

where the potential ϕ and the distance x are normalized by the plasma parameters, electron temperature T_e and Debye length λ_D.
Assuming that the potential drop ϕ_0 is $\phi_0 >> kT_e/e$ and $d\phi/dx \simeq 0$ at sheath edge, x=d, we obtain the expression of the electric field as follows,

$$d\eta/d\xi = [\sqrt{2}(\sqrt{(\eta_0-\eta)}-a_n\sqrt{\eta_0}-\sqrt{\eta})]^{1/2},$$

$$(d\eta/d\xi)_{\xi=0} = [\sqrt{(2\eta_0)}(1-a_n)]^{1/2} \qquad (2)$$

and Eq.(2) shows that a_n must be $a_n \leq 1$. If $a_n=1$, the electric field on the cathode surface vanishes and then the sheath becomes a double layer (DL). The space-charge density varies with the potential as shown in Fig.2. In the cold-cathode discharges, $a_n<<1$, we can approximate the solution of Eq.(2) as follows,

$$\xi^2 = (4\sqrt{2}/9)\eta^{3/2} \qquad (3)$$

This is Child-Langmuir law of three-halves.
In hot-cathode discharge, $a_n \lesssim 1$, however, it is too difficult to solve it analytically. Some typical profiles of the potential and the electric field are calculated by using a computer (see Fig.3). The relationship between the potential drop and the sheath thickness is demonstrated in Fig.4.

3. MAGNITUDE OF POTENTIAL DROP AT FIXED DISCHARGE CURRENT: Now we consider the plasma sustained by only primary electrons accelerated through the sheath. The equations of particle balance and energy balance in the plasma of length L are expressed as follows /6/,

$$J_i = J-J_e = en(1/\tau_g-1/\tau_1)L \qquad (4)$$

$$\phi_0(J_e-J_e') = e\phi_i(n/\tau_g)(1+\Delta_E)L \qquad (5)$$

where τ_g, τ_1 are characteristic times of generation and recombination loss and ϕ_i, Δ_E are the ionization potential and a fraction of all energy losses except energy loss by ionization process. From Eqs.(4) and (5), we can directly obtain the magnitude of potential drop,

$$\phi_0 = \phi_i(1+\Delta_E)(J_i+enL/\tau_1)/(J_e-J_e') \quad (6)$$

$$\simeq \phi_i(1+\Delta_E)(1/\gamma)/(1-J_e'/J_e) > \phi_i \quad \text{for } \gamma=J_e/J_i<1 \quad (6.1)$$

$$\simeq \phi_i(1+\Delta_E)(enL/J_e\tau_1)/(1-J_e'/J_e) \gtrsim \phi_i \quad \text{for } \sqrt{(M/m)}\geq\gamma=J_e/J_i<1 \quad (6.2)$$

where $(1-J_e'/J_e)$ means the effect of trapping of primary electrons and varies with p.L and fields for electron confinement /7/. Thus we find some limits for the diode discharge as shown in Fig.5.

When the emission current is large enough compared with J_e, the potential barrier ϕ_m for the emitted electrons occurs near the cathode. The excess electrons trapped by the barrier may be neutralized by the ion current from the plasma and then a dark plasma is formed between the cathode and the sheath /1/. The sheath (DL) separates from the cathode. The location of DL may be determined by using the principle of energy minimum. The energy loss of primary electrons is given by Eq.(5) in which L (>>d) is replaced with L-x. In order to fill the space from x=0 to x with the plasma, ions must be sufficiently slow down and lose their energies. The total energy loss could be written by

$$P = \phi_0(J_e+J_i) = e\phi_i[(n_A/\tau_{gA})(1+\Delta_E)(L-x)/(1J_e,/J_e) + (n_B/\tau_{1B})x/(1-J_i'/J_i)]$$

$$= e\phi_i\{A^*(L-x)+B^*x\} \quad (7)$$

where suffix A and B denote the parameter in the glowing plasma A and the dark plasma B. The density in dark plasma is estimated to be $\sim(1/2)(T_{eA}/T_{eB})^{1/2}n_A$ from divJ=0.

The double layer could occur at $x=x_{DL}$ where Eq.(7) takes the minimum value. If the energy loss rates, A^* and B^*, are independent of x, Eq.(7) takes minimum value at x=0 for $A^*<B^*$ or x=L for $A^*>B^*$. In this situation, DL occurs near the cathode (Langmuir mode) or the anode (anode glow mode)/1/. When the loss rates depend on x, DL appears at $0<X_{DL}<L$. In

this case we could find the position of DL by calculating dP/dx=0 under an appropriate assumption, but stational DL at $0<x_{DL}<L$ rarely occurs when A^* and B^* satisfies very restricted conditions. This simplified theory may be applied to stepwise varying potential structure in striated discharges.

4. CONCLUSION: The cathode sheath structure $\{\phi(x), \phi_0(J_i/J_e), d, X_{DL}\}$ in low pressure diode discharges is able to be explained by using collisonless sheath theory and plasma balance equations. Sheath-plasma transition is not discussed here. To clear joining plasma and sheath, more detailed discussions are required.

REFERENCES:

1. L.Malter, E.O.Johson, W.M.Webster, RCA Rev.12(1951), 415.
2. E.O.Johnson, W.M.Webster, RCA Rev. 16(1955),82.
3. R.N.Franklin, Plasma Phenomena in Gas Discharge (ClarendonPress, 1976), Chap.5.
4. H.Mase, T.Tanabe, T.Ikehata, Trans. IEE Japan 111-A(1991),383.
5. I.Langmuir, Phys. Rev. 33(1929), 954.
6. H.Mase, T.Tanabe, I.Ikehata, Proc. XVIII ICPIG (Swansea,1987) IV, 746.
7. H.Mase, T.Tanabe, T.Ikehata, N.Y.Sato, Proc. 1992ICPP,in press.

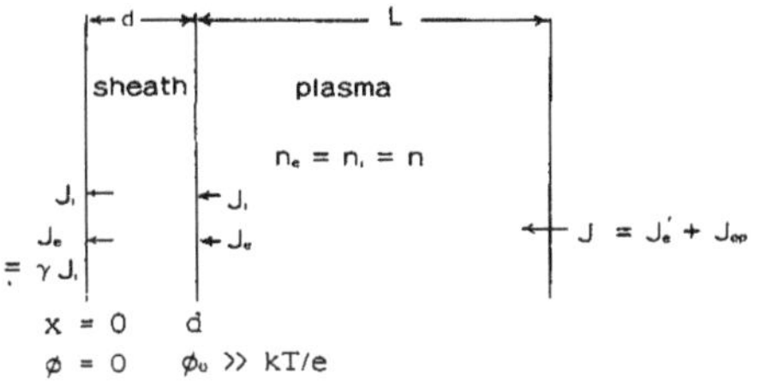

Fig.1.One-dimensinal sheath and plasma.(J is the total current density, $J=J_i+J_e$. J_e' is the current carried by primary electrons which are not de-celerated in the plasma volume. J_{ep} is the current due to plasma electrons.

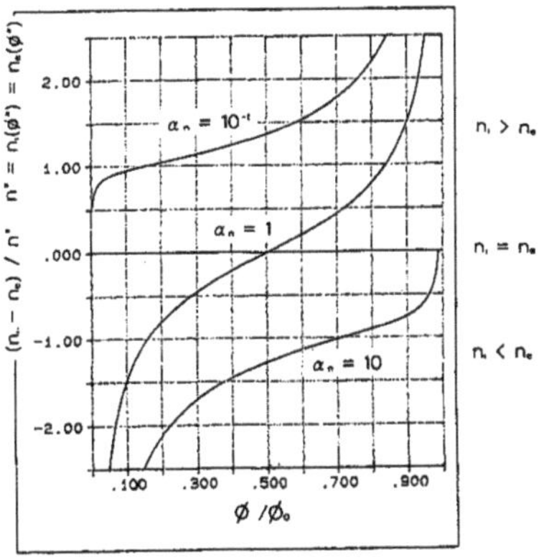

Fig.2.Variation of chaege density with potential, including a case of vacuum diode (a_n=10).

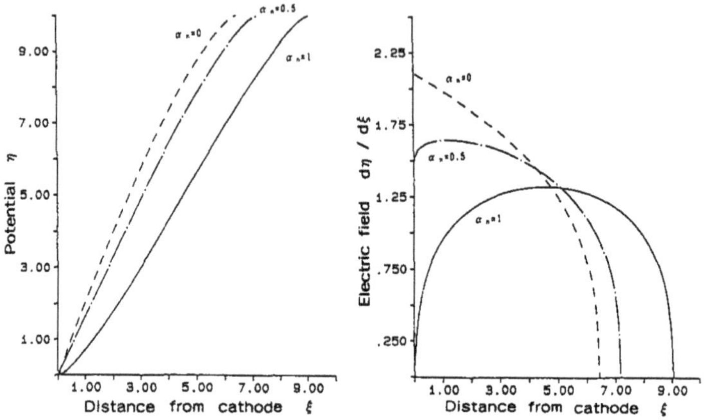

Fig.3.Typcal profiles of potential and electric field.

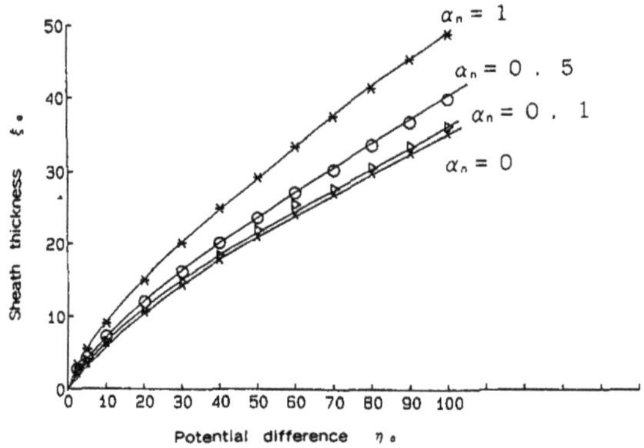

Fig.4.Variation of sheath thickness with potential drop. When $\alpha_n=0$, $\xi_0 \propto \eta_0^{3/4}$.

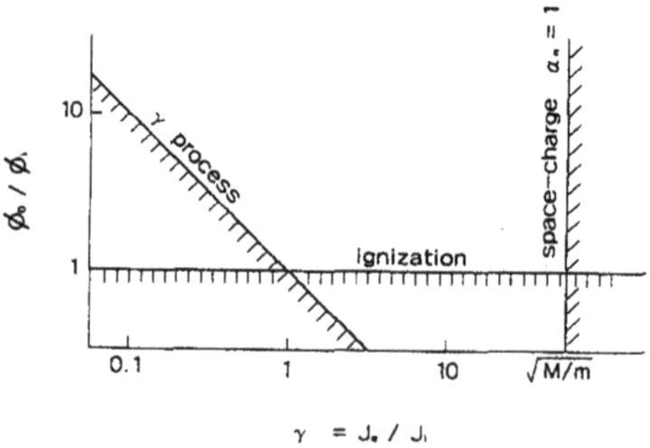

Fig.5.Limits for low pressure diode discharge. Hatched area is forbidden.

THE EXCITATION OF A THREE-DIMENSIONAL NONMONOTONOUS DOUBLE LAYER IN A MAGNETIZED PLASMA WAVEGUIDE

V.I.Maslov Kharkov Institute of Physics & Technology, 310108 Kharkov,Ukraine

The energy exchange between particle and nonhomogeneous oscillations as well as the nonlinear generation of hogher harmonics by these oscillations, if their dispersion relation is near linear, can lead to the formation of a localized wave structure. In particular, the possible formation of the correlated perturbation - the potential well - has been investigated with ion-acoustic instability development in /1-4/. Also the solitary perturbation can form in a magnetized plasma-loaded waveguide /5-10/. Instead of treating this potential well emergence we describe its evolution and properties when it is excited by the "hot" electron beam in a magnetized plasma-loaded waveguide. We consider the case of the strong external longitudinal magnetic field. Then the electron dynamics is one-dimensional.

We choose the initial perturbation in the form of the potential well $\varphi = J_0(K_\perp r)\,\varphi_w(z,t)$ of small amplitude (depth)-φ_0, and of width Δz smaller than the system length L, moving with a velocity V_0, that satisfy the inequalities $V_b - V_{bth} < V_0 < V_b$. Here $J_0(K_\perp r)$ is the zeroth order Bessel function, $K_\perp R_0$ is its zero, Ro is the radius of waveguide, V_b is the beam velocity, V_{bth} is the thermal velocity of beam electrons. The potential well is unstable in a plasma with electron beam. It reflects the resonant electrons, absorbing its energy and the well amplitude is growing. Owing to the reflection from the potential well of resonant electrons, the velocity distribution of which is asymmetric relatively to Vo, the quasi-neutrality is broken near the well: the electrons are short in the first front of the well and they are surplus on the back one. The restoration of the latter is due to the acceleration or to the deceleration of electrons. This is caused by the self-consistent formation of the potential shock $\Delta\varphi$ near the well. Hence, the resonant region in electron velocity space is wider in front of the well, $\xi = z - V_0 t$, by the value of potential shock, than behind the well, $\xi < 0$.

When the well amplitude increases up to certain value, the slow evolution of well starts as compared with the dynamics of general part of electrons. It is true when the increase time of interaction $\tau^{-1} = V_{tr}/\Delta z = (2e\varphi_0/m)^{1/2}/\Delta z$ of the general part of resonant electrons with the well exceeds the growth rate $\gamma(\varphi_0) = \partial \ln \varphi_0/\partial t$ of the well amplitude $V_{tr}(\varphi_0) > \gamma(\varphi_0)\,\Delta z(\varphi_0)$. In this case the changing velocity distribution of resonant electrons moves from the well with the relative velocity equal to V_{tr}.

We describe the well evolution using the expanding of small parameter $\alpha=\gamma\Delta\tau/V_{tr}$

Below we neglect the nonadiabatic dynamics of electrons with the velocities $|V-V_0| \leq V_{tr}$ in the field of potential tails. The latter moves with the velocity V_{tr}. Consideration of the nonadiabatic dynamics leads to the additional terms of the higher order as compared to the used ones.

Taking into account that the resonant electrons are reflected from the well, we obtain from the Vlasov equation the expression for the velocity distribution function of beam electrons

$$f_b(t,\xi,u) = \begin{cases} f_b((u^2-2e\varphi/m)^{1/2}+V_0-V_b), & u > A(\varphi)\,\mathrm{sign}(\xi), \quad A(\varphi)=(2e(\varphi_{00}+\varphi)/m)^{1/2} \\ f_b((u^2-2e\varphi/m)^{1/2}-V_0+V_b), & u < A(\varphi)\,\mathrm{sign}(\xi), \quad \varphi_{00}=\varphi_0 J_0(k_\perp r) \end{cases} \qquad (1)$$

In order to describe the slow evolution of a system we shall seek for the solution of Vlasov and Poisson equations in the form (1) and the potential distribution in the form of a well with amplitude $\varphi_0(t)$ slowly varying in time. Integrating (1) over velocities, we obtain the expression for the density perturbation of beam electrons in the first order of $(V_b-V_0)/V_{bth}$ and the second order of $e\varphi/T_b (T_b=mV^2_{bth})$

$$\delta n_b(\xi)=(en_{b0}/T_b)\{\varphi(1+e\varphi/2T_b)-$$
$$-\mathrm{sign}(\xi)\sqrt{2/\pi}((V_b-V_0)/V_{bth})[\sqrt{\varphi_{00}(\varphi_{00}+\varphi)}+\varphi \ln(\sqrt{-\varphi}/(\sqrt{\varphi_{00}}+\sqrt{\varphi_{00}+\varphi}))]\} \qquad (2)$$

Finding from the Vlasov equation the expression for the density perturbation of plasma electrons, and substituting it and (2) in the Poisson equation, we derive the evolution equation

$$2\dot{\varphi}_w/V_0^3+\omega_p^{-2}\varphi_w'''+\varphi_w'(V_0^{-2}-V_c^{-2}-V_m^{-2})-\varphi_w'\{(e/m)\beta\varphi_w(3V_0^{-4}+V_m^{-2}V_{bth}^{-2})-$$
$$-\mathrm{sign}(\xi)\sqrt{2/\pi}((V_b-V_0)/V_{bth}V_m^2)[(1+\varphi_w/\varphi_0)^{-1/2}+\ln(\sqrt{-\varphi_w}/(\sqrt{\varphi_0}+\sqrt{\varphi_0+\varphi_w}))]\}=0 \qquad (3)$$

of KdV-type but with nonquadratic nonlinearity. Here "$\cdot$"$=\partial/\partial t$, "$'$"$=\partial/\partial\xi$; V_c is the maximum phase velocity of a wave in waveguide with no beam, $V_m=V_{bth}(n_0/n_{b0})^{1/2}$ is the maximum phase velocity of electron acoustic mode /12/, $\omega_p^2=4\pi n_0 e^2/m$, $\beta=\int_0^{R_0} J_0^3(k_\perp r)r\,dr/\int_0^{R_0} J_0^2(k_\perp r)r\,dr$. The equation (3) describes the space distribution of nonhomogeneous potential, its deformation, the increase of $\varphi_0(t)$ in time and also the motion of the potential well with the velocity V_0. The last term 3,4,11, in (3) is determined by the resonant electrons.

Let us derive from (3) the expressions for properties of potential well. For this we choose symmetric $\Phi(\xi)$ and asymmetric $\delta\varphi(\xi)$ parts of potential distribution $\varphi_w(\xi)=\Phi(\xi)+\delta\varphi(\xi)$. $\delta\varphi(\xi)$ describes the asymmetric potential, added in presence of a beam, to the symmetric well. $\delta\varphi(\xi)$ changes from $-\Delta\varphi$ at $\xi<-\Delta z$ to $\Delta\varphi$ at $\xi>\Delta z$. Assuming $\Delta\varphi \ll \varphi_0$ we obtain from (3) the equation describing the space distribution of the potential:

$$(\varphi')^2=\omega_p^2\varphi^2(V_c^{-2}+V_m^{-2}-V_0^{-2})+\varphi^3\beta(e/m)(\omega_p^2/V_0^4+d_B^{-2}V_{Bth}^{-2}/3), \qquad d_B = V_m/\omega_p \tag{4}$$

From (4) and $\varphi'|_{\varphi=-\varphi_0}=0$ we obtain the expression for well velocity Vo

$$V_0 = V_{00}+\Delta V(\varphi_0), \quad V_{00}=V_mV_c/(V_m^2+V_c^2)^{1/2}$$
$$\Delta V(\varphi_0) = (e\varphi_0\beta/2T_B)(V_{00}^3/3V_m^2+V_{Bth}^2/V_{00}) \tag{5}$$

V_0 increases with increasing φ_0 and is equal to V_m at $V_c \gg V_m$ and V_c at $V_m \gg V_c$. From (4),(5) it may be concluded that the well is determined by two modes with the linear despersion relation : the mode in magnetized plasma-loaded waveguide and the electron-acoustic mode. In the quasi-stationary approximation from (3) at $\xi > \Delta z$ we obtain the rough expression for the potential shock near the well

$$e\Delta\varphi/m = \sqrt{2/\pi}((V_B-V_0)/V_{Bth})/\beta(V_{Bth}^{-2}/3+V_m^2/V_{00}^4) \tag{6}$$

In the case being considered the shock is greater by a large factor $T_b/e\varphi_0$ in comparison with the case, when fronts of electrons reflected from the well reach the boundaries of the system 4.

From (4) we determine roughly the well width:

$$\Delta z=\{(e\varphi_0\beta/8m)[(\omega_p/V_{00}^2)^2+(d_BV_{Bth})^{-2}/3]+ +\mathrm{sign}(\xi)\sqrt{2/\pi}\,d_B^{-2}((V_B-V_0)/V_{Bth})[1-3/2\sqrt{2}+0.25\ln(1+\sqrt{2})]\}^{-1/2} \tag{7}$$

The well is asymmetric in the presence of the beam: the first front is steeper than the backone. The well width decreases with increasing amplitude.

To describe the well evolution in time we determine from (3),(4) the well amplitude growth rate $\gamma = \partial \ln\varphi_0/\partial t$

$$\gamma=(1+\sqrt{2})((V_B-V_0)/V_{th}^6\,2\sqrt{\pi})\sqrt{e\varphi_0\beta/m}\,(V_{00}^3/V_m^2)(\omega_p^2/V_{00}^4+d_B^{-2}V_{Bth}^{-2}/3)^{1/2} \tag{8}$$

Nonlinear growth rate increases with increasing well amplitude.

In a magnetized plasma-loaded waveguide two types of solitary structures may exist. The other type - soliton - moves with a thermal velocity of plasma electrons, V_{th}. Let us consider its properties. In the case of a small soliton amplitude, φ_s ($<< T_e/e = mV^2_{th}/e$), from the Vlasov equation we obtain the expression for the velocity distribution function of electrons. Integrating the latter over velocities, we obtain the expression for the electron density perturbation in the second order of $\Phi = e\varphi/T_e$

$$\delta n' = \dot{\Phi}[y+(1-2y^2)(1-Q(y))/y] + \Phi' Q(y) + \Phi\cdot\Phi'[1-y^2+(3/2-y^2)(Q(y)-1)] \quad (9)$$

Here "$\cdot$" $=\partial/\partial(t\omega_p\sqrt{2})$, "$'$" $=\partial/\partial(\xi\omega_p/V_{th})$, $\xi = x - tV_s$, $Q(y)=1+(y/\sqrt{\pi})\int_{-\infty}^{\infty}dt\,exp(-t^2)/(t-y)$, $y=V_0/V_{th}\sqrt{2}$; V_s, φ are the velocity and potential of the soliton. Substituting (9) in the Poisson equation, we derive the KdV evolution equation

$$\dot{\Phi}[y+(1-2y^2)(1-Q(y))/y] + \Phi' Q(y) + \Phi\cdot\Phi'[1-y^2+(3/2-y^2)(Q(y)-1)] - \Phi''' = 0 \quad (10)$$

From (10) we derive the equation describing the space distribution of the potential:

$$(\Phi')^2 = \Phi^2 Q(y) - [1+(2y^2-3)Q(y)]\Phi^3/6 \quad (11)$$

From (11) and $\Phi'|_{\Phi=\Phi_s} = 0$ we obtain (similar to 13) the expression for V_s

$$V_s = 1.32\, V_{th}(1 - e\varphi_s/6T_e) \quad (12)$$

V_s is close to V_{th} and decreases with increasing φ_s.
We determine roughly the soliton width from (11):

$$\Delta X = \sqrt{48 T_e/e\varphi_s} \quad (13)$$

The soliton width decreases with increasing amplitude. The soliton is the "hole" in the electron phase space, the area of which, $\Delta X\sqrt{e\varphi_s/m}$, is independent of φ_s.
In the case of large amplitudes, $e\varphi_s/T_e > 1$, from the Vlasov equation we have the expression for the velocity distribution function of plasma electrons (without electrons trapped by a soliton field) $f = f_0[(u^2 - 2e\varphi/m)^{1/2} + V_s\, sign(u)]$ for $|u| = |V - V_s| > \sqrt{2e\varphi/m}$. Here f_0 is the Maxwellian dustribution function. Thus we derive the equation for the soliton shape

$$(\Phi')^2 = -\Phi + (2/\sqrt{\pi})\int_{-\infty}^{\infty}dt\,(t-y)^2 exp(-t^2)\{[1+\Phi/(y-t)^2]^{1/2} - 1\} \quad (14)$$

From (14)

$$\Delta X = (2e\varphi_s / T_e(\sqrt{2}-1))^{1/2} \qquad (15)$$

we conclude that the soliton width increases with φ_s . Hence, taking into account the trapped electrons is important. Assuming for their density distribution $n_{t2}(x) = n_2 \exp(e\varphi(x)/T_{t2})$, we obtain similarly to (15) that the width and velocity of the soliton increase with amplitude growing (in contrast to the case of small amplitude φ_s).

These properties of the solitary pulses and their amplitude dependences were observed in experiments and numerical simulations 6.

REFERENCES:

1. H.Schamel, Phys. Rep. 140 (1986), 163.
2. T.Sato, H.Okuda, Phys.Rev.Lett., 44 (1980) 740.
3. K.Nishihara et al., Riso-R-472 (1982) 41.
4. V.I.Maslov, Plasma Physics. 16 (1990) 759.
5. S.M.Krivoruchko et al., JETP 67 (1974) 2092.
6. I.P.Lynov, P.Michelsen, H.L.Pecseli, J.J.Rasmussen, K.Saeki, V.A.Turikov, Physica Scripta 20 (1979) 328.
7. H.Ikezi et al., Phys. Fluids 14 (1971) 1997.
8. B.N.Rutkevich et al., JTP 42 (1972) 493.
9. V.I.Kurilko, A.P.Tolstolushskij, JTP 44 (1974) 1418.
10. V.K.Sayal, S.R.Sharma, Phys. Lett., 149 (1990) 155.
11. J.E.Allen, Plasma Phys. and Contr. Fus. 27 (1985) 1343.
12. V.K.Jain, N.T.Tsintsadze IC/89/ 150.
13. H.Schamel, Plasma Physics 14 (1972) 905.

A DOUBLE LAYER INDUCED BY A RELATIVISTIC ELECTRON BEAM

V.I.Maslov
Institute of Physics & Technology, Kharkov 310108, Ukraine

When an electron beam of density, n_b , comparable with the plasma density, n_i, is injected from the isolated gun into the plasma, the plasma electrons are accelerated to the electron beam source in the field occurring between the source and the beam. In this case their density falls and they do not compensate a positive charge of the source. Therefore the potential shock attains the beam energy value. The beam returns to the source and neutralizes its charge. Then, the potential separates from the source and moves inside the plasma with a velocity V_1. Thus, the double layer (DL) may form (i.e., a potential jump from the boundary value $\varphi_0 = \varphi(z=0)$ to zero in a small space interval L). The DL reflects the beam from the plasma. We investigate the properties of such a DL in the case of relativistic beam injection into one-dimensional plasma. The latter assumption holds true for a plasma in a strong magnetic field.

The DL formation needs a charge separation. In the case of the electron monotonous DL formation by the electron - beam - plasma mode, as will be shown, two groups of injected electrons are necessary. Let us consider the half - space plasma, z>0, into which a high energy beam, $V_b >> V_{th}$, V_{bth}, and a slowly moving beam, $V_c < V_{tho} << V_b$, of densities n_b and n_o arebeing injected from some time. $V_{bth} = (T_b/m_e)^{\frac{1}{2}}$, $V_{tho} = (T_o/m_e)^{\frac{1}{2}}$ are the thermal velocities of the electron beams. As the distribution function of electrons of the slowly moving beam is symmetric relative to V_1, after they are reflected from the DL, the current velocity of this beam can be chosen to equal V_1.

The density of the slow - moving beam exponentially falls off

$$n_o(z) = n_o \exp((\varphi - \varphi_o)e/T_o), \qquad (1)$$

resulting in positive charge at $\varphi_a < \varphi < \varphi_o$; while the density of both the high energy beam and plasma electrons

$$n_e(z) = n_e(V_1/c)[1-(1+(y_0-1)\varphi/\varphi_0)^{-2}]^{-1/2} \quad (2)$$

$$n_b(z)=n_b(1-y_0^{-2})^{\frac{1}{2}}(1-(1+e\varphi/mc^2)^{-2})^{-\frac{1}{2}}, \quad y_0=(1-(V_b+V_{bth})^2/c^2)^{-\frac{1}{2}}$$

grows in accordance with the power law. This leads to negative charge at $0<\varphi\leq\varphi_a=\varphi_0(y_0-1)^{-1}[((y_0+1)/2)^{\frac{1}{2}}-1]$. At $\varphi_c = \varphi_0(2V_{bth}/c)y_0$ the beam starts to be reflected. As a result, the quasineutrality is restored outside the DL.

Since the nonresonant beam electrons propagate into the plasma, where they are decelerated, their density grows. Therefore, the quasineutrality condition in front of the DL (at z>>L) requires V_1 to be lower than the thermal velocity of plasma electrons $V_1<V_{th}$, and the density of the beam electrons passing through the DL should be small $n_{bo}<<n_b$. Hence the approximate expression for the potential shock follows

$$\varphi_0 = (mc^2/e)(y_0-1) \quad (3)$$

All electrons transfer their momentum to the DL. In the DL field only ions gain the momentum. The plasma electrons and ions take away the energy from the DL. The electrons of beams interacting with the DL lose their energy. Using the momentum- and energy - balance equations and also the quasineutrality condition at the boundary of beam injection, we obtain the following expressions

$$V_1/c = (n_{bo}/n_e)(1-y_0^{-2})^{\frac{1}{2}}<<1 \quad (4)$$

$$n_b = (n_i/2 - n_0(T_0/2e\varphi_0))y_0/(y_0+1)$$

$$n_0 = [1+(T_0/e\varphi_0)y_0/(y_0+1)]n_i/(y_0+1)$$

It is seen from (4) that n_0 is inversely proportional to y_0, and n_b tends to $n_i/2$ if y_0 is large.

From (1) we have that in the region, where the slow - moving beam is reflected, $\delta n(\varphi_b)= -n_0$, where φ_b is defined by $dn(\varphi_b)/d\varphi=0$ and equals $\varphi_b/\varphi_0= 1-T_0/e\varphi_0$. It is clear from the latter that the potential drop is small in the region, where the density perturbation is defined by the change of the slow - moving beam density. From (2) it follows that in the region, where the density perturbation is defined by the change in the density of the high energy beam at its slowdown δn grows up to the potential $\varphi = \varphi_c$. δn reaches its maximum value in the region of strong beam deceleration and equals $\delta n(\varphi_c) = n_b(2V_b/V_{bth})^{\frac{1}{2}} y_0^{\frac{1}{2}}$. Integrating Poisson equation over φ ,we obtain

$$(\partial\varphi/\partial z)^2/8\pi e = \int_{\varphi_0}^{\varphi} d\varphi \cdot \delta n = \tag{5}$$

$$= n_i(\varphi_0 \cdot \varphi)^{\frac{1}{2}}[[(\varphi/\varphi_0)(y_0+1)+2]^{\frac{1}{2}}/(y_0+1)^{\frac{1}{2}} - (\varphi/\varphi_0)^{\frac{1}{2}}]$$

From (5) it follows that the potential profile of the DL is described at $y_0 \gg 1$ by the equation

$$\varphi/\varphi_0 = 1 - (z2^{\frac{1}{2}} w_p / y_0 c) \tag{6}$$

We determine roughly the DL width from (5)

$$L = (c/4e)(2m/\pi n_i)^{\frac{1}{2}}(y_0-1)/[1-(2/(y_0+1))^{\frac{1}{2}}]^{\frac{1}{2}} \tag{7}$$

At $y_0 \gg 1$, $L = (c/2^{\frac{1}{2}} w_p) y_0$, where $w_p = (4\pi n_i e^2/m)^{\frac{1}{2}}$, the DL width is proportional to y_0.

Although this electron DL, formed by electron beam injection into the plasma, is determined by electron velocities and times, the momentum- and energy - exchange of the DL with ions is essential. So, the ions, passing through the DL, are accelerated in its field to an energy $E_i = mc^2(y_0-1)$, comparable with the beam energy.

The DL forms at $y_0 \gg 1$ during the time $y_0 w_p^{-1}/2^{\frac{1}{2}}$. But the time of plasma electron response to the field initiation, according (5) is

$$t_0 = (y_0/w_p 2^{\frac{1}{2}})(\varphi_0/\varphi_0(t))^{\frac{1}{2}} \tag{8}$$

Hence, it may be concluded that during DL formation time, when $\varphi_0(t) < mc^2(y_0-1)/e$, the plasma electrons have no time for the response to the field initiation, and thus are trapped by the DL fitld. In this way the slow electron group is formed, which is necessary, as stated above, for the DL formation on injecting only the high energy beam. But now the DL is non-monotonous: first, the potential slowly rises from φ_0 to φ_1 and then abruptly drops from φ_1 to zero. After the DL formation is completed, the plasma electrons incoming to the DL region are accelerated opposite to the beam propagation.

Since the DL slowly drifts downstream, the density of the trapped electrons n_0 in the nonmonotonous DL decreases because of the growth of the region, where these electrons are localized. The stability analysis of the electron fluxes relative to the low frequency perturbations shows that if the density of the trapped electrons falls below the critical value

$$(V_{tho}/V_b)^2/y_0^3 > n_0/n_i \qquad (9)$$

the DL becomes unstable relative to the perturbation with the phase velocity equal to V_1. The computer simulation[1] has shown that the DL, that has penetrated deep into the plasma, decays with the generation of a single vortex structure in the electron phase space. At the same time, a new DL arises at the boundary. It is necessary to note that, according to (4),(9), in the case of a relativistic beam, $y_0 \gg 1$, as opposed to the nonrelativistic case[2] a small part of slow - moving electrons is sufficient for the forming DL to become quasistationary.

Similar processes for the potential distribution and beam behaviour have been observed in a full - scale experiment[3] and numerical simulations[1,4]. The beam injection into the plasma may lead to the virtual cathode formation. But in numerical simulations[4] at certain conditions the virtual cathode transforms to the potential hump. The DL, under study, separates from the hump and moves inside the plasma at a thermal velocity of plasma electrons.

References

1. H.Okuda, R.Horton, M.Ashour-Abdalla, Phys.Fluids, 30(1987), 200.
2. V.I.Maslov, Ukr. Fiz. Zh., 33(1988), 1342.
3. T.Obayashi, N.Kawashima, K.Kuriki, et al., Science, 225(1984), 195.
4. N.Singh, R.W.Schunk, Plasma Phys. and Contr. Fus., 26(1984), 859.

DOUBLE LAYER AT A SLOW ION MODE

V.I.Maslov, Kharkov Institute of Physics & Technology, 310108 Kharkov Ukraine

It is shown in 1 that the ion-acoustic instability development in one-dimensional current-carrying plasma results in ion-acoustic turbulence splitting into widely spaced short perturbations. The width of the latter is of the order of the wavelength. The perturbation of this type may be transformed in the nonlinear stage into solitary perturbation, viz. the potential well. The possible formation of the correlated perturbation - the potential well - has been investigated with ion-acoustic instability development in 2-9. The selfconsistent evolution of a nonmonotonous double layer (potential well φ with potential shock $\Delta\varphi$ in its vicinity) is considered in this paper in one-dimensional current-carrying bounded plasma, 0<x<L. Initially electrons move with respect to ions with a current velocity V_d satisfying $\sqrt{T_e/m_i}=V_s<V_d<V_{th}=$ $=\sqrt{T_e/m_e}$. Here, T_e is the temperature of electrons; m_e, m_i are the electron and ion masses. At boundaries, the injected electrons have a Maxwellian distributions of velocities (with $V_d\neq0$), while the distribution of leaving electrons is determined selfconsistently.

The reflection of resonant electrons of a current-carrying plasma by the potential well causes the potential shock formation near the well and the well amplitude growth, $\varphi_0(t)$. When $\varphi_0(t)$ increases up to certain value $V_{tr}>\gamma\cdot\Delta X$ ($\gamma=\partial \ln\varphi_0/\partial t$, Δx is the well width, $V_{tr}=(2e\varphi_0/m_e)^{1/2}$), the slow (as compared with the dynamics of electrons) evolution of the well starts. In this approximation and at $e\varphi/T_e=\varphi<1$, $V_d/V_{th}=V_c<1$ we obtain from the Vlasov equation the expression for the perturbation of electron distribution function. From the latter the expression for the electron density perturbation follows

$$\delta n=\varphi+\varphi^2/2-\mathrm{sign}(\xi)\{2\Delta\varphi\sqrt{(\varphi_0+\varphi)/\pi}+V_c\sqrt{2/\pi}\,[\sqrt{\varphi_0(\varphi_0+\varphi)}+ + \varphi\ln(\sqrt{-\varphi}/(\sqrt{\varphi_0}+\sqrt{\varphi_0+\varphi}))]\}, \qquad (1)$$

$\xi=x-tV_0$, V_0 is the well velocity.

Using the quasineutrality condition outside the well we obtain

$$\Delta\varphi/\varphi_0=V_c\sqrt{2/\pi}\,(1+2\sqrt{\varphi_0/\pi}) \qquad (2)$$

The potential shock is proportional to the amplitude φ_0 and the electron current velocity V_e.

From the Vlasov equation we obtain the expression for the perturbation of ion distribution function. From this the expression for the ion density perturbation of ion distribution function. From this the expression for the ion density perturbation is obtained:

$$\delta n_i' = \varphi' + 2\dot{\varphi} - 2(V_e - 1)\varphi' + 3\varphi\cdot\varphi' \tag{3}$$

Here "." $=\partial/\partial t$, "," $=\partial/\partial\xi$, $V_e = V_0/V_s$. Substituting (1), (3) in the Poisson equation, we derive the evolution equation

$$\dot{\varphi} - (V_e - 1)\varphi' + \varphi\cdot\varphi' + \varphi'''/2 + \mathrm{sign}(\xi)(\Delta\varphi/2)\varphi'[(1-\sqrt{\varphi_0/\pi})/\sqrt{\varphi_0(\varphi_0+\varphi)} + \varphi_0^{-1}(1-2\sqrt{\varphi_0/\pi})\ln(\sqrt{-\varphi}/(\sqrt{\varphi_0}+\sqrt{\varphi_0+\varphi}))] = 0 \tag{4}$$

of KdV-type but with nonquadratic nonlinearity 4,7,9. From (4) we derive the equation describing the space distribution of the potential:

$$(\varphi')^2/2 = (V_e-1)\varphi^2 - \varphi^3/3 - \Delta\varphi\{-(\varphi_0+\varphi) + (\varphi^2/2\varphi_0)(1-2\sqrt{\varphi_0/\pi})\ln(\sqrt{-\varphi}/(\sqrt{\varphi_0}+\sqrt{\varphi_0+\varphi})) + (2/3\sqrt{\pi})\varphi_0^{3/2} + \sqrt{1+\varphi/\varphi_0}\,[\varphi_0(1-(2/3)\sqrt{\varphi_0/\pi}) + (\varphi/2)(1+(2/3)\sqrt{\varphi_0/\pi})]\} \tag{5}$$

From (5) and $\varphi'|_{\varphi=-\varphi_0} = 0$ we obtain the expression for V_e

$$V_e = 1 - \varphi_0/3 + (2/3\sqrt{\pi})(\Delta\varphi/\sqrt{\varphi_0}) \tag{6}$$

The profile of the well is characterized by its width. We determine roughly the width frim (5):

$$\Delta X = \{V_c\sqrt{2/\pi}\,[1 - 3/2\sqrt{2} + 0.25\ln(1+\sqrt{2}) + (1-2\sqrt{2}/3)\sqrt{\varphi_0/\pi}\,]\}^{-1/2} \tag{7}$$

To describe the well evolution in time we determine from (4), (5) the well amplitude growth rate $\gamma = \partial\ln\varphi_0/\partial t$

$$\gamma = (2/\pi)^{3/4}(V_c^{3/2}/\sqrt{2})[1 + (8/3)\sqrt{\varphi_0/\pi} - \sqrt{\pi}/2\,(\varphi_0/3V_c) \tag{8}$$

There exists a threshold in V_c for the formation of a large-amplitude well 3. Moreover, the stationary ion-acoustic perturbation in the form of the potential well doesn't exist in equilibrium plasma.

Hence, let us consider the properties and the evolution of a potential well at slow ion mode in current-carrying plasma, $V_{th}^i < V_d < V_{th}$. In other words, we describe the well, propagating with the velocity V_0 approximately equal to the ion thermal velocity, $V_{th}^i = (Ti/M)^{1/2}$. Taking into account that the resonant electrons are reflected from the well, we obtain from the Vlasov equation the expression for the velocity distribution function of electrons

$$f_e(t,\xi,u)=\begin{cases} f_0^e((u^2-2e(\varphi-\Delta\varphi)/m)^{1/2}+V_d), & u\le A(\varphi)\,\mathrm{sign}(\xi)\\ f_0^e((u^2-2e(\varphi+\Delta\varphi)/m)^{1/2}-V_d), & u>A(\varphi)\,\mathrm{sign}(\xi)\end{cases} \tag{9}$$

Here, $A(\varphi)=(2e(\varphi_0+\varphi)/m)^{\frac{1}{2}}$, $u=V-V_0$. Integrating (9) over velocities, we obtain the expression for the density perturbation of electrons in the first order of $(V_d-V_0)/V_{th}=V_c$ and in the second order of $e\varphi/T_e=\varphi$

$$\delta n_e=\varphi(1+\varphi/2)-\mathrm{sign}(\xi)\{2\Delta\varphi\sqrt{(\varphi_0+\varphi)/\pi}+\sqrt{2/\pi}\,V_c[\sqrt{\varphi_0}(\varphi_0+\varphi)+$$
$$+\varphi\ln(\sqrt{-\varphi}/(\sqrt{\varphi_0}+\sqrt{\varphi_0+\varphi}))]\} \tag{10}$$

From the Vlasov equation we obtain the expression for the perturbation of ion distribution function. From the latter the expression for the ion density perturbation follows

$$\delta n_i=\delta n_1+\delta n_2,\quad \delta n_2'=-\theta^{3/2}\dot{\varphi}[1-z^2-R(z)(1-2z^2)]/z\sqrt{2}$$
$$\delta n_1=-\theta[\varphi-\Delta\varphi\,\mathrm{sign}(\xi)]R(z)+\theta^2(\varphi/2)^2[R(z)(3-2z^2)-1] \tag{11}$$
$$z=V_0/V_{th}^i\sqrt{2},\quad \theta=T_e/T_i,\quad R(z)=1+(z/\sqrt{\pi})\int_{-\infty}^{\infty}dt\,(t-z)^{-1}\exp(-t^2)$$

In the quasi-stationary approximation from (10),(11) at $\xi>\Delta x$ we obtain (2). Substituting (10),(11) in the Poisson equation, we derive the evolution equation

$$\dot{\varphi}\theta^{3/2}[1-z^2+(2z^2-1)R(z)]/z\sqrt{2}+\varphi'[1+\theta R(z)]+\varphi\cdot\varphi'\{1+[1+(2z^2-3)R]\theta^2/2\}-$$
$$-\varphi'''+\sqrt{2/\pi}\,V_c|\varphi'|[(1+\sqrt{\varphi_0/\pi})/\sqrt{\varphi_0+\varphi}+\ln(\sqrt{-\varphi}/(\sqrt{\varphi_0}+\sqrt{\varphi_0+\varphi}))]=0 \tag{12}$$

From (12) we derive the equation describing the space distribution of the potential:

$$(\varphi')^2=\varphi^2(1+\theta R(z))+(\varphi^3/6)\{2+\theta^2[1+(2z^2-3)R(z)]\}-\mathrm{sign}(\xi)(\Delta\varphi/\varphi_0)\times$$
$$\times\{\varphi^2(1-2\sqrt{\varphi_0/\pi})\ln(\sqrt{-\varphi}/(\sqrt{\varphi_0}+\sqrt{\varphi_0+\varphi}))+\sqrt{\varphi_0(\varphi_0+\varphi)}(2\varphi_0+\varphi)-2\varphi_0^2+ \tag{13}$$
$$+2\varphi_0\varphi\theta R(z)+\varphi_0[5(\varphi_0+\varphi)^{3/2}-3\varphi_0\sqrt{\varphi_0+\varphi}-2\varphi_0^{3/2}]/3\sqrt{\pi}\}$$

From (13) and $\varphi'|_{\varphi=-\varphi_0}=0$ we obtain (similar to 10) the expression for V_0

$$z=z_0[1+1/\theta-(\varphi_0/6)(\theta+3-2z_0^2+2/\theta)],\quad z_0=0.924 \tag{14}$$

The well velocity is close to V_{th}^i and decreases with increasing φ_0.
From (13),(14) it may be concluded that the potential well at slow ion mode (in contrast to the ion-acoustic mode case) may exist in the current-free plasma.
We determine roughly the width of the well from (13)

$$(\Delta x)^{-2}=(\varphi_0/48)\{2+\theta^2[1+(2z^2-3)R(z)]\}+\mathrm{sign}(\gamma)(\Delta\varphi/\varphi_0)\times$$
$$\times\{0.25(1-2\sqrt{\varphi_0/\pi})\ln(1+\sqrt{2})+2-3/2\sqrt{2}+(\varphi_0/3\pi)^{1/2}(3/2+3/\sqrt{2}-5/2\sqrt{2})\} \quad (15)$$

The well is asymmetric: the first front is steeper than the back one. The well width decreases with increasing amplitude. The area of the "hole" $\Delta x\sqrt{\varphi_0}=\{48/[2+\theta(\theta+3-2z_0^2)]\}^{\frac{1}{2}}$ in the ion phase space is independent of φ_0 . Now we determine from (12),(13) the well amplitude growth rate

$$\gamma=\dot{\varphi}_0/\varphi_0=(z/\sqrt{3\theta})/[1-z^2+(2z^2-1)R(z)] \quad (16)$$

There is a threshold in θ . The growth rate is not zero for $\theta>\theta_{th}=\frac{1}{2}(1-Z_0)^{-1}$. For large amplitudes, $\varphi_0\gtrsim 1$, we consider the properties of the well in the current-free plasma. In the latter case we have

$$n_e=n_0\exp(\varphi),\quad n_i=(n_0/\sqrt{\pi})\int_{-\infty}^{\infty}dt\,\exp(-t^2)/[1-\theta\varphi/(z-t)^2]^{1/2} \quad (17)$$

From (17) and the Poisson equation we obtain

$$[\varphi'(\varphi=-\varphi_0/2)]^2=\sqrt{2}-2+2\exp(-\varphi_0/2)-\sqrt{2}\exp(-\varphi_0) \quad (18)$$

φ_0 cannot exceed the critical value φ_a determined by

$$1-\sqrt{2}+\sqrt{2}\exp(-\varphi_a/2)-\exp(-\varphi_a)=0 \quad (19)$$

The width of the well increases with growing φ_0 . Therefore we take into account the trapped ions. We assume their distribution function $f^i_{tr}=\mathrm{const}=f_0(v_0)$ Then their density equals

$$n^i_{tr}=2\sqrt{-\theta\varphi}\,(n_0/\sqrt{\pi})\exp(-z_0^2) \quad (20)$$

From (17),(20) and Poisson equations we find, that the width and velocity of the well increase with growing φ_0 (in contrast to the case of small amplitude φ_0).

REFERENCES

1. A.S.Bakai, V.I.Maslov, Plasma Physics, 13 (1987) 1019.
2. H.Schamel, Phys. Rep., 140 (1986) 163.
3. T.Sato, H.Okuda, Phys.Rev.Lett., 44 (1980) 740.
4. K.Nishihara et al., Riso-R-472 (1982) 41.
5. T.H.Dupree, Phys. Fluids 29 (1986) 1820.
6. C.Chan, M.H.Cho, N.Hershkowitz, T.Intrator, Phys.Rev.Lett. 52 (1984) 1782.
7. J.E.Allen, Plasma Phys. and Contr. Fus. 27 (1985) 1343.
8. A.N.Sekar, Y.C.Saxena, Plasma Phys. and Contr. Fus. 27 (1985) 181.
9. V.I.Maslov, Plasma Physics, 16 (1990) 759.
10. H.Schamel, Plasma Physics, 14 (1972) 905.

OBTAINING DOUBLE LAYERS IN FREE EXPANSION PLASMA PROBLEMS

Stefano Mola, Giovanni Manfredi, Marc Feix, PMMS/CNRS, 3A Av. de la Recherche Scientifique, 45071 Orléans, France.

1. INTRODUCTION

This paper is devoted to the study of the expansion of a two-component plasma (TCP) into vacuum, in plane and spherical geometry. From a practical point of view, a plasma expansion into vacuum can be realized in several ways. An interesting case, analyzed in [1], is given by the drift of polarization charges along the magnetic field lines.

Our approach to the problem will be a combination of analytical and numerical tools. In fact, the main difficulty that is found in simulating an expansion into vacuum is the need to work with free boundary conditions. Moreover, since the plasma expands to infinity, the choice of the computational grid becomes an actual problem.

We shall show that, by performing some *analytical* work on the equations (rescaling), we can "absorb" automatically the expansion term. The transformed equations, which take into account the "residual" motion, can subsequently be integrated by standard numerical methods.

2. MATHEMATICAL MODEL AND RESCALING TECHNIQUES.

First, let us consider a TCP in plane geometry (one dimension). The two species have identical charge (+q and -q) but different masses. The evolution of the TCP is described by the Vlasov-Poisson system :

$$\begin{cases} \dfrac{\partial f_e}{\partial t} + p\dfrac{\partial f_e}{\partial x} - E\dfrac{\partial f_e}{\partial p} = 0 \\ \dfrac{\partial f_i}{\partial t} + \dfrac{p}{M}\dfrac{\partial f_i}{\partial x} + E\dfrac{\partial f_i}{\partial p} = 0 \\ \dfrac{\partial E}{\partial x} = \int (f_i - f_e)\,dp \end{cases} \tag{1}$$

where f_e and f_i are the phase space distribution functions for electrons and ions, and

$$q_i = -q_e = m_e = \varepsilon_0 = 1 \quad ; \quad M = m_i/m_e$$

The rescaling methods [2,3] consist in introducing a "new space time" (ξ,θ) defined by

$$
\begin{aligned}
x &= C(t)\xi \\
dt &= A^2(t)d\theta
\end{aligned}
\tag{2}
$$

where A(t) and C(t) are two regular, positive functions of t, so far left at our disposal. The interest of relations (2) for our expansion problem is obvious. If one could "guess" the right law of expansion, and accordingly choose C(t), then, in the ξ space, the plasma would experience no expansion at all. Secondly, since the plasma frequency changes during the expansion, the factor A(t) can be chosen in order to adapt the new time θ to the plasma natural time scale.

To complete the relations (2), we define the new momentum π by the following

$$
p = \frac{\pi}{C} + m_{i,e}\dot{C}\xi \tag{3}
$$

Note that the new phase space (ξ,π) is entirely determined by C(t), whereas A(t) only changes the time scale. Their choice is then totally independent.

Here, we formulate the conjecture that the plane geometry expansion is of the form x~t (free particle motion). The choice of C(t) is thus imposed :

$$
C(t) = 1 + \Omega t
$$

where Ω is a parameter of the transformation.

A(t) is chosen so that :

$$
A^2(t) = \omega_p^{-1}(t) = (1+\Omega t)^{1/2}
$$

Substituting (2) and (3) in (1) gives :

$$
\begin{cases}
\dfrac{\partial F_i}{\partial \theta} + \dfrac{\pi}{M\mu(\theta)}\dfrac{\partial F_i}{\partial \xi} + \mu(\theta)\varepsilon\dfrac{\partial F_i}{\partial \pi} = 0 \\
\dfrac{\partial F_e}{\partial \theta} + \dfrac{\pi}{\mu(\theta)}\dfrac{\partial F_e}{\partial \xi} - \mu(\theta)\varepsilon\dfrac{\partial F_e}{\partial \pi} = 0 \\
\dfrac{\partial \varepsilon}{\partial \xi} = \int (F_i - F_e)d\pi
\end{cases}
\tag{4}
$$

where $\mu(\theta) = (1+\Omega\theta/2)^3$ and $F_{i,e}(\xi,\pi,\theta) = f_{i,e}(x,p,t)$; $\varepsilon(\xi,\theta) = E(x,t)$

3. NUMERICAL RESULTS.

The system (4) is solved numerically in the variables ξ and π, then coming back to (x,p) through (2) and (3). The values of Ω and M used in the simulation are :

$$\Omega = 0.5 \quad ; \quad M = 2$$

The initial condition is uniform in x, both for f_i and f_e (local neutrality), and maxwellian in p, with different temperatures :

$$T_e = 1 \quad ; \quad T_i = 0.3$$

In figure 1, we show the phase space portrait for ions (1a) and electrons (1b), when the solution is already asymptotical. The relation $v \sim xt$ is verified both for electrons and ions. The two species then expand at constant and equal velocities. In other words, the TCP behaves asymptotically as a neutral gas.

Figure 2 confirms this hypothesis : it shows the net charge density in the rescaled space

$$n(\xi) = \int (F_i - F_e) d\pi$$

at the same time.

Sheets of negative and positive charge rapidly alternate, expressing the fact that the plasma is achieving neutrality over a smaller and smaller scale. As time passes, the oscillations of $n(\xi)$ become more and more violent.

Finally, figure 3 shows the time evolution of the energy. We see that the potential energy, which is zero at $t=0$ (since we prepared a neutral initial condition), reaches a maximum, and then falls back to zero for large values of t. The kinetic energies of the ions and the electrons become asymptotically constant, and proportional to the respective masses : this confirms that the two species are moving with the same velocity.

4. EXPANSION IN SPHERICAL GEOMETRY.

We treat the case of a system with radial symmetry and we ignore all tangential velocities. The phase space is then represented by the radius x and the radial momentum p. The price we have to pay for this simplification is that the origin becomes a singular point : in fact, particles with inward bound velocities can accumulate at the origin, consequently generating an infinite density. Therefore, in the simulation we shall restrict ourselves to those initial conditions for whiwh all particles have positive (outward bound) radial velocities.

It turns out that the Vlasov equations in spherical geometry, within the above approximations, are identical to the 1D plane geometry ones. The difference lies in the Poisson equation, which is now written :

$$\frac{\partial}{\partial x}(x^2 E) = \int (f_i - f_e) dp \tag{5}$$

where E is the radial field and $f_{i,e}$ the distribution function multiplied by x^2.

Once again, we use the rescaling (2), with

$$C(t) = 1+\Omega t \quad ; \quad A^2(t) = (1+\Omega t)^{3/2} \propto \omega_p^{-1}(t)$$
$$f_{i,e}(x,p,t) = F_{i,e}(\xi,\pi,\theta) \quad ; \quad E(x,t) = \varepsilon(\xi,\theta) \tag{6}$$

Under (6), the Vlasov equations get transformed as in (4), where now

$$\mu(\theta) = (1-\Omega\,\theta/2)^{-1} \tag{7}$$

The Poisson equation (5) is left invariant.
The new time θ varies on the finite interval $[0, 2/\Omega[$ ("time renormalization").

5. NUMERICAL RESULTS.

The rescaling (6) absorbs entirely the expansion, which is then once again linear in t (free particle motion). Though, the phenomenon of local neutralization found in plane geometry no longer takes place, as it is apparent in figure 4. Here, we show the net charge density as a function of the rescaled radius ξ : a multiple layer, non-neutral structure is created, which seems to be very stable over long times. Figure 5 shows the evolution of the root of the mean square value of the radius : ions and electrons move asymptotically with constant but different velocities.

In fact, the stability of the non-neutral structure of figure 4 can be proved by means of an easy calculation in the rescaled space. The equations of motions corresponding to the Vlasov Poisson system are :

$$\begin{cases} \dfrac{d\xi}{d\theta} = \dfrac{\pi}{\mu(\theta)} \\[2mm] \dfrac{d\pi}{d\theta} = \mu(\theta)\,\varepsilon(\xi,\theta) \end{cases} \tag{8}$$

where $\mu(\theta)$ is given by (7).

We integrate (8) from θ to $\theta_{\lim} = 2/\Omega$, assuming that ε varies slowly as compared to μ. The displacement between θ and $\theta_{\lim}$ is found to be :

$$\Delta\,\xi \propto \theta_{\lim} - \theta$$

and tends to zero from $\theta \to \theta_{\lim}$, that is for $t \to \infty$.

REFERENCES

[1] M. Galvez and J. E. Borowsky, Phys. Fluids B 3(8), August 1991.

[2] J. R. Burgan, J. Gutierrez, A. Munier, E. Fijalkow, M. R. Feix, in *Strongly coupled plasmas* (ed. G. Kalman), Plenum Press, 1978.

[3] G. Manfredi, S. Mola, M.R. Feix, to appear in Phis. Fluids B, Plasma Physics.

fig. 1a

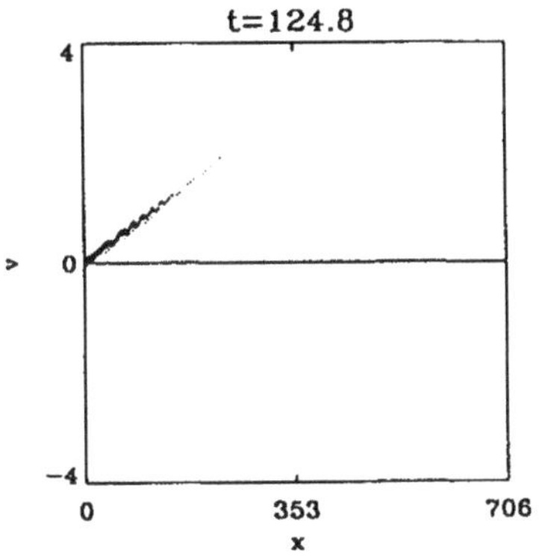

fig. 1b

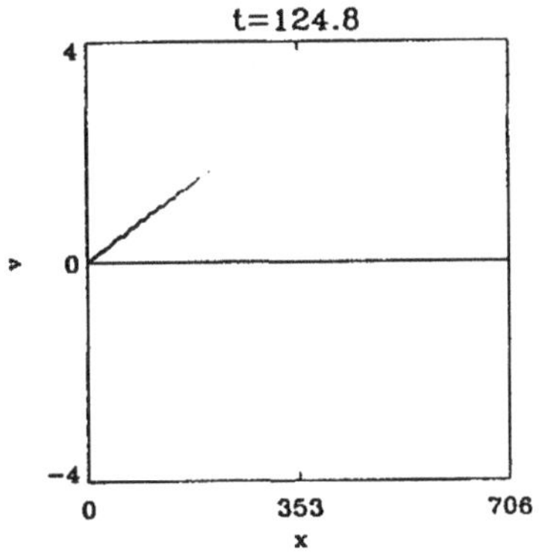

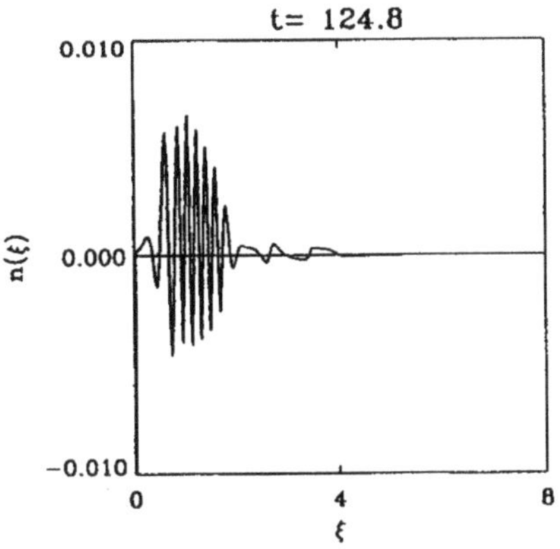

fig. 2

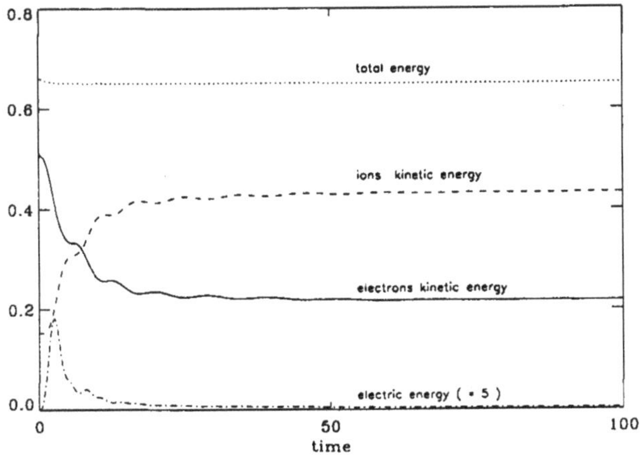

fig. 3

fig. 4

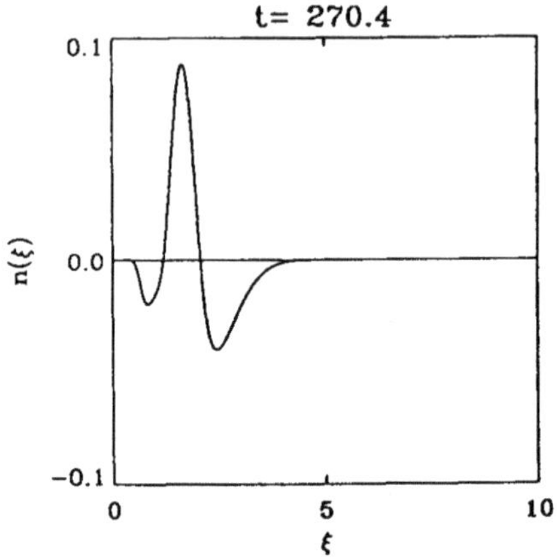
t= 270.4
0.1
0.0
-0.1
n(ξ)
0
5
10
ξ

fig. 5

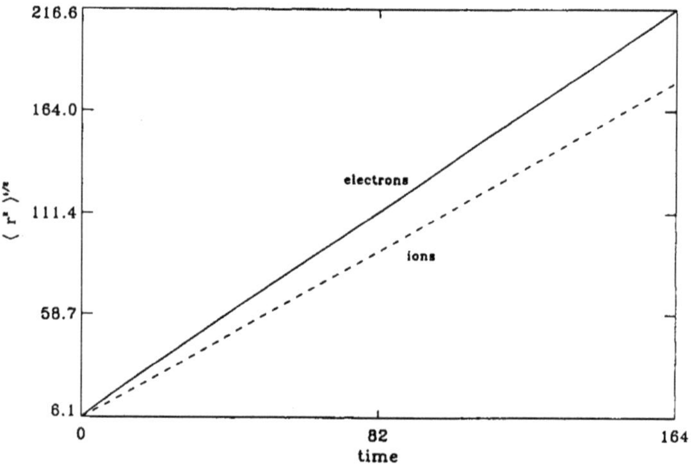
216.6
164.0
111.4
58.7
6.1
electrons
ions
0
82
164
time

ELECTRON INERTIA EFFECTS IN ION-ACOUSTIC SOLITONS AND DOUBLE LAYERS

W.K.M. Rice*, M.A. Hellberg*, S. Baboolal*†, R.L. Mace*, G. Gray*
*Plasma Physics Research Institute, Department of Physics, University of Natal, Durban, South Africa
†Department of Computer Science, University of Durban-Westville, Durban, South Africa

1 Introduction

In investigations of ion-acoustic double layers and solitons in fluid theory, it is common to model the electron density by a Boltzmann distribution[1-5], i.e. the mobility of the electrons is assumed to be so large that they can respond instantaneously - their mass is neglected.

In this paper we present results from some studies of the effects on the amplitudes and existence domains of ion-acoustic solitons and double layers of taking into account finite electron mass.

Following the technique first discussed by Baboolal, *et al*[1], we obtain the Sagdeev pseudopotential numerically from the particle densities, which are found by numerical solution of the fluid equations. The existence of a soliton or double layer is deduced from the form of the Sagdeev potential[2,3].

Using this numerical procedure we can study nonlinear structures of arbitrary amplitude and are not restricted by the use of the small-amplitude Korteweg-de Vries (KdV) equation to structures for which $e\phi \ll \kappa T$. This technique has been applied to a variety of situations[1-8].

In particular we consider a plasma consisting of cold hydrogen ions and two electron fluids at different temperatures. The results can be compared with those of Baboolal, *et al*[3], who considered a similar plasma, but with the two electron distributions being of Boltzmann type.

Our main result is that in the case of rarefactive solitons, the Boltzmann electron model overestimates considerably the existence domain in parameter space. A similar effect is observed by Mace & Hellberg[6] who have considered the effects of finite ion mass on dust-acoustic double layers, in which one would normally ignore the ion mass by comparison with that of the massive dust particles.

Thus, just as we have warned earlier about the indiscriminate application of the KdV and mKdV equations and similar expansions to larger amplitude structures[4,7], we now suggest that great care should be taken in deducing the existence of solitons and double layers from models in which the light particle mass is ignored.

2 The Fluid Model

We consider the fluid equations for an infinite, collisionless, unmagnetised plasma

$$\frac{\partial n_j}{\partial t} + \frac{\partial (n_j u_j)}{\partial x} = 0, \tag{1}$$

$$\frac{\partial u_j}{\partial t} + u_j \frac{\partial u_j}{\partial x} + \frac{1}{m_j n_j}\frac{\partial p_j}{\partial x} = \frac{-Z_j}{m_j}\frac{\partial \phi}{\partial x}, \tag{2}$$

$$\frac{\partial^2 \phi}{\partial x^2} = n_h + n_c - n_i, \tag{3}$$

where standard ion-acoustic normalisations have been used and the temperature is normalised with respect to an effective electron temperature $T_{eff} = T_h T_c/(n_{co}T_h + n_{ho}T_c)$. The subscript $j = i, c, h$ refers to the ions, cool and hot electrons. Equation (2) explicitly introduces the electron mass.

We consider two possible models: in Case A both electron species are isothermal, with $\gamma_{c,h} = 1$, while in Case B the hot electrons are assumed isothermal, while the more sluggish cool electron component is taken to be adiabatic ($\gamma_c = 3$), i.e. we assume that the wave speed is greater than the typical cool electron speed.

We assume that the plasma is undisturbed at $| x | \rightarrow \infty$, with normalised boundary conditions:

$$\phi, \frac{\partial \phi}{\partial x}, u_i, u_c, u_h \rightarrow 0;\ n_i, n_c, n_h \rightarrow 1, n_{co}, n_{ho};\ p_i, p_c, p_h \rightarrow 0, n_{co}T_c, n_{ho}T_h. \quad (4)$$

3 Arbitrary-amplitude theory

Following Baboolal, *et al.*[1], we seek stable, time-independent solitons and double layers, transforming to the moving frame through $\xi = x - \mu t$, where the Mach number μ represents the normalised soliton (double layer) speed.

We can then integrate (1) directly, yielding $u_j = \mu(1-(n_{jo}/n_j))$ for the particle flow velocities. With the aid of the equation of state $p_j n_j^{-\gamma_j} = p_{jo} n_{jo}^{-\gamma_j}$, one finds $p_i = 0$ for the ions; $p_{c,h} = n_{c,h}T_{c,h}$ for the isothermal electrons; $p_c = n_c^3 T_c/n_{co}^2$ for the adiabatic electrons. Using these expressions for u_j and p_j in equation (2) and integrating, we obtain the ion density explicitly,

$$n_i = \frac{1}{(1 - \frac{2\phi}{\mu^2})^{\frac{1}{2}}}. \quad (5)$$

The densities of the two electron components are expressed in terms of equations that depend on the value of the adiabatic constant γ_j. In Case A the equations for both species take the form

$$m_e \mu^2 \frac{n_{jo}^2}{n_j^2} - m_e \mu^2 + 2T_j \ln(\frac{n_j}{n_{jo}}) - 2\phi = 0, \quad (6)$$

which, because of their transcendental nature, have to be solved numerically. Care needs to be taken in finding the appropriate root, i.e. the one that matches to that obtained for the assumption of Boltzmann electrons.

In Case B the hot electrons satisfy equation (6), but the cold electron density equation is

$$m_e \mu^2 \frac{n_{co}^2}{n_c^2} + 3T_c(\frac{n_c^2}{n_{co}^2}) - 2\phi - m_e \mu^2 - 3T_c = 0, \quad (7)$$

which can be solved as a quadratic equation in n_c^2, yielding

$$(\frac{n_c}{n_{co}})^2 = \frac{1}{6T_c}\{(3T_c + 2\phi + m_e\mu^2) \pm [(3T_c + 2\phi)^2 - 2(3T_c - 2\phi)m_e\mu^2 + m_e^2\mu^4]^{\frac{1}{2}}\}. \quad (8)$$

The densities are then substituted in Poisson's equation (3) to get

$$\frac{d^2\phi}{d\xi^2} = n_h + n_c - n_i = N(\phi, \mu) = -\frac{d\psi(\phi, \mu)}{d\phi}, \quad (9)$$

where $\psi(\phi,\mu)$ is the Sagdeev pseudopotential, defined as

$$\psi(\phi,\mu) = -\int_o^{\phi} N(\xi,\mu)d\xi. \tag{10}$$

In seeking soliton solutions for fixed μ one applies the conditions: $\psi(0) = \partial\psi(0)/\partial\phi = 0$, $\psi(\phi_o) = 0$ for some ϕ_o and $\psi(\phi) < 0$ for $0 < |\phi| < |\phi_o|$. Equation (9) can then be integrated with initial value $\phi(0) = \phi_o$ to yield a soliton solution.

On the other hand, for double layers the conditions are $\psi(0) = \partial\psi(0)/\partial\phi = 0$ for all μ; $\psi(\phi_{DL},\mu_{DL}) = \partial\psi(\phi_{DL},\mu_{DL})/\partial\phi = 0$ for some ϕ_{DL},μ_{DL}; and $\psi(\phi) < 0$ for $\mu = \mu_{DL}$ and $0 < |\phi| < |\phi_{DL}|$. An initial value problem is then solved as above[1].

4 Results and Discussion

In our calculations we have concentrated on solitons - both compressive (positive electrostatic potential) and rarefactive (negative potential) - and briefly considered double layers. Unless stated otherwise, the results are for an electron temperature ratio of $T_h/T_c = 30$.

4.1 Case A: Isothermal Hot and Cool Electrons

4.1.1 Solitons

In Figure 1 we consider a typical Sagdeev potential for a rarefactive soliton ($\mu = 1.1; n_{co} = 0.2$). We note that the Boltzmann model (essentially the results of Baboolal, *et al*[3]; dashed line) yields a shallower pseudopotential well, which gives rise to a slightly lower amplitude soliton.

While the effect of electron inertia on amplitude is small, it plays a very important role in governing the soliton existence domain in parameter space (amplitude versus cool electron density). In Figure 2 we consider the existence regions of rarefactive solitons in this space, the parameter labelling the curves being the Mach number. Confirming the indications from Figure 1 for a particular value of n_{co}, we see that the magnitude of the soliton amplitude is generally found to be a little larger when finite electron mass is included. The small increase may be ascribed to the fact that there is a small change in the sound speed and hence in the effective Mach number from the earlier calculations. The phase speed of the wave in the presence of finite mass electrons is decreased by an amount $\sim (m_e/M_i)^{\frac{1}{2}}$ below that for Boltzmann electrons. Thus when we retain a fixed value of μ in our calculations, we are actually using an increased value of effective Mach number, and as we can see from the figure, amplitude increases with Mach number.

More important than this small change in amplitude, however, is the fact that the inclusion of finite electron inertia dramatically reduces the range of values of n_{co} and ϕ_o over which solitons are predicted[3], the relevant region decreasing rapidly with Mach number.

In particular, whereas Baboolal, *et al.*[3] found solitons with normalized amplitudes as large as $\chi = -14$, the inclusion of electron inertia in the calculations now restricts the range to $|\phi_o| \leq 1$. Thus very large amplitude solitons appear not to occur. We believe that the reason for this is probably similar to that discussed by Mace & Hellberg[6], who show that the solution to equation (6) for the cool light component (ions in their case, electrons in this case) is very sensitive and breaks down for larger values of ϕ_o. The function in equation (6) has a global minimum. If the negative potential (positive electron potential energy) is sufficiently large, this minimum may be raised sufficiently to prevent real roots from occurring.

The existence diagram for rarefactive solitons when $T_h/T_c = 15$ is shown in Figure 3 and we see that the results are similar to those of Figure 2.

We draw particular attention to these results, as, guided by the Boltzmann electrons model, one may expect solitons in a part of parameter space which is totally inappropriate.

We have carried out similar calculations for compressive solitons. No cut-off equivalent to that occurring for rarefactive solitons is found. In general the pairs of curves coincide, implying that the soliton amplitude is well-represented by the 'massless' model over the whole range of cool electron density and Mach number values.

If we bear in mind the reason for the reduction in the existence domain in parameter space for rarefactive solitons this is understandable. One is now dealing with a positive potential (negative electron potential energy), and as a result the global minimum in equation (6) cannot be raised above the zero-line as may be the case for the rarefactive soliton.

We note that these results for compressive solitons correct erroneous preliminary results reported earlier[8], which appeared to indicate a limited reduction in existence domain for compressive solitons, too.

4.1.2 Double Layers

Using the Boltzmann electron model, Baboolal, *et al.*[1,3] found large amplitude double layers for electron temperature ratios of 15 and 30. On the other hand, it was also shown that the presence of negative ions is deleterious to the existence of double layers[4]. This applied to a wide range of mass-values for the negative ions, including a 'negative ion' mass of m_e. In view of that, we would expect electron inertia to have an effect on double layers.

In a preliminary search for double layers in a two electron component plasma, including the effects of electron inertia, we have so far been unable to find any for $T_h/T_c = 15$ or 30.

The reason for this negative result might be related to the situation reported by Mace & Hellberg[6], where no dust-acoustic double layers are found for an ion/dust mass ratio larger than 10^{-4}. In our present calculations, of course, the mass ratio is 1/1836, as we have assumed hydrogen ions.

It was earlier argued[4] that symmetry between the two light species was required for the formation of double layers. Thus one could get double layers when both species have Boltzmann distributions, but would not expect them when there was a hot Boltzmann electron distribution and a non-Boltzmann distribution of the second negative species. In the present case we have both the light negative species having the same mass, but there is a lack of symmetry in mobility because of the temperature difference.

4.2 Case B: Isothermal Hot and Adiabatic Cool Electrons

In this case one may calculate the phase speed to be $C_s = (\kappa T_{eff}/M_i)^{\frac{1}{2}}$, where $T_{eff} = (3T_cT_h)/(3n_{ho}T_c + n_{co}T_h)$. For $n_{co} \simeq 0$ this phase speed is the same as that for the case of double isothermal ions, while for $n_{co} \simeq 1$ it introduces a factor $\simeq \sqrt{3}$ relative to the earlier normalizing velocity.

The assumption of adiabatic cool electrons should apply if their thermal velocity, $(\kappa T_c/m_e)^{\frac{1}{2}}$, is less than the phase speed of the wave. Clearly, this requirement is most easily satisfied if n_{co} is small (in fact $\sim m_e/M_i$), and if $T_c/T_h < m_e/M_i$. We have thus considered our standard temperature ratio of 30, as well as the more appropriate (for the model), but somewhat extreme case of $T_h/T_c = 2000$.

Equation (8) has two possible solutions. The negative surd is the one that one would expect to use with the adiabatic model ($T_c/m_e \ll C_s^2$), as it is continuous with the solution for the cold electron case ($T_c = 0$). Using that choice of solution for the cool electron density, preliminary calculations reveal the existence of solitons for a temperature ratio of 2000, for very small cool electron density ($\sim 10^{-4} - 10^{-3}$).

However, that solution does not yield solitons when we use $T_h/T_c = 30$. But on the other hand, for that ratio solitons are found when the positive surd in equation (8) is considered. This branch of the solution of the cool electron density equation is continuous with the choice m_e tending to zero. Thus it relates more closely to the double Boltzmann electron model and the double isothermal electron model, but not to the adiabatic assumption for the cool electrons.

5 Conclusions

In summary, the inclusion of electron inertia reduces considerably the existence domain of rarefactive solitons and double layers in two electron temperature plasmas, while having a negligible effect on compressive solitons. This reduction is due to the limit on negative soliton potential imposed by the presence of a global minimum in the function in equation (6) and is analogous to that found by Mace & Hellberg for dust-acoustic double layers[6].

However, where solitons and double layers are shown to exist in the model including electron inertia, the Boltzmann model is a good one for evaluating their amplitudes, the small difference being due to the difference in sound speed.

Our calculations have dealt with the case of the smallest possible ion/electron mass ratio, viz. a hydrogen plasma. One might believe that with more massive ions the neglect of electron mass is more easily justified. However, we should note the results of Mace & Hellberg[6], who show that for an ion/dust mass ratio as small as 10^{-6}, finite ion mass still has an effect on the existence domain of the double layers.

6 References

1. S. Baboolal, R. Bharuthram & M.A. Hellberg, *J. Plasma Phys.* **40**, 163 (1988)
2. S. Baboolal, R. Bharuthram & M.A. Hellberg, *J. Plasma Phys.* **41**, 341 (1989)
3. S. Baboolal, R. Bharuthram & M.A. Hellberg, *J. Plasma Phys.* **44**, 1 (1990)
4. S. Baboolal, R. Bharuthram & M.A. Hellberg, *J. Plasma Phys.* **45**, 247 (1991)
5. R.L. Mace, S. Baboolal, R. Bharuthram & M.A. Hellberg, *J. Plasma Phys.* **45**, 323 (1991)
6. R.L. Mace & M.A. Hellberg, *Proc Fourth Symp on Double Layers and other Nonlinear Potential Structures in Plasmas, Innsbruck, Austria* (1992)
7. M.A. Hellberg, S. Baboolal, R.L. Mace & R. Bharuthram *IEEE Trans. on Plasma Science* (To be published, 1992)
8. W.K.M. Rice, M.A. Hellberg, R.L. Mace & S. Baboolal *Proc. International Conference on Plasma Physics, Innsbruck, Austria* **III**, 1819 (1992)

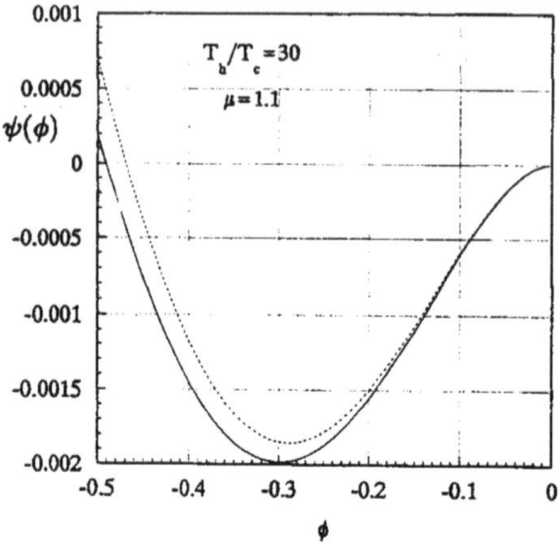

Figure 1: Typical Sagdeev potential ($T_h/T_c = 30; n_{co} = 0.2$).

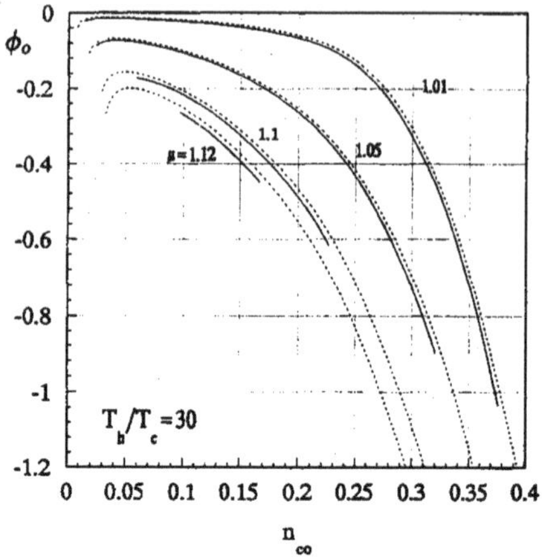

Figure 2: Existence domains for rarefactive solitons ($T_h/T_c = 30$).

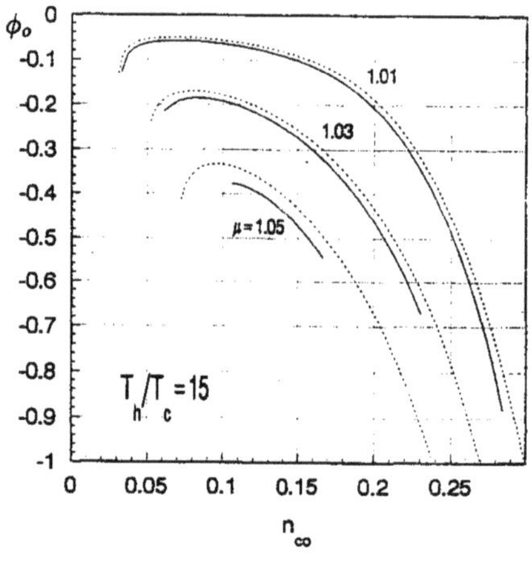

Figure 3: Existence domains for rarefactive solitons ($T_h/T_c = 15$).

SOLITONS FOR FREE ELECTRON LASER APPLICATIONS

A.Sen[1] and G.L. Johnston[2]
[1]Institute for Plasma Research, Bhat, Gandhinagar 382424, India
[2]Plasma Fusion Center, Massachussetts Institute of Technology, Cambridge, MA 02139, U.S.A.

A new class of nonlinear traveling wave solutions of the relativistic cold fluid model are presented for a system consisting of a relativistic cold electron beam propagating through a helical wiggler magnetic field. The solutions, which are in the form of isolated soliton pulses of coupled electromagnetic and plasma waves, are obtained numerically. They represent possible nonlinear saturated states of the free electron laser instability and may also have useful applications in particle acceleration schemes.

1. INTRODUCTION: The free electron laser (FEL), which generates coherent electromagnetic radiation using an intense relativistic electron beam propagating through a static transverse magnetic field, has been the subject of several theoretical and experimental investigations[1]. While many simple model calculations exist to analytically estimate the linear growth rate of the instability, very few calculations other than those dependent on large computer simulations and numerical codes, have addressed the nonlinear development and saturation of the instability. In particular, there are no simple model descriptions of the nonlinear saturated state. A notable exception is the work of Lane and Davidson[2] where large amplitude traveling wave solutions were obtained for the nonlinear Vlasov-Maxwell equations. These BGK type solutions are stationary in a frame of reference moving with the ponderomotive phase velocity. An interesting question that arises is whether there are simpler fluid analogs of these BGK type solutions, especially finite pulse solutions like solitons. To the best of our knowledge such solutions have not been found for the one dimensional relativistic cold fluid model of the FEL, although the model has been extensively employed for linear calculations. This is somewhat surprising since traveling wave solutions (including soliton solutions) of the cold plasma model have been widely investigated in the context of the nonlinear theory of intense laser-plasma interactions[3-6]. A possible reason for this lacuna in the FEL problem could be the complexity introduced in the fluid model by the presence of the beam and the wiggler field which make a traveling wave analysis somewhat difficult. In this paper we carry out such an analysis and find that soliton solutions do indeed exist for the one dimensional cold fluid model of the FEL. Our findings are based on numerical solutions of the relativistic fluid equations modelling the nonlinear evolution

of perturbations to a relativistic cold electron beam propagating through a helical magnetic wiggler. The solutions are in the form of isolated soliton pulses which occur with a discrete velocity spectrum. This new class of traveling wave solutions, consisting of intense light pulses coupled to large amplitude plasma waves, represent possible saturated states of the FEL. They might also form interesting candidates for particle accelerator schemes like the inverse free electron laser (IFEL)[7] in which the large electrostatic potential of the plasma wave may be used to get significant acceleration of electrons.

2. MODEL EQUATIONS: We begin with a brief discussion of the relativistic cold fluid model that has been previously developed in Davidson et al[8] for describing the nonlinear evolution of the free electron laser instability. For a relativistic cold electron beam of uniform density propagating in the z direction through a helical wiggler magnetic field $\vec{B}^{(0)}(\vec{x}) = -\hat{B}[\cos(k_0 z)\hat{e}_x + \sin(k_0 z)\hat{e}_y]$, the cold fluid equations consist of the continuity equation

$$\frac{\partial n}{\partial t} + \frac{\partial (nV_z)}{\partial z} = 0 \tag{1}$$

and the z component of the equation of motion

$$\frac{\partial p_z}{\partial t} = e\frac{\partial \phi}{\partial z} - mc^2\frac{\partial \gamma}{\partial z} \tag{2}$$

where $\gamma = \sqrt{(1 + p^2/m^2c^2)}$ is the relativistic factor, and p_z is the longitudinal momentum. Note that for one dimensional variations (in z) the symmetry of the problem permits elimination of the transverse momentum by the exact integration $\vec{p}_\perp = (e\vec{A}/c)$, where $\vec{A}$ is the total vector potential. The electrostatic potential ϕ can be eliminated from (2) by a further differentiation in time and making use of Poisson's equation and the z component of Ampere's law, to get

$$\frac{\partial^2 p_z}{\partial t^2} = -mc^2\frac{\partial \gamma}{\partial t \partial z} - \omega_{p0}^2\left(\frac{n}{n_0}\frac{p_z}{\gamma} - \frac{p_{0z}}{\gamma_0}\right) \tag{3}$$

The evolution of the transverse electromagnetic perturbations are given by the inhomogeneous wave equation

$$\left(\frac{\partial^2}{\partial z^2} - \frac{1}{c^2}\frac{\partial^2}{\partial t^2}\right)\delta\vec{A} = \frac{\omega_{p0}^2}{c^2}\left(\frac{n\vec{A}}{n_0\gamma} - \frac{\vec{A}^{(0)}}{\gamma_0}\right), \tag{4}$$

Here γ_0,$\vec{A}_0$,p_0 and $\omega_{p0}^2 = 4\pi e^2 n_0/m$ denote unperturbed values due to the wiggler magnetic field and the beam. In (3) and (4) the nonlinear terms consist of the striction (n/n_0) and relativistic (γ) nonlinearities. In the cold plasma model these two nonlinearities oppose each other[3,4] and their competition determines whether solutions of the soliton type are possible. To investigate such solutions it is convenient first to transform Eqn(4) to the so-called wiggler coordinates defined by $\hat{e}_1 = \cos(k_0 z)\hat{e}_x + \sin(k_0 z)\hat{e}_y$, $\hat{e}_2 = -\sin(k_0 z)\hat{e}_x + \cos(k_0 z)\hat{e}_y$ and $\hat{e}_3 = \hat{e}_z$. The components of Eqn(4) in the new coordinate system take the form

$$\left(\frac{\partial^2}{\partial z^2} - k_0^2 - \frac{1}{c^2}\frac{\partial^2}{\partial t^2}\right)\delta A_1 - 2k_0\frac{\partial \delta A_2}{\partial z} = \frac{\omega_{p0}^2}{c^2}\left(\frac{nA_1}{n_0\gamma} - \frac{A_1^{(0)}}{\gamma_0}\right)$$

$$\left(\frac{\partial^2}{\partial z^2} - k_0^2 - \frac{1}{c^2}\frac{\partial^2}{\partial t^2}\right)\delta A_2 - 2k_0\frac{\partial \delta A_1}{\partial z} = \frac{\omega_{p0}^2}{c^2}\frac{nA_2}{n_0\gamma} \qquad (5)$$

We now make the traveling-wave ansatz that all the dynamical quantities depend on z and t through the combination $\xi = z - ut$ where u = constant is the signal speed. In the wave frame the continuity equation can be readily integrated to give

$$\frac{n}{n_0} = \frac{(\beta - \beta_b)\gamma}{(\gamma\beta - \rho_3)} \qquad (6)$$

where $\beta = u/c$, $\beta_b = V_0/c$ and $\rho_3 = p_z/mc$. Using (6) and some straightforward algebra it is now possible to reduce (3),(5) to a set of three coupled nonlinear differential equations for the variables ρ_1, ρ_2 and $Z = \beta\rho_3 - \gamma$

$$(1-\beta^2)\ddot{\rho}_1 - 2\dot{\rho}_2 - \rho_1\{1 + F(\rho_1,\rho_2,Z)\} = -\hat{\omega}_c\gamma_0(1+\hat{\omega}_p^2) \qquad (7)$$

$$(1-\beta^2)\ddot{\rho}_2 + 2\dot{\rho}_1 - \rho_2\{1 + F(\rho_1,\rho_2,Z)\} = 0 \qquad (8)$$

$$(1-\beta^2)\ddot{Z} - F(\rho_1,\rho_2,Z)Z - \hat{\omega}_p^2\gamma_0(1-\beta\beta_b) = 0 \qquad (9)$$

$$F(\rho_1,\rho_2,Z) = \frac{\hat{\omega}_p^2\gamma_0|\beta-\beta_b|}{[(\beta^2-1)(1+\rho_1^2+\rho_2^2)+Z^2]^{1/2}} \qquad (10)$$

where the overdot denotes $d/d\zeta$ and the normalisations are $\hat{\omega}_c = \omega_c/ck_0$, $\hat{\omega}_p = \omega_p/ck$, $\zeta = k_0\xi$, $\rho_\alpha = p_\alpha/mc$ where $\alpha = 1,2,3$. Equations (7-10) are the final set of coupled equations that we must solve for studying the nonlinear evolution of the FEL. These equations exhibit many similarities to, and important differences from, traveling wave equations studied in the past for intense laser-plasma interactions [3-6]. The most striking difference is the inhomogeneous term in Eq.(7) for the ρ_1 oscillator, which results from the presence of the wiggler field. The β_b contribution arises from the beam velocity. These equations admit of one exact integral of motion which may be written as

$$Q = \frac{1}{2}[(\beta^2-1)(\dot{\rho}_1^2+\dot{\rho}_2^2)+\dot{Z}^2] + W(\rho_1,\rho_2,Z) = const \qquad (11)$$

where

$$W(\rho_1,\rho_2,Z) = \frac{1}{2}(\rho_1^2+\rho_2^2) + \frac{\hat{\omega}_p^2\gamma_0(1-\beta\beta_b)}{(\beta^2-1)}Z - \hat{\omega}_c\gamma_0\rho_1(1+\hat{\omega}_p^2)$$

$$+\frac{\hat{\omega}_p^2\gamma_0|\beta-\beta_b|}{(\beta^2-1)}[(\beta^2-1)(1+\rho_1^2+\rho_2^2)+Z^2]^{1/2} \qquad (12)$$

The equilibrium solutions ($d/d\zeta = 0$) of (7-10) are given by $\rho_{01} = \hat{\omega}_c\gamma_0$, $\rho_2 = 0$ and $Z_0 = \gamma_0(\beta\beta_b - 1)$. A small signal analysis around the equilibrium yields a traveling wave dispersion relation [8] which can be formally related to the familiar cold fluid normal mode dispersion relation for a helical wiggler FEL. As shown in Davidson et al[8], the FEL instability corresponds to the condition $\beta < \beta_b < 1$. We note from (11) that in this case the problem is similar to a Hamiltonian of coupled anharmonic oscillators with three degrees of freedom (ρ_1, ρ_2, Z) where the effective mass for two of the anharmonic oscillators is negative. In general there can be a rich variety of solutions including infinite wave trains, solitons and even aperiodic chaotic solutions[9].

3. SOLITON SOLUTIONS: Our primary interest is to investigate soliton solutions for which we have solved equations (7-10) numerically and looked for solutions that decay exponentially as $\zeta \rightarrow \pm\infty$. We find that such solutions do indeed exist and occur for discrete values of $\hat{\omega}_p$ and β. In other words, for a fixed value of β (and $\hat{\omega}_c$ and γ_0) finding soliton solutions turns out to be an eigenvalue problem in $\hat{\omega}_p$. The sizes and shapes of these solitons also vary as a function of $\hat{\omega}_p$. Figure 1 shows a typical soliton solution for $\beta = 0.5$, $\hat{\omega}_c = 0.5$, $\gamma_0 = 10.$ and $\hat{\omega}_p = 0.501935$. Typically the $\delta Z = (Z - Z_0)$ profile (Fig. 1a)has a characteristic bell shape whereas $\delta\rho_1 = (\rho_1 - \rho_{01})$ (Fig. 1b) and ρ_2 (Fig. 1c) have a number of nodes. The $n/n_0\gamma$ profile (Fig. 1d) shows that the density is reduced in most of the soliton region but piles up at the edges. This suggests that physically these solitons correspond to trapped light bubble solutions in the beam fluid, in which standing light waves are set up in an 'effectively empty' cavity. We see further evidence of this as larger amplitude solitons appear with increasing $\hat{\omega}_p$ (keeping other quantities fixed). These solitons have a larger spatial extent in ζ and accomodate larger units of the standing wave pattern in them. Table I lists some of the eigenvalues (of $\hat{\omega}_p$) at which solitons appear for $\beta = 0.5$, $\hat{\omega}_c = 0.5$, $\gamma = 10.$ and their maximum amplitudes in δZ.

Table I

$\hat{\omega}_p$	0.5019	0.5118	0.5138	0.5151
δZ_{max}	-6.079	-18.74	-31.25	-71.98

Similar discrete spectra of solitons can be obtained at other values of β as well (as long as $\beta < \beta_b$).

4. DISCUSSION: These soliton solutions constitute a new class of nonlinear stationary solutions for the FEL which can be viewed as fluid analogs of the BGK type solutions. Unlike the kinetic solutions, the saturation mechanism here is quite different and occurs from a balance of the nonlinearities due to striction and relativity effects. This results in a trapping of the electromagnetic wave in a plasma wave created in the beam due to bunching of the beam density. Since the electrostatic field plays an important role in the formation of the soliton, these solutions can only occur in the Raman regime where collective effects are significant.
To conclude, we have obtained soliton solutions for the first time in the context of the free electron laser. These one dimensional nonlinear traveling wave solutions have been obtained by a numerical solution of the relativistic cold fluid model equations and are in the form of isolated pulses of coupled intense electromagnetic and plasma waves. We suggest that this new class of solutions can have a large number of applications. They could provide simple model descriptions of the nonlinear saturated state of the free electron instability in the Raman regime. The possibility of obtaining large electrostatic fields also makes them interesting from the point of view of particle and photon accelerators. The inverse free electron laser (IFEL) is one such scheme where the large electric fields of the soliton solutions could in principle be used to transfer energy from a laser to a relativistic electron beam in the presence of the magnetic field of an undulator.

ACKNOWLEDGEMENT: It is a pleasure to thank Professor P.K. Kaw for many stimulating discussions during the course of this work. One of us (AS) is grateful to Professor Abdus Salam, the IAEA ,UNESCO and SAREC for hospitality at the ICTP, Trieste where a part of this work was carried out.

REFERENCES:

1. T.C. Marshall, *Free Electron Lasers* (Macmillan, New York, 1985)
2. B. Lane and R.C. Davidson, Phys. Rev. A **27**, 2008 (1983)
3. A.I. Akhiezer and R.V. Polovin, Zh. Eksp. Teor. Fiz. **30**, 915 (1956) [Sov. Phys. JETP **3**, 696 (1956)]
4. V.A. Kozlov, A.G. Litvak and E.V. Suvorov, Zh. Eksp. Teor. Fiz. **76**, 148 (1979) [Sov. Phys. JETP **49**, 75 (1979)]
5. P. Sprangle, E. Esarey and A. Ting, Phys. Rev. Lett. **64**, 2011 (1990)
6. S.C. Wilks, J.M. Dawson, W.B. Mori, T. Katsouleas, M.E. Jones Phys. Rev. Letts. **62**, 2600 (1989)
7. C. Pellegrini, Proc. ECFA-RAL *The Challenge of Ultrahigh Energies* (ed: J. Mulvey) 1982
8. R.C. Davidson, G.L. Johnston and A. Sen, Phys. Rev. A **34**, 392 (1986)

9. P.K.Kaw, A. Sen and E. Valeo, Physica D **9D**, 96 (1983)

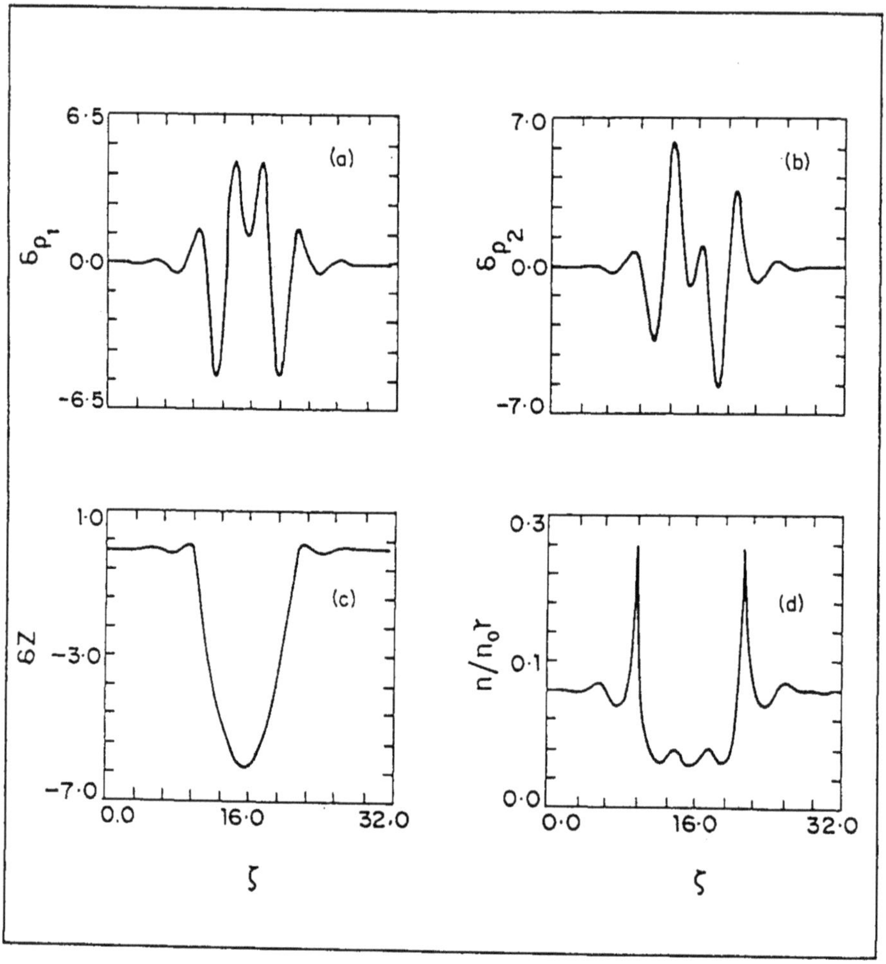

Figure 1. Profiles of (a)$\delta\rho_1$, (b)$\delta\rho_2$, (c) δZ and (d)$n/n_0\gamma$ for a soliton pulse with $\beta = 0.5$,$\gamma = 10.$, $\hat{\omega}_c = 0.5$ and $\hat{\omega}_p = 0.501935$.

STABILITY ISSUES OF DOUBLE LAYER STRUCTURES

J. A. Tataronis, University of Wisconsin - Madison, Madison, WI 53706 USA

1. INTRODUCTION: Strong local electric fields associated with double layers accelerate charged particles and produce thereby ion and electron beams that are potential sources of plasma waves, instabilities, turbulence, and anomalous resistivity.[1] In this paper, we derive equations that govern the electrostatic dynamics and wave stability of one dimensional double layer structures. Our analysis will be based on WKB theory, and therefore our considerations will be limited to short wavelength fluctuations. WKB analyses of electrostatic waves in hot inhomogeneous plasmas have been made in the past.[2] This paper adopts an alternate expansion method. We assume the existence of a static double layer equilibrium that is a solution of the coupled Poisson-Vlasov equations. Solutions of this system depend upon the free and trapped particle velocity distributions that must be specified at boundaries of the double layer plasma. We assume that the equilibrium problem has been solved and proceed to the stability problem by linearizing the time dependent Poisson-Vlasov equations.

2. GOVERNING EQUATIONS AND PERTURBATION ANALYSIS: We assume a one dimensional, unmagnetized, double layer structure that extends parallel to the x-axis over the infinite range, $-\infty < x < \infty$. Under the electrostatic approximation, the double layer dynamics is governed by the coupled Poisson-Vlasov system of equations,

$$\frac{\partial^2 \Phi}{\partial x^2} = \frac{q n_o}{\varepsilon_o}\left[\int_{-\infty}^{\infty} dv\, f_e(v,x,t) - \int_{-\infty}^{\infty} dv\, f_i(v,x,t)\right] \tag{1}$$

$$\frac{\partial f_i}{\partial t} + v\frac{\partial f_i}{\partial x} - \frac{q}{m_i}\frac{\partial \Phi}{\partial x}\frac{\partial f_i}{\partial v} = 0 \tag{2}$$

$$\frac{\partial f_e}{\partial t} + v\frac{\partial f_e}{\partial x} + \frac{q}{m_e}\frac{\partial \Phi}{\partial x}\frac{\partial f_e}{\partial v} = 0 \tag{3}$$

where $f_i(v,x,t)$ and $f_e(v,x,t)$ designate respectively the ion and electron velocity distributions, $\Phi(x,t)$ is the electric potential, q represents the positive electric charge of the ions, m_i and m_e are respectively the ion and electron masses, ε_o is the vacuum permittivity, and n_o is a constant density that is used to normalize the velocity distributions. MKS units are used throughout this paper.

We solve Eqs. (1) - (3) by a perturbation expansion about a static double layer state that is characterized by the ion and electron velocity distributions, $f_{io}(v,x)$ and $f_{eo}(v,x)$, respectively, and the electric potential, $\Phi_o(x)$. The time independent Vlasov equation implies that $f_{io}(v,x)$ and $f_{eo}(v,x)$ depend on v and x through a dependence on the total particle energy,

$$f_{io}(v,x) \equiv f_{io}\left[E_i(v,x)\right],\ E_i(v,x) = \frac{1}{2}m_i v^2 + q\Phi_o(x) \tag{4}$$

$$f_{eo}(v,x) \equiv f_{eo}\left[E_e(v,x)\right],\ E_e(v,x) = \frac{1}{2}m_e v^2 - q\Phi_o(x) \tag{5}$$

The perturbation expansion consists in representing each time dependent field as a superposition of a static equilibrium field and a fluctuation that varies harmonically in time with frequency ω,

$$\Phi(x,t) = \Phi_o(x) + \Phi_1(x)\, e^{i\omega t} \tag{6}$$

$$f_i(v,x,t) = f_{io}(v,x) + f_{i1}(v,x)\, e^{i\omega t} \tag{7}$$

$$f_e(v,x,t) = f_{eo}(v,x) + f_{e1}(v,x)\, e^{i\omega t} \tag{8}$$

The amplitude functions of the fluctuations are labeled with a subscript one. Substituting Eqs. (6) - (8) in Eqs. (1) - (3) and linearizing with respect to the fluctuations result in the following system of coupled partial differential equation for the fluctuation amplitudes,

$$\frac{\partial^2 \Phi_1}{\partial x^2} = \frac{q n_o}{\varepsilon_o}\left[\int_{-\infty}^{\infty} dv\, f_{e1}(v,x) - \int_{-\infty}^{\infty} dv\, f_{i1}(v,x)\right] \tag{9}$$

$$i\omega f_{s1} + v\frac{\partial f_{s1}}{\partial x} - \frac{q_s}{m_s}\frac{d\Phi_o}{dx}\frac{\partial f_{s1}}{\partial v} = \frac{q_s}{m_s}\frac{\partial \Phi_1}{\partial x}\frac{\partial f_{so}}{\partial v} \tag{10}$$

In Eq. (10), the subscript s labels particle species, with s replaced by i and e for ions and electrons, respectively. Equations (9) and (10) form the basis of our stability analysis. If the plasma equilibrium were uniform, plane wave solutions of the form, $\exp[i(\omega t - kx)]$, exist. The wave frequency ω is connected to the wavenumber k via the well known dispersion relation,

$$D(\omega,k) \equiv k^2 - \sum \omega_{ps}^2 \int_{-\infty}^{\infty} dv \frac{df_{so}(v)/dv}{v - \omega/k} = 0 \tag{11}$$

where the summation extends over the particle species, and ω_{ps} is the plasma frequency of particle species s. In the following section, we consider a spatially inhomogeneous double layer equilibrium, and solve Eqs. (9) and (10) via the WKB method.

3. WKB SOLUTION: The WKB method requires that the effective wavelength of the fluctuations be much smaller than the spatial scale length of the double layer inhomogeneity. We introduce this constraint in our analysis by assuming that the equilibrium potential depends on x through the scaled variable , αx, where α is an arbitrarily small parameter,

$$\Phi_o(x) \equiv \Phi_o(\alpha x) \tag{12}$$

Effectively, Eq. (12) implies that the double layer electric potential depends weakly on x, while the electric field is small, scaling with the parameter α,

$$\frac{d\Phi_o(\alpha x)}{dx} = \alpha \frac{d\Phi_o(\alpha x)}{d(\alpha x)} \tag{13}$$

The associated WKB representations of the amplitudes of the fluctuating fields read as follows,

$$\Phi_1(x) = \hat{\Phi}(\alpha x)\, e^{-i\psi(x)} \tag{14}$$

$$f_{s1}(v,x) = \hat{f}_s(v,\alpha x)\, e^{-i\psi(x)} \tag{15}$$

where $\psi(x)$ is the eikonal. The amplitude functions, $\hat{\Phi}(\alpha x)$ and $\hat{f}_s(v,\alpha x)$, are slowly varying functions of x. Substitute Eqs. (14) and (15) in Eqs. (9) and (10). After differentiation, and rearrangement of terms, the following system of coupled equations result,

$$i(\omega - kv)\hat{f}_s + i\frac{q_s}{m_s}k\frac{\partial f_{so}}{\partial v}\hat{\Phi} = \alpha\left(\frac{q_s}{m_s}\frac{d\Phi_o}{d\tilde{x}}\frac{\partial \hat{f}_s}{\partial v} - v\frac{\partial \hat{f}_s}{\partial \tilde{x}} + \frac{q_s}{m_s}\frac{f_{so}}{\partial v}\frac{\partial \hat{\Phi}}{\partial \tilde{x}}\right) \tag{16}$$

$$k^2 \hat{\Phi} - \frac{q n_o}{\varepsilon_o}\left[\int_{-\infty}^{\infty} dv\, \hat{f}_i - \int_{-\infty}^{\infty} dv\, \hat{f}_e\right] = -i\alpha\left(2k\frac{\partial \hat{\Phi}}{\partial \bar{x}} + \frac{\partial k}{\partial \bar{x}}\hat{\Phi}\right) + \alpha^2 \frac{\partial^2 \hat{\Phi}}{\partial \bar{x}^2} \quad (17)$$

In Eqs. (16) and (17), $\bar{x}$ represents the scaled spatial coordinate, (αx), and k is the derivative of the eikonal with respect to x,

$$k \equiv \frac{\partial \psi}{\partial x} \quad (18)$$

It should be noted that k, which can be interpreted as an effective wavenumber, is a function of $\bar{x}$, $k \equiv k(\bar{x})$, implying that it depends weakly on x. The solution of Eq. (16) and (17) for the WKB amplitudes is obtained iteratively by expressing $\hat{\Phi}(\bar{x})$ and $\hat{f}_s(v,\bar{x})$ as power series in the small parameter α,

$$\hat{\Phi}(\bar{x}) = \hat{\Phi}_o(\bar{x}) + \alpha\, \hat{\Phi}_1(\bar{x}) + \ldots\ldots \quad (19)$$

$$\hat{f}_s(v,\bar{x}) = \hat{f}_{so}(v,\bar{x}) + \alpha\, \hat{f}_{s1}(v,\bar{x}) + \ldots\ldots \quad (20)$$

Substitute these expansions in Eqs. (16) and (17), group terms according to powers of α, then solve the equations for the expansion functions to each order in α. We summarize the results that we find to order α.

Order α^o

To order α^o, we find an equation that connects $\hat{f}_{so}(v,\bar{x})$ and $\hat{\Phi}_o(\bar{x})$,

$$\hat{f}_{so}(v,\bar{x}) = \frac{q_s}{m_s}\frac{\partial f_{so}(v,\bar{x})/\partial v}{v - \omega/k}\hat{\Phi}_o(\bar{x}) \quad (21)$$

and the following homogeneous algebraic equation,

$$\left[k^2 - \sum \omega_{ps}^2 \int_{-\infty}^{\infty} dv \frac{\partial f_{so}(v,\bar{x})/\partial v}{v - \omega/k}\right]\hat{\Phi}_o(\bar{x}) = 0 \quad (22)$$

Since we require that $\hat{\Phi}_o(\bar{x})$ be non zero, we set the bracketed expression in Eq. (22) equal to zero,

$$D(\omega,k,\bar{x}) \equiv k^2 - \sum \omega_{ps}^2 \int_{-\infty}^{\infty} dv \frac{\partial f_{so}(v,\bar{x})/\partial v}{v - \omega/k} = 0 \tag{23}$$

Equation (23) is the local WKB dispersion relation and the eikonal equation that governs the phase $\psi(x)$. If the plasma is spatially uniform, $D(\omega,k,\bar{x})$ is identical to the dispersion function given by Eq. (11).

Order α

To order α, a first order, linear, ordinary differential equation for $\hat{\Phi}_o(\bar{x})$ results,

$$\frac{\partial \hat{\Phi}_o}{\partial \bar{x}} + Q(\omega,k,\bar{x})\,\hat{\Phi}_o = 0 \tag{24}$$

where the coefficient $Q(\omega,k,\bar{x})$ has the following structure,

$$Q(\omega,k,\bar{x}) = \frac{N(\omega.k.\bar{x})}{D_k(\omega,k,\bar{x})} \tag{25}$$

The denominator of Eq. (25) represents the partial derivative of $D(\omega,k,\bar{x})$ with respect to k, $D_k(\omega,k,\bar{x}) \equiv \partial D(\omega,k,\bar{x})/\partial k$, while the numerator is a function that is defined in terms of velocity space integrals,

$$N(\omega.k.\bar{x}) \equiv \left[1 + \frac{\omega}{k} \sum \frac{\omega_{ps}^2}{k^2} \int_{-\infty}^{\infty} dv \frac{v\,\partial f_{so}(v,\bar{x})/\partial v}{(v - \omega/k)^3}\right] \frac{\partial k}{\partial \bar{x}}$$

$$- \frac{d\Phi_o}{d\bar{x}} \sum \frac{q_s}{m_s} \frac{\omega_{ps}^2}{k} \int_{-\infty}^{\infty} dv \frac{\partial f_{so}(v,\bar{x})/\partial v}{(v - \omega/k)^3}$$

$$+ \sum \frac{\omega_{ps}^2}{k} \frac{\partial}{\partial \bar{x}} \int_{-\infty}^{\infty} dv \frac{v\,\partial f_{so}(v,\bar{x})/\partial v}{(v - \omega/k)^2} \tag{26}$$

Equation (24) is readily solved for $\hat{\Phi}_o(\bar{x})$,

$$\hat{\Phi}_o(\bar{x}) = \mathcal{A} \exp\left[- \int^{\bar{x}} dx' \, Q(\omega,k(x'),x') \right] \tag{27}$$

where $\mathcal{A}$ is an integration constant.

4. DISCUSSION OF RESULTS: The WKB solution of the linearized Poisson-Vlasov system is given by Eq. (27), where Eqs. (25) and (26) define $Q(\omega,k,\bar{x})$. There is one such solution for each wavenumber, $k(\bar{x})$, that satisfies the local dispersion relation, $D(\omega,k,\bar{x}) = 0$, defined by Eq. (23). It should be noted that WKB solutions are not uniformly valid throughout space. Specifically, they are not valid about the spatial singularities of $Q(\omega,k,\bar{x})$, which occur at zeros of the denominator of Eq. (25),

$$D_k(\omega,k,\bar{x}) \equiv 2k + \omega \sum \frac{\omega_{ps}^2}{k^2} \int_{-\infty}^{\infty} dv \, \frac{\partial f_{so}(v,\bar{x})/\partial v}{(v - \omega/k)^2} = 0 \tag{28}$$

Solutions of Eq. (28) are generally complex. A boundary layer analysis or phase integral methods are required to continue the WKB solutions across the singular points in the complex $\bar{x}$ plane. Issues of double layer stability can be addressed with the WKB solutions. Local stability can be assessed by solving the dispersion relation, Eq. (23), for complex ω in terms of real k. If the imaginary part of ω is negative, instability is implied. An important question concerns the nature of any unstable mode, i.e. whether the instability is convective or absolute. The dangerous instabilities are absolute because they imply that fluctuations grow in time throughout the plasma. Convective instabilities can be regarded as transient since the growing mode convects or propagates out of the system. The classification of instabilities depends on the positions of the Stokes' and anti-Stokes' lines in the complex $\bar{x}$ plane. The classification of double layer instabilities will be the subject of a forthcoming paper.

ACKNOWLEDGMENTS: This work was supported by the U. S. National Science Foundation, grant PHY-8921300.

REFERENCES:

1. H. Schamel, Z. Naturforsch. **38a** (1983) 1170.
2. H. L. Berk, M. N. Rosenbluth and R. N. Sudan, Phys. Fliuds **9** (1966) 1606.

NONLINEAR THEORY OF SHEATHS IN COAXIAL AND CONCENTRIC GEOMETRIES

H.-B. Valentini
Institute for Physical High Technology (e.V.)
Helmholtzweg 4, O-6900 Jena, Germany

ABSTRACT
Using a two fluid plasma model space charge sheaths in coaxial and concentric geometries are treated. The generation of charge carriers and elastic and charge exchange collisions are included. Boundary conditions for various physical conditions, including external electric fields and electric currents, are considered. To solve the nonlinear set of differential equations involving isolated singular points various analytical methods combined with numerical integrations are used. Results are given for the potential and plasma structures and the thickness of the sheaths in coaxial and concentric models describing double probes and their characteristics or plasma columns with an internal wall. The results are more detailed than those ones obtained by circuit diagram calculations. Especially, it is shown that in the considered geometries the floating potentials of the inner and outer electrode can be different.

1. INTRODUCTION: In plasmas the potential and space charge structures before walls and electrodes substantially determine the properties of plasmas and their technical applications. It is well known that external electric fields are strongly changed within the plasma (e.g. /1/-/7/). The plasma diagnostics require to know these structures as well /8/,/9/,/10/.

In this paper plasmas of low pressure discharges in coaxial and concentric geometries are taken into consideration /11/. Ionization by electron impacts in the volume and recombination at the walls are included. The space charge density in the whole plasma and, in contrast to the well-known theories of positive column, an electric current between the inner and outer walls are taken into account. On this way, improved models of double and single probes are obtained allowing to calculate the electric field, the densities and drift velocities of electrons and ions in detail.

2. PHYSICAL MODEL AND BASIC EQUATIONS: The plasma is considered to consist of electrons and of single charged positive ions in a motionless neutral gas. Charged particles are generated within the plasma, flow to the walls or probes, recombine there and re-enter as neutral atoms.

A steady-state model in an one-dimensional plane ($\beta=0$) or cylindrical (coaxial) ($\beta=1$) or spherical (concentric) ($\beta=2$) geometry is considered. If $\beta=0$ walls are assumed at $x=\pm d$. In contrast to the existing theories in the case $\beta=1$ and $\beta=2$ we have two walls, i.e., an inner wall at $r=r_1$ and an outer wall at $r=r_2$.

It is well known that insulating walls are charged negatively with respect to the plasma. Here, conducting walls, i. e., probes or electrodes, also considered, but without emission of electrons or ions. If an external electric voltage is applied between two probes the potentials of both remain negatively with respect to the plasma in a large region of the voltage. This region is treated.

The well known basic equations for the ions (index s=i), the electrons (s=e), the electric field intensity E and the potential Φ are the equations of continuity and momentum transfer and the Poisson equation. They read

$$(d/dr)N_s v_s = Q-\beta N_s v_s/r, \quad s=i,e, \qquad (1)$$

$$N_s M_s v_s dv_s/dr+kT_s dN_s/dr=N_s Z_s eE-P_s \qquad (2)$$

$$dE/dr=(e/\varepsilon_0)(N_i-N_e)-\beta E/r \qquad (3)$$

$$d\Phi/dr=-E \qquad (4)$$

and further $$j=e(N_i v_i-N_e v_e) \qquad (5)$$

where k, e, ε_0 are used as usual and constant temperatures T_s are supposed. M_s is the mass, N_s is the number density, v_s the drift velocity and $Z_i=-Z_e=1$.

$$Q=\nu_{ni}N_e, \quad P_i=(Q+N_i\nu_k)M_i v_i, \quad P_e=(\nu_{ni}+\nu_{en})N_e M_e v_e \qquad (6)$$

describe the source term for the charge carrier generation and the momentum transfer, respectively, with the frequencies of ionization ν_{ni}, elastic electron-atom collisions ν_{en} and an effective frequency ν_k for charge exchange and elastic impacts between ions and atoms.

From (1) and (5) can be derived $r^{\beta}j=\text{const}.$ (7)

The differential equation (1) for the electrons (or the ions) can be replaced by (7). It can be seen that the remaining set of five differential equations (1)-(4) is nonlinear and involves singular points at the sound speeds

$$v_s = \pm v_{sc} \quad \text{where} \quad v_{sc} = (kT_s/M_s)^{1/2}. \qquad (8)$$

The index c denotes the singular points. In the used model the ion velocity attains (8) within the plasma. The singularity at this point can be removed by the constraint /3/, /5/, /8/

$$(Q_c - \beta N_{ic} v_{ic}/r_c) M_i v_{ic} = N_{ic} Z_i e E_c - P_{ic} \qquad (9)$$

derived from (1) and (2). Further, since the space charge density within the bulk of the plasma is extremely small the set of governing equations represents a singularly perturbated problem.

3. BOUNDARY CONDITIONS: Under our assumptions a point r_0 exists where

$$v_i(r_0) = 0, \quad r_1 < r_0 < r_2. \qquad (10)$$

The ions generated in $r_1 < r < r_0$ flow to the inner wall, those from $r_0 < r < r_2$ to the outer wall. The boundary of the spheres of influence of the inner and the outer wall on the ion flux is determined by r_0. An appreciable mathematical simplification is obtained if only cold ions, i. e., $T_i=0$ and $v_{ic}=0$, are taken into account, because $r_c=r_0$ follows. Then, from (9) $E(r_0)=0$ (11) results. $\Phi(r_0)=0$ (12) can be stated. Expanding (1) and (2) for the ions and (3) into power series of $r-r_0$ or using (9) gives

$$w_0^2(w_0-1)/a_0 - Kw_0 - 2 = 0 \qquad (13)$$

where $1/a_0 = (\omega_{pi}/\nu_{ni})^2$ with the ion plasma frequency $\omega_{pi} = (Z_i e^2 N_e(r_0)/\varepsilon_0 M_i)^{1/2}$, $K = \nu_k/\nu_{ni}$ and $w_0 = N_i(r_0)/N_e(r_0)$. Here, $N_e(r_0)$ can be stated and $N_i(r_0)$ is given by (13). At the walls a simplified electron kinetics gives

$$v_e = \pm v_{ew}, \quad v_{ew} = (2kT_e/\pi M_e)^{1/2} \qquad (14)$$

if $r=r_2$ and $r=r_1$, respectively.

We have 5 boundary conditions at r_0 and one at r_1 and one at r_2 for the 5 remaining differential equations. The electron temperature determining the collision frequencies occurs as eigen value. The ratio r_2/r_1 following from (14) depends on r_0. At the beginning r_0 is unknown. Approximating a given value of r_1/r_2 by use of an iteration scheme one can determine r_0.

4. NORMALIZATION AND METHODS OF SOLUTION: It can be easily seen that $v_0 = (kT_e/M_i)^{1/2}$, $l_{ni} = v_0/\nu_{ni}$, $l_D = (\varepsilon_0 kT_e/e^2 N_e(r_0))^{1/2}$,

$$E_0 = kT_e/el_{ni}, \quad a_0 = (l_D/l_{ni})^2, \quad K = \nu_K/\nu_{ni}, \quad K_e = \nu_{en}/\nu_{ni} \qquad (15)$$

are characteristic parameters, where l_D is the Debye length, a_0 describes the influence of space charge, K that of ion-atom collisions, and K_e that of elastic electron-atom collisions, where $K_e \ll K$. The equations can be written in the dimensionless variables

$$s=r/l_i,\ V_{sr}=v_n/v_0,\ w_s=N_n/N_c(r_0),\ f=E/E_0,\ \eta=-e\Phi/kT_e. \quad (16)$$

If $T_i>0$ two boundary value problems have to be solved, the first in the subsonic region of the ion flow, the second in the hypersonic region. A shooting method was used. Restricting to $T_i=0$ leads to a hypersonic ion flow in the whole interval /5/.
Setting $s=s_c+u$ or $V_i=V_{ic}+\chi$ the set (1)-(7) can be solved by an expansion into series of powers of u or χ near the singular point of the ion gas. This solution is continued by numerical integration. If $0<T_i<<T_e$ an expansion into power series of T_i/T_e is useful /3/,/5/. This leads to an regular perturbation scheme. If $a_0K<<1$ an expansion into power series of a_0 can be used. One obtains an singular perturbation scheme /3/,/5/,/6/.
5.RESULTS AND CONCLUSIONS: For the case without net current between the walls, i.e. C=0, and coaxial cylindrical geometry, i.e. $\beta=1$, results are shown in Fig.1 and 2 for argon. The radial profiles of the ion and electron densities w_i, w_e and of the electric potential η are plotted. Under the considered conditions the ion density has a minimum close to the inner wall. It can be seen that the particle densities and the potentials at the inner and outer wall are not equal. This result also holds for the drift velocities and the ion currents /11/. Similar results follow for other values of a_0, K and r_2/r_1.
The Fig.3-8 represents results for a double probe configuration, i.e., if a net current flows between inner and outer wall. In Fig.3 a typical current-voltage characteristic is plotted for a coaxial cylindrical model where $r_2/r_1=100$. A saturation occurs in the ionic branch (C<0) as well as in the electronic branch (C>0). In this model, saturation takes place if all created ions flow to the first wall and all created electrons to the second wall or vice versa. Fig.4 shows that in the saturation regions one of the two normalized wall potentials η_1 or η_2 strongly increases with the current density C. It can be seen that in both the branchs the walls remain negatively charged with respect to the plasma. Fig.5 shows the boundaries of the regions from where ions (r_0) and electrons (r_{0e}) flow to the inner or to the outer wall.
A very small spherical probe ($\beta=2$) in a large vessel ($r_2/r_1=10^4$) approximates to a single probe (see Fig.6-8). In contrast to the

foregoing situation, in the electronic branch the potential difference between probe and plasma vanishes (Fig.6). The characteristics (Fig.7) show no saturation up to this point. Under these conditions the saturation of the electron current can be only reached if the externally imposed potential drop is so high that the probe is positively biased with respect to the plasma. Further, it can be seen from the profiles of particle densities that at the end of the ionic branch the space charge sheath before the inner wall is much smaller than that before the outer wall. Considering the here given end points of the characteristics the different dimensions of the regions from where electrons and ions flow to the walls can be taken from the plots of the drift velocities depicted in Fig.8. In this cases the boundaries of the regions of particle collection, i.e., of sheath and presheath, are exactly calculated. In Fig.8 r_0-r_i, $r_{0e}-r_i$ are substantially smaller than in Fig.1 and 5.

The given model can be easily generalized.

ACKNOWLEDGEMENTS: I am grateful to Dr. D. Wolff for discussions and performing a part of numerical calculations and drawings.

REFERENCES:

1. P.D. Prewett, J.E. Allen, *Proc. Roy. Soc.* **A348**(1976),p.435
2. G.J. Andrew, E.J. Allen, *Proc. Roy. Soc.* **A320**(1971),p.459
3. R.N. Franklin, *"Plasma Phenomena in Gas Discharges"* Clarendon, Oxford, 1976
4. R. Chodura, *"Physics of Plasma Wall Interaction in Controlled Fusion"* (ed. by D.E.Post and K.Behrisch) Plenum, London, 1986
5. H.-B. Valentini, *J.Phys.D:Appl Phys.* **21**(1988),p.311 and *Contrib.Plasma Phys.* **31**(1991),p.211
6. K.-U. Riemann, *J.Phys.D:Appl.Phys.* **24**(1991),p.493
7. C.K. Birdsall, A.B. Langdon, *"Plasma Physics via Computer Simulation*, Adam Hilger, Bristol, (1991)
8. S.H. Lam, *8th Intern. Conf. on Phen. in Ionized Gases*, Wien, (1967), Invited Pap. p.545
9. P.M. Chung, L. Talbot, K.J. Touryan, *"Electric Probes in Stationary and Flowing Plasmas"*, Springer, Berlin 1975
10. I.M. Cohen, *Phys. Fluids* **8**(1965),p.2097
11. H.-B. Valentini, *Proc. 10th Intern.Symp. on Plasma Chemistry, Bochum, Germany*, 4.-9.8.1991 Vol.2, p.2.1-51

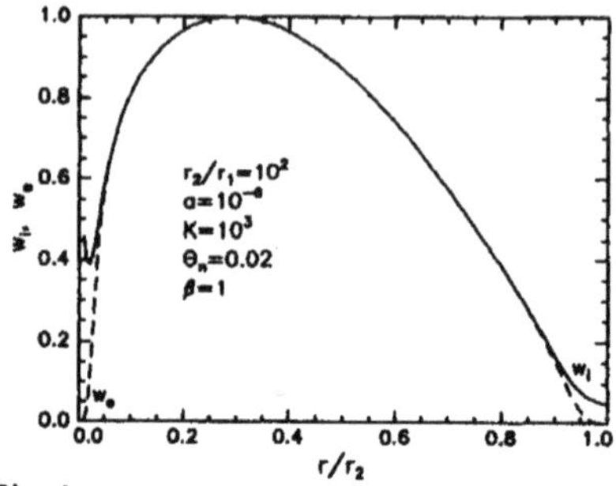

Fig.1

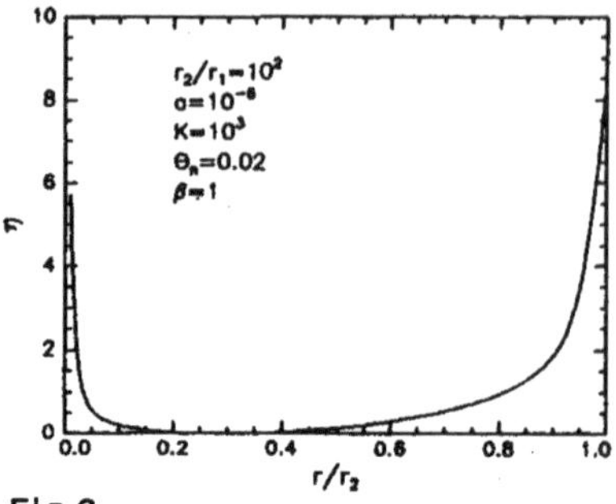

Fig.2

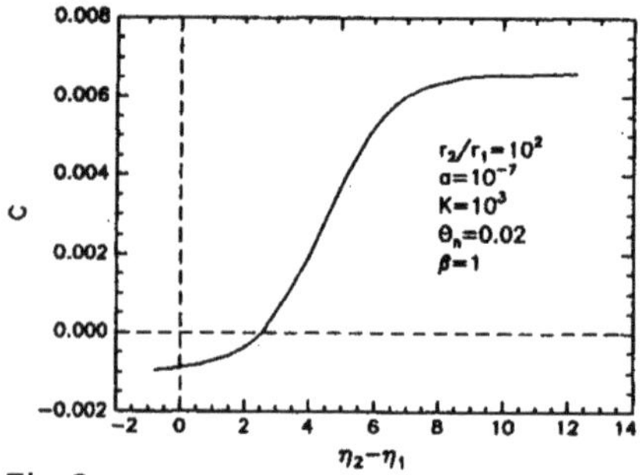

Fig.3

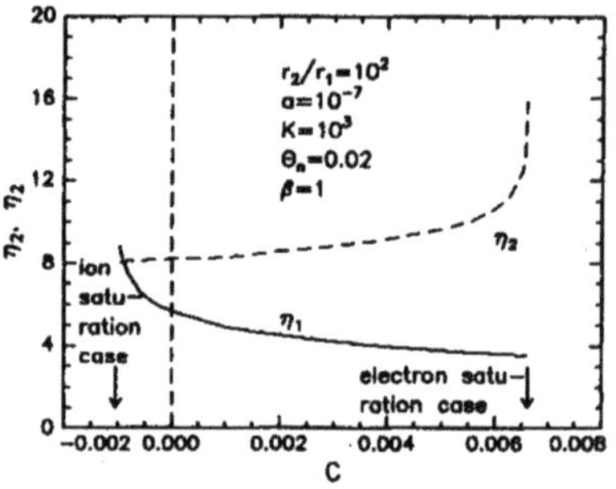

Fig.4

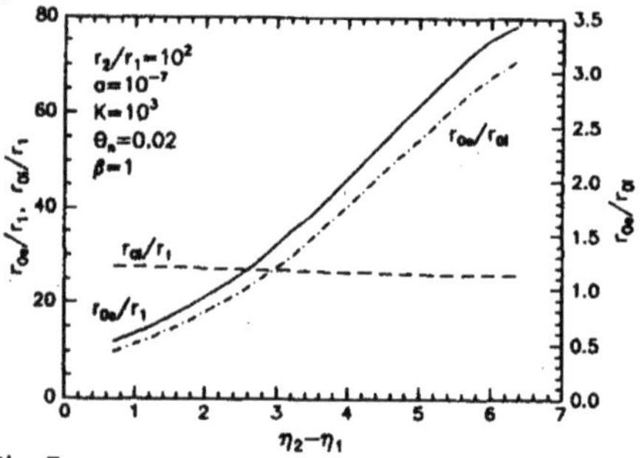

Fig.5

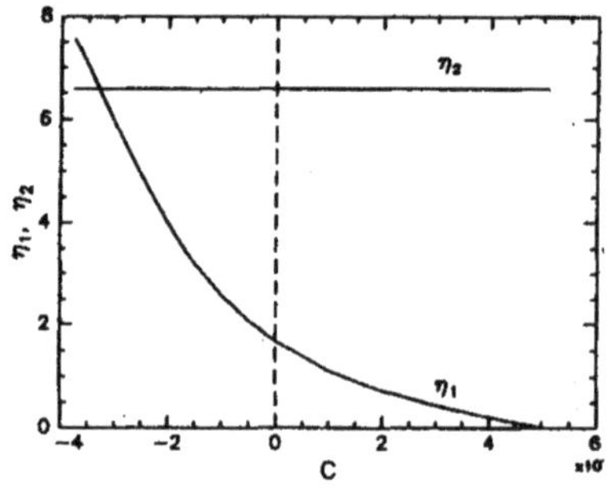

Fig.6

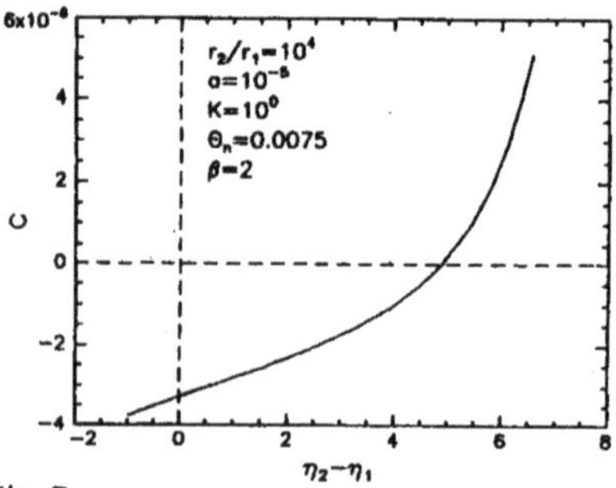

Fig.7

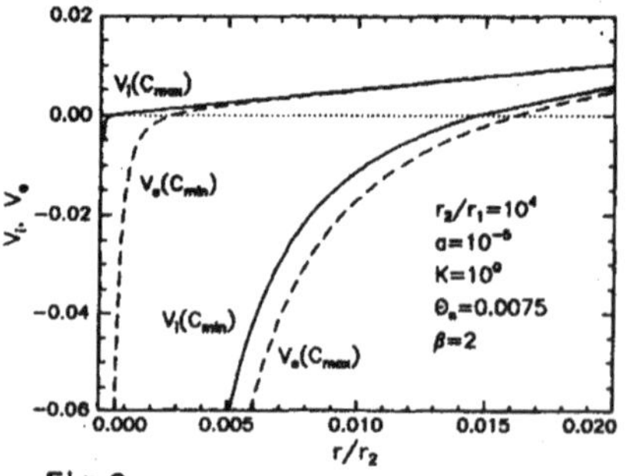

Fig.8

Analysis of Problems of Trapped Particles in Boundary Kinetic Problem

Zhykharsky A.V., Shmykov V.N.
Scientific-technical Centre "Kinetika",
P.O. Box 36, 630078, Novosibirsk-78, Russia

The paper presents the analysis of the trapped particle problems with thermionic diodes since this model permits obtaining sufficiently detailed information about the electric potential in the diode without immediately solving the Poisson equation. This, in its turn, makes it possible within the range of this model to make an investigation into potential wells which are necessary for trapped particles to exist.

Formulation of the boundary kinetic problem for thermionic diodes in the absence of an external magnetic field has the form /1/

$$V_r \cdot \frac{\partial f_e}{\partial r} - \frac{\partial U_e}{\partial r} \cdot \frac{\partial f_e}{\partial V_r} = 0\,, \tag{1}$$

$U_e = \Psi$ -- plane diode;

$U_e = \dfrac{M^2}{2 \cdot r^2} + \Psi$ -- cylindrical diode;

$U_e = \dfrac{S}{2 \cdot r^2} + \Psi$ -- spherical diode;

$$f_e(R_1)\Big|_{V_r > 0} = g\left(\frac{|V|^2}{2}\right), \quad f_e(R_2)\Big|_{V_r < 0} = 0 \tag{2}$$

Here $\Psi = -\dfrac{e}{m_e} \cdot \Phi$, Φ is electrical potential; M and S are independent variables (see /1/), R_1 is a coordinate of the cathode, R_2 is a coordinate of the anode surface.

Fig.1 presents the profiles of the function Ψ, characteristic of the electron diodes. Here and further we assume that $\Phi(R_1) = 0$. Analysis of the expression for generalized potentials showed that in cylindrical and spherical diodes there may exist potential wells on the segment $r \in [R_1, r^*]$ in some interval of the values M and S.

We'll further consider a cylindrical diode. Fig.2 presents the generalized potential profiles of this diode for the values M providing the existence of potential wells.

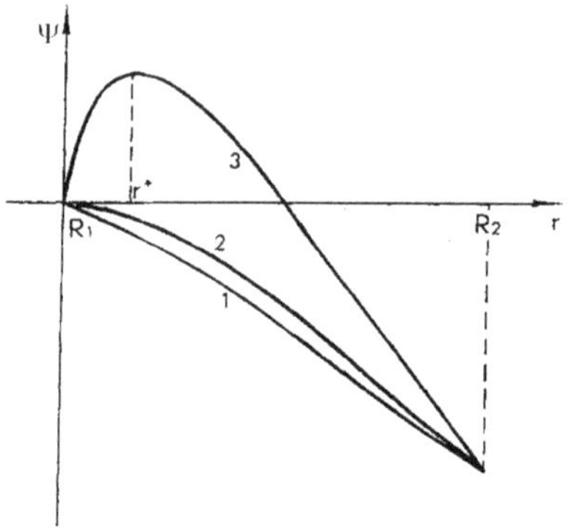

Fig.1.Profiles of function Ψ for electron thermionics diodes.

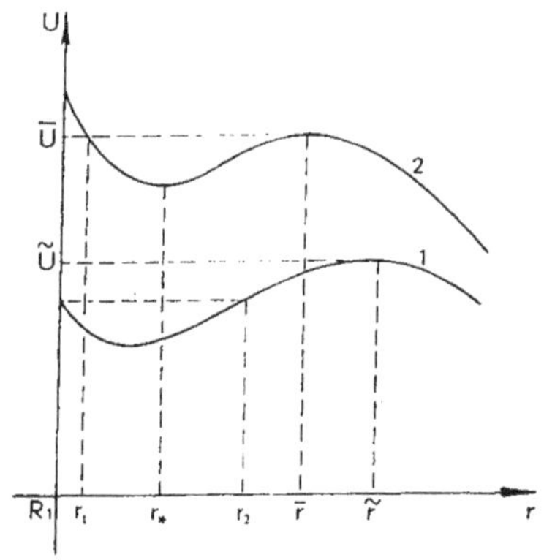

Fig.2. Profiles of generalized potential of the cylindrical diode operating as "virtual cathode".

A solution to the boundary problem (1), (2), in our case, has the form /1/

$$f_e = H(\lambda)\cdot g\left(\frac{|V|^2}{2}\right),$$

$$\lambda = V_r - sign(r - \tilde{r}_e)\cdot\left[2\cdot(\tilde{U}_e - U_e)\right]^{1/2} \qquad (3)$$

$$U_e = \frac{M^2}{2\cdot r^2} + \Psi, \quad \tilde{U}_e = \max_{R_1 \le r \le R_2}(U_e), \quad U_e(\tilde{r}_e) = \tilde{U}_e .$$

To determine whether DF (3) presents the trapped particles we'll estimate the electron density produced by this function inside the potential wells (see Fig.2). The expression for electron density in the cylindrical diode has the form /1/

$$n(r) = \frac{1}{r}\cdot\int_{-\infty}^{\infty} dV_z\, dM\cdot\int_{-\infty}^{\infty} dV_r\cdot H(\lambda)\cdot g\left(\frac{V_z^2}{2} + \frac{M^2}{2\cdot r^2} + \frac{V_r^2}{2} + \Psi\right).$$

(a) Let's estimate the density of trapped particles at the point $r = r_*$ for a potential well on curve "1" (see Fig.2). Since in this case $\tilde{r}_e > r_*$ we'll obtain from the expressions (3)

$$\lambda = V_r + \left[2\cdot\left(U_e(\tilde{r}_e) - U_e(r_e)\right)\right]^{1/2} = V_r + V^*,\quad H(\lambda) = 1 \quad \text{for } V_r > -V^*.$$

Consequently: $n(r_*) \sim \int_{-V^*}^{\infty} dV_r\cdot g$.

The interval of the velocities $V_r \in [-\varepsilon, \varepsilon]$, is characteristic of trapped particles where $\varepsilon = \left[2\cdot\left(U_e(R_l) - U_e(r_*)\right)\right]^{1/2} < V^*$. Consequently the density estimate of these particles has the form : $n(r_*)\Big|_{|V_r| \le \varepsilon} \sim \int_{-\varepsilon}^{\varepsilon} dV_r\, g > 0$, since $\varepsilon < V^*$ the interval $V_r \in [-\varepsilon, \varepsilon]$ falls into a full interval of integration by V_r.

(b) In the case of a potential well on curve "2" (Fig.2) we have $r_* > r_e = R_l$. Consequently:

$$\lambda = V_r - \left[2\cdot\left(U_e(R_l) - U_e(r_*)\right)\right]^{1/2} = V_r - V^*, \quad H(\lambda) = 1 \quad \text{for } V_r > V^*.$$

In this case the estimate of the overall particles density at the point $r = r_*$ has the form: $n(r_*) \sim \int_{V^*}^{\infty} d V_r \cdot g$. Since in this case the interval of the velocities V_r for trapped particles is determined by the expression:

$$\varepsilon = \left[2\cdot\left(\bar{U_e} - U_e(r_*)\right)\right]^{1/2} < V^*,$$

the density estimate of these particles has the form: $n(r_*)\Big|_{|V_r| \le \varepsilon} \sim \int_{-\varepsilon}^{\varepsilon} d V_r\, g = 0$, since in this case the interval $V_r \in [-\varepsilon, \varepsilon]$ doesn't fall into a full interval of integration by V_r. Thus we can draw the conclusion that the DF (3) fills only those potential wells which are located between the particles source (cathode) and the maximum of generalized potential.

For analysis of the possibility of filling a potential well on curve "2" (Fig.2) we'll treat the function of the kind

$$f = g\left(\frac{|V|^2}{2} + \Psi\right)\cdot H(r - r_l)\cdot H(\bar{r} - r)\cdot H\left(\bar{U} - U - \frac{V_r^2}{2}\right).$$

This function fills the potential well on curve "2" and satisfies the boundary conditions (2). However before taking this function as a solution to the boundary kinetic problem (1), (2) we must prove that the function

$$\Lambda = H(r - r_l)\cdot H(\bar{r} - r)\cdot H\left(\bar{U} - U - \frac{V_r^2}{2}\right) \tag{4}$$

is a solution to the equation (1). This statement can be proved mathematically and here we'll just present an "illustrative" proof

$$V_r\cdot\frac{\partial\Lambda}{\partial r} - \frac{dU_e}{dr}\cdot\frac{\partial\Lambda}{\partial V_r} = V_r\cdot H\left(\bar{U_e} - U_e - \frac{V_r^2}{2}\right)\cdot\left\{\delta_{r_l}\cdot H(\bar{r} - r) - \delta_{\bar{r}}\cdot H(r - r_l)\right\} = 0,$$

since on the one hand, with $r_l < r < \bar{r}$ the expression in the brackets is equal to zero, on the other hand, we have:

$$V_r H\left(\bar{U}_e - U_e - \frac{V_r^2}{2}\right)\Bigg|_{\substack{r\to r_1\\ r\to \bar{r}}} \longrightarrow V_r H\left(-\frac{V_r^2}{2}\right) = 0.$$

Thus we can consider it established that the function of the kind

$$f_e = g\left(\frac{|V|^2}{2} + \Psi\right)\cdot\left\{H(\lambda) + H(r - r_1)\cdot H(\bar{r} - r)\cdot H\left(\bar{U} - U - \frac{V_r^2}{2}\right)\right\}$$

is a solution to the boundary problem (1), (2) and fills with trapped particles all the potential wells in the cylindrical thermionic diode. Further investigations showed that the "Λ"-type functions can serve to construct distribution functions for any plasma components, to arbitrarily fill potential wells with trapped particles.

In particular, one can show that in the case of cylindrical diode a distribution function of the kind (see Fig.2)

$$f_e = g\left(\frac{|V|^2}{2} + \Psi\right)\cdot\left\{H(\lambda) - H(r_2 - r)\cdot H\left(U(R_1) - U - \frac{V_r^2}{2}\right)\right\} +$$

$$+ F\left(\frac{|V|^2}{2}\right)\cdot\left\{H(r_2 - r)\cdot H\left(U(R_1) - U - \frac{V_r^2}{2}\right) + H(r - r_1)\cdot H(\bar{r} - r)\cdot H\left(\bar{U} - U - \frac{V_r^2}{2}\right)\right\}, \quad (5)$$

where F is arbitrary function, is a solution to the boundary problem (1), (2). The quantities: $r_1, r_2, \bar{r}, U(R_1), \bar{U}$ included in the expression (5) are the functions of the invariable "M". This fact should be taken into account when solving applied problems making use of the functions of the type (5), i.e. in calculating the right side of the Poisson equation.

We will write the Poisson equation as an example for the case of completely filling all the potential wells in the cylindrical thermionic diode. The potential wells are assumed to have been filled with electrons "originating" on the cathode. In this case DF (5) is transformed into the form

$$f_e^* = \left\{H(\lambda) + H(r - r_1)\cdot H(\bar{r} - r)\cdot H\left(\bar{U} - U - \frac{V_r^2}{2}\right)\right\}\times$$

$$\times\, n\cdot\left(\frac{m_e}{2\pi kT}\right)^{3/2}\cdot\exp\left(-\frac{m_e}{kT}\cdot\frac{|V|^2}{2} + \frac{e}{kT}\cdot\Phi\right)$$

Further, substituting this function on the right side of the Poisson equation and integrating by V_z and V_r we will obtain the following form of the equation

$$\frac{d^2\Phi}{d r^2} + \frac{1}{r}\cdot\frac{d\Phi}{d r} = \frac{e}{\varepsilon}\cdot\frac{n}{\sqrt{\pi}}\cdot\exp\left(\frac{e\Phi}{kT}\right)\times$$

$$\times\left\{\int_0^\infty \exp(-\tau^2)\cdot\left[1 - sign(r - \bar{r})\, erf(\,[\bar{U} - U\,]^{1/2})\right]d\tau +\right.$$

$$+ 2\cdot\int_{\tau_1}^{\tau_2} H(r-r_I)\cdot H(\bar{r}-r)\cdot\exp(-\tau^2)\cdot erf\left([\bar{U}-U]^{1/2}\right) d\tau \Big\},$$

where $U = \tau^2 - \dfrac{e}{kT}\cdot\Phi, \quad U(\bar{r}) = \bar{U},$

$$\bar{U} = \max_{R_I \le x \le R_2}\left(\tau^2\cdot\frac{r^2}{x^2} - \frac{e}{kT}\cdot\Phi(x)\right).$$

Here substitution has been made for the variables of the form $\tau = \left(\dfrac{m_e}{2kT}\right)^{1/2}\cdot\dfrac{M}{r}$. The values "$\tau_1$ " and "τ_2 " should be determined at every iteration step (they are "functions" of the electric potential and its derivatives). These values determine the interval "τ", for which there exist empty potential wells (see curve "2" in Fig.2).

The results obtained permit drawing the following conclusions:

(1) A new class of accurate solutions to the boundary kinetic problem has been obtained (functions of the type (4)) that permitted filling potential wells arbitrarily with trapped particles. Thus, it has been established that within tha range of the stationary boundary kinetic problem it is impossible to uniquely determine the DF (distribution function) of trapped particles.

(2) The choice of the DF of trapped particles (explicit, i.e. analytical or implicit, i.e., pre-set with the aid of some computational algotithm) determines the character of filling potential wells with these particles and is actually an external parameter of the problem to be solved. This fact results in two significant consequences to be taken into account in solving applied problems.

(a) When solving Poisson equations with the aid of iterative process at every iteration step calculation of the right side of this equation is performed with the electric potential obtained in the previous iteration. So when calculating the right side of the equation the electric potential (and its derivatives) is considered to have been pre-set.

It can be shown that if it is possible for potential wells to exist in the system the shape and the dimensions of these wells are determined by electric potantial and its derivatives. Thus, we come to the conclusion that in the process of an iterative solution of the Poisson equation in the space area considered there may appear an odd (or vanish a true) charge whose quantity is determined by the variation character of electric potential. Moreover since at the outset of the iterative process the form of electric potential may be rather wide of the "true one" it is quite possible that there may arise charge oscillations even when solving those problems for which the absence of trapped particles is evident (the nature giving rise to these oscillations is probably related to the fact that realization of an iterative computing process disturbs a cause-effect relation accepted in classic electrodynamics: so the field seems to give rise to charges and not vice versa).

It should be noted that from the standpoint of the Poisson equation the odd charges are quite natural and it is impossible to get rid of them by means of conventional computing techniques (For example, by decreasing the pitch size of a differential grid).

(b) A thorough analysis of the fact related to the absence of solution uniqueness for the stationary boundary kinetic problem indicated that the classical formulation of the stationary kinetic problem ($\frac{\partial}{\partial t} = 0$) needs to be essentially precised. Particularly, it turned out that a stationary state of the system should be considered as the result of a limiting transition from a pre-set initial state of the system. Moreover there may arise a situation where a statioanary state of the system can depend on the method of this transition.

It is evident that the results obtained should be taken into account when solving applied problems. Indeed any specific iterative process realized while solving the stationary problem actually simulates a transition of the system from a pre-set initial state to a final stationary state. Consequently, one must take account of the possible situation where a final state of the system will depend on the specific algorithmic realization of this process. Therefore numerically when simulating the stationary boundary kinetic problems precautions should be taken.

In addition it should be noted that even introducing into the equation the time-dependent terms in pronciple saves no situation in the case where at every time step an actual solution of the stationary problem is required.

/1/ Zhykharsky A.V. Physica Scripta, v.44:6, p.606.

CONTRIBUTED PAPERS

SIMULATION

PARTICLE SIMULATIONS OF STRONG DOUBLE LAYERS IN A COLLISIONLESS BOUNDED PLASMA: EFFECTS OF BOUNDARY CONDITIONS AND HF FIELDS

S.V. Bulanov,[a)] S. Kuhn,[b)] and A.S. Sakharov[a)]

[a)] Institute of General Physics, Russian Academy of Sciences
117924 Moscow, Russia

[b)] Institute for Theoretical Physics, University of Innsbruck
A–6020 Innsbruck, Austria

Abstract. The evolution of electrostatic double layers in a collisionless plasma diode with injection of current–carrying plasma from both ends is systematically investigated by means of particle–in–cell simulations performed with the one–dimensional electrostatic bounded–plasma codes PDW1 and PDP1. It is found that asymmetry in the injection conditions can shift the double layer toward one of the boundaries and may either stabilize or destabilize it. If a given dc potential difference is applied between the electrodes and a time–independent double layer exists, the latter is always found to be adjacent to the low–potential boundary of the diode. Application of a strong hf field (modeled by an ac bias) can lead to stabilization or destabilization of the double layer, depending on the driving frequency. For boundary conditions corresponding to a given constant external–circuit current, the double layers observed in the simulations are transient and decay on the time scale on which the diode fills with plasma.

In this paper we present results of PIC simulations of electrostatic double layers (DL's) in a diode with injection of plasma from both ends. The effects of (i) boundary and external–circuit conditions and (ii) an applied hf field on the DL dynamics are investigated.

The following model is used. The plane diode fills up with plasma injected from the electrodes with half–Maxwellian electron and ion distribution functions with the same temperatures T on both sides. The injection current densities satisfy the condition $j_{il}/j_{el} = j_{ir}/j_{er} = (m_e/m_i)^{\frac{1}{2}}$, so that the injected plasma beams are neutral. Particles hitting boundaries are absorbed. Two types of boundary conditions are considered: (i) a given dc potential difference between the electrodes, and (ii) a given dc current in the external circuit. The first one can easily be generalized by applying to the diode an additional ac voltage to study the effect of an hf field.

The following normalized variables are used: $x = \hat{x}/\hat{L}$, $t = \hat{t}/\hat{t}_0$, $j = \hat{j}/\hat{j}_0$, $\Phi = \hat{\Phi}/\hat{\Phi}_0$, $\rho = \hat{\rho}/\hat{\rho}_0$, and $T = \hat{T}/\hat{T}_0$. Here, $\hat{}$ denotes the usual (dimensional) variables, $\hat{L}$ is the distance between the electrodes, $\hat{t}_0 = (m_e\hat{L}^2/e\hat{\Phi}_0)^{\frac{1}{2}}$, $\hat{\rho}_0 = \hat{j}_0\hat{t}_0/\hat{L}$, $\hat{T}_0 = e\hat{\Phi}_0$, and $\hat{j}_0$, $\hat{\Phi}_0$ and $\hat{L}$ satisfy the Child–Langmuir–type condition

$$\hat{j}_0 = \left[\frac{e}{16\pi^2 m_e}\right]^{\frac{1}{2}} \frac{\hat{\Phi}_0^{3/2}}{\hat{L}^2}. \qquad (1)$$

The total current density is defined as $J(t) = j_e(x,t) + j_i(x,t) + \partial E(x,t)/\partial t$.

For a given dc bias we choose $\hat{\Phi}_0 = \hat{\Phi}(\hat{L})$, so that $\Phi(1) = 1$ and $\hat{j}_0$ is defined by Eq. (1). For a given dc external–circuit current $\hat{I}$ we set $\hat{j}_0 = \hat{I}/\hat{A}$, the electrode area being $\hat{A} = 1$. Then $I = J = 1$ and $\hat{\Phi}_0$ is defined by Eq. (1).

A theoretical analysis shows that the structure of a "strong" $[\Phi(L) \gg T]$ time–independent DL in the plasma diode depends strongly on the parameter

$$R \equiv \frac{j_{ir}}{j_{el}}\sqrt{\frac{m_i}{m_e}}. \qquad (2)$$

For our injection conditions, $R = j_{er}/j_{el}$. If $R = 1$ (symmetric injection), a time–independent solution with the DL in the middle of the diode can exist. For $j_{el} \gg 1$ its width is much less than L and the DL separates two extended regions of quasineutral plasma, so that we are dealing with a "long" diode. If the symmetry of injection is violated ($|R-1| \geq T^{\frac{1}{2}}$), the DL is adjacent to the right ($R < 1$) or to the left ($R > 1$) boundary. The simulation results show that in a long diode the DL's stability also depends strongly on R.

First we demonstrate results for a given dc bias. In all simulations, $T = 0.01$ and $m_i/m_e = 100$. The simulations were performed with the PIC codes PDW1[1] and PDP1[2].

Figure 1 corresponds to $R = 1$ (symmetric injection) and $j_{el} = 10$ ("long" diode). Figures 1(a) and 1(b) show the time history of $|J|$ and the level lines of $\Phi(x,t)$, which exhibit strongly nonlinear DL oscillations. During one cycle, the DL moves from the left electrode to the middle of the diode and is then destroyed due to a Pierce–type instability. Then the ions are reflected to the left–hand boundary. After the destruction of the old DL, the formation of a new one begins due to the injection of plasma from the left boundary.

An asymmetry in injection can stabilize ($R > 1$) or destabilize ($R < 1$) the DL. This is demonstrated by Figs. 2 and 3. In both cases, $j_{el} = 10$. Figure 2 corresponds to $R = 0.5$. The dynamics of the plasma diode are similar to that observed in simulations of a Q–machine with injection of plasma from the low–potential boundary only.[3] The time dependence of J reflects inharmonic oscillations with larger amplitude and period than for $R = 1$.

Figure 3 shows results for $R = 1.5$. After some time the system settles down at a quasi–dc state. Figure 3(c) ($t = 100$) shows that the DL is shifted towards the left electrode, in accordance with the theoretical analysis for $R > 1$.

The case of a given dc external–circuit current differs basically from that of a dc potential bias. In this case no immediate feedback between the electrodes is provided, and thus no Pierce–type instability may be expected. Here we present simulation results on the long–time evolution of a diode which is initially empty and subsequently fills with plasma injected from boundaries.

Figure 4 corresponds to $j_{el} = 10$ and $R = 1$. The DL exists only during some finite time interval. The right–hand potential history, $\Phi(L,t)$, exhibits four stages: (i) The diode is still largely empty and $\Phi(L,t)$ grows linearly up to the value corresponding to the "pure electron diode". (ii) A transition to the regime of the "Langmuir diode" occurs, in which accelerated ions and electrons move in opposite directions and $\Phi(L,t)$ decreases. (iii) The diode fills with two streams of quasineutral plasma, the effective width of the DL decreases and $\Phi(L,t)$ decreases as well. (iv) The whole diode is filled with quasineutral plasma and only fluctuations of $\Phi(L,t)$ are observed.

A strong hf electric field applied to the plasma can result in aperiodic instability when $\omega < \omega_p$.[4] If $\omega > \omega_p$, the hf field can be a stabilizing factor. Our results show that DL's may also be stabilized or destabilized by an hf electric field. Our first simulation results on the subject, performed with a flat–charged–sheet model, were reported in Ref. 5. In the present PIC simulations, the boundary conditions are the same as those for the case of a given dc bias, but with an additional ac voltage: $\Phi(L,t) = 1 + \varphi\sin(\omega t)$, with period $\tau \equiv 2\pi/\omega$. Two parameter sets were used: (i) $\varphi = 1$, $\tau = 1.25$ and (ii) $\varphi = 5$, $\tau = 0.25$. The electron quiver energy $[(\varphi/\omega)^2/2 \ll 1)$ is the same for both sets, but in case (i) τ is larger than the electron transit time, $t_e \simeq 1$, and ω is less than the typical plasma frequency in the DL region, $\bar{\omega}_p \equiv 2\pi/\tau_e$. In case (ii) we have $\tau < \tau_e$ and $\omega > \bar{\omega}_p$. As clearer evidence of hf effects may be expected for symmetrical particle emission from the boundaries, we have chosen $j_{el} = 10$ and $R = 1$ for all runs.

Figure 5 ($\tau = 1.25$, $\omega < \bar{\omega}_p$) shows that the "low–frequency" hf field destabilizes the DL. The total current is averaged over $\Delta t = \tau$. The frequency of the relaxation oscillations decreases and their amplitude becomes larger as compared with the dc–bias case (see Fig. 1).

Figure 6 shows that for $\omega > \bar{\omega}_p$ ($\tau = 0.25$) the DL is actually stabilized. Both the potential distribution and the electron and ion charge–density distributions (Fig. 6(c)) are averaged over the interval $145 \lesssim t \lesssim 150$. The stabilized DL is adjacent to the left boundary, which in some sense is similar to the time–independent situation with $R > 1$.

This work was partly supported by the Austrian Science Foundation (FWF) under contracts P7005 and P8405 and the Austrian Academy of Sciences.

[1]W.S. Lawson, J. Comp. Phys. **80**, 253 (1989).
[2]V. Vahedi, M.A. Lieberman, M.V. Alves, J.P. Verboncoeur, and C.K.Birdsall, J. Appl. Phys. **60**, 2008 (1991).
[3]P.C. Gray, S. Kuhn, T.L. Crystal, and C.K. Birdsall, *2nd Symp. on Plasma Double Layers and Related Topics*, Innsbruck, 1984, ed. by R. Schrittwieser and G.Eder, p.266.
[4]V.P. Silin, Sov. Phys. JETP 21, 1127 (1965).
[5]S.V. Bulanov, A.S. Sakharov, and S. Kuhn, *2nd Symp. on Plasma Double Layers and Related Topics*, Innsbruck, 1984, edited by R. Schrittwieser and G. Eder, p. 327.

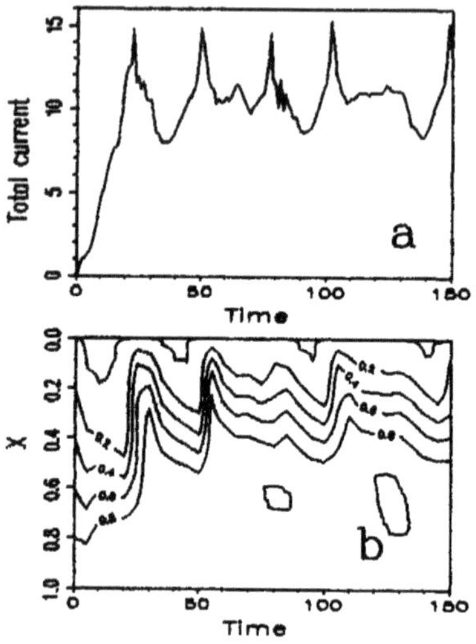

Fig.1

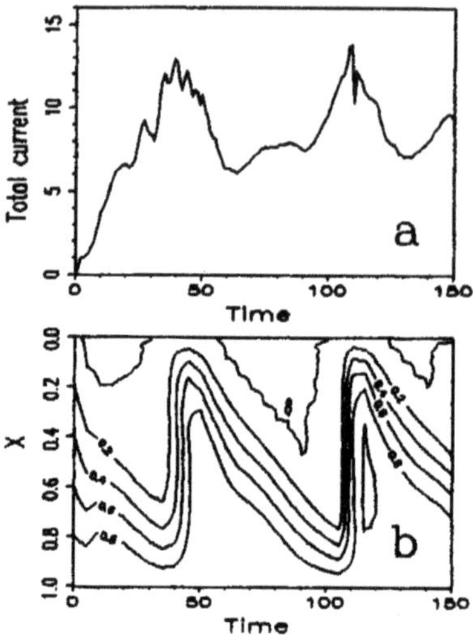

Fig.2

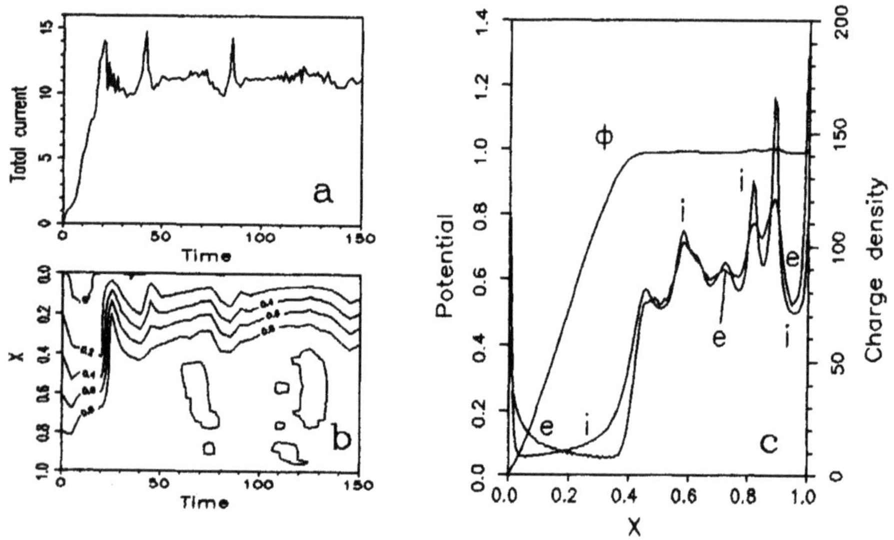

Fig.3

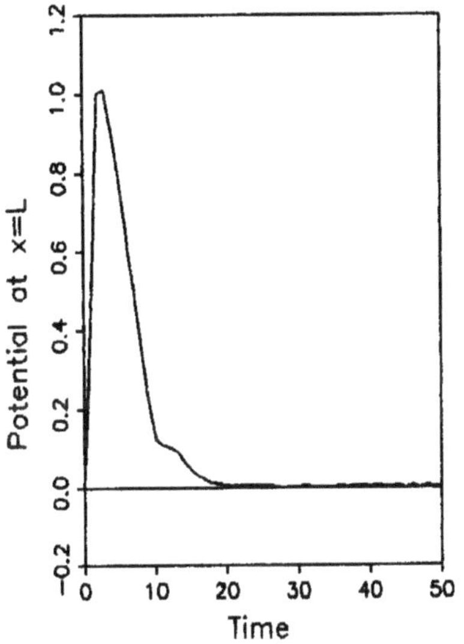

Fig.4

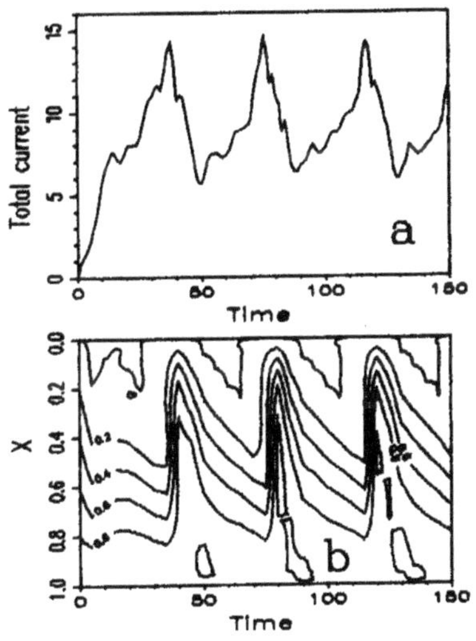

Fig.5

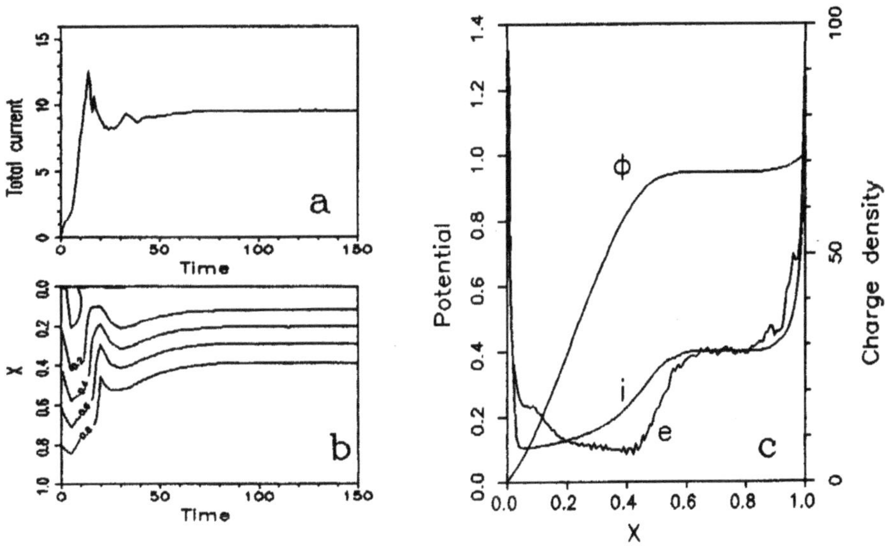

Fig.6

AN EULERIAN VLASOV PLASMA SIMULATION CODE FOR ION PHASE SPACE PHENOMENA IN ION BEAM-PLASMA SYSTEMS

R. Cherkhani, G. Bonhomme, A. Ghizzo, P. Bertrand
Lab P.M.I., U.R.A. 835 du CNRS, Université de Nancy I
BP 239, F-54506 Vandoeuvre Cedex, France

Abstract
A one-dimensional electron eulerian Vlasov simulation code[1] has been modified to model plasmas in which ions are described by the Vlasov equation while electrons are Boltzmann distributed. Electric fields are introduced through the Poisson equation, which is now non linear in the electric potential. The simulation code has a time step of the order of the ion plasma period and allows an accurate study of the plasma dynamics over long times (i.e. several thousands time steps). The simulations are performed using two types of spatial boundary conditions:

- for periodic boundary conditions the results obtained on unstable ion beam-plasma systems exhibit the ion phase vortices formation previously observed with P.I.C. simulation codes[2,3]. Moreover careful comparisons of the Fourier modes in the two approaches have been done. This allows to point out the crucial importance of the choice of the initial perturbation, particularly in the case of linearly unstable systems.

- non periodic boundary conditions may be used as well and allow the study of the long time behaviour of electrostatic shocks induced by a transient perturbation of the ion beam velocity.

In both cases the results are shown to be in close agreement with those obtained with a P.I.C. simulation code. However as the Vlasov code describes better the tail of the ion distribution function and has a low noise level, more complex phase space correlations can be displayed.

1 INTRODUCTION

A number of papers have been devoted during the last years to the non-linear saturation of various instabilities. One proposed mechanism is the formation of vortices in phase-space. P.I.C. simulations have shown [2], [3] the occurrence of such structures in the non-linear saturation regime of the ion beam driven acoustic instability. The need for a low noise level and a very good phase-space resolution to study the long time behaviour of these structures leads us to prefer an eulerian Vlasov simulation code. Thus we have modified a one-dimensional electron Vlasov code [1] to model ion beam-plasma systems.

In this paper the first results obtained with this code are discussed. Namely detailed comparisons between the P.I.C. and the eulerian Vlasov simulation approaches will be reported.

2 THE THEORETICAL MODEL AND SIMULATION METHODS

We consider a one-dimensional collisionless unmagnetized plasma in which ions are described by the Vlasov equation and electrons are assumed to be in a Boltzmann equilibrium. The ion distribution function $F(x,v,t)$ and the normalised self-consistent electrostatic potential $\Phi(x,t)$ are thus described by the following dimensionless Vlasov-Poisson system:

$$\begin{cases} \dfrac{\partial F}{\partial t} + v\dfrac{\partial F}{\partial x} - \dfrac{\partial \Phi}{\partial x}\dfrac{\partial F}{\partial v} = 0 \\ \dfrac{\partial^2 \Phi}{\partial x^2} = -\int F dv + \exp(\Phi / T) \end{cases}$$

Here $T = T_e / T_i$ and $x \in [0, L]$, $v \in (-\infty, +\infty)$ and t are the space, velocity and time variables respectively measured in units of ion Debye length, ion thermal velocity and inverse ion plasma period.

Eulerian Vlasov simulation code: This code is based on the splitting scheme algorithm described by Cheng & Knorr (1976) [4], or Gagné & Shoucri (1977) [5]. It consists of alternately integrating the free streaming term and the acceleration term in Vlasov equation, using cubic spline interpolations.

But, contrary to the problem of electron dynamics in a background of static ions, the Poisson equation is now non linear. To solve it, a regular grid with grid spacing ΔX and N_G grid points is introduced. Thus the Poisson equation can be written:

$\Phi_{j+1} - 2\Phi_j + \Phi_{j-1} = (\exp(\Phi_j / T) - \rho_j)\Delta X^2$, where $\rho(x) = \int F(x,v)dv$ is the ion charge density. The resulting set of equation in matrix form, i.e. $\mathbf{A\Phi} = \mathbf{B} + \mathbf{C}(\mathbf{\Phi})$, can be solved iteratively by the formula $\mathbf{A\Phi}^{(n)} = \mathbf{B} + \mathbf{C}(\mathbf{\Phi}^{(n-1)})$.

The x-v phase-space is discretized using 256×128 mesh points (for L=150).

P.I.C. simulation code: The ion distribution function is discretized using Np=512000 superparticles and the leap-frog scheme is chosen to solve the simulation particle trajectories. The self-consistent electric field is obtained from the Poisson equation solved with the same iterative method as above, with a grid of N_G=256 points. The ion charge density at a mesh point is calculated according to the formula:

$\rho_j(X_j) = \sum_{i=1}^{N_p} \frac{q_i}{\Delta X} W(x_i - X_j)$ where the charge assignment function $W(x)$ is constructed using quadratic spline interpolation.

3 THE NON LINEAR REGIME OF THE ION BEAM-PLASMA INSTABILITY

Periodic boundary conditions are used throughout this study

3.1 Initial data: We consider the following initial distribution function: $f(x,v,t=0)$ =

$$(2\pi)^{-1/2}\left[\alpha (T_{ib} / T_i)^{-1/2} \exp\left(-\frac{(v-v_b)^2}{2T_{ib}/T_i}\right) + (1-\alpha)\exp\left(-\frac{v^2}{2}\right)\right] \times \left(1 + \sum_n a_n \sin(k_n x + \varphi_n)\right) \quad (1)$$

where a_n (resp. φ_n) is the amplitude (resp. phase) of the initial perturbation of the mode with wave vector $k_n = n(2\pi / L)$, and v_b is the beam velocity.

The following plasma parameters are chosen: $T = 20,\ T_{ib} / T_i = 0.05,\ v_b = 6,\ \alpha = 0.1$

For this data set the n=4 mode (four wavelengths in a box of length L=150) is the most linearly unstable, as predicted by the linear dispersion relation (see table 1) [2]. All the simulations are performed with a time step $\Delta t = 0.25$. P.I.C. simulations need no initial perturbation and thus $a_n = 0, \forall n$. On the contrary the intrinsically noiseless Vlasov code needs some $a_n \neq 0$. We will discuss the results obtained with the following data sets:

Vlasov1: $a_1 = a_2 = 10^{-5}, a_3 = 10^{-3}$

Vlasov2: $a_2 = 10^{-5}, \quad a_{n\neq 2} = 0$

Vlasov3: all modes initially excited with amplitudes a_n distributed according to the thermal levels in a maxwellian plasma and randomly distributed φ_n

$a_1 = 0.4928 \times 10^{-8}, a_2 = 0.3543 \times 10^{-7}$
$a_3 = 0.9253 \times 10^{-6}, a_4 = 0.1471 \times 10^{-7}, \ldots$

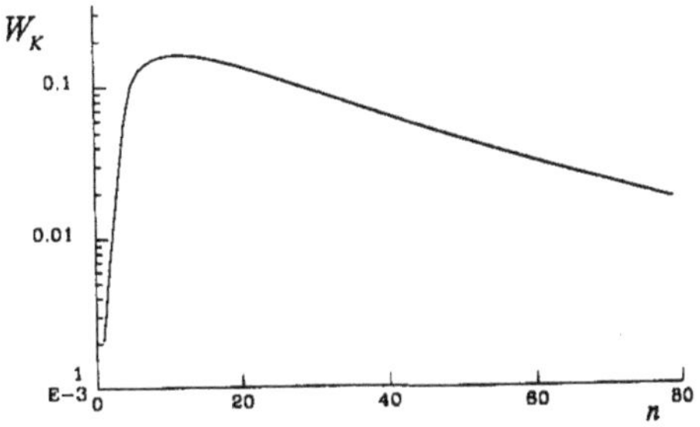

Fig. 1 Noise modelling

These values are obtained from the following expression for the spectral power density as derived using a test-particle method:

$$W_{K,\Omega} \propto \frac{F_i(\Omega/K)+F_e(\Omega/K)}{K^3|\varepsilon(\Omega,K)|^2}$$

The related k-dependence is displayed on the figure 1. The k-spectrum of the fluctuations in a PIC simulation of a maxwellian plasma shows a good agreement with this expression.

3.2 Growth of the ion acoustic instability: In the corresponding simulations we have observed the growth and the non-linear saturation of the ion acoustic instability. The growth rates for the various Fourier modes supported by the box have been obtained from their time evolution (fig. 2) and the main results are summarised in table 1

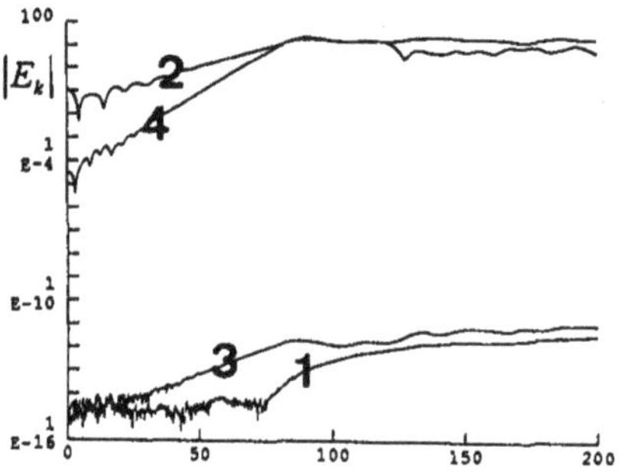

Fig. 2 Time evolution of the first Fourier modes for Vlasov2 simulation

	growth rates				
	γ_2	γ_3	γ_4	γ_5	$\gamma_{n>5}$
linear theory	0.081	0.101	0.103	0.048	≈ 0
P.I.C.	0.079	0.100	**0.102**	0.044	≈0.2*
Vlasov1	0.081	**0.101**	*0.18**	*0.19**	≈0.2*
Vlasov2	**0.078**	0.102	*0.17**		
Vlasov3	0.081	0.101	**0.103**	0.044	≈0.2*

Table 1

* these values correspond to non linear growth rates related to the non linear resonant coupling of the modes actually supported by the system.

In the early linear regime it is seen that in Vlasov2 simulations only the growth rate of the mode n=2 which is initially excited is in agreement with the predictions of the linear theory. Nonetheless, before saturation takes place, the modes not present in the initial perturbation are excited by non-linear resonant coupling processes. The great efficiency of this mode coupling process can be explained by the almost linear ω-k dependence for the ion beam-plasma system, so that the resonance condition is fullfilled for the excited modes.

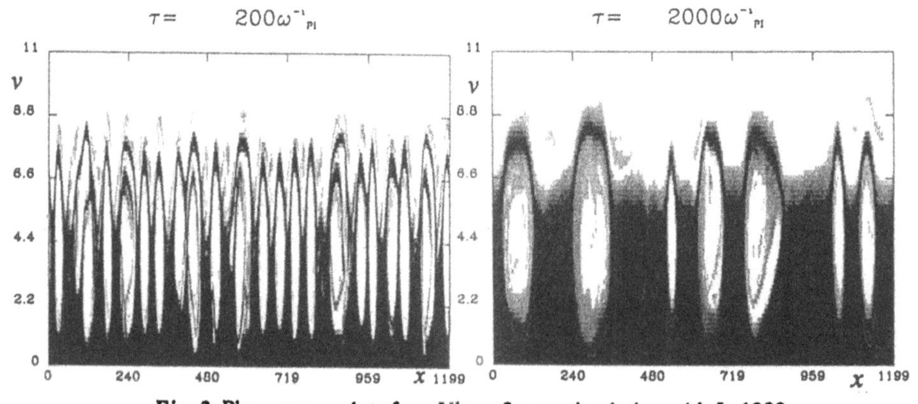

Fig. 3 Phase-space plots for a Vlasov3-type simulation with L=1200

Now the number of vortices observed in the early saturated regime is fixed by the dominant mode. Thus, four vortices are observed in the P.I.C. and in the Vlasov3 simulations, instead of two in the Vlasov2 simulation and three in the Vlasov1 simulation.

Although these results have been obtained using small boxes where the most unstable modes correspond to say n=1,2,3,4,
a clear understanding of the saturation process points to the need for much larger system. The figure 3 shows phase space plots corresponding to Vlasov3-type initial condition where the most unstable mode is n=32.

Finally it appears that the actual evolution of the system depends strongly on the choice of the initial perturbation.

3.3 Non linear evolution after saturation: The excellent phase-space resolution of the Vlasov simulation code enables to study the long time behaviour of the trapped particles.
The following plasma parameters are chosen:
$T = 30, T_{ib} / T_i = 0.05, v_b = 6, \alpha = 0.1$
The only excited mode is the n=1 mode (one wavelength in a box of length L=50) which is the most linearly unstable.

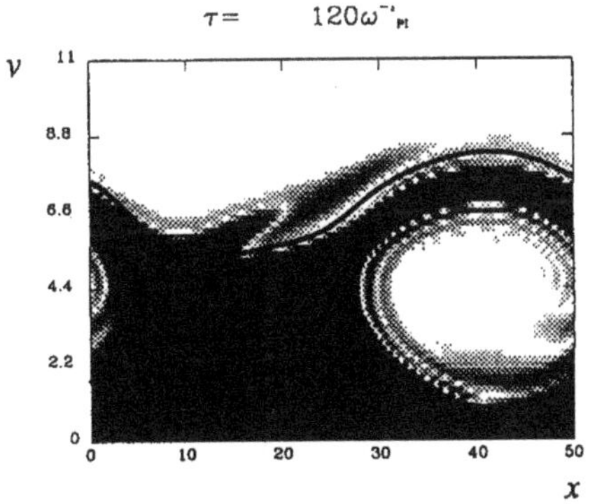

Fig. 4 Phase-space plot showing detrapped particles above the separatrix

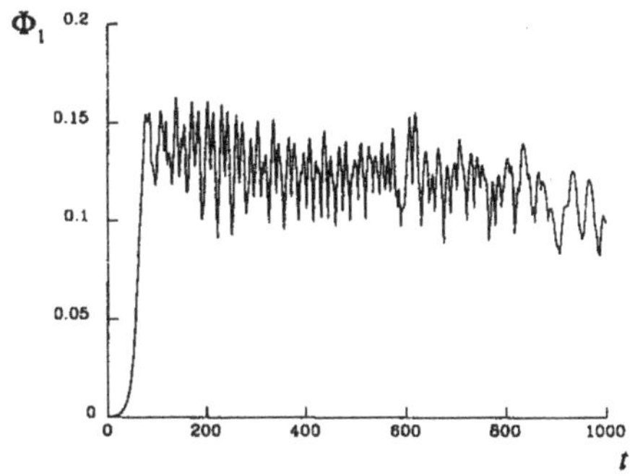

Fig. 5 Time evolution of Φ_1 showing large amplitude side-band oscillations

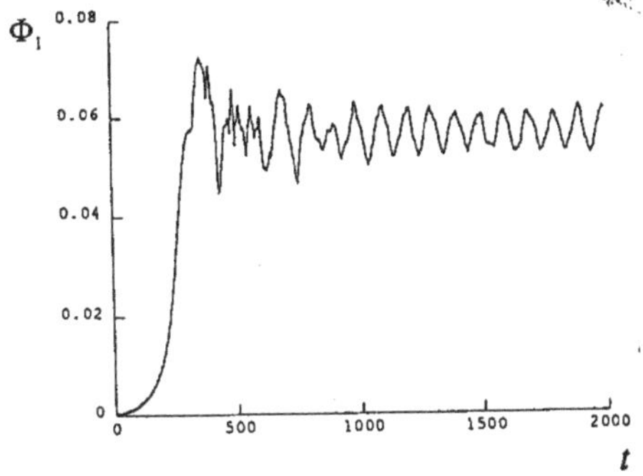

Fig. 6 Time evolution of Φ_1 showing no side-band oscillations

It has been observed (fig. 4 and 5) as previously reported for the electron beam-plasma instability [6] that growing side-bands can destroy the trapped particle oscillations which follow the saturation of the ion acoustic instability. Fig.4 shows the detrapping of particles above the marginally trapped orbit computed from potential data given by the code. The phase velocity of the side-band oscillations (fig.5) is in agreement with the velocity of detrapped particles.On the contrary, fig.6 shows the results of a simulation performed with a lower beam density (α=0.01 and L=101 corresponds to the n=1 most unstable mode). Only trapped particles oscillation can be observed ; indeed no detrapping mechanism is present.

4 THE ION BEAM-PLASMA SYSTEM IN THE PRESENCE OF A TRANSIENT PERTURBATION

For this study non periodic boundary conditions are assumed.

4.1 Initial data: The following plasma parameters are chosen:

$T = 30, T_{ib} / T_i = 0.05, \alpha = 0.1$

We consider again the initial distribution function as given by (1) where the initial Fourier perturbation is replaced by a step-like beam velocity discontinuity.

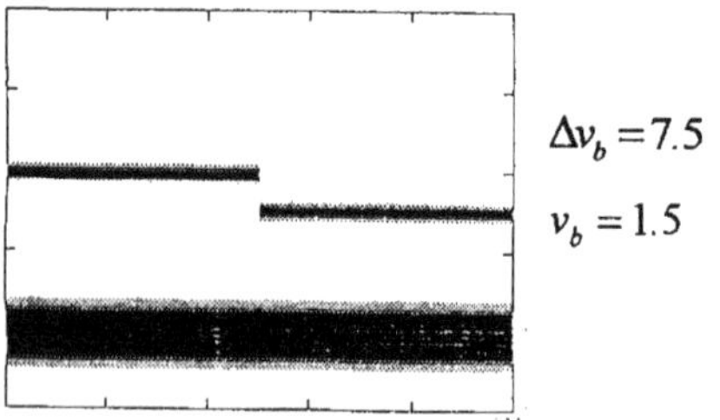

These parameters are close to the experimental situation discussed in ref.2

4.2 Results:

Some preliminary simulations are presented here allowing to recover the previous results discussed in [2]. Moreover the good phase-space resolution has enabled to compare more accurately the simulations with experimental data. As a matter of fact figs. 8 and 9 show the shock structure evolution. The good resolution of the Vlasov code allows to study the associated phase-space phenomena (see fig.10).

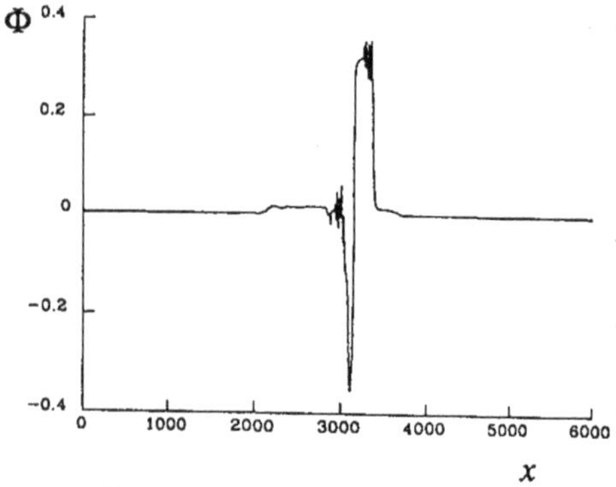

Fig. 8 Electric potential at t=100

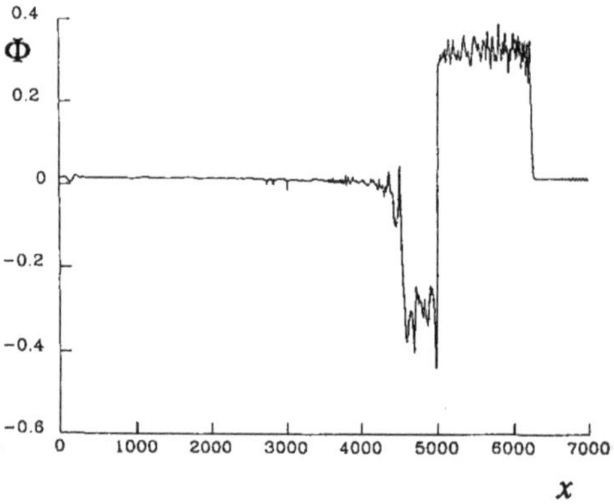

Fig. 9 Electric potential at t=600

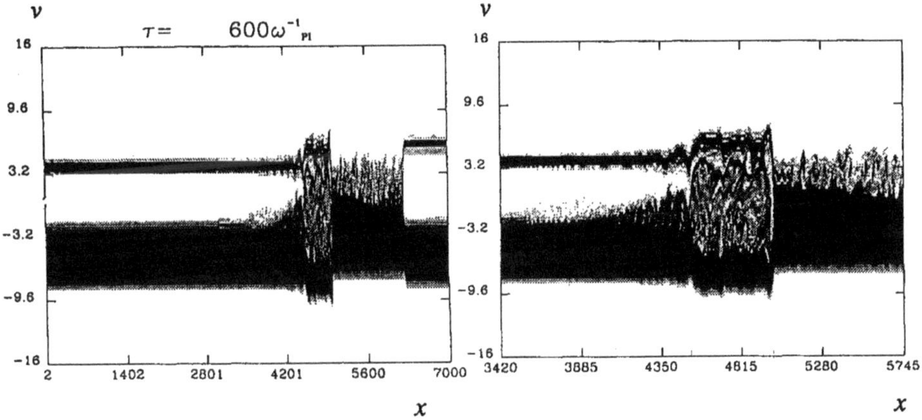

Fig. 10 Phase-space plots at *t=600*

5 CONCLUSION

A one-dimensional eulerian Vlasov simulation code has been developed to model ion beam-plasma systems. The results obtained on unstable ion beam-plasma systems, using periodic boundary conditions, exhibit the ion phase-space vortices formation previously observed in P.I.C. simulations. Moreover careful comparisons of the Fourier modes in the two approaches has been done. This allows to point out the crucial importance of the choice of the initial perturbation, particularly in the case of linearly unstable systems. Indeed, in the case of P.I.C. simulations, unstable modes grow spontaneously from the numerical noise, while Vlasov simulations are noiseless and then require some initial perturbation. When this initial perturbation is distributed among the different modes according to the thermal levels in a maxwellian plasma, the results obtained with Vlasov simulations are in close agreement with those of P.I.C. simulations.

Improvements of this code are also presented and permit to study the long time behaviour of more realistic non periodic ion beam-plasma systems. Again the results agree with P.I.C. simulations.

However the Vlasov code displays more complex phase-space correlations and permit to observe the long time behaviour of non linear phenomena. Preliminary examples are discussed.

References:

[1] A. GHIZZO, *et al.*, Phys. of Fluids, **31**$_1$ (1987) 72
[2] G. BONHOMME, *et al.*, Plasma Physics and Controlled Fusion, **33**$_5$, 507,(1991).
[3] H.L. PECSELI, *et al.*, Phys. Scripta **29** (1984), 241.
[4] C.J. CHENG, G. KNORR, Journal of Computational Physics, **22**, 330, (1976).
[5] R.J. GAGNE, M.M. SHOUCRI, Journal of Computational Physics, **24**, 445, (1977).
[6] M.M. SHOUCRI, Phys. of Fluids, **23**$_{10}$ (1980) 2030

A CONTRIBUTION TO THE CONTROVERSY ON THE EXISTENCE OF COLLISIONLESS SPACE CHARGE DOUBLE LAYERS

R. Schrittwieser, Institute for Ion Physics, University of Innsbruck, Technikerstr. 25, A–6020 Innsbruck, Austria

In this contribution a number of arguments are presented which support the assumption that Double Layers (DL) need *no* collisions to be formed. Recently the claim has been made that for the formation of DL's inelastic collisions are essential, even in very thin plasmas. This was based on the argument that a DL contains energy and that a part of this energy needs to be "stored" in excitation and ionization processes. Though ionization collisions are certainly necessary for the creation of DL's in low pressure discharges, a DL can also form in an absolutely collisionless plasma, since one of the necessities for its formation is simply the existence of two different plasmas which get in touch with each other. In addition, the results of one–dimensional numerical simulations are presented which have been performed by use of the computer code "PDP1". Also in this case collisions are entirely neglected.

1. INTRODUCTION: It is usually assumed that space charge Double Layers (DL) can form under special conditions both in collisional and in collisionless plasmas. Several different types of DL's are known. As a general rule, a DL is found at the boundary between two plasmas of different species, densities, temperatures or potentials. These two plasmas can be produced by various processes and thereby create different types of DL's. In the following an incomplete list of DL–types is given:

A) "Free" DL's[1] in collisional low–pressure discharge plasmas are usually ascribed to electrons which, under the influence of the external electric field, are accelerated and produce new plasma on a higher space potential by collisional ionization of the background neutral gas atoms. The DL will then form between the "old" and the "new" plasma. One prerequisite is that the mean free path for ionization collisions be smaller than the length of the plasma column, but also not too small. Only then the DL is created at a definite position, namely there where the kinetic energy of the electrons just surpasses the ionization potential Φ_i of the atoms.[1-2] This is also the reason why such DL's cannot have a potential drop Φ_d higher than Φ_i. When the applied voltage exceeds Φ_i, multiple DL's can form, each having a potential drop of Φ_i approximately. This process is connected to the formation of stationary or moving striations. When the ionization mean free path is much smaller than the length of the plasma column, a DL cannot form because the collisions are more frequent and randomized.

B) In double–ended Q–machines, DL's of extremely high potential drops up to $e\Phi_d \simeq 2\mathrm{x}10^3\, \kappa T_e$ (where κ is the Boltzmann constant and T_e the electron temperature) are obtained by applying a potential V_d between the two plasma sources.[3-4] On the other hand, it is important to note that also arbitrarily small DL's can easily be produced in such devices with a Φ_d *lower* than Φ_i of the alkali metal, not to mention Φ_i of the background gas.

C) Also in discharge plasmas under certain conditions ultrastrong DL's, with Φ_d greatly exceeding Φ_i of the background gas, could be produced, so for instance by Torvén[5] in a triple plasma device where two Penning discharge plasma sources feed two plasmas of different potentials into a third chamber. The pressure in the third chamber is kept sufficiently low as to prevent ionization collisions.
D) Special forms of DL's are seen in DP–machines, for instance by reflecting a strong ion beam from the source chamber by a positively biased plane probe in the target chamber.[6-9] Also these DL's can have arbitrary heights and can be understood as boundaries between two plasmas since the reflected beam ions form a cloud of plasma on a higher space potential in front of the probe.
E) Small DL's have been found to appear at the boundary between two magnetized plasmas of different T_e and different ion species: a K^+ plasma, formed by contact ionization, and an Ar^+ plasma, produced by a low pressure discharge.[10] The plasmas are produced at the two ends of a vacuum cylinder and *no* current is drawn between the plasma sources.
F) DL's of kilovolts Φ_d have been postulated in the ionosphere[11] and in the auroral plasma of the magnetosphere.[12] These DL's are probably the reason for the appearance of high energetic electrons which precipitate into the ionosphere, there exciting the auroras. This idea has, however, recently been contested by Bryant et al.;[13] but their view has subsequently been strongly disputed by Borovsky[14] and Boström et al.[15] A lively discussion on the existence of DL's in the surroundings of the earth has evolved since; these ideas are well represented in the present proceedings.

2. THEORIES OF DOUBLE LAYERS: The number of theoretical papers on DL's is abundant. Nevertheless, no full analytical theory has been developed as yet, obviously because of the fact the DL's are found solely in *bounded* systems, and the theoretical treatment of bounded plasma systems is still at its beginning. Note that also the auroral acceleration region, where type–F DL's appear, is necessarily a bounded system.[15] An excellent review of some theoretical aspects of DL's has recently been published by Raadu.[16]

Despite the absence of a comprehensive theory, at least two arguments for the existence of collisionless DL's can yet be found in the existing theories:

- A DL can be described as a BGK solution[17] of the Vlasov equation which *neglects collisions entirely*.
- According to Hershkowitz[18] and Chen,[19] a DL can also be understood as an ion–acoustic shock wave which is Doppler shifted down to zero velocity in the laboratory frame by the plasma flow. Ion acoustic waves are a typical phenomenon of a *collisionless* plasma.

It should be emphasized that from all theoretical investigations, which do not explicitely take into account collisions, the existence of space charge DL's can be derived *despite* the neglect of Coulomb collisions and of inelastic collisions like exitation and ionization processes. These theories do of course not *contradict* the existence of collisionally created DL's, like those described above as type–A DL's, but they make clear that also *collisionless* DL's can exist.

On the other hand, the existence of purely collisionless DL's has recently been disputed by Sanduloviciu[20] and Sanduloviciu and Popa[21] based on the argument that a DL contains energy and that part of this energy has to be – and can only be – "stored" in excitation and ionization processes, which constantly take place in DL's due to the particle acceleration by the electric field. The authors argue that the cross section for inelastic collisions changes across the DL (which is of course true since the particle velocity changes across the DL) and that therefore the creation of DL's *as such* is caused by inelastic collisions.

As mentioned above, inelastic collisions are certainly necessary for the creation of type–A DL's and might occur, depending on the density, in other types of DL's too. But according to Ref. 20 and 21, even in very thin plasmas, where the mean free paths for all types of collisions are many times larger than the width of the DL and usually even larger than the dimensions of the entire plasma volume, inelastic collisions should be essential for the DL formation.

However, in addition to this discrepancy, we point out the following obvious contradiction: were the production of DL's in any way caused by or connected to excitation or ionization processes with neutrals of the background gas and/or the atoms forming the plasma ions, then we would expect the height of any DL, even of those which we understand as purely collisionless (from type–B to type–E DL's, but probably also type–F DL's), to be correlated to either the excitation levels of the specific atom and/or ion or the ionization potential of the atom.

Such a correlation could, for instance, be one of the following effects:

- No DL should be observed below the lowest exitation level of the atom and/or ion.
- DL's would form with a special preference at ecitation levels of the atom and/or ion.
- As mentioned above, no DL could exists with a voltage drop higher than the ionization potential of the atom.

Such a correlation has, however, never been found, but experiments for instance in double–ended Q–machines (type–B DL's) and triple plasma devices (type–C DL's) made it clear that in these cases Φ_d solely depends on the voltage V_d between the two plasma sources. I.e., both, DL's with a arbitrarily small and arbitrarily high Φ_d could easily be achieved.

The reason is, to our understanding, very simple: in three of the above listed cases (B, C and E) the two different plasmas are produced by two independent plasma sources, and the DL simply forms as a natural boundary between the two plasmas. Even for type–D DL no inelastic collisions need be invoked to explain its formation since it appears between the background plasma and a small plasma population which is created locally. In none of these cases (B – E) collisional interactions are needed to produce another type of plasma to form the DL.

3. SIMULATIONS OF DOUBLE LAYERS: By use of the numerical code PDP1[22] a brief investigation of numerical DL's has been performed, showing very clearly that no collisions of whatsoever type are necessary for their formation. The code has been developed at the University of California at Berkeley for the purpose to simulate the simplest case of a one–dimensional bounded plasma system, namely a one–emitter or two– emitter plasma diode. This system consists of two plane electrodes, facing each other, which both can emit plasma. An arbitrary voltage V_d can be applied between the two plates. Thus the conditions are very similar to those of a double–ended Q–machine and type–B DL's are to be expected. Although collisions of various kinds can be taken into account in the code, for the present example all collisions have of course been ignored.

In the following, one example of a numerical DL's is shown (Figure 1). In this case the two plates both emit plasma of the same density, temperature and ion species. The areas of both plates are 1 cm^2 each. The voltage of the LHS plate with respect to the one on the RHS is $V_d = +10$ V, the length of the system is $L = 2$ cm, the ratio between ion and electron mass is 500. Both plates emit an ion and electron current with a current density of $j_i = 0.115$ A/m^2 and a thermal speed of $v_{ith} = 8.9\times10^3$ m/sec, and $j_e = 2.56$ A/m^2 and a thermal speed of $v_{eth} = 2\times10^5$ m/sec, respectively. These thermal velocities correspond to a plasma source temperature of 0.2 eV and the conditions are electron–rich. Though PDP1 is able to take into account collisions of various kinds, during these investigations *all collisions have been ignored.* The system starts empty and needs about 4×10^{-7} sec to be filled with plasma. The figures on the axes are in SI–units. The two topmost inserts both show the particle phase space in two different ranges of the velocity v_x at the time $t = 3.32\times10^{-5}$ sec. Thus the upper insert shows mainly the ion phase space and the lower one the electron phase space. The two lower inserts show on the LHS the axial profile of the space potential at $t = 3.32\times10^{-5}$ sec and one the RHS the temporal evolution of the current through the system from $t = 0$ to $t = 3.3\times10^{-5}$ sec.

The numerical results clearly show the formation of a stable one–dimensional DL between the two plasma emitters. The width of the DL is 0.53 cm, the potential drop is $\Phi_d = 4.68$ V, thus fairly close to the applied potential of $V_d = 5$ V. These results agree qualitatively with earlier investigations in double–ended Q–machines.[7-8,23] The DL between the two plates is subject to strong, however quite irregular oscillations, whose frequency seems to peak around 130 MHz.

For higher values of V_d similar results are obtained, the amplitude of the oscillations, however, being generally smaller. In some cases the oscillations tend to be more regular. Further investigations will show whether these oscillations can evolve into a real chaotic behaviour, and if yes, under which conditions. Once again we emphasize that in these numerical investigations no collisions were taken into account. They can therefore be considered as another proof for the fact that *collisionless DL's are possible.*

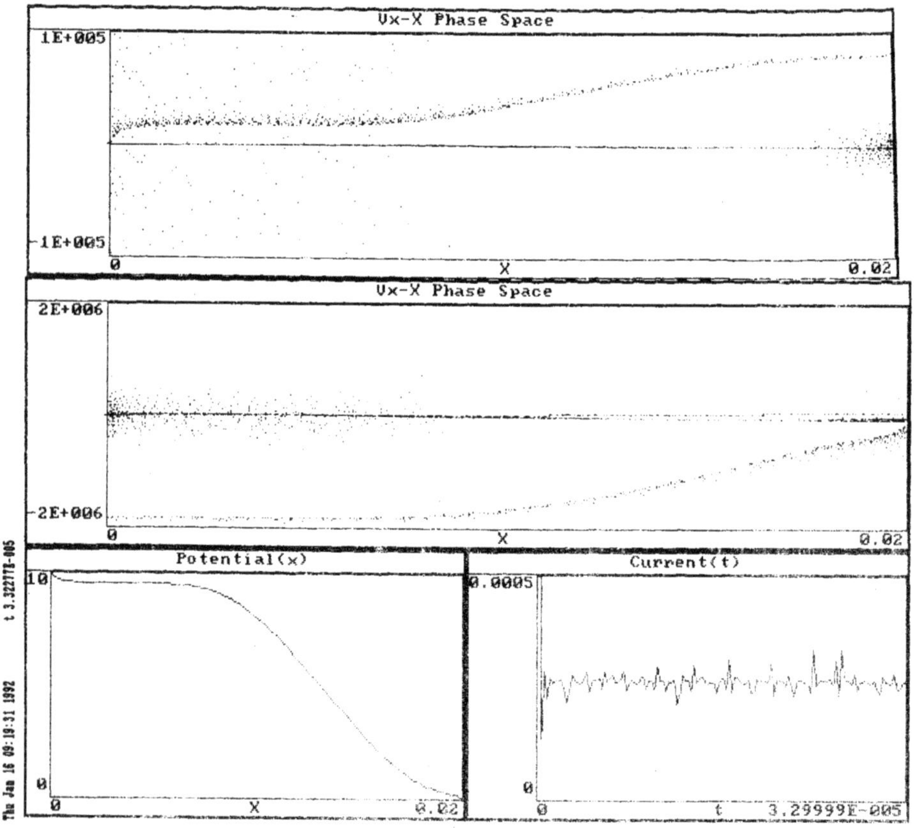

Figure 1: Numerical DL shown at $t = 1.88125\times10^{-6}$ sec after start. The two topmost inserts show the phase space with the horizontal axis as the coordinate x along the system in m: (a) in the interval $10^5 \leq v_x \leq -10^5$ m/sec, thus mainly the ions are shown; (b) in the interval $2\times10^6 \leq v_x \leq -2\times10^6$ m/sec, thus mainly the electrons are shown; (c) LHS lower figure shows the axial profile of the potential in volts at this time; (d) RHS lower figure shows the temporal evolution of the current through the system in amperes from the start to the time t.

Also in an experimental investigation[23] oscillations of the DL have been observed. But these were usually more regular and had a frequency given by the ion transit time through the system. Nevertheless, under certain conditions, either close to the on–set or far beyond it, also in the experiment the oscillations tended to show a hysteretic, intermittent or even chaotic behaviour. This will be the subject of further thorough investigations.

Acknowledgements

The author would like to thank C.K. Birdsall, S. Kuhn, G. Popa and M. Sanduloviciu for enlighting discussions. This work has been supported by the Fonds zur Förderung der wissenschaftlichen Forschung (Austria) under grant No. P8709–PHY.

References:

1. H.S. Maciel and J.E. Allen, *J. Plasma Phys.* **42** (1989), 321; see also J.E. Allen and H.S. Maciel, *these Proceedings.*
2. D. Anderson, *J. Phys. D: Appl. Phys.* **14** (1981), 1403.
3. N. Sato, R. Hatakeyama, S. Iizuka, T. Mieno, K. Saeki, J.J. Rasmussen and P. Michelsen, *Phys. Rev. Lett.* **46** (1980), 1330.
4. N. Sato, R. Hatakeyama, S. Iizuka, T. Mieno, K. Saeki, J.J. Rasmussen, P. Michelsen and R. Schrittwieser, *J. Phys. Soc. Japan* **52** (1983), 875.
5. S. Torvén, *J. Phys. D: Appl. Phys.* **15** (1982), 1943.
6. W. Weber, R.J. Armstrong and J. Trulsen, *J. Appl. Phys.* **50** (1979), 4545.
7. R.L. Stenzel, M. Ooyama and Y. Nakamura, *Phys. Fluids* **24** (1981), 708.
8. Å. Skøelv, R.J. Armstrong and J. Trulsen, *Phys. Fluids* **27** (1984), 2744.
9. R. Armstrong and R. Schrittwieser, *Plasma Phys. Contr. Fus.* **33** (1991), 1407.
10. R. Hatakeyama, Y. Suzuki and N. Sato, *Phys. Rev. Lett.* **50** (1983), 1203.
11. L.P. Block, *Cosmic Electrodyn.* **3** (1972), 349.
12. M. Temerin, K. Cerny, W. Lotko and F.S. Mozer, *Phys. Rev. Lett.* **48** (1982), 1175.
13. D.A. Bryant, R. Bingham and U. de Angelis, *Phys. Rev. Lett.* **68** (1992), 37; see also *these Proceedings.*
14. J.E. Borovsky, *Phys. Rev. Lett.* **69** (1992), 1054.
15. R. Boström, P.–O. Dovner, A.I. Erikson and G. Holmgren, *these Proceedings.*
16. M.A. Raadu, *Phys. Rep.* **178** (1989), 25; see also *these Proceedings.*
17. I. Bernstein, J.M. Green and M.D. Kruskal, *Phys. Rev.* **108** (1957), 546.
18. N. Hershkowitz, *J. Geophys. Rev.* **86** (1981), 3307.
19. F.F. Chen, *Plasma Physics and Controlled Fusion, Vol. 1: Plasma Physics* (Plenum Press, New York and London), 1984, p. 305.
20. M. Sanduloviciu, *Analele Ştiinţifice ale Universităţii "Al. I. Cuza" din Iaşi* **XXXI** (1985), 5; *Plasma Phys. Contr. Fus.* **28** (1986), 585.
21. M. Sanduloviciu and G. Popa, *Proc. of the Third Symposium on Plasma Double Layers* (Bucharest, Romania, 1987), *Analele Ştiinţifice ale Universităţii "Al. I. Cuza" din Iaşi* **XXXIV** (1988), 83.
22. J.P. Verboncoer, V. Vahedi and M.V. Alves, PDP1, PDC1, PDS1 Plasma device 1 dimensional bounded electrostatic codes, *Reference Manual*, Version 2.0, Univ. of California, Berkeley, 1990.
23. S. Iizuka, P. Michelsen, J. Juul Rasmussen, R. Schrittwieser, R. Hatakeyama, K. Saeki and N. Sato, *J. Phys. Soc. Japan* **54** (1985), 2516.

CONTRIBUTED PAPERS

SPACE

VIKING OBSERVATIONS OF WEAK DOUBLE LAYERS AND SOLITARY STRUCTURES IN THE AURORAL ACCELERATION REGION: LATEST RESULTS ON WDL POLARITIES

Anssi Mälkki, Finnish Meteorological Institute, Department of Geophysics, P.O.Box 503, SF-00101 Helsinki, Finland

Rolf Boström and Gunnar Holmgren, Swedish Institute of Space Physics, Uppsala Division, S-75591 Uppsala, Sweden

Abstract. The latest results of the analysis of Viking Weak Double Layer (WDL) data are presented. The results show that the potentials of the structures are very weak resulting in a net potential drop at most of the order of 0.5 keV over 5000 km along the auroral field lines. The potential drop of a single structure is shown to be consistent with asymmetric reflection of passing electrons from the potential hole.

1. INTRODUCTION: The Weak Double Layers (WDLs) and Solitary Structures (SSs) observed with the Swedish Viking satellite[1] initiated a discussion of their possible role for auroral particle acceleration. The initial analysis showed that when the structures had a net potential drop over the structure, it was up to 1 volt and the polarity was predominantly such that it pointed upwards along the geomagnetic field lines. Consequently, it was suggested that several thousands of WDLs could build up a potential needed for accelerating observed keV regime field-aligned beams of upward flowing ions, and downward flowing auroral electrons.

Already before the Viking observations, similar kinds of structures were observed by the U.S. S3–3 satellite[2], but from those observations it was not possible to definitely determine the polarity of the potential. With the instrumentation of S3–3 it was possible to conclude that either the structures had a negative potential and were propagating upwards, or positive potential and propagating downwards. In those days much work was done on solitary structures in current-driven instabilities (for a review, see Ref. 3.). The authors of Ref. 3. concluded that the ion-hole instability[4,5] is in best agreement with observations both in its assumptions and predicitions. Later, Tetreault[6,7] extended previous work on ion-hole instabilities to a three-dimensional model and also showed that this model agrees well with known observational properties of the structures.

In this work we present our latest results of the analysis of the Viking low-frequency wave data in search for solitary structures. We have carried out more detailed analyses on the electrical polarities of the structures and extended the initial analysis to new regions of the auroral oval.

2. DATA ANALYSIS: The Viking spacecraft was equipped with four 40 m long radial wire booms extended perpendicular to each other within the orbital plane and also within the magnetic meridian plane. Each boom carried a spherical Langmuir probe. Signals from two channels could be simultaneously digitized (853.3 samples/s) and telemetred to ground.

Two of the probes could in one of the operational modes be used for measurements of $\Delta n/n$ simultaneously at two points in space separated by 80 m. From these measurements, we can deduce the propagation velocities of the structures. Solitary density depletions of up to 50 % of the background plasma density, are recorded first by the lower of the two probes, then by the upper one. The structures evidently propagate upwards with speeds varying from 5 km/s up to the limit of resolution of the instrument (approximately 68 km/s), having a mean value of 18 km/s. A typical characteristic growth/decay time of the structures is estimated to be of the order of 10 ms.[8]

In another operational mode, these two probes could be used for measurements of fluctuations of the plasma potential difference $\phi_1 - \phi_2$ between the two probes separated by 80 m. The other pair of probes was only used for measurements of fluctuations of the potential difference $\phi_3 - \phi_4$ over 80 m in a direction perpendicular to the first one. When one of the first pair of probes was measuring density, and the other pair potential difference, we can combine the two different signals and make correlations between the potential and particle density of the structures. In the polarity analysis, which will be described in more detail in section 3, we have used data from both this mode and from the third operational mode with both pairs of probes measuring potential difference.

At the moment, we have some 50 analysed orbits. The analysis has been done by visual inspection of the time-series data on a terminal screen, as we have not succeeded in developing a reliable pattern recognition program for computerized analysis.

We have found solitary structures at an altitude range from 2800 km up to the apogee of the spacecraft at 13500 km. In Figure 1, we see normalized occurrence frequency of the structures versus satellite altitude for the analysed orbits. The normalization reduces the measure for the highest altitude bins, as the satellite spends relatively more time closer to its apogee. We see that the most typical altitude to meet SS/WDLs is in the middle of the range of the typically accepted acceleration region, i.e. between 7000 and 10000 km. The gaps at 8500 to 9000 km and 11000 to 12000 altitude are most probably due to a too small dataset, as we have not found any physical reason for expecting such a phenomenon. In the lowest altitude bins we have not found many structures, but we also have very little data from the lowest altitudes due to the spacecraft orbit.

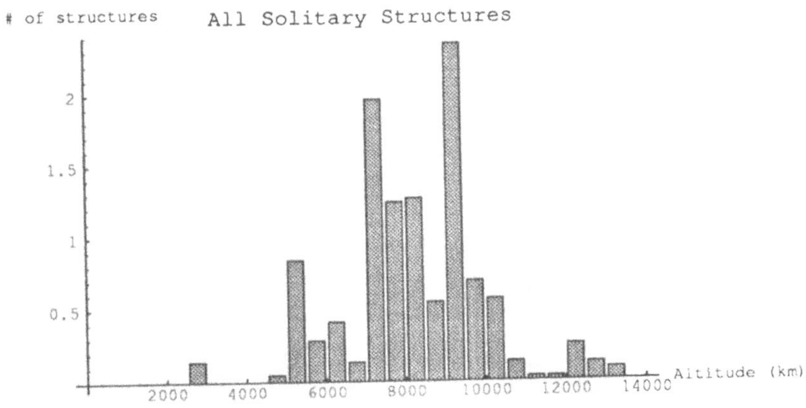

Figure 1. Number of WDLs found (arbitrary units) plotted versus altitude. The number is normalized to correspond to the probability of finding WDL/SSs in an altitude bin by multiplying the measured occurrence for each bin with a factor proportional to 1/(time spent in the altitude bin).

The WDL/SSs are typically seen in regions of upward flowing ion beams (UFI), as reported earlier[9]. There are a few exceptions to this, but their number is not representative. The correlation is such that when we see these structures, we are in an ion beam region with larger than 90 % probability, but not vice versa: we have seen several occasions of UFI beams without WDL/SSs.

3. RESULTS OF POLARITY ANALYSIS: Some statistics on the polarity of the potential structures is shown in Figure 2. The polarities of the structures are determined from the time series for the electric field by time integrating the signal and converting the time integral to the corresponding space integral along a magnetic field line by assuming a constant velocity of 15 km/s along the field line for each structure. The mean value of the measured velocities (from the measurements in the mode with two probes measuring particle density) is 18 km/s with the distribution of the structure velocities peaking between 10 and 20 km/s; we consider 15 km/s a suitable value.

As a measure of particle energy in Figure 2 we have used the sum of ion beam bulk drift energy and the maximum energy of the trapped electron population. When this sum is greater than 2 keV, the case is classified to high energy, and a sum smaller than or equal to 2 keV gives low energy classification.

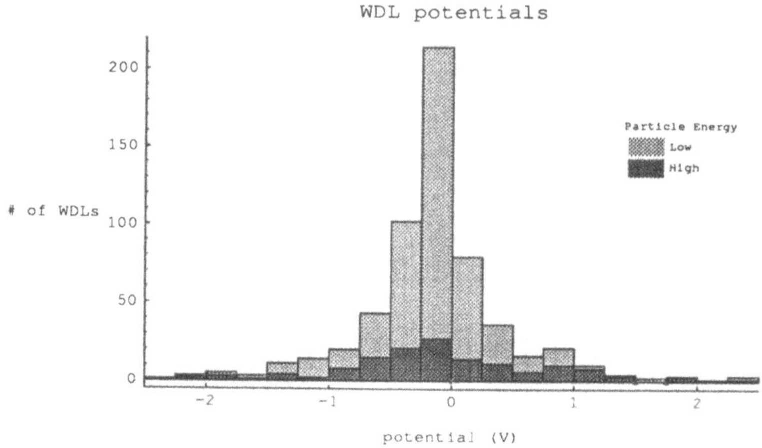

Figure 2. WDL net potentials on analysed orbits. Sign convention is as follows: structures having net electric field pointing upwards in the Earth's frame of reference (i.e., opposite to the geomagnetic field direction) have negative sign. For description of net potential determination from time series and particle energy classification see text.

We see that the polarity histogram peaks strongly on the bin containing structures with polarities between −0.25 V and 0 V. Of a total of 616 structures in this statistics, about two thirds (419) have a net potential pointing upwards. With high particle energy the distribution is broader, but also has a smaller mean value for the potential of −0.06 V, whereas for the narrower low energy distribution the mean potential is −0.13 V. However, both numbers are small, and structures with small or negligible net potential drops dominate, with an overall mean potential of −0.09 V.

When we see WDL/SSs, they are intermittently distributed. I.e., we may see some tens of these structures during an interval of some seconds, sometimes only a couple of them during a fraction of a second, then nothing for some seconds and so on. Knowing the spacecraft velocity to be about 5 km/s and practically perpendicular to the geomagnetic field, we may estimate that these structures appear in localized regions with a perpendicular (to **B**) extension of the order of at most a couple of kilometers. When seen, the "packing density" of the structures along a field line is estimated to be typically 10 structures/s assuming an isotropic packing (taking a safe upper limit). With a mean speed of 18 km/s and an average net potential of 0.2 V (again an upper limit) this would give a net magnetic field-aligned electric field of 0.1 mV/m. Thus for a filament of length 5000 km we get a net potential drop of only 0.5 kV along the field line, but we must note that our estimates are relatively crude.

Assuming that we already have a potential hole, the net potential drop over a single structure can be explained by the asymmetric reflection of electrons from the structure. Electrons with velocities within velocity range $\frac{1}{2}m(v - v_{\text{hole}})^2 < e|\phi_{\text{min}}|$ are reflected from a negative potential hole with potential $\phi_{\text{min}} < 0$.

Using the approach of Dupree[4] we can calculate the potential drop over a single structure from

$$\lambda^{-2}\Delta\phi = \frac{n_e e}{\epsilon_0}\int dv\,(f_r - f_0) \tag{1}$$

where $\Delta\phi$ is the net potential drop over the structure, f_0 is the unperturbed velocity distribtion, f_r is the distribution function of the reflected particles, and

$$\lambda^{-2} = -\sum_\alpha \omega_{p\alpha}^2 \int_{\text{passing}} \frac{dv\ \partial f_{0\alpha}/\partial v}{v-u}, \tag{2}$$

is a shielding length. The sum over α in Eq. (2) runs over the particles which are not reflected and u is the drift velocity of each particle population (subscript α omitted). Assuming equal temperatures for electrons and ions ($T_e \approx T_i$) the shielding length is of the order of the Debye length λ_D.

Setting in numerical values $T_e = T_i = 10$ eV, $\phi_{min} = -2$ V, $j = 10^{-7}\text{A/m}^2$ gives for the electron drift velocity $v_{de} \approx 0.33\ v_{te}$ (v_{te} being the electron thermal velocity) and reflection boundary $\Delta v_e \approx 0.45\ v_{te}$. Using these values in integral (1), assuming only electron reflection and the unperturbed distributions to be drifting Maxwellians for ions and electrons, we get a net potential drop of 0.2 V, which is in very good agreement with the observed value. Taking into account only electron reflection is a reasonable assumption as the electrons are strongly bound to the magnetic field lines, whereas for the ions the perpendicular motion must also be taken into account.

4. CONCLUSION: We have extended the initial analysis of the data to new regions in the auroral acceleration region. WDL/SSs are found in all MLT sectors between 60 and 90 degrees invariant latitude and 2800 to 13500 km altitude (limited by the Viking orbit). Most structures are found between 7000 and 10000 km altitude. Our new data shows solitary structure characteristics in agreement with the initial analysis.

We have made new analyses on the polarities of the electric field of the structures. We find that although numerous cases of reversed polarity (downward pointing electric fields) are inferred, the cases with upward pointing fields dominate when moderate energy ion beams are present. Under other conditions, for instance, with higher particle energy, upward and downward fields are equally frequent. We must note that false net potentials may be inferred in the presence of perpendicular electric fields when the probes are separated perpendicular to the magnetic field, even if we have excluded observations with too large perpendicular separation of the probes. Such false net potentials should show both polarities of equal frequency. We also note that structures with very small or negligible net potential drop clearly outnumber those with clear upwards or downwards accelerating potential drops.

The small net potential drops over the solitary structures can be explained using a simple one-dimensional calculation of asymmetric reflection of drifting electrons. The resulting numerical value agrees very well with observational results. We do not know how the initial disturbance in the potential is created, but, for instance, for the nonlinear ion hole instability[4] thermal fluctuations are sufficient. Once created, a solitary wave can then develop to a weak double layer, which locally accelerates passing particles.

The structures appear only intermittently in localized regions, while the ion beams may have a spatial extension of several hundreds of kilometers at the same altitudes. Inside ion beam regions, however, the E-field and $\Delta n/n$ data signals are often quite disturbed and thus identification of SS/WDLs is impossible, even if they exist. Consequently, inside the ion beams we may have more double layers which we cannot identify in the data. Still the resulting net potential drop caused by the WDLs is hardly more than of the order of one kilovolt. It is thus likely that energetic ion beams with energies up to several keV have been accelerated, not only by the weak double layers, but also by some other means.

ACKNOWLEDGEMENTS: The authors thank L. Eliasson for providing particle data and A.I. Eriksson, B. Holback, A. Roux and H. Koskinen for useful discussions.

REFERENCES:

1. R. Boström, G. Gustafsson, B. Holback, G. Holmgren, H. Koskinen, and P. Kintner, *Phys.Rev.Lett.* **61** (1988), 82.
2. M. Temerin, K. Cerny, W. Lotko, and F. Mozer, *Phys.Rev.Lett.* **48** (1982), 1175.
3. A. Mälkki, H. Koskinen, R. Boström, and B. Holback, *Physica Scripta* **39** (1989), 787.
4. T.H. Dupree, *Phys.Fluids* **25** (1982), 277.
5. R.H. Berman, D.J. Tetreault and T.H. Dupree, *Phys.Fluids* **28** (1985), 155.
6. D.J. Tetreault, *Geophys.Res.Lett* **15** (1988), 164.
7. D.J. Tetreault, *J.Geophys.Res.* **96** (1991), 3549.
8. R. Boström, P.-O. Dovner, A.I. Eriksson, and A. Mälkki, *these proceedings* (1992).
9. H. Koskinen, R. Lundin, and B. Holback, *J.Geophys.Res.* **95** (1990), 5921.

Pulsar Core Emission and Double Layers

Martin Volwerk and Jan Kuijpers
Sterrekundig Instituut Utrecht

Radiation from relativistic double layers

The large scale electric field in a relativistic Double Layer (DL) accelerates charged particles. These particles will emit radiation by the process of linear acceleration emission, as first proposed by Melrose [3]. This mechanism is comparable to that of synchrotron emission but it has the disadvantage of being independent of the particles Lorentz-factor.

We show (Volwerk and Kuijpers [5]) that in the case of a relativistic DL a maser process can enhance the emitted radiation and produce high brighness temperatures, which are needed if we want to apply this mechanism to a pulsar radiation mechanism.

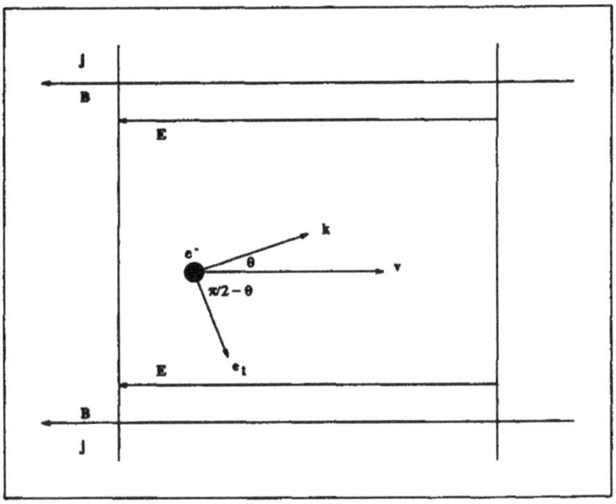

Figure 1: the emission topology in a relativistic DL

In Fig. 1 we show the topology of the radiation emitting DL. The radiation is emitted in a cone around the direction of the velocity of the electron (relativistic beaming) with a half angle:

where γ is the Lorentz-factor of the emitting particle. The emitted power is calculated by:

$$U(\vec{k}) = 4\pi R_E(\vec{k})|\hat{e}^* \cdot \vec{j}(\vec{k})|^2, \tag{2}$$

with $\vec{j}$ the current carried by the emitting particle, R_E the ratio of electric to total energy of the emittind wave and $\hat{e}$ a unit vector in the direction of the electric field of the emitted wave.

Amplified linear acceleration emission in a DL

In order to find the emitted power we calculate the current carried by one particle inside the DL, using a Fourier transformation to get the result in frequency space. Because of the very strong magnetic field at the polar cap of a pulsar ($\sim 10^12$ Gauss) the particles are in the lowes Landau orbit, which means we can assume a one dimensional particle distribution function (the case of a 3D distribution function is also considered in the paper by Volwerk and Kuijpers [5]):

$$f(\vec{p}) = f(p_x)\delta(p_y)\delta(p_z). \tag{3}$$

The emitted energy U combined with the particle distribution function and the transition time over the DL $t_{\rm tr}$ can define an absorption (μ) and an emission (α) coefficient:

$$\alpha = \int_0^\infty f\, U\, t_{\rm tr}^{-1}\, d\epsilon, \tag{4}$$

$$\mu = \text{constant} \int_0^\infty \frac{\partial f}{\partial\, \epsilon}\, U\, t_{\rm tr}^{-1}\, d\epsilon. \tag{5}$$

α and μ can be used to form an equation fo radiative transport:

$$\frac{dI}{ds} = \alpha - \mu\, I, \text{boundary condition}: I(s = 0) = 0. \tag{6}$$

The solution to this differential equation is just exponential variation in the case that α and μ are independent of s, which is the case in our paper. To obtain enhanced emission we need a inverted population of electrons th enter the DL. We therefor assume that the distribution function can be written as a power law:

A pulsar polar cap model

The combination of hte observational results by Rankin [4] and the theoretical results of Beskin et al. [1] seems to be very fruitful to provide for a model of pulsar radio emission in the core radiation.

In Fig. 2 we show the topology at the polar cap of a pulsar. Rankins observational results showed that the core emission from pulsars had to be created very near the stellar surface to be in agreement with the fact that the opening angle of the core emission can be written as the simple relation:

$$W \approx 2.5^\circ P^{-0.5} \tag{8}$$

as was found experimentally and can also be found theoretically assuming small enission height.

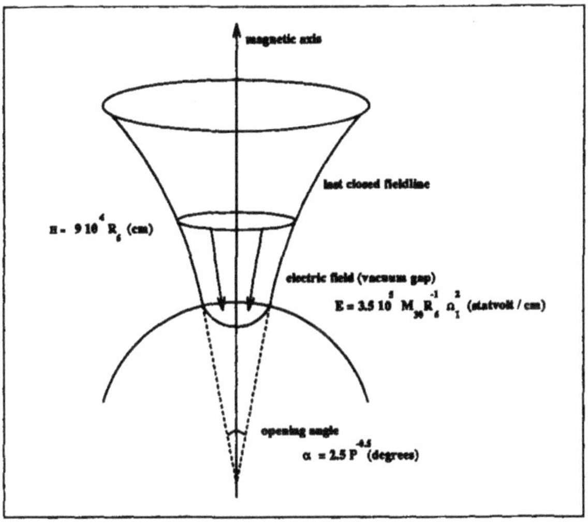

Figure 2: The geometry at the polar cap of a pulsar.

Beskin et al. presented a pulsar magnetosphere model in which they showed that at the polar cap a Langmuir DL exists. The DL has the following properties:

$$\text{height}: \quad H \approx 9.0 \cdot \beta^{0.5} R_6 \text{ cm}, \tag{10}$$

where M_{30} is the magnetic moment of the pulsar in units of 10^30 Gauss cm^3, Ω_1 is the angular velocity in units of 10 s^{-1}, R_6 is the radius of the pulsar in units of 10^6 cm and β is a small number (put equal to 0.01).

Results

We have appied our calculations t a standard pulsar with the following parameters:

$$R = R_6 \, 10^6 \text{ cm} \tag{11}$$

$$\Omega = \Omega_1 \, 10\text{s}^{-1} \tag{12}$$

$$M = M_{30} \, 10^{30} \text{ G cm}^3 \tag{13}$$

The emission takels place at a frequency which is given by:

$$\omega = \gamma_0 k_{DL} c, \text{ with } k_{DL} = \frac{2\pi}{L_{DL}}, \tag{14}$$

with a bandwidth of:

$$\frac{\Delta\omega}{\omega} \approx \frac{1}{2} \frac{\ln 2}{\mu s}, \tag{15}$$

which we see is inversely proportional to the absorption coefficient.

The brightness temperature of the emitted radaitin is found by using the intensity and making an assumption of the area of the polar cap. Making a Raleigh-Taylor apporximation of the Planck-function we find:

$$T_b = \frac{2I\pi^2 c^2}{k_B \omega^2}. \tag{16}$$

For our pulsar model this results in:

$$10^{21}(\text{K}) \leq T_b \leq 10^{32}(\text{K}) \tag{17}$$

$$7 \leq \mathcal{C} \leq 15 \tag{18}$$

$$\mathcal{C} = \beta_{-2}^3 \; M_{30} \; \Omega_1^5 R_6^2 \gamma_{0,2}^{-4} \tag{19}$$

while observed is:

Discussion

The theoretical model we present here describes the radiation emitted by relativistic DLs. This model is a relativistic extension of the model presented by Kijpers [2].

The one dimensional calculation we have made for the pulsar core radiation is one of the possible applications of this model, but also complete three dimensional calculations can be performed, though this complicates matters substantially.

References

[1] Beskin,V.S., Gurevich, A.V. and Istomin, Ya.N. 1983, Sov. Pys. JETP, 58, 235.

[2] Kuijpers, J. (1990) in: Plasma Phenomena in the Solar Atmosphere, eds. M.A. Dubois, F. Bély-Dubau and D. Grésillon, les Edition de Physique, Les Ulis, France.

[3] Melrose, D.B. 1978, Ap. J., 225, 557.

[4] Rankin, J.M. 1990, Ap. J., 352, 247.

[5] Volwerk, M. and Kuijpers, J. 1992, in preparation for A&A.

CONTRIBUTED PAPERS

OTHERS

DOUBLE LAYERS OF CELL MEMBRANE AND DISCHARGE PLASMA

N. Tanizuka
Integrated Arts & Sciences, University of Osaka Pref., Gakuen-cho 1-1, Sakai, 593 Japan

The paper deals with a problem of double layers and related noise from an analogy of the function of cell membrane and by a consideration of experimental results of DC magnetron discharge which produces noise. A proposition is given that double layers must be considered as a controller of a selforganization system in which a nonlinear computation principle functions in noise fluctuations.

1. PLASMA AND SYSTEM

A plasma is one state of the matter and one of events; energy, material and information are focussed on the production and the maintenance of the plasma. We consider a plasma on a macroscopic view, as a plasma system, then double layers (we indicate them by DL, including a plasma sheath)[1] and the outside domain of plasma come into our sight. When energy and material are appropreate, a plasma is generated in the system, coincident with self-organization of DL through which charged particles (electrons and ions) come in and out for the maintenance of the plasma. We call the event as the self-organization of a structure (the plasma and DL). Generally, the energy is poured into the plasma through DL (static potential) by the charged particles. DL knows the amount of the energy and the particle numbers, and DL controls them according to system conditions. System condiions are the density and the temperature of the plasma, the intensities of the currents of charged particles, energy source, walls(electrodes, dielectrics) and so on.

2. MEMBRANE POTENTIAL

A neuron (nerve cell) is an element of the nerve (=biological computation) system. "Plasma" was named by an analogy from the biological cell plasma, and besides that, it has been found that a function of DL is analogous to that of plasma membrane,[1] across which carriers go in and out and polarization arizes. The cell plasma and the cell membrane are not in existence independent of each other but are necessary each other for the existence. The cell membrane is a controller by which the transport of plasma materials are controlled. Flicker noise was observed from the pulse series of membrane potential of a snail giant neuron(autoactive neuron)[2] and from the resting membrane potential of a frog neuron(axon).[3] Both are noise of the membrane potential.

Alfvén wrote that DL always produces noise and fluctuations, and noise is essential for the formation and maitenance of DL.[1,4]

A biological system operates nonlinear. The membrane potential ϕ equilibrates at

$$\phi=\frac{RT}{zF}\ln\frac{C2}{C1} \text{ (Nernst equation)}, \qquad (1)$$

R:gas constant,T:temperature,z:charge number,F:Faraday constant,C1,C2:density of an ion species at 1 and 2.

The numbers 1 and 2 indicate plasma side and outside of the membrane, respectively. ϕ is the potential of 1 relative to 2. The difference of ion density is kept constant by ion pumps across the membrane.

For a neuron(axon) of squid, C1=400mM and C2=20mM for K^+, and C1=50mM and C2=440mM for Na^+.[5] K^+ and Na^+ ion channels form nonlinear conductance across the membrane for K^+ flux and Na^+ flux, respectively. Below a threshold of the membrane potential ϕ, the gate open probability of the Na^+ channel is zero and the conductance for Na^+ flux is null; then eq.(1) stands for K^+ (K^+ leak free through the membrane) and ϕ is the resting membrane potential ϕ_K, for the squid neuron ϕ_K=-77mV (T=300K) from eq(1). The gate open probability of the K^+ channel is high; the patch clamp recording method tells us that one ion gate opens at random interval and the conductance is 10^{-11}S when the gate opens.[5] The conductance of the membrane is the integral of these ion channels.

In Fig.1 the nonlinearity of the conductance of the ion channel is shown. When the ion channel is excited by an electrical or chemical signal, the gate of Na^+ channel opens and the gate open probability of Na^+ channel increases rapidly above the threshold potential ϕ_t with a positive feed back (depolarization), while the gate open probability of the K^+ channel becomes null, and eq.(1) stands for Na^+ (Na^+ leak free) and ϕ becomes the action membrane potential ϕ_{Na}, for the squid neuron $\phi_{Na}=+56mV$ (T=300K). And then, the Na^+ channels enter inactive and the K^+ channels active, and the membrane potential rapidly decreases to $\phi_K(<0)$ from $\phi_{Na}(>0)$. This is the mechanism of single impulse which is the carrier of neuron signal.

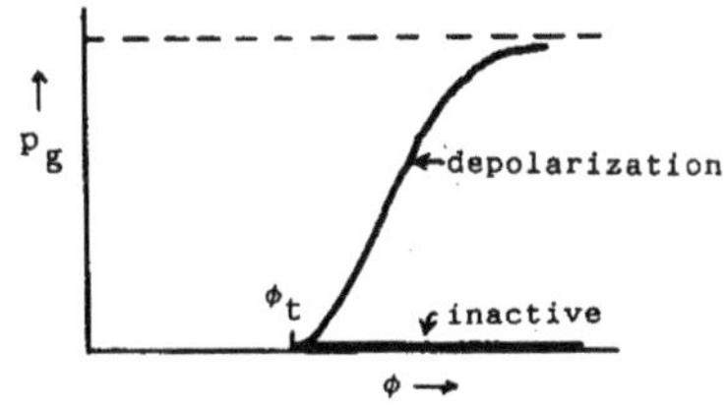

Fig.1 Nonlinearity of conductance.[5]
ϕ: membrane potential
ϕ_t: threshold potential
p_g: gate open probability of Na^+ channel

The equation of ion currents through the membrane is given as follows:[5]

$$\sum_x g_x(\phi-\phi_x)+I_p+I_i=0 \tag{2}$$

g_x:conductance for x ion ($x=K^+, Na^+$, nonlinear)
I_p:Na^+-K^+·ATPase pump current
I_i:ion leak current except by x.

I_p and I_i are very small compared to the term of Σ, and g_x is the nonlinear conductance which senses the membrane potential ϕ (the electric field across the membrane is very strong, about 10^5V/cm), ion species and its density.[5] It can be said that the experimental solution of eq.(2) for the frog neuron is $\phi=\phi_K(<0$, resting potential) with flicker noise $\tilde{\phi}^2\propto\frac{1}{f}$ in the frequency domain $0.15<f<1500$Hz,[3] and that for the snail autoactive neuron it is the pulse series ϕ (resting potential $\phi=\phi_K<0$, action potential $\phi=\phi_{Na}>0$) with flicker noise $\tilde{\phi}^2\propto\frac{1}{f}$ in the frequency domain $0.0005<f<0.5$Hz.[2] $\tilde{\phi}^2$ is the power spectral density of the membrane potential. Eq.(2) is a system equation of the biological plasma system. ϕ is nonlinear because g_x is nonlinear and ϕ is with flicker noise. The membrane functions as a controller and it is for maintenance of the system which is a structure of the plasma, the membrane and the outside of environment. In the next section we discuss similar points experimentally on a discharge plasma system.

3. DC DISCHARGE SYSTEM AND NOISE

The author made an experiment on the discharge with a hot-cathode DC magnetron cell.[6] The test gas was argon. When the discharge became active by increasing the axial magnetic field under a proper anode voltage (U: a few hundred volts) and thermionic current (one to a hundred milliamperes), noise was detected with a Langmuir probe under a certain amount of secondary electron current; then, it is noted that the magnetic field B passed the usual magnetron cut-off field $B_C (B>B_C)$:

$$B_C \propto \sqrt{U/(R^2-r^2)}. \quad (3)$$

R and r are the radii of the cylinder anode and the coaxal cathode, respectively. A result is shown in Fig.2.

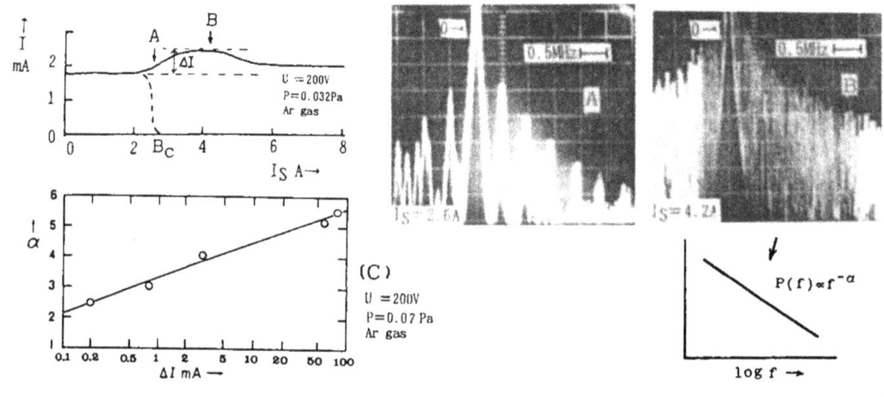

Fig.2 Discharge current vs. solenoid current characteristics.[4] A: Ar^+ ion oscillation in electron tube, B: noise with self-organization of potential structure, C: spectrum index α and discharge current ΔI. $(B \propto I_S)$

The power spectral density P(f) of noise followed the power law in the frequency domain $0.1<f<2MHz$:[6]

$$P(f) \propto f^{-\alpha} \tag{4}$$

with α the spectrum index ($2.5<\alpha<5.5$). α depends on the secondary current ΔI (see Fig.2C):[4]

$$2.9\alpha = \log_2 \Delta I/\Delta I_0 \text{ (bits)}. \tag{5}$$

The shift of the cut-off field ($B>B_C$) forces us to image that the anode shifted inward to a virtual anode $R'(R'<R)$ or the cathode outward to a virtual cathode $r'(r'>r)$. Then from eq.(3) $R'$and r' make the shift of B_C to a larger cut-off field ($B>B_C$). In Fig.3 is outlined the self-organizaton of a new potential structure by an active discharge with noise.

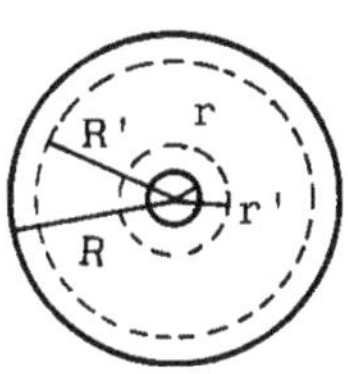

Fig.3 Self-organization of a structure with noise.

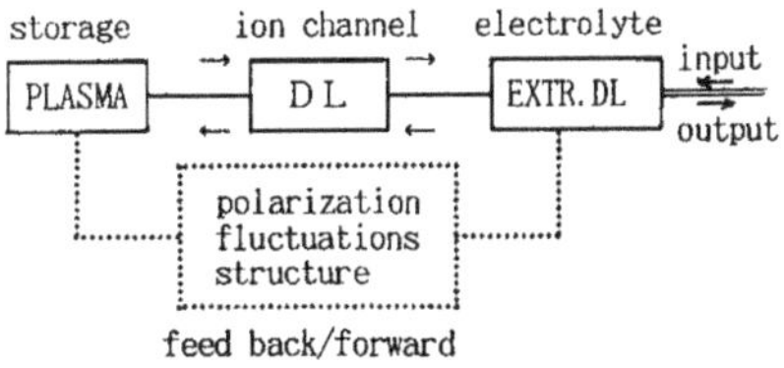

Fig.4 A plasma system, DL as a controller with noise.

An analogy of eq.(2) to the discharge plasma system leads to the system equation (ϕ:potential, ΔI:discharge current):

$$\Sigma g\phi + I_p + \Delta I = 0, \tag{6}$$

g: conductance(nonlinear)
I_p:current by pump effects
ΔI:discharge current.

Under $\Delta I=0$, eq.(6) is for an usual electron tube with I_p the anode current and with no noise (the experimental fact); but under $\Delta I\neq 0$, eq.(6) is for the discharge system in which the plasma and DL are self-organized by which a new potential structure appears with the nonlinear conductance and with noise of eq.(4). The nonlinear eq.(6) includes noise fluctuations in the potential ϕ ($-\alpha$ noise) and the relation between the index α

and the discharge current ΔI (eq.(5)) is a boundary condition.

4. DISCUSSION AND CONCLUSION

In Fig.4 is shown a conceptual diagram of the plasma system. The system is stabilized by DL where polarization takes place simultaniously with noise (fluctuations of ϕ, stabilizing process). By noise a feedback of extraDL information to plasma is executed and vice versa, through DL the controllar of the system. We consider the plasma system as a nonlinear computation system: 1. the plasma is a storage of information (energy, material), 2. DL is the controller, 3. extraDL is an input/ output connector to the another system, and 4. the whole system is an arithmetic unit. Eq.(6) is the equation of the system.

A neuron produces a double layer across its membrane as an element of neural computing. An integration of neurons (neural network, self-organization) is like a formation of plasmas. Both are produced according to the input of energy, material and information and maintained by a computing in the systems.

α is an index characteristic to the system, for the nerve system α=1 and for the discharge system it is in eq.(5).

The author presented an analogical consideration of DL in the plasma system from the cell membrane, and showed an evidence of forming a potential structure with noise from a result of experimental study of a DC magnetron discharge.

REFERENCES:

1. H.Alfvén, Double Layers and Circuits in Astrophysics, TRITA-EPP-86,Royal Inst.Technol.(1986,Stockholm).
2. T.Musha,H.Takeuchi and T.Inoue, IEEE Trans.Biom.Engin.,**BME-30**(1983)3,p.194.
3. H.E.Derksen and A.A.Verveen, Science,**151**(1966)p.1388.
4. N.Tanizuka, Proc.Int.Seminar on Reactive Plasmas,(Nagoya Univ.,1991)p.121.
5. B. Alberts, D. Bray, J. Lewis, M. Raff, K. Roberts and J.D. Watson, Molecular Biology of the Cell(Garland Publishing, New York, 1983)Chaps.6 and 18.
6. N.Tanizuka, Jpn.J.Appl.Phys.,**30**(1991)p.171.

www.ingramcontent.com/pod-product-compliance
Lightning Source LLC
La Vergne TN
LVHW082254241026
850861LV00008B/42
* 9 7 8 9 8 1 0 2 1 4 3 2 6 *